Theory and Applications of Numerical Analysis

Theory and Applications of Numerical Analysis

SECOND EDITION

G. M. PHILLIPS
University of St Andrews

and

P. J. TAYLOR
formerly of the University of Strathclyde

ACADEMIC PRESS
Harcourt Brace & Company, Publishers
London San Diego New York
Boston Sydney Tokyo Toronto

ACADEMIC PRESS LIMITED
24–28 Oval Road
LONDON NW1 7DX

U.S. Edition published by
ACADEMIC PRESS INC.
San Diego, CA 92101

This book is printed on acid free paper

Copyright © 1996 ACADEMIC PRESS LIMITED
First published 1973
Second edition 1996

All rights reserved

No part of this book may be reproduced or transmitted in any form or by any means, electronic or mechanical, including photocopying, recording, or any information storage and retrieval system without permission in writing from the publisher

A catalogue record for this book is available from the British Library
ISBN 0-12-553560-0

1001232626

Typeset by Mathematical Composition Setters Ltd, Salisbury, UK
Printed in Great Britain by WBC, Bridgend, Mid Glamorgan

CONTENTS

PREFACE

Although no text of reasonable length can cover all possible topics, it seemed to us that the first edition of this text would be improved by including material on two further items. Thus the main change in the second edition is the addition of two entirely new chapters, Chapter 6 (Splines and other approximations) and Chapter 11 (Matrix eigenvalues and eigenvectors). In the preface to the first edition we stated that the material was equivalent to about sixty lectures. This new edition has adequate material for two separate semester courses.

When this book first appeared in 1973, computers were much less 'friendly' than they are now (1995) and, in general, it was only specialists who had access to them. The only calculating aids available to the majority of our readers were mathematical tables and desk calculators; the latter were mainly mechanical machines, which are now museum pieces. In contrast, the readership of our second edition will be able to appreciate the power and elegance of the algorithms which we discuss by implementing them, and experimenting with them, on the computer. (Sounding like our parents of old, we can say to our students 'We never had it so easy: you are lucky!') To encourage the active pursuit of the algorithms in the text, we have included computing exercises amongst the problems at the end of each chapter.

Despite the changes made in the second edition, the flavour of this text remains the same, reflecting the authors' tastes: in short, we like both theorems and algorithms and we remain in awe of the masters of our craft who discovered or created them. The theorems show how the mathematics hangs together and point the way to the algorithms.

We are deeply grateful to our readership and to our publishers for keeping this text in print for such a long time. We also owe much to the many colleagues from many countries with whom we have shared the sheer fun of studying mathematics in our research collaborations, and we mention their names here as a sign of our respect and gratitude: G. E. Bell, L. Brutman, B. L. Chalmers, E. W. Cheney, F. Deutsch, D. K. Dimitrov, D. Elliott, Feng Shun-xi, D. M. E. Foster, H. T. Freitag, R. E. Grundy, A. S. B. Holland, Z. F. Koçak, S. L. Lee, W. Light, J. H. McCabe, A R. Mitchell, D. F. Paget, A. Sri Ranga, B. N. Sahney, S. P. Singh, E. L. Wachspress, M. A. Wolfe, D. Yahaya.

G. M. PHILLIPS
Mathematical Institute
University of St Andrews
Scotland

P. J. TAYLOR
formerly of the
Department of Mathematics
University of Strathclyde
Scotland

December 1995

FROM THE PREFACE TO THE FIRST EDITION

We have written this book as an introductory text on numerical analysis for undergraduate mathematicians, computer scientists, engineers and other scientists. The material is equivalent to about sixty lectures, to be taken after a first year calculus course, although some calculus is included in the early chapters both to refresh the reader's memory and to provide a foundation on which we may build. We do not assume that the reader has a knowledge of matrix algebra and so have included a brief introduction to matrices. It would, however, help the understanding of the reader if he has taken a basic course in matrix algebra or is taking one concurrently with any course based on this book.

We have tried to give a logical, self-contained development of our subject *ab initio*, with equal emphasis on practical methods and mathematical theory. Thus we have stated algorithms precisely and have usually given proofs of theorems, since each of these aspects of the subject illuminates the other. We believe that numerical analysis can be invaluable in the teaching of mathematics. Numerical analysis is well motivated and uses many important mathematical concepts. Where possible we have used the theme of approximation to give a unified treatment throughout the text. Thus the different types of approximation introduced are used in the chapters on the solution of non-linear algebraic equations, numerical integration and differential equations.

It is not easy to be a good numerical analyst since, as in other branches of mathematics, it takes considerable skill and experience to be able to leap nimbly to and fro from the general and abstract to the particular and practical. A large number of worked examples has been included to help the reader develop these skills. The problems, which are given at the end of each chapter, are to be regarded as an extension of the text as well as a test of the reader's understanding.

Some of our colleagues have most kindly read all or part of the manuscript and offered wise and valuable advice. These include Dr J. D.

Lambert, Prof. P. Lancaster, Dr J. H. McCabe and Prof. A. R. Mitchell. Our thanks go to them and, of course, any errors or omissions which remain are entirely our responsibility We are also much indebted to our students and to our colleagues for the stimulus given by their encouragement.

G. M. PHILLIPS
Department of Applied Mathematics
University of St Andrews
Scotland

P. J. TAYLOR
Department of Computing Science
University of Stirling
Scotland

December 1972

Chapter **1**

INTRODUCTION

1.1 What is numerical analysis?

Numerical analysis is concerned with the mathematical derivation, description and analysis of methods of obtaining numerical solutions of mathematical problems. We shall be interested in *constructive methods* in mathematics; these are methods which show how to construct solutions of mathematical problems. For example, a constructive proof of the existence of a solution to a problem not only shows that the solution exists but also describes how a solution may be determined. A proof which shows the existence of a solution by *reductio ad absurdum*† is not constructive. Consider the following simple example.

Example 1.1 Prove that the quadratic equation

$$x^2 + 2bx + c = 0 \tag{1.1}$$

with real coefficients b and c satisfying $b^2 > c$ has at least two real roots.

Constructive proof For all x, b and c,

$$\begin{aligned} x^2 + 2bx + c &= x^2 + 2bx + b^2 + c - b^2 \\ &= (x+b)^2 + c - b^2. \end{aligned}$$

Thus x is a root of (1.1) if and only if

$$(x+b)^2 + c - b^2 = 0$$

that is,

$$(x+b)^2 = b^2 - c.$$

† In such a proof we assume that the solution does not exist and obtain a contradiction.

On taking square roots, remembering that $b^2 - c > 0$, we see that x is a root if and only if

$$x + b = \pm\sqrt{(b^2 - c)}$$

that is

$$x = -b \pm \sqrt{(b^2 - c)}. \tag{1.2}$$

Thus there are two real roots and (1.2) shows how to compute these. □

Non-constructive proof Let

$$q(x) = x^2 + 2bx + c.$$

Suppose first that there are no roots of (1.1) and hence no zeros of q. As q is a continuous function (see § 2.2), we see that $q(x)$ is either always positive or always negative for all x. Now

$$\begin{aligned} q(-b) &= b^2 - 2b^2 + c \\ &= c - b^2 < 0 \end{aligned}$$

by the condition on b and c. Thus $q(x)$ must always be negative. However, for $|x|$ large,

$$x^2 > |2bx + c|$$

and thus $q(x) > 0$ for $|x|$ large. We have a contradiction.

Secondly, suppose that q has only one zero. Then from the continuity of q we must have $q(x) > 0$ for x large and $q(x) < 0$ for $-x$ large or vice versa. This contradicts the fact that $q(x) > 0$ for $|x|$ large, regardless of the sign of x.

This proof does not show how to compute the roots. □

The word 'algorithm' has for a long time been used as a synonym for 'method' in the context of constructing the solution of a mathematical problem. Measured against the long history of mathematics it is only relatively recently that a proper mathematical definition of algorithm has been devised. Much of this important work is due to A. M. Turing (1912–54), who gave a definition based on an abstract concept of a computer. For the purposes of this book we define an algorithm to be a complete and unambiguous description of a method of constructing the solution of a mathematical problem. One difficulty in making a formal definition is deciding precisely what operations are allowed in the method. In this text we shall take these to be the basic arithmetical operations of addition, subtraction, multiplication and division. In general, the solution and method occurring in an algorithm need not be numerical. For example, the early Greek geometers devised algorithms using ruler, pencil and compass. One such algorithm is that for constructing the perpendicular to a straight line at a given point. However, the greatest of the Greek mathematicians, Archimedes (287–212 BC), constructed a most elegant numerical algorithm for approximating the area of a circle (thus estimating π), by bounding it above and below by the areas of regular polygons with an increasingly large number of sides. In this book we shall deal only with numerical algorithms.

It is interesting to note that until the beginning of the twentieth century much mathematical analysis was based on constructive methods, and some of the finest mathematicians of every era have produced important algorithms and constructive proofs, for example Isaac Newton (1642–1727), L. Euler (1707–83) and K. F. Gauss (1777–1855). It is perhaps ironic that the trend in more recent years has been towards non-constructive methods, although the introduction of calculating devices made the execution of algorithms much easier.

The rapid advances in the design of digital computers have, of course, had a profound effect on numerical analysis. At the time the first edition of this book was published in 1973, computers could already perform over a million arithmetic operations in a second, while around 1960 speeds had been nearer one thousand operations a second. As we prepare the second edition of this text we reflect that the most impressive and important development in computers over the past two decades has been the dramatic decrease in their cost, leading to their near universal availability. Yet, while lengthier and more complex calculations may be tackled, the scope for error has correspondingly increased. Unfortunately, the expectations of users of numerical algorithms have more than matched the improvements in computers. However, one cannot select the best algorithm for the problem in hand without having a sound understanding of the theoretical principles involved.

Although most numerical algorithms are designed for use on digital computers, the subject of numerical analysis should not be confused with *computer programming* and *information processing* (or *data processing*). Those who design and use numerical algorithms need a knowledge of the efficient use of a digital computer for performing calculations. Computer programming is primarily concerned with the problem of coding algorithms (not necessarily numerical) in a form suitable for a computer. Information processing is concerned with the problems of organizing a computer so that data can be manipulated. The data is not necessarily numerical.

1.2 Numerical algorithms

For many problems there are no algorithms for obtaining exact numerical values of solutions and we have to be satisfied with approximations to the values we seek. The simplest example is when the result is not a rational number and our arithmetic is restricted to rational numbers. What we do ask of an algorithm, however, is that the error in the result can be made as small as we please. Usually, the higher the accuracy we demand, the greater is the amount of computation required. This is illustrated in the following algorithm† for computing a square root, based on *bisection*, where a demand for higher accuracy in the result means completing more basic arithmetical operations, as well as increasing the accuracy with which these are performed.

Example 1.2 Given $a>1$ and $\varepsilon>0$, we seek an approximation to $\sqrt{a}$ with error not greater than ε. At each stage we have two numbers x_0 and x_1. In the calculation, we change x_0 and x_1, keeping $x_0 \leqslant \sqrt{a} \leqslant x_1$ and at the same time repeatedly halving the interval $[x_0, x_1]$. Initially we take $x_0 = 1$ and $x_1 = a$, as $1 < \sqrt{a} < a$. We compute the midpoint $x = (x_0 + x_1)/2$ and then decide whether x is greater than $\sqrt{a}$. Since we do not know $\sqrt{a}$, we compare x^2 and a. If $x^2 > a$ we replace the number x_1 by x; otherwise we replace the number x_0 by x. We stop when $x_1 - x_0 \leqslant 2\varepsilon$, since then we must have $0 \leqslant x - \sqrt{a} \leqslant \varepsilon$, where ε was chosen in advance to give the accuracy we desire in the final approximation x. □

Algorithm 1.1 (square root by bisection) We begin with $a>1$ and $\varepsilon>0$, and $x_0 = 1$, $x_1 = a$.

```
repeat
   x := (x0 + x1)/2
   if x^2 > a then x1 := x
           else x0 := x
until x1 - x0 <= 2ε
```
□

In the above algorithm, the symbol := should be interpreted as 'becomes equal to'. The 'loop' of instructions between **repeat** and **until** is called an *iteration*. If we make the positive number ε smaller, the number of iterations required increases. We must also make the arithmetic more accurate as ε is decreased, as otherwise the test whether $x^2 > a$ will give incorrect results and x_0 and x_1 will not remain distinct. If we are prepared to do the arithmetic to arbitrary accuracy, we can make ε as small as we please, provided it is positive. We cannot take $\varepsilon = 0$ as this would require an infinite number of iterations. □

In Algorithm 1.1 we compute the first few members of a sequence of values (the midpoints x) which converge (see § 2.3) to $\sqrt{a}$, the solution of the problem. As we can calculate only a finite number of members of the sequence, we do not obtain the precise value of $\sqrt{a}$. Many algorithms

† This algorithm is not recommended; it is given merely as a simple illustration. A more efficient square root algorithm is described in § 8.6.

consist of calculating a sequence and in this context we say that the algorithm is *convergent* if the sequence converges to the desired solution of the problem.

In the analysis of an algorithm we shall consider the error in the computed approximation to the solution of our problem. This error may be due to various factors, for example, rounding in arithmetic and the termination of an infinite process, as in Example 1.2. Errors, which are a deliberate but often unavoidable part of the solution process, should not be confused with *mistakes* which are due to failure on the part of the human user of an algorithm, particularly when writing a computer program. For each algorithm we look for an *error bound*, that is, a bound on the error in the computed solution. A bound which is not directly dependent on the solution and therefore can be computed in advance, is called an *a priori error bound.* When directly dependent on the solution, a bound cannot be computed until after the completion of the main part of the calculation and is called an *a posteriori bound.* If, as is often the case, an error bound is very much larger than the actual error, the bound cannot be used to provide a sensible estimate of the error. For this reason we also devise alternative methods of estimating the error for some problems. Error bounds and estimates are often very valuable in telling us about the asymptotic behaviour of the error as we improve approximations made in deriving an algorithm.

There are two ways in which we measure the magnitude of an error. If a is some real number and a^* is an approximation to a we define the *absolute error* to be $|a-a^*|$ and we define the *relative error* to be $|a-a^*|/|a|$. We can see that the absolute error may be quite meaningless unless we have some knowledge of the magnitude of a. We often give relative errors in terms of percentages.

One important point we must consider in looking at an algorithm is its efficiency. We usually measure this in terms of the number of basic arithmetical operations required for a given accuracy in the result. Sometimes we also consider other factors such as the amount of computer storage space required.

1.3 Properly posed and well-conditioned problems

There is one important question which we must investigate before applying a numerical method to a problem and that is 'How sensitive is the solution to changes in data in the formulation of the problem?' For example, if we are solving a set of algebraic equations, the coefficients in the equations form data which must be provided. There will certainly be difficulties if small changes in these coefficients have a large effect on the solution. It is very unlikely that we can compute a solution without perturbing these coefficients if only because of errors due to rounding in the arithmetic.

Suppose that $S(d)$ represents† a solution of a problem for a given set of data d. Now suppose that $d+\delta d$ is a perturbed set of data and $S(d+\delta d)$ is a corresponding solution. We write $\|\delta d\|$ to denote a non-negative number which is a measure of the magnitude of the perturbation in the data and we write $\|S(d+\delta d)-S(d)\|$ to denote a non-negative number which is a measure of the magnitude of the difference in the solutions. We say that the problem is *properly posed* for a given set of data d if the following two conditions are satisfied.

(i) A *unique* solution exists for the set of data d and for each set of data 'near' d. That is, there is a real number $\varepsilon>0$ such that $S(d+\delta d)$ exists and is unique for all δd with $\|\delta d\|<\varepsilon$.

(ii) The solution $S(d)$ is continuously dependent on the data at d, that is

$$\|S(d+\delta d)-S(d)\|\to 0 \qquad \text{whenever} \qquad \|\delta d\|\to 0.$$

If there were more than one solution for a given set of data, then we would not know which solution (if any) our numerical method obtains. Usually we can avoid existence of more than one solution by imposing extra conditions on the problem. For example, if we wish to solve a cubic equation we could stipulate that the largest real root is required.

If the condition (ii) above is not satisfied, it would be difficult to use a numerical algorithm as then there are data which are arbitrarily close to d and such that the corresponding solutions are quite different from the desired solution $S(d)$.

We say that a problem which is properly posed for a given set of data is *well-conditioned* if every small perturbation of that data results in a relatively small change in the solution. If the change in the solution is large we say that the problem is *ill-conditioned*. Of course these are relative terms.

Often the solution of a properly posed problem is *Lipschitz* dependent on the data (see § 2.2). In this case there are constants $L>0$ and $\varepsilon>0$ such that

$$\|S(d+\delta d)-S(d)\|\leqslant L\|\delta d\| \tag{1.3}$$

for all δd with $\|\delta d\|<\varepsilon$. In this case we can see that the problem is well-conditioned if L is not too large. The restriction, $\|\delta d\|<\varepsilon$, only means that there are limits within which the data must lie if (1.3) is to hold. This restriction should not worry us unless ε is small. If L has to be made large before (1.3) is satisfied or if no such L exists, then a small change in the data can result in a large change in the solution and thus the problem is ill-conditioned.

---

† $S(d)$ and d may be numbers, functions, matrices, vectors or combinations of these depending on the problem.

As we remarked earlier, an ill-conditioned problem may be difficult to solve numerically as rounding errors may seriously affect the solution. In such a case we would at least need to use higher accuracy in the arithmetic than is usual. If the problem is not properly posed, rounding errors may make the numerical solution meaningless, regardless of the accuracy of the arithmetic. Sometimes the data contains *inherent* errors and is only known to a certain degree of accuracy. In such cases the amount of confidence we can place in the results depends on the conditioning of the problem. If it is ill-conditioned, the results may be subject to large errors and, if the problem is not properly posed, the results may be meaningless.

The next example illustrates that the properties properly posed and well-conditioned depend on both the problem and the data.

Example 1.3 We consider the computation of values of the quadratic

$$q(x) \equiv x^2 + x - 1150.$$

This is an ill-conditioned problem if x is near a root of the quadratic. For example, for $x = 100/3$, $q(100/3) = -50/9 \simeq -5.6$. However, if we take $x = 33$, which differs from $100/3$ by only 1%, we find that $q(33) = -28$ which is approximately five times the value of $q(100/3)$. To obtain a relation of the form (1.3) when $x = 100/3$ we need to take $L \simeq 70$ (see Problem 1.5). □

Suppose that we wish to solve a problem which we will denote by P. We often find it necessary to replace P by another problem P_1 (say) which approximates to P and is more amenable to numerical computation. We now have two types of error. Firstly, there is the difference between P and P_1 and we write E_P to denote some measure of this error. Often this results from cutting off an infinite process such as forming the sum of an infinite series (see § 2.3) after only a finite number of steps, in which case we call it the *truncation* error. Secondly, there is the error consisting of the difference between the solutions of P and P_1 and we denote some measure of this by E_S. This second type of error is sometimes called the *global* error. Of course there is a relation between E_P and E_S, but it does not follow that if E_P is small then so is E_S. We shall derive relations between E_P and E_S for different types of problem and, for a satisfactory method, will require E_S to become arbitrarily small as the approximation is improved, so that E_P tends to zero. If this property does hold we say that the resulting algorithm is *convergent*. Unfortunately there are problems and approximations such that E_S does not tend to zero as E_P is decreased.

Example 1.4 Suppose that f is a given real-valued differentiable function of a real variable x and that we seek values of the derivative f' for given particular values of x in the interval $a \leq x \leq b$. We will assume that f' cannot be obtained explicitly as f is a very complicated function but that we can compute $f(x)$ for any given value of x.

Our algorithm consists of replacing f by another function g which can easily be differentiated. We then compute values of $g'(x)$ instead of $f'(x)$. In Chapter 5 we will describe one way in which g can be chosen and that provided f satisfies certain conditions we can make the error $E_P(x) = f(x) - g(x)$ arbitrarily small for all x with $a \leq x \leq b$. Unfortunately it is possible for $E_S(x) = f'(x) - g'(x)$ to be arbitrarily large although $E_P(x)$ is arbitrarily small. For example, if $E_P(x) = (1/n)\sin(n^2x)$, where $n = 1, 2, 3, \ldots$ corresponds to different choices of g, then

$$E_P(x) \to 0 \qquad \text{as} \qquad n \to \infty,$$

for all values of x. However,

$$E_S(x) = n\cos(n^2x)$$

and thus $E_S(x) \to \infty$ as $n \to \infty$, for $x = 0$. □

If we replace a problem P by another, more convenient problem P_1 which is an approximation to P, we shall normally expect both P and P_1 to be properly posed and reasonably well-conditioned. If P were not properly posed, in general we would not obtain convergence of the solution of P_1 to that of P. We are almost certain in effect to perturb the data of P in making the approximation P_1. We therefore require, as a necessary condition for the convergence of our algorithm, that the original problem P be properly posed. The convergence will be ‘slow’ unless P is well-conditioned.

We also have to solve P_1 using numerical calculations and we are almost certain to perturb the data of P_1, which should therefore be properly posed and well-conditioned. If P is well-conditioned, we normally expect P_1 to be well-conditioned; if not, we should look for a better alternative than P_1.

In some numerical algorithms, especially those for differential equations, we compute elements of a sequence as in Example 1.2 but are interested in each element of the sequence and not just the limit. In both types of problem we use a *recurrence relation*, which relates each element of a sequence to earlier values. For example, we may compute elements of a sequence $(y_n)_{n=0}^{\infty}$ using a recurrence relation of the form

$$y_{n+1} = F(y_n, y_{n-1}, \ldots, y_{n-m}) \tag{1.4}$$

for $n = m,\ m+1,\ m+2, \ldots$, where F is a known function of $m+1$ variables. We need starting values $y_0, y_0, \ldots, y_m$ and normally these must be given. In using (1.4) we are certain to make rounding errors. Now an error introduced in any one step (one value of n) will affect all subsequent values and, therefore, we need to investigate the propagated effect of rounding errors. These effects may be very important, especially as computers allow us to attempt to compute very many elements of a sequence and so introduce many rounding errors. In this context we use the term *stability* in referring to whether the process is properly posed and well-conditioned.

Example 1.5 If $(y_n)_{n=0}^{\infty}$ satisfies

$$y_{n+1} = 100.01 y_n - y_{n-1}, \qquad n \geq 1, \tag{1.5}$$

with $y_0 = 1$ and $y_1 = 0.01$ then $y_5 = 10^{-10}$, whereas if $y_0 = 1 + 10^{-6}$ and $y_1 = 0.01$ then $y_5 \simeq -1.0001$. The large difference in the value of y_5 shows that the process is very ill-conditioned. In fact we say that the recurrence relation is *numerically unstable*. Further details are given in §13.11, particularly Example 13.18. □

Problems

Section 1.1

1.1 Give both constructive and non-constructive proofs of the following statements.

(i) Every quadratic equation has at most two distinct real roots.
(ii) There are at least two real values of x such that

$$2 \cos(2 \cos^{-1} x) = 1.$$

(iii) Given a straight line ℓ and a point P not on ℓ, there is a perpendicular to ℓ which passes through P.

Section 1.2

1.2 Show that an *a priori* error bound for the result of the bisection method for finding $\sqrt{a}$ with $0 < a < 1$ is

$$|x - \sqrt{a}| \leq (1 - a)/2^N,$$

where N is the number of iterations completed.

1.3 In Algorithm 1.1 we have at each stage

$$|x - \sqrt{a}| \leq \frac{x_1 - x_0}{2}.$$

Show that if

$$(x_1 - x_0)^2 < 4a\varepsilon^2,$$

then the relative error in x as an approximation to $\sqrt{a}$ satisfies

$$\left|\frac{x-\sqrt{a}}{\sqrt{a}}\right|<\varepsilon.$$

Hence modify the algorithm so that the iterations stop when the relative error is less than ε.

Section 1.3

1.4 Show that the equations

$$\begin{aligned} x+y&=2\\ \tfrac{1}{6}x+\tfrac{1}{6}y&=\tfrac{1}{3}\\ x+2y&=3 \end{aligned}$$

have a unique solution, but that the problem of computing this solution is not properly posed, in the sense given in § 1.3. Notice that, if we express coefficients as rounded decimal fractions, there is no solution.

1.5 If $q(x)$ is the quadratic of Example 1.3, show that

$$q(x+\delta x)-q(x)=(2x+\delta x+1)\delta x$$

and thus that, if $33\leqslant x\leqslant 34$ and $|\delta x|\leqslant 1$, then

$$|q(x+\delta x)-q(x)|\leqslant 70|\delta x|.$$

Chapter 2

BASIC ANALYSIS

2.1 Functions

Definition 2.1 A *set* is a collection of objects. The number of objects may be finite or infinite. □

We write

$$S = \{a, b, c, \ldots\},$$

where S denotes a set and $a, b, c, \ldots$ denote the objects belonging to the set. We call $a, b, c, \ldots$ the *elements* or *members* of the set S and we write $a \in S$ as a shorthand for 'a belongs to S'. If S and T are sets such that† $x \in T \Rightarrow x \in S$, we say that T is a subset of S and write $T \subset S$.

Example 2.1 The set of all numbers x such that $a \leqslant x \leqslant b$ is denoted by $[a, b]$ and is called a *closed interval.* The set of x such that $a < x < b$ is denoted by (a, b) and is called an *open interval.* We write $(a, b]$ to denote the set of x such that $a < x \leqslant b$ and $(-\infty, \infty)$ to denote the set of all real numbers. □

Definition 2.2 If, given two sets X and Y, there is a relation between them such that each member of X is related to exactly one member of Y, we call this relation a *function* or *mapping* from X to Y. □

Example 2.2 If, at some instant, X denotes the set of all London buses in use and Y the set of all human beings, the relation 'y is the driver of x, with $x \in X$, $y \in Y$' is a function from X to Y. □

---

† The symbol $\Rightarrow$ means 'implies'.

Example 2.3 If X is the set $[0, 1]$ and Y is the set of all real numbers y, then the relation

$$y = x + 2, \qquad 0 \leqslant x \leqslant 1,$$

defines a function from X to Y. □

We sometimes write $y = f(x)$, where $x \in X$ and $y \in Y$, to denote a function by which x is mapped into y. We then refer to this as 'the function f'. Note that not every $y \in Y$ need be such that $y = f(x)$, for some $x \in X$. X is called the *domain* of f and the set of all $y \in Y$ such that $y = f(x)$, for some $x \in X$, is called the *range* of f or ran(f). Thus we have

$$\text{ran}(f) \subset Y.$$

In Example 2.3, the domain is the closed interval $[0, 1]$ and the range is $[2, 3]$.

Example 2.4 Let X denote the set of all vectors $\mathbf{x}$ with n real elements and $\mathbf{A}$ denote an $n \times n$ matrix with all elements real. Then (see Chapter 9) $\mathbf{Ax} \in X$ and the relation $\mathbf{y} = \mathbf{Ax}$ is a function from X to X. □

We may depict a function schematically as in Fig. 2.1, which shows the sets X, Y and ran(f) and shows a 'typical element' $x \in X$ being mapped into an element $y \in Y$ by f.

If X and Y are subsets of the real numbers and $y = f(x)$ denotes a function from X to Y, the reader will be used to representing the function by a graph, as in Fig. 2.2. Such a function is called a function of one (real) variable.

Example 2.5 Let X denote the set of *ordered pairs* (x, y) where x and y are reals and let Y denote the set of all real numbers. A mapping from X to Y is called a function of two real variables. Similarly, we may define a function of n variables. The set of all ordered pairs (x, y) is called the xy-plane. Geometrically, the function $z = f(x, y)$ may be represented by a *surface* in xyz-space. Thus, for instance, the function $z = 1 + x + y$ is represented by the plane which passes through the three points with coordinates $(-1, 0, 0)$, $(0, -1, 0)$, $(0, 0, 1)$. □

Definition 2.3 A function f from X to ran$(f) \subset Y$ is said to be *one–one* if to each $y \in \text{ran}(f)$ there corresponds *exactly one* $x \in X$ such that $y = f(x)$. □

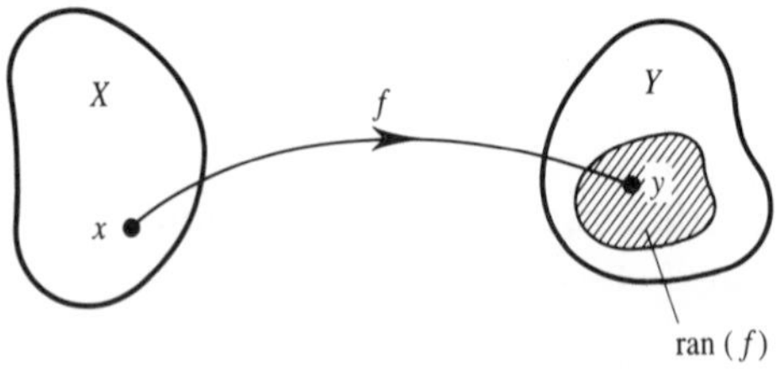

Fig. 2.1 Schematic representation of a function.

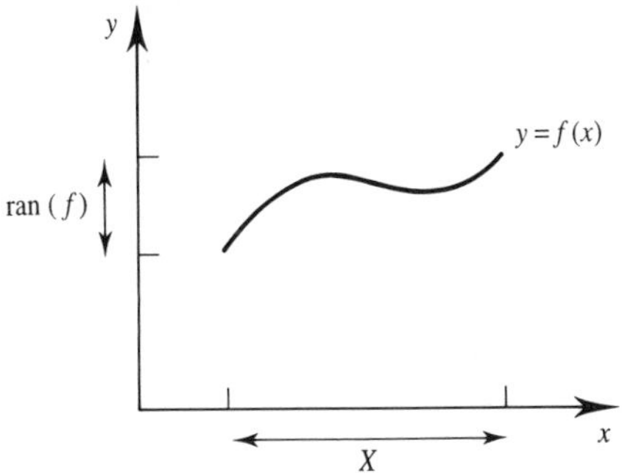

Fig. 2.2 Graph of a function from the reals to the reals.

Thus the function of Example 2.3 is one–one. The function defined by $y = x^2$, $-1 \leqslant x \leqslant 1$, is not one–one since each y such that $0 < y \leqslant 1$ corresponds to two values of x.

Definition 2.4 If f is a one–one function from X to $\mathrm{ran}(f) \subset Y$, we may define the *inverse function* of f by

$$x = f^{-1}(y).$$

The inverse function f^{-1} maps $y \in \mathrm{ran}(f)$ into the unique $x \in X$ such that $y = f(x)$. Note that $\mathrm{ran}(f^{-1}) = X$. □

Example 2.6 If $y = f(x)$, $x \in X$, is the function $y = x + 2$, $0 \leqslant x \leqslant 1$, of Example 2.3, then the inverse function $x = f^{-1}(y)$ exists and is given by $x = y - 2$, $2 \leqslant y \leqslant 3$. □

For the remainder of this chapter we shall be concerned mainly with functions from the reals although most of the functions may also be defined for a complex variable. We start by briefly considering polynomials and circular functions. Later in the chapter we look at logarithmic and exponential functions.

Polynomials

The function

$$y = a_0 + a_1 x + \cdots + a_n x^n, \tag{2.1}$$

where x belongs to the reals and the *coefficients* $a_0, a_1, \ldots, a_n$ are fixed real numbers with $a_n \neq 0$, is called a *polynomial* in x of degree n.

The function in Example 2.3 is thus a polynomial of degree 1. Throughout the text, we use P_n to denote the set of all polynomials of degree n or less. If p is a polynomial of degree $n \geq 1$ and α is a *zero* of p, that is $p(\alpha) = 0$, then we have (see Problem 2.2)

$$p(x) = (x - \alpha)q(x) \tag{2.2}$$

where q is a polynomial of degree $n - 1$. If p is a polynomial of degree $n \geq r > 1$ and

$$p(x) = (x - \alpha)^r q(x),$$

where q is a polynomial of degree $n - r$ such that $q(\alpha) \neq 0$, we say that α is a multiple zero of p of order r. If $r = 2$, we say that α is a double zero. For example, $p(x) \equiv (x - 1)^2(x - 2)$ has three real zeros, a single zero at $x = 2$ and a double zero at $x = 1$.

We can use (2.2) to prove that a polynomial of degree n cannot have more than n real zeros unless the polynomial is identically zero (that is, all its coefficients are zero). This result is clearly true for a polynomial of degree one (whose graph is a straight line) and using (2.2) and the principle of induction (see §2.3) we can extend this to the general case.

Circular functions

In Fig. 2.3, AB is an arc of a circle of radius 1 and centre O. In radian measure, the *angle* θ is defined as the length of the circular arc AB. Thus the angle of one complete revolution is 2π and a right angle is $\frac{1}{2}\pi$. Evidently BC and OC are functions of θ and these are denoted by

$$\text{BC} = \sin\theta, \qquad \text{OC} = \cos\theta, \qquad 0 < \theta < \tfrac{1}{2}\pi.$$

Since the sum of the angles of triangle OBC is π, we have

$$\cos\theta = \sin(\tfrac{1}{2}\pi - \theta), \tag{2.3}$$

where $0 < \theta < \frac{1}{2}\pi$.

The domain of the sine function may be extended from $(0, \frac{1}{2}\pi)$ to all real θ as follows:

$$\sin 0 = 0, \ \sin\tfrac{1}{2}\pi = 1, \tag{2.4}$$

$$\begin{aligned} \sin(\pi - \theta) &= \sin\theta, & 0 \leq \theta \leq \tfrac{1}{2}\pi, \\ \sin(-\theta) &= -\sin\theta, & 0 \leq \theta \leq \pi, \\ \sin(\theta + 2n\pi) &= \sin\theta, & n = 0, \pm 1, \pm 2, \ldots. \end{aligned} \tag{2.5}$$

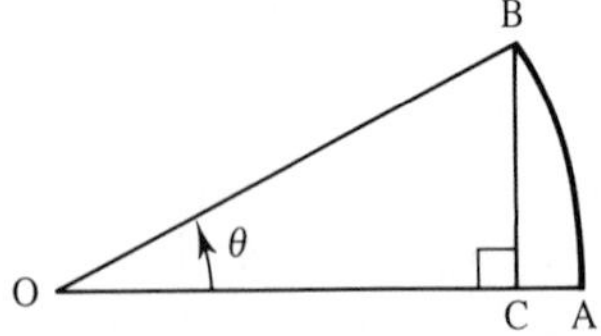

Fig. 2.3 OA = OB = 1, $\sin\theta = \text{BC}$, $\cos\theta = \text{OC}$

The domain of $\cos\theta$ is then extended to all real θ by allowing (2.3) to hold for all θ. Note that the values in (2.4) are chosen to make $\sin\theta$ *continuous* (see §2.2). Equation (2.5) demonstrates the *periodic* nature of $\sin\theta$. Both $\sin\theta$ and $\cos\theta$ are said to be *periodic*, with period 2π (see Fig. 2.4). We note some identities satisfied by the sines and cosines:

$$\sin^2\theta + \cos^2\theta = 1, \qquad \text{for all } \theta, \tag{2.6}$$

$$\sin\theta + \sin\phi = 2\sin\left(\frac{\theta+\phi}{2}\right)\cos\left(\frac{\theta-\phi}{2}\right), \tag{2.7}$$

$$\cos\theta + \cos\phi = 2\cos\left(\frac{\theta+\phi}{2}\right)\cos\left(\frac{\theta-\phi}{2}\right). \tag{2.8}$$

The identities (2.7) and (2.8) hold for all θ and ϕ. Other circular functions (for example, $\tan\theta$) may be defined in terms of the sine and cosine.

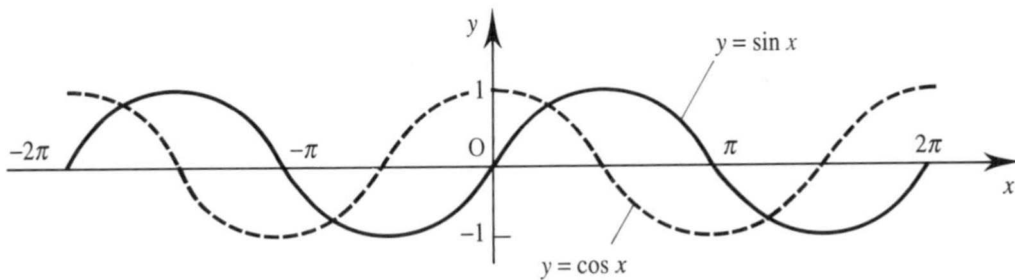

Fig. 2.4 The sine and cosine functions.

Further classes of functions

Definition 2.5 A function f from the reals to the reals is said to be *even* if $f(-x) = f(x)$ and to be *odd* if $f(-x) = -f(x)$, for all x. □

Thus $\sin x$ is odd, $\cos x$ is even and $1 + x$ is neither.

Definition 2.6 A function f defined on $[a, b]$ is said to be *convex* if

$$f(\lambda x_1 + (1-\lambda)x_2) < \lambda f(x_1) + (1-\lambda)f(x_2)$$

for all $x_1, x_2 \in [a, b]$ and any $\lambda \in (0, 1)$. □

Geometrically, this means that a chord joining any two points on the curve $y = f(x)$ lies above the curve, as for the function x^2 on $(-\infty, \infty)$. We can similarly define a *concave* function by reversing the inequality so that points on a chord must lie below the curve.

Definition 2.7 A function f is said to be *bounded above* on an interval I (which may be finite or infinite) if there exists a number M such that

$$f(x) \leqslant M, \qquad \text{for all } x \in I.$$

Any such number M is said to be an *upper bound* of f on I. The function is said to be *bounded below* on I if there exists a number m such that

$$f(x) \geqslant m, \qquad \text{for all } x \in I.$$

If a function is bounded above and below, it is said to be *bounded.* □

If f is bounded above on I, it can be shown that there exists a *least upper bound*, say M^*, such that $M^* \leqslant M$, for every upper bound M. The least upper bound, which is unique, is also called the *supremum*, denoted by

$$\sup_{x \in I} f(x).$$

The supremum is thus the smallest number which is not less than any $f(x)$, $x \in I$. Similarly, if f is bounded below on I, we have the *greatest lower bound* or *infimum*, denoted by

$$\inf_{x \in I} f(x),$$

which is the greatest number which is not greater than any $f(x)$, $x \in I$.

Example 2.7 For $\sin x$ on $[0, 2\pi]$, we have

$$\inf_{0 \leqslant x \leqslant 2\pi} \sin x = -1, \qquad \sup_{0 \leqslant x \leqslant 2\pi} \sin x = 1,$$

the infimum and supremum being attained at $x = 3\pi/2$ and $\pi/2$ respectively. □

Example 2.8 For $1/x^2$ on $(-\infty, \infty)$ there is no supremum, since $1/x^2$ is not bounded above. However, we have

$$\inf_{-\infty < x < \infty} 1/x^2 = 0$$

although the infimum is *not attained*, since there is no value of x for which $1/x^2 = 0$. □

If a supremum is attained, we write 'max' (for maximum) in place of 'sup'. If an infimum is attained, we write 'min' (for minimum) in place of 'inf'.

2.2 Limits and derivatives

We define

$$g(x) = \frac{\sin x}{x}, \qquad 0 < x \leqslant \tfrac{1}{2}\pi.$$

Table 2.1 Limit of $(\sin x)/x$ as $x \to 0$

x	0.50	0.20	0.10	0.05	0.02
$(\sin x)/x$	0.959	0.993	0.998	1.000	1.000

Note that we have not defined $g(0)$. What can we say about $g(x)$ *near* $x = 0$? Inspection of a table of values of $\sin x$ (see Table 2.1) suggests that $g(x)$ approaches the value 1 as x approaches 0. To pursue this, we see from Fig. 2.5 that

$$\text{area of } \triangle\text{OAB} < \text{area of sector OAB} < \text{area of } \triangle\text{OAC}$$

or (see Problem 2.11)

$$\tfrac{1}{2}\sin x < \tfrac{1}{2}x < \tfrac{1}{2}\tan x, \qquad 0 < x < \tfrac{1}{2}\pi.$$

This yields

$$1 < \frac{x}{\sin x} < \frac{1}{\cos x}$$

and thus

$$\cos x < \frac{\sin x}{x} < 1. \tag{2.9}$$

Since $\cos x$ approaches the value 1 as x approaches 0, (2.9) shows that $(\sin x)/x$ also approaches 1 as x approaches 0. More precisely, we deduce from (2.9) (see Problem 2.12) that

$$0 < 1 - \frac{\sin x}{x} < \tfrac{1}{2}x^2. \tag{2.10}$$

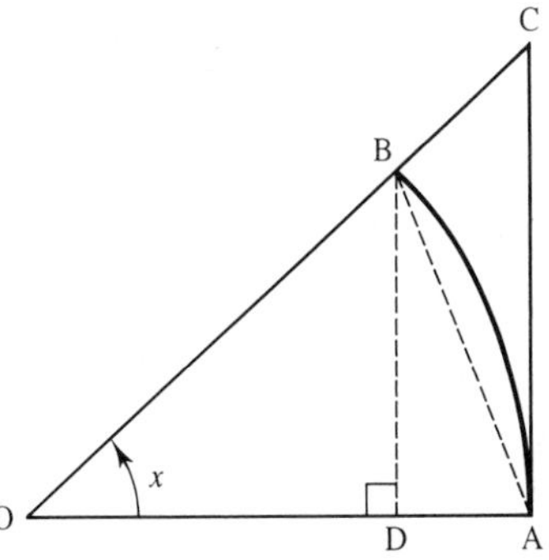

Fig. 2.5 OA = OB = 1; AB is a circular arc; OAC is a right angle.

This shows that $(\sin x)/x$ may be made as near to 1 *as we please* for all values of x sufficiently near to zero. This is an example of a *limit.*

Definition 2.8 Suppose that $x_0 \in [a, b]$ and the function g is defined at every point of $[a, b]$ with the possible exception of x_0. We say that g has limit c at x_0 if to each $\varepsilon > 0$ there corresponds a number $\delta > 0$ such that

$$x \in [a, b] \text{ with } |x - x_0| < \delta \text{ and } x \neq x_0 \Rightarrow |g(x) - c| < \varepsilon.$$

As a short-hand for the classical ε–δ terminology, we write

$$\lim_{x \to x_0} g(x) = c. \qquad \square$$

Example 2.9 The function $(\sin x)/x$ has the limit 1 at $x = 0$. Given $\varepsilon > 0$, we may take $\delta = (2\varepsilon)^{1/2}$ since, for all x such that $x \in (0, \pi/2)$ with $|x| < (2\varepsilon)^{1/2}$, we have $\frac{1}{2}x^2 < \varepsilon$ and thus from (2.10)

$$\left| \frac{\sin x}{x} - 1 \right| < \varepsilon,$$

as required by Definition 2.8. $\square$

We say that

$$\lim_{x \to \infty} g(x) = c$$

if to any $\varepsilon > 0$ there corresponds a number M such that

$$|g(x) - c| < \varepsilon, \qquad \text{for all } x > M.$$

Properties of limits

If the functions g and h are both defined on some interval containing x_0, a point at which both have a limit, then

(i) $\lim_{x \to x_0} (\alpha g(x) + \beta h(x)) = \alpha \lim_{x \to x_0} g(x) + \beta \lim_{x \to x_0} h(x)$,

where α and β are any constants;

(ii) $\lim_{x \to x_0} g(x)h(x) = \left(\lim_{x \to x_0} g(x) \right) \left(\lim_{x \to x_0} h(x) \right)$;

(iii) $\lim_{x \to x_0} \dfrac{1}{g(x)} = 1 / \lim_{x \to x_0} g(x)$

provided $\lim_{x \to x_0} g(x) \neq 0$.

Continuity

Consider the function $g(x)$ defined by

$$g(x) = \begin{cases} -1, & -1 \leqslant x \leqslant 0 \\ 1, & 0 < x \leqslant 1. \end{cases}$$

We say that g has a *discontinuity* at $x=0$, where the value of $g(x)$ jumps from -1 on the left to $+1$ on the right.

Definition 2.9 A function g defined on $[a, b]$ is said to be *continuous* at $x_0 \in [a, b]$ if g has a limit at x_0 and this limit is $g(x_0)$. □

Thus g is continuous at a point $x_0 \in [a, b]$ if, to each $\varepsilon > 0$, there corresponds a number $\delta > 0$ such that

$$x \in [a, b] \text{ with } |x - x_0| < \delta \quad \Rightarrow \quad |g(x) - g(x_0)| < \varepsilon.$$

We say that g is continuous on $[a, b]$ if g is continuous at each $x_0 \in [a, b]$. The concept of continuity is applied also to functions defined on open and infinite intervals.

Definition 2.10 A function g, defined on $[a, b]$, is said to be *uniformly continuous* on $[a, b]$ if, to each $\varepsilon > 0$, there corresponds a number $\delta > 0$ such that

$$x \in [a, b] \text{ with } |x - x_0| < \delta \quad \Rightarrow \quad |g(x) - g(x_0)| < \varepsilon \tag{2.11}$$

for *every* $x_0 \in [a, b]$. □

The point of this definition is that, given an $\varepsilon > 0$, the corresponding δ which appears in (2.11) is *independent* of x_0, so that the same δ will serve for each point of $[a, b]$. We state without proof that if a function g is continuous on the closed interval $[a, b]$ it is also uniformly continuous on $[a, b]$.

We quote the following theorems concerning continuous functions. The first of these is a simple consequence of the properties of limits.

Theorem 2.1 If f and g are continuous on an interval I, so also are fg and $\alpha f + \beta g$, where α and β are constants. If f is continuous on I and $f(x) \neq 0$ for all $x \in I$ then $1/f$ is continuous on I. □

Theorem 2.2 If f is continuous on $[a, b]$ and η is a number lying between $f(a)$ and $f(b)$, then there is a number $\xi \in [a, b]$ such that $f(\xi) = \eta$. □

Thus a function f which is continuous on $[a, b]$ assumes every value between $f(a)$ and $f(b)$. As a special case of this, we have the following result.

Theorem 2.3 If f is continuous on $[a, b]$, there exists a number $\xi \in [a, b]$ such that

$$f(\xi) = \tfrac{1}{2}(f(a) + f(b)). \qquad \square$$

The notion of continuity is easily extended to functions of several variables. For example, for two variables we state the following.

Definition 2.11 A function $f(x, y)$ defined on some region R of the xy-plane is said to be continuous at a point (x_0, y_0) of R if, to each $\varepsilon > 0$, there corresponds a number $\delta > 0$ such that

$$(x, y) \in R \text{ with } |x - x_0| < \delta \text{ and } |y - y_0| < \delta \quad \Rightarrow \quad |f(x, y) - f(x_0, y_0)| < \varepsilon.$$
$\square$

Lipschitz condition

Definition 2.12 A function f, defined on $[a, b]$, is said to satisfy a *Lipschitz condition* on $[a, b]$ if there exists a constant $L > 0$ such that

$$|f(x_1) - f(x_2)| \leqslant L|x_1 - x_2|, \tag{2.12}$$

for all $x_1, x_2 \in [a, b]$. L is called the *Lipschitz constant.* $\square$

We may deduce from (2.12) that if f satisfies a Lipschitz condition on $[a, b]$, then f is uniformly continuous on $[a, b]$.

Example 2.10 Consider $f(x) \equiv x^2$ on $[-1, 1]$. Then

$$f(x_1) - f(x_2) = x_1^2 - x_2^2 = (x_1 - x_2)(x_1 + x_2).$$

Thus on $[-1, 1]$

$$|f(x_1) - f(x_2)| \leqslant 2|x_1 - x_2|,$$

showing that $f(x) \equiv x^2$ satisfies a Lipschitz condition on $[-1, 1]$ with Lipschitz constant 2. $\square$

Example 2.11 Consider $f(x) \equiv x^{1/2}$, with $x \geqslant 0$. Then

$$f(x_1) - f(x_2) = x_1^{1/2} - x_2^{1/2} = (x_1 - x_2)/(x_1^{1/2} + x_2^{1/2}).$$

By taking x_1 and x_2 sufficiently close to 0 we can make $1/(x_1^{1/2} + x_2^{1/2})$ as large as we please. Thus the function $x^{1/2}$ does not satisfy a Lipschitz condition on any interval which includes $x = 0$. $\square$

This last example demonstrates that not every continuous function satisfies a Lipschitz condition.

Derivatives

We are often interested in how much $f(x)$ changes for a given change in x. As a measure of the *rate of change* as x varies from, say, x_0 to $x_0 + h$, we may calculate the quotient

$$\frac{f(x_0+h)-f(x_0)}{(x_0+h)-x_0}=\frac{\text{QR}}{\text{PR}},$$

as shown in Fig. 2.6. Geometrically, this ratio is the gradient of the chord PQ. The number h (shown as positive in Fig. 2.6) may be positive or negative. The rate of change of $f(x)$ with respect to x at x_0 is defined as

$$\lim_{h\to 0}\frac{f(x_0+h)-f(x_0)}{h},$$

assuming that this limit exists. Geometrically, the limit gives the gradient of the tangent to the curve at P. Replacing x_0 by x, we write

$$f'(x)=\lim_{h\to 0}\frac{f(x+h)-f(x)}{h} \tag{2.13}$$

and call this limit the *derivative* of f with respect to x. Instead of $f'(x)$ we sometimes write $(\mathrm{d}/\mathrm{d}x)f(x)$, $\mathrm{d}f/\mathrm{d}x$ or $\mathrm{d}y/\mathrm{d}x$ if $y=f(x)$. If the derivative of f exists for all x in some interval I, we say that f is differentiable on I and write f' for the resulting function on I.

Properties of derivatives

(i) $$\frac{\mathrm{d}}{\mathrm{d}x}(\alpha f(x)+\beta g(x))=\alpha\frac{\mathrm{d}}{\mathrm{d}x}f(x)+\beta\frac{\mathrm{d}}{\mathrm{d}x}g(x). \tag{2.14}$$

(ii) $$\frac{\mathrm{d}}{\mathrm{d}x}(f(x)g(x))=f(x)\frac{\mathrm{d}}{\mathrm{d}x}g(x)+g(x)\frac{\mathrm{d}}{\mathrm{d}x}f(x). \tag{2.15}$$

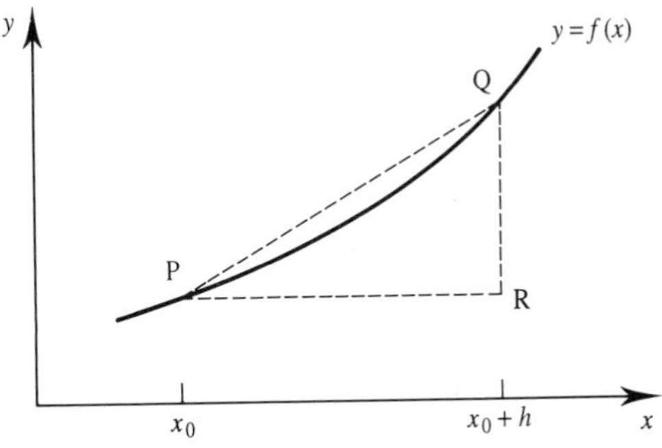

Fig. 2.6 Measurement of rate of change.

(iii) If we write $u = g(x)$, then

$$\frac{\mathrm{d}}{\mathrm{d}x} f(g(x)) = \frac{\mathrm{d}f}{\mathrm{d}u}\frac{\mathrm{d}u}{\mathrm{d}x} = f'(g(x))g'(x). \tag{2.16}$$

It is assumed that all the derivatives occurring in the right sides of these equations exist. The verification of these properties uses the basic properties of limits.

Property (iii) is called the *function of a function rule* or *chain rule*. One application of this rule concerns inverse functions. If the function $y = f(x)$ has an inverse function $x = \phi(y)$, we may write

$$y = f(\phi(y))$$

and deduce from (2.16) that

$$1 = \frac{\mathrm{d}y}{\mathrm{d}y} = \frac{\mathrm{d}y}{\mathrm{d}x}\frac{\mathrm{d}x}{\mathrm{d}y},$$

so that

$$\frac{\mathrm{d}x}{\mathrm{d}y} = 1 \Big/ \frac{\mathrm{d}y}{\mathrm{d}x}. \tag{2.17}$$

The derivative of an arbitrary polynomial is obtained by first showing directly from the definition of a derivative (see Problem 2.16) that

$$\frac{\mathrm{d}}{\mathrm{d}x} x^n = nx^{n-1}, \qquad n \text{ an integer.} \tag{2.18}$$

(This result holds also if n is any real number.) Applying (2.14) repeatedly and using (2.18) we find

$$\frac{\mathrm{d}}{\mathrm{d}x}(a_0 + a_1 x + a_2 x^2 + \cdots + a_n x^n) = a_1 + 2a_2 x + \cdots + na_n x^{n-1}.$$

For the derivatives of the circular functions, we find (Problem 2.18) that

$$\frac{\mathrm{d}}{\mathrm{d}x}\sin x = \cos x. \tag{2.19}$$

Making the substitution $t = \pi/2 - x$, using the chain rule and finally replacing t by x, we deduce from (2.19) that

$$\frac{\mathrm{d}}{\mathrm{d}x}\cos x = -\sin x.$$

We now discuss two simple but very important theorems involving derivatives.

Theorem 2.4 (Rolle's theorem) If f is continuous on $[a, b]$ and is differentiable in (a, b) and $f(a)=f(b)=0$, then there exists at least one point $\xi \in (a, b)$ such that $f'(\xi)=0$. □

We omit the proof. Notice that the same result holds if we substitute the single condition 'f is differentiable on $[a, b]$' in place of the more complicated conditions 'f is continuous on $[a, b]$ and is differentiable in (a, b)'. The reason why this simplification is not usually adopted is because it weakens the theorem's generality.

Let f be continuous on $[a, b]$ and differentiable in (a, b) and define

$$g(x)=f(x)-f(a)-(x-a)\left(\frac{f(b)-f(a)}{b-a}\right)$$

on $[a, b]$. (The function g is the difference between f and the polynomial which interpolates f at a and b. See (4.3).) It follows from the above conditions on f that g also is continuous on $[a, b]$ and differentiable in (a, b). In addition, we have $g(a)=g(b)=0$ and

$$g'(x)=f'(x)-\left(\frac{f(b)-f(a)}{b-a}\right)$$

in (a, b). Thus g satisfies the conditions of Rolle's theorem and we deduce that there exists a $\xi \in (a, b)$ such that $g'(\xi)=0$. Hence we obtain the following result.

Theorem 2.5 (mean value theorem) If f is continuous on $[a, b]$ and differentiable in (a, b), then there exists at least one point $\xi \in (a, b)$ such that

$$f(b)-f(a)=(b-a)f'(\xi). \tag{2.20}$$ □

Recall the remark after Theorem 2.4 on how the conditions on f could be simplified, at a cost of loss of generality of the theorem. This remark applies to Theorem 2.5 also.

Rolle's theorem is a special case of the mean value theorem, which is in turn a particular case of Taylor's theorem 3.1. We shall need Rolle's theorem in the proof of Theorem 3.1.

From the mean value theorem, if $x_1, x_2 \in [a, b]$ then

$$f(x_1)-f(x_2)=(x_1-x_2)f'(\xi),$$

where ξ depends on the choice of x_1 and x_2 and thus

$$|f(x_1)-f(x_2)| \leq L|x_1-x_2|,$$

where

$$L = \sup_{a \leq x \leq b} |f'(x)|.$$

Hence

differentiability $\Rightarrow$ Lipschitz condition $\Rightarrow$ uniform continuity.

We cannot reverse these implication symbols, as may be seen from Example 2.11 and Problem 2.19.

Higher derivatives

If the derivative of f' exists at x, we call this the *second derivative* of f and write it as $f''(x)$. Similarly, we define third, fourth and higher order derivatives from

$$f^{(n)}(x) = \frac{\mathrm{d}}{\mathrm{d}x}(f^{(n-1)}(x)), \qquad n = 3, 4, \ldots.$$

The nth derivative is sometimes written as $\mathrm{d}^n f/\mathrm{d}x^n$ or $\mathrm{d}^n y/\mathrm{d}x^n$ if $y=f(x)$. To unify the notation, we sometimes write $f^{(0)}(x)$ and $f^{(1)}(x)$ to denote $f(x)$ and $f'(x)$ respectively.

We have already noted the geometrical interpretation of $f'(x)$ as the gradient of the tangent to $y=f(x)$ at the point $(x, f(x))$. Thus if $f'(x)>0$, $f(x)$ is increasing and if $f'(x)<0$, $f(x)$ is decreasing. The second derivative too is easily related to the graph of $y=f(x)$: if $f''(x)>0$, f' is increasing and therefore f is convex (see Problem 2.20). If $f''(x)<0$, $-f$ is convex and thus f is concave.

We now consider a simple extension of Rolle's theorem which involves higher derivatives. Suppose that f is continuous on $[a, b]$ and differentiable in (a, b), and is zero at $n+1$ points, $t_0 < t_1 < \cdots < t_n$, of $[a, b]$. (For the special case of $n=1$ with $t_0=a$ and $t_1=b$, this just reproduces the conditions of Theorem 2.4.) Let us apply Theorem 2.4 to the function f on each of the n intervals $[t_{i-1}, t_i]$, $1 \leq i \leq n$. We deduce that f' has at least n zeros in (a, b) since there is a zero of f' in each of the subintervals (t_{i-1}, t_i). We now assume further that f' is continuous on $[a, b]$ and is differentiable in (a, b). Then if $n \geq 2$ we may apply Theorem 2.4 to the function f' on each of the intervals between consecutive pairs of its zeros to show that f'' has at least $n-1$ zeros in (a, b). This line of reasoning may clearly be extended and we deduce the following result.

Theorem 2.6 (extended Rolle theorem) If $f(t_i)=0,\ i=0, 1, \ldots, n$, where $a \leqslant t_0 < t_1 < \cdots < t_n \leqslant b$, and $f^{(n-1)}$ is continuous on $[a, b]$ and is differentiable in (a, b), then there exists at least one point $\xi \in (a, b)$ such that $f^{(n)}(\xi)=0$. □

Note (cf. the remarks made after Theorems 2.4 and 2.5) that the latter condition could be simplified by merely asking that $f^{(n)}$ exists on $[a, b]$, resulting in a loss of generality of the theorem.

Maxima and minima

Definition 2.13 We say that $f(x_0)$ is a local maximum of a function f if there exists a number $\delta > 0$ such that $f(x_0) > f(x)$ for all x such that $x \neq x_0$ and $|x - x_0| < \delta$. □

We often drop the adjective 'local' and speak simply of a maximum of $f(x)$ at $x = x_0$, meaning that $f(x_0)$ is the greatest value of $f(x)$ in the vicinity of x_0. Similarly we say we have a minimum of $f(x)$ at $x = x_0$ if $f(x_0)$ is the smallest value of $f(x)$ in the vicinity of x_0. If $f(x_0)$ is the greatest (least) value of $f(x)$ for *all* $x \in [a, b]$, we call $f(x_0)$ the *global* maximum (minimum) on $[a, b]$.

Theorem 2.7 If f'' is continuous at x_0 and $f'(x_0) = 0$ then $f(x_0)$ is a local maximum (minimum) if $f''(x_0) < 0$ (>0). □

The truth of this result is intuitively obvious from our remark above that $f''(x_0) < 0$ (>0) implies that the function $f(x)$ is concave (convex). The theorem is easily proved (Problem 3.5) with the aid of Taylor's theorem. Notice that Theorem 2.7 tells us nothing if $f'(x_0) = f''(x_0) = 0$.

Partial derivatives

For a function of two variables, $f(x, y)$, it is convenient to analyse the rate of change of f with respect to x and y *separately*. This involves the use of a *partial derivative*, defined by

$$\frac{\partial f}{\partial x} = \lim_{h \to 0} \frac{f(x+h, y) - f(x, y)}{h} \tag{2.21}$$

if the limit exists. Note that $\partial f/\partial x$ is obtained by treating y as a constant and differentiating with respect to x in the usual way. Similarly we obtain $\partial f/\partial y$ on differentiating f with respect to y, treating x as a constant.

For given values $x = x_0$, $y = y_0$, $\partial f/\partial x$ is the gradient of the tangent *in the x-direction* to the surface $z = f(x, y)$ at the point (x_0, y_0). Sometimes we write f_x and f_y for $\partial f/\partial x$ and $\partial f/\partial y$ respectively. We also have higher partial derivatives, for example

$$f_{xx} = \frac{\partial^2 f}{\partial x^2} = \frac{\partial}{\partial x}\left(\frac{\partial f}{\partial x}\right),$$

$$f_{xy} = \frac{\partial^2 f}{\partial x\,\partial y} = \frac{\partial}{\partial x}\left(\frac{\partial f}{\partial y}\right).$$

Example 2.12 If $f(x, y) = x^2 \cos y$, we have

$$\begin{aligned} f_x &= 2x \cos y, & f_y &= -x^2 \sin y,\\ f_{xx} &= 2 \cos y, & f_{xy} &= -2x \sin y,\\ f_{yx} &= -2x \sin y, & f_{yy} &= -x^2 \cos y. \end{aligned}$$

□

If f is a function of two variables x and y and each of these variables is in turn a function of say t, we have the function of a function (or chain) rule

$$\frac{\mathrm{d}f}{\mathrm{d}t} = \frac{\partial f}{\partial x}\frac{\mathrm{d}x}{\mathrm{d}t} + \frac{\partial f}{\partial y}\frac{\mathrm{d}y}{\mathrm{d}t}. \tag{2.22}$$

The definition of a local maximum or minimum of $f(x, y)$ at (x_0, y_0) is similar to that for a function of one variable. For a local maximum or minimum, we require $f_x = f_y = 0$ at (x_0, y_0) but these are not sufficient conditions.

2.3 Sequences and series

Sequences

Consider a function whose domain is the set of positive integers such that the integer r is mapped into an element u_r. The elements $u_1, u_2, u_3, \ldots$ are called collectively a *sequence*, which we will denote by (u_r) or $(u_r)_{r=1}^{\infty}$. The sequences in which we will be most interested are those with real members (elements).

Definition 2.14 We say that the sequence of real numbers $(u_r)_{r=1}^{\infty}$ *converges* if there is a number, say u, such that

$$\lim_{r\to\infty} u_r = u, \tag{2.23}$$

and u is called the limit of the sequence. □

By the limit (2.23) we mean that to each $\varepsilon > 0$ there corresponds a number N such that

$$r > N \quad \Rightarrow \quad |u_r - u| < \varepsilon.$$

If a sequence does not converge, it is said to *diverge*.

Principle of induction

Let S_r denote a statement involving the positive integer r. If

(i) S_1 is true,
(ii) for each $k \geqslant 1$, the truth of S_k implies the truth of S_{k+1},

then *every* statement in the sequence $(S_r)_{r=1}^{\infty}$ is true. This is known as the *principle of induction*.

Example 2.13 If $0 \leqslant \alpha \leqslant 1$, show that

$$(1-\alpha)^r \geqslant 1 - r\alpha, \qquad r = 1, 2, \ldots. \tag{2.24}$$

If S_r denotes the inequality (2.24), we see that S_1 is obviously true. Assuming S_k is true, we have

$$(1-\alpha)^k \geqslant 1 - k\alpha$$

and, on multiplying each side of this inequality by $1 - \alpha \geqslant 0$, we have

$$\begin{aligned}(1-\alpha)^{k+1} &\geqslant (1-\alpha)(1-k\alpha) \\ &= 1 - (k+1)\alpha + k\alpha^2 \geqslant 1 - (k+1)\alpha,\end{aligned}$$

showing that S_{k+1} is true. By induction, the result is established. □

Convergence of sequences

Having defined convergence of a sequence (Definition 2.14) we now state a result which is often referred to as Cauchy's principle of convergence.

Theorem 2.8 (Cauchy) The sequence $(u_r)_{r=1}^{\infty}$ converges if, and only if, to each $\varepsilon > 0$, there corresponds a number N such that

$$n > N \Rightarrow |u_{n+k} - u_n| < \varepsilon, \qquad \text{for } k = 1, 2, \ldots. \quad \square$$

This result enables us to discuss convergence of a sequence without necessarily knowing the value of the limit to which the sequence converges.

Series

A sum

$$c_1 + c_2 + \cdots + c_n,$$

where $c_1, c_2, \ldots, c_n$ are real numbers, is called a *finite* series. Suppose that we have an infinite sequence of real numbers $(c_r)_{r=1}^{\infty}$ and that

$$u_n = c_1 + c_2 + \cdots + c_n. \tag{2.25}$$

If the sequence $(u_n)_{n=1}^{\infty}$ converges to a limit u, we write

$$u = c_1 + c_2 + \cdots \tag{2.26}$$

and say that the *infinite series* on the right of (2.26) is convergent with sum u. We call u_n the nth *partial sum* of the infinite series.

Example 2.14 Consider the infinite series

$$1 + \frac{1}{2} + \frac{1}{3} + \cdots + \frac{1}{n} + \cdots.$$

From the inequalities

$$\frac{1}{3} + \frac{1}{4} > 2 \times \frac{1}{4} = \frac{1}{2},$$

$$\frac{1}{5} + \frac{1}{6} + \frac{1}{7} + \frac{1}{8} > 4 \times \frac{1}{8} = \frac{1}{2},$$

and so on, we have

$$1 + \frac{1}{2} + \frac{1}{3} + \cdots + \frac{1}{2^n} \geqslant 1 + \frac{1}{2}n.$$

Thus the partial sums can be made as large as we please by taking enough terms and the series is divergent. □

Uniform convergence

Consider now a sequence $(u_r(x))_{r=0}^{\infty}$ of which each member $u_r(x)$ is a function of x, defined on the interval $[a, b]$.

Definition 2.15 The sequence $(u_r(x))_{r=0}^{\infty}$ is said to converge *pointwise* on $[a, b]$ if, for each $x \in [a, b]$, the sequence converges. □

Definition 2.16 The sequence $(u_r(x))_{r=0}^{\infty}$ is said to converge *uniformly* to a function $u(x)$ on $[a, b]$ if the sequence

$$\left(\sup_{a \leqslant x \leqslant b} |u_r(x) - u(x)| \right)_{r=0}^{\infty} \tag{2.27}$$

converges to zero. □

It is clear that uniform convergence to $u(x)$ implies pointwise convergence to $u(x)$. We now see that the converse does not always hold.

Example 2.15 Consider the sequence (x^r) on $[0, 1]$. For any $x \in [0, 1)$, $x^r \to 0$ as $r \to \infty$ and at $x = 1$, $x^r \to 1$ as $r \to \infty$. Thus (x^r) converges pointwise on $[0, 1]$ to the limit function $u(x)$ defined by

$$u(x) = \begin{cases} 0, & 0 \leqslant x < 1 \\ 1, & x = 1. \end{cases}$$

However, for any $r \geqslant 0$,

$$\sup_{0 \leqslant x \leqslant 1} |x^r - u(x)| = \sup_{0 \leqslant x < 1} |x^r| = 1$$

and therefore the sequence (x^r) does *not* converge uniformly (see Fig. 2.7). □

In Example 2.15, each member of the sequence is continuous on $[0, 1]$ but the limit function $u(x)$ is *not* continuous. This could not have occurred had the convergence been uniform, as the following theorem shows.

Theorem 2.9 If each element of the sequence $(u_r(x))_{r=0}^{\infty}$ is continuous on $[a, b]$ and the sequence converges uniformly to $u(x)$ on $[a, b]$, then $u(x)$ is continuous on $[a, b]$. □

If all the $u_r(x)$ and $u(x)$ are continuous on $[a, b]$ we can replace the 'sup' in (2.27) by 'max'.

As an alternative definition of uniform convergence of $(u_r(x))_{r=0}^{\infty}$, we may say that to each $\varepsilon > 0$ there corresponds a number N such that

$$n > N \Rightarrow |u_n(x) - u(x)| < \varepsilon, \qquad \text{for all } x \in [a, b].$$

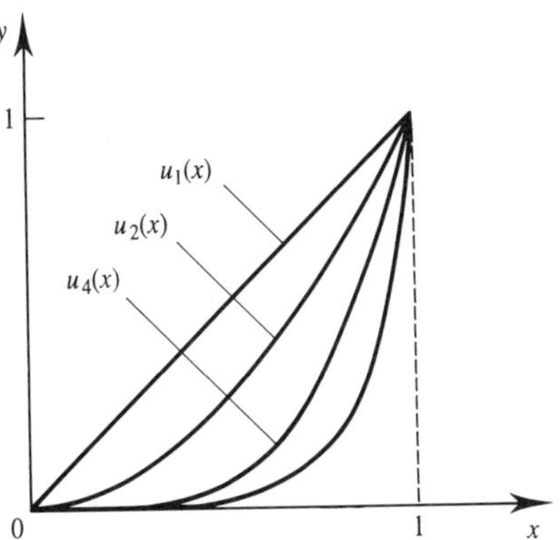

Fig. 2.7 Non-uniform convergence.

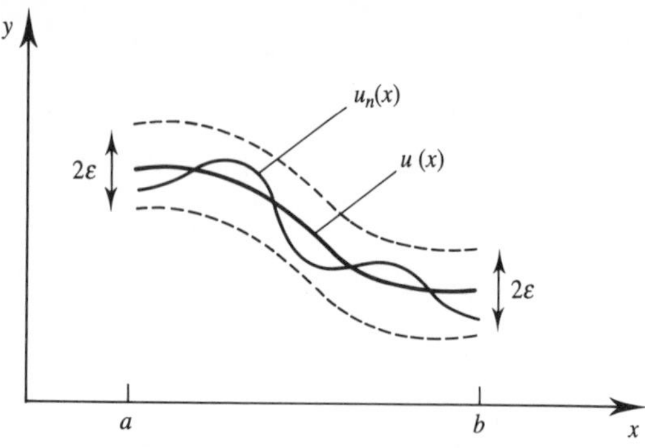

Fig. 2.8 Uniform convergence.

Notice that N does not depend on x. This statement is represented in Fig. 2.8. Given ε, we draw a strip about $u(x)$ of width 2ε: for $n>N$, the graph of $u_n(x)$ lies entirely inside this strip. We can make the strip as narrow as we please.

The following theorem (cf. Theorem 2.8) is also useful.

Theorem 2.10 The sequence $(u_r(x))_{r=0}^{\infty}$ converges uniformly on $[a, b]$ if, to each $\varepsilon>0$, there corresponds a number N such that

$$n>N \Rightarrow |u_{n+k}(x)-u_n(x)|<\varepsilon$$

for all $x \in [a, b]$ and $k=1, 2, \ldots$. □

Let $c_1, c_2, \ldots$ denote an infinite sequence of functions defined on $[a, b]$. We will form partial sums (as we did for a sequence of real numbers in (2.25)) and write

$$u_n(x)=c_1(x)+c_2(x)+\cdots+c_n(x)$$

for $a \leqslant x \leqslant b$. If the sequence of functions $(u_n(x))_{n=1}^{\infty}$ converges uniformly on $[a, b]$ to a limit function $u(x)$, we say that the infinite series $c_1(x)+c_2(x)+\cdots$ converges uniformly to $u(x)$ on $[a, b]$ and write

$$u(x)=c_1(x)+c_2(x)+\cdots. \tag{2.28}$$

2.4 Integration

We consider a function f, continuous on an interval $[a, b]$ which we divide into N equal sub-intervals with points of sub-division $x_r=x_0+rh$, $r=0, 1, \ldots, N$, with $x_0=a$, so that $x_N=b$ and $h=(b-a)/N$. Then the sum

$$S_N=\sum_{r=1}^{N} hf(x_r) \tag{2.29}$$

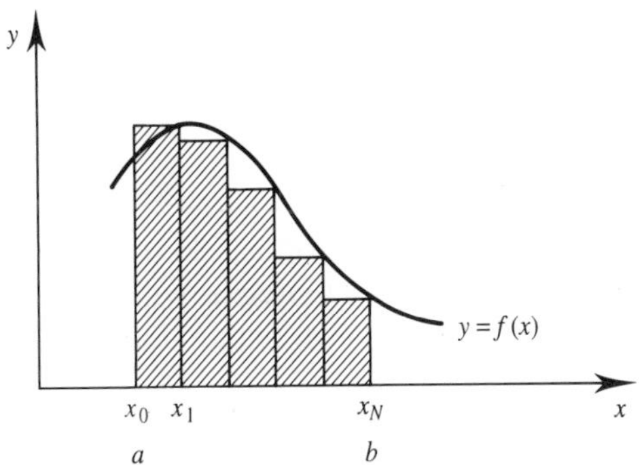

Fig. 2.9 Area approximation by a 'sum' of rectangles.

is the sum of the areas of the rectangles depicted in Fig. 2.9. If $(S_N)_{N=1}^{\infty}$ converges, we write

$$\lim_{N\to\infty} S_N = \int_a^b f(x)\,\mathrm{d}x. \tag{2.30}$$

We call this limit a *definite integral* and say that f is *integrable* on $[a, b]$. It can be shown that the limit always exists if f is continuous and the definite integral is thus interpreted as the area under the curve $y = f(x)$, above the x-axis (assuming f is positive) and between the limits $x = a$ and $x = b$. If we allow the rectangles, which contribute to the sum S_N, to vary in width and ask that the longest width tends to zero as $N \to \infty$, the sequence (S_N) still converges to the same limit. Also, we remark that continuity of f is not a necessary condition for convergence of the sequence (S_N).

In (2.30), the right side is a constant and x is therefore called a *dummy variable*. Thus we could equally express (2.30) as

$$\lim_{N\to\infty} S_N = \int_a^b f(t)\,\mathrm{d}t.$$

Now we allow the upper limit b to vary, so that the integral becomes a function of b. Writing x for b, we define

$$F(x) = \int_a^x f(t)\,\mathrm{d}t$$

and note that $F(a) = 0$. We have the following result.

Theorem 2.11 If f is continuous on $[a, b]$, then

$$F'(x) = f(x), \qquad a \leqslant x \leqslant b. \qquad \square$$

We write $\int f(x)\,\mathrm{d}x$, called an *indefinite integral*, to denote any function ϕ such that $\phi'(x) = f(x)$. The function is sometimes called an *antiderivative* of f. From Theorem 2.11 we deduce the following.

Theorem 2.12

$$\int_a^b f(t)\,\mathrm{d}t = \phi(b) - \phi(a), \tag{2.31}$$

where ϕ is any anti-derivative of f. □

We usually abbreviate the right side of (2.31) to $[\phi(t)]_a^b$. This important theorem allows us often to evaluate an integral without explicitly relying on the limiting process (2.30). We shall also require the following results.

Theorem 2.13 (mean value theorem for integrals) If f is continuous and g is integrable on $[a, b]$ and $g(x) \geqslant 0$ for $a \leqslant x \leqslant b$, then there exists a number $\xi \in (a, b)$ such that

$$\int_a^b f(x)g(x)\,\mathrm{d}x = f(\xi)\int_a^b g(x)\,\mathrm{d}x.$$ □

Theorem 2.14 (integration by substitution) If g' is continuous on $[a, b]$,

$$\int_a^b f(g(t))g'(t)\,\mathrm{d}t = \int_{g(a)}^{g(b)} f(u)\,\mathrm{d}u.$$

(We obtain the integral on the right from that on the left by making the substitution $u = g(t)$.) □

Theorem 2.15 (integration by parts) If f' and g' are continuous on $[a, b]$,

$$\int_a^b f(x)g'(x)\,\mathrm{d}x = [f(x)g(x)]_a^b - \int_a^b f'(x)g(x)\,\mathrm{d}x. \tag{2.32}$$

This is proved by integrating (2.15). □

Suppose that a given infinite series $c_1(x) + c_2(x) + \cdots$ converges uniformly to $u(x)$ on $[a, b]$ and that all of these functions can be integrated to give, say,

$$U(x) = \int_a^x u(t)\,\mathrm{d}t \quad \text{and} \quad C_j(x) = \int_a^x c_j(t)\,\mathrm{d}t, \qquad j = 1, 2, \ldots \tag{2.33}$$

for $a \leqslant x \leqslant b$. Then it can be proved that

$$U(x) = C_1(x) + C_2(x) + \cdots, \tag{2.34}$$

where the series on the right of (2.34) converges uniformly to $U(x)$ on $[a, b]$. Thus, in view of (2.33), we have integrated the given infinite series *term-by-term* to give (2.34). This result is summarized as follows.

Theorem 2.16 (term-by term integration) If an infinite series of functions converges uniformly to a given limit function $u(x)$ on $[a, b]$, then the series can be integrated term-by-term to give a new series which converges uniformly on $[a, b]$ to the integral of the limit function $u(x)$. □

For an application of this theorem, see Example 7.1 of Chapter 7.

2.5 Logarithmic and exponential functions

We define the function

$$\log x = \int_1^x \frac{\mathrm{d}t}{t} \tag{2.35}$$

for $x>0$ and call $\log x$ the *natural logarithm* of x. The integral in (2.35) exists for all $x>0$ as the integrand is continuous. From the definition and Theorem 2.11 we have

$$\frac{\mathrm{d}}{\mathrm{d}x} \log x = 1/x. \tag{2.36}$$

From (2.35) we also deduce (Problem 2.28) that

$$\log x_1 x_2 = \log x_1 + \log x_2 \tag{2.37}$$

for $x_1>0$, $x_2>0$. (It is this property of the logarithm, allowing multiplications to be replaced by additions, which motivated the construction of logarithm tables in the early part of the seventeenth century.) If $N>2$ is a positive integer,

$$\log N = \int_1^N \frac{\mathrm{d}t}{t} = \sum_{r=2}^{N} \int_{r-1}^{r} \frac{\mathrm{d}t}{t}$$

and thus

$$\log N > \frac{1}{2} + \frac{1}{3} + \cdots + \frac{1}{N}. \tag{2.38}$$

From this inequality and Example 2.14, it follows that $\log x \to \infty$ as $x \to \infty$. Also putting $x_1 = N$, $x_2 = 1/N$ in (2.37) and noting that $\log 1 = 0$, we have

$$\log(1/N) = -\log N$$

and we see that $\log x \to -\infty$ as $x \to 0$ from above.

Since the logarithm function is evidently one–one, we may define its inverse function. If $x = \log y$, $0 < y < \infty$, we define the inverse function

$$y = \exp(x), \tag{2.39}$$

which is thus defined for $-\infty < x < \infty$. We see that $\exp(x) \to \infty$ as $x \to \infty$ and $\exp(x) \to 0$ as $x \to -\infty$. We deduce from (2.37) (see Problem 2.29) that

$$\exp(x_1)\exp(x_2) = \exp(x_1 + x_2), \tag{2.40}$$

where $-\infty < x_1, x_2 < \infty$.

To obtain the derivative of $y = \exp(x)$, we use (2.17) to give

$$\frac{dy}{dx} = 1 \bigg/ \frac{dx}{dy}$$

and

$$\frac{dx}{dy} = \frac{d}{dy}(\log y) = 1/y = 1/\exp(x).$$

Thus

$$\frac{d}{dx}(\exp(x)) = \exp(x). \tag{2.41}$$

Hereafter, we shall write e^x in place of $\exp(x)$. The number $e = e^1$ evidently satisfies the equation

$$\int_1^e \frac{dt}{t} = 1$$

and, in fact, $e = 2.71828\ldots$. For any $a > 0$ we define

$$a^x = e^{x \log a} \tag{2.42}$$

and this is also called an *exponential function*. It follows from (2.41) that

$$\frac{d}{dx} a^x = a^x \log a,$$

and it is essentially the simplification which occurs for $a = e$, namely

$$\frac{d}{dx} e^x = e^x,$$

which makes e^x *the* exponential function.

Problems

Section 2.1

2.1 For the following functions f, state the range of f and say which functions have an inverse.

(i) $1+x+x^2$, $x \in (-\infty, \infty)$,
(ii) $(1+x)/(2+x)$, $x \in [-1, 1]$,
(iii) $1+x+x^2$, $x \in [0, \infty)$.

2.2 Show that, for any real numbers x and α,

$$x^n - \alpha^n = (x-\alpha)(x^{n-1} + \alpha x^{n-2} + \alpha^2 x^{n-3} + \cdots + \alpha^{n-1}).$$

Hence, if $p(x) = a_0 + a_1 x + \cdots + a_n x^n$, $a_n \neq 0$, show that

$$p(x) - p(\alpha) = (x-\alpha)q(x),$$

where q is a polynomial of degree $n-1$, and thus obtain (2.2).

2.3 Suppose that $x_1 < x_2 < \cdots < x_k$ and that p is a polynomial which satisfies the conditions $(-1)p(x_1) > 0$, $(-1)^k p(x_k) > 0$ and $p(x_i) = 0$ for $2 \leqslant i \leqslant k-1$. Show that p has at least $k-1$ zeros in (x_1, x_k). (*Hint*: if k is even, $p(x_1).p(x_k) < 0$ and p has therefore an *odd* number of zeros in (x_1, x_k). Investigate also the case where k is odd.)

2.4 With the notation of the previous problem, show that if p satisfies the conditions $(-1)^i p(x_i) \geqslant 0$, $1 \leqslant i \leqslant k$, then p has at least $k-1$ zeros in $[x_1, x_k]$.

2.5 If p and q are polynomials and

$$p(x) = (x-\alpha)^{r+1} q(x),$$

show that $p^{(k)}(\alpha) = 0$ for $k = 1, 2, \ldots, r$.

2.6 Assuming that (2.3) (for all values of θ) and (2.7) hold, deduce that (2.8) holds.

2.7 Assuming that (2.3) holds for all values of θ, deduce from the behaviour of the sine function that $\cos\theta$ satisfies the properties $\cos(\pi - \theta) = -\cos\theta$ and $\cos(-\theta) = \cos\theta$.

2.8 By writing $1 - \cos x = \cos 0 + \cos(\pi - x)$ and using (2.8), or otherwise, deduce that $1 - \cos x = 2\sin^2 \tfrac{1}{2}x$.

2.9 Show directly from Definition 2.6 that $y = x^2$ is convex on $(-\infty, \infty)$.

2.10 Investigate which of the following suprema and infima exist and, of those that exist, find which are attained.

(i) $\sup_{1<x<\infty} (2-x-x^2)$, (ii) $\inf_{-\infty \leqslant x < \infty} (2+\sin x)e^x$,

(iii) $\inf_{0<x<\infty} |\cos x \log x|$ (iv) $\sup_{0 \leqslant x < \infty} (1-x-x^2)$.

Section 2.2

2.11 In Fig. 2.5 show that

(i) area of ΔOAB $= \frac{1}{2}$ OA.BD $= \frac{1}{2}\sin x$,
(ii) area of ΔOAC $= \frac{1}{2}$ OA.AC $= \frac{1}{2}\tan x$,
(iii) area of sector OAB $= \frac{1}{2}x$, assuming that the area of a circle with radius unity is π.

2.12 Deduce from (2.9) that

$$0 < 1 - \frac{\sin x}{x} < 1 - \cos x, \qquad 0 < x < \tfrac{1}{2}\pi.$$

Use the result of Problem 2.8 and the right inequality of (2.9) to show that

$$0 < 1 - \cos x < \tfrac{1}{2}x^2, \qquad 0 < x < \tfrac{1}{2}\pi,$$

thus verifying (2.10).

2.13 Which of the following functions are continuous on $[-1, 1]$?

$$\text{(i) } x^n, \quad \text{(ii) } |x|, \quad \text{(iii) } \sin\left(\frac{1}{1+x}\right), \quad \text{(iv) } \cot x.$$

2.14 Verify Theorem 2.1.

2.15 Show that if f satisfies a Lipschitz condition on $[a, b]$, f is uniformly continuous on $[a, b]$.

2.16 Using the identity

$$a^n - b^n = (a-b)(a^{n-1} + a^{n-2}b + \cdots + ab^{n-2} + b^{n-1}),$$

n a positive integer, verify (2.18). (Treat separately the cases n positive and n negative in (2.18).)

2.17 From the *definition*, obtain the derivatives of

(i) $2x^2 + 3x$, (ii) $1/(x+1)$, (iii) $x^{1/2}$.

2.18 Writing $\sin(x+h) - \sin x = \sin(x+h) + \sin(-x)$ and using (2.7) and the result of Example 2.9, find the derivative of $\sin x$ directly from the definition of a derivative.

2.19 If $f(x) \equiv |x|$ on $[-1, 1]$, show that f satisfies a Lipschitz condition with Lipschitz constant $L = 1$ on $[-1, 1]$, but that $f'(x)$ does not exist at $x = 0$.

2.20 Suppose that f'' is continuous on $[x_1, x_2]$ and that $p_1(x) = 0$ denotes the equation of the chord joining the points $(x_r, f(x_r))$, $r = 1, 2$, so that

$p_1(x_r) = f(x_r)$. Then, anticipating the result of (4.13), we have

$$f(x) - p_1(x) = \tfrac{1}{2}(x - x_1)(x - x_2)f''(\xi),$$

for some $\xi \in [x_1, x_2]$. Deduce that if $x \in [x_1, x_2]$ and $f''(x) > 0$ on $[x_1, x_2]$ then f is convex on $[x_1, x_2]$.

Section 2.3

2.21 Show that the sequence $(u_n)_{n=1}^{\infty}$ converges, where $u_n = (n+1)/2n$.

2.22 Prove by induction that

$$1^2 + 2^2 + \cdots + n^2 = n(n+1)(2n+1)/6$$

and hence find the limit of the sequence (s_n), where

$$s_n = \frac{1}{n}\left(\left(\frac{1}{n}\right)^2 + \left(\frac{2}{n}\right)^2 + \cdots + \left(\frac{n}{n}\right)^2\right).$$

2.23 If $u_n(x) = a_0 + a_1 x + \cdots + a_n x^n$ and $|a_r| \leqslant K$ for all r, use Theorem 2.10 to show that $(u_n(x))_{n=0}^{\infty}$ converges uniformly in any interval $[-\rho, \rho]$, where $0 < \rho < 1$.

2.24 Show that the infinite series

$$1 + \tfrac{1}{2} + \tfrac{1}{4} + \tfrac{1}{8} + \tfrac{1}{16} + \cdots$$

has partial sums

$$u_r = 2 - \frac{1}{2^r}, \qquad r = 0, 1, 2, \ldots,$$

and deduce that the series is convergent with sum 2.

2.25 Show that the sequence $(u_n(x))_{n=1}^{\infty}$ with

$$u_n(x) = \begin{cases} nx, & 0 \leqslant x \leqslant 1/n \\ 2 - nx, & 1/n < x < 2/n \\ 0, & 2/n \leqslant x \leqslant 2 \end{cases}$$

is pointwise convergent to $u(x) \equiv 0$ on $[0, 2]$ but that it is not uniformly convergent.

Section 2.4

2.26 Evaluate $\int_0^1 x^2\,dx$ in two different ways:

(i) by finding an anti-derivative (Theorem 2.12),
(ii) by treating the integral as the limit of a sum, as in (2.29), using the result of Problem 2.22.

2.27 Deduce from Theorem 2.13 that, if f is continuous on $[a, b]$,

$$m(b-a) \leqslant \int_a^b f(x)\, \mathrm{d}x \leqslant M(b-a)$$

where m and M denote the minimum and maximum values of $f(x)$ on $[a, b]$.

Section 2.5

2.28 For $x_1, x_2 > 0$, write $\log x_1 x_2$ as an integral, as in the definition (2.35). Split the range of integration $[1, x_1 x_2]$ into $[1, x_1]$ and $[x_1, x_1 x_2]$ and make the change of variable $t = x_1 u$ in the second integral. Thus establish (2.37).

2.29 If $y_i = \exp(x_i)$, $i = 1, 2$, use (2.37) to deduce (2.40).

Chapter 3

TAYLOR'S POLYNOMIAL AND SERIES

3.1 Function approximation

One very important aspect of numerical analysis is the study of approximations to functions. Suppose that f is a function defined on some interval of the reals. We now seek some other function g which 'mimics' the behaviour of f on some interval. We say that g is an *approximation* to f on that interval. Usually we make such an approximation because we wish to carry out some numerical calculation or analytical operation involving f, but we find this difficult or impossible because of the nature of f. For example, we may wish to find the integral of f over some interval and there may be no explicit formula for such an integral. We replace f by a function g which may easily be integrated. Other problems include:

- evaluating f for a particular x, especially if f is defined to be the solution of some equation;
- differentiation of f;
- determining zeros of f (that is, finding roots of $f(x) = 0$);
- determining extrema of f.

In all such cases we may need to replace f by some approximation g. We choose g so that it is more amenable than f to the type of operation involved. We can rarely then determine the exact solution of the original problem, but we do hope to obtain an approximation to the solution. There are, of course, many different ways in which an approximating function g may be chosen but we would normally restrict g to a certain class of functions. The most commonly used class is the polynomials. These are easily integrated and differentiated and are well-behaved, in that all derivatives exist and are continuous.

One very desirable property, which we expect of any class of approximating functions, is that, for any function f with certain basic properties on some finite interval, it is possible to approximate to f to within arbitrary accuracy on that interval. The class of all polynomials satisfies this property, with only the requirement that f be continuous. This remarkable result is embodied in Weierstrass's theorem† in Chapter 5. It must be stressed, however, that polynomials are not the only functions with this property but are certainly the most convenient for general manipulation.

Even if the approximating function g is restricted to be a polynomial of a given degree, there are still many ways of choosing g and the question of determining the most suitable g is very complicated. We will return to this point in Chapter 5.

3.2 Taylor's theorem

In this chapter we shall consider a polynomial approximation which mimics a function f near one given point. We will restrict our choice of polynomial to an element of P_n, the set of all polynomials of degree not exceeding n. We will seek coefficients $a_0, a_1, \ldots, a_n$ such that the polynomial

$$p_n(x) = a_0 + a_1 x + \cdots + a_n x^n$$

approximates to f near $x = x_0$. Assuming that f is n times differentiable at $x = x_0$, we try to choose the coefficients of p_n so that

$$p_n^{(j)}(x_0) = f^{(j)}(x_0), \qquad j = 0, 1, 2, \ldots, n, \tag{3.1}$$

where $p_n^{(0)}(x) \equiv p_n(x)$ and $f^{(0)}(x) \equiv f(x)$. Thus we seek p_n such that the values of p_n and f are equal at $x = x_0$ and the values of their first n derivatives are equal at $x = x_0$.

The equations (3.1) are

$$\left.\begin{aligned}
a_0 + a_1x_0 + a_2x_0^2 + \quad \cdots \quad + a_{n-1}x_0^{n-1} + a_n x_0^n &= f(x_0)\\
a_1 + 2a_2x_0 + \cdots + (n-1)a_{n-1}x_0^{n-2} + na_nx_0^{n-1} &= f'(x_0)\\
&\;\;\vdots\\
(n-1)!a_{n-1} + n(n-1)\ldots 2a_nx_0 &= f^{(n-1)}(x_0)\\
n!a_n &= f^{(n)}(x_0).
\end{aligned}\right\} \tag{3.2}$$

These are $n+1$ equations‡ in the $n+1$ unknown coefficients $a_0, a_1, \ldots, a_n$ and we see that these coefficients are determined uniquely by (3.2). The last equation in (3.2) gives a_n and then, from the penultimate equation, we may determine a_{n-1}. By working through the equations in reverse order we may determine $a_n, a_{n-1}, a_{n-2}, \ldots, a_0$ uniquely. Hence p_n is unique, that is, there is only one member of P_n satisfying (3.1).

† Due to the German mathematician K. Weierstrass (1815–97).

‡ These are triangular equations, and their solution is discussed in detail in Chapter 9.

We do not usually bother to obtain $a_0, a_1, \ldots, a_n$ explicitly (see Problem 3.1) but instead write p_n in the form

$$p_n(x) = f(x_0) + \frac{(x-x_0)}{1!} f'(x_0) + \frac{(x-x_0)^2}{2!} f''(x_0) + \cdots + \frac{(x-x_0)^n}{n!} f^{(n)}(x_0). \quad (3.3)$$

It should be noted that (3.3) is a polynomial in x, as each term is a polynomial in x. (Remember that x_0 is a constant.) We call $p_n(x)$ of (3.3) the *Taylor polynomial*† of f constructed at the point $x = x_0$. The reader is probably familiar with (3.3) as the first $n+1$ terms of the *Taylor series* of $f(x)$; however, it is instructive to verify that the Taylor polynomial does indeed satisfy (3.1).

Example 3.1 For $f(x) = e^x$ (the exponential function), $f^{(r)}(x) = e^x$ and $f^{(r)}(0) = 1$ for $r = 0, 1, 2, \ldots$. Thus the Taylor polynomial of degree n constructed at $x = 0$ is

$$p_n(x) = 1 + x + \frac{x^2}{2!} + \frac{x^3}{3!} + \cdots + \frac{x^n}{n!}. \qquad \square$$

Example 3.2 For $f(x) = \sin x$,

$$f^{(2r)}(x) = (-1)^r \sin x \quad \text{and} \quad f^{(2r+1)}(x) = (-1)^r \cos x, \qquad r = 0, 1, 2, \ldots.$$

Thus, for example, the Taylor polynomial of degree 4 constructed at $x = 0$ is

$$p_4(x) = x - \frac{x^3}{3!}.$$

In this case the coefficients of 1, x^2 and x^4 are zero.

At $x = \pi/4$ we obtain

$$p_4(x) = \frac{1}{\sqrt{2}}\left(1 + \frac{(x-\pi/4)}{1!} - \frac{(x-\pi/4)^2}{2!} - \frac{(x-\pi/4)^3}{3!} + \frac{(x-\pi/4)^4}{4!}\right). \qquad \square$$

The function $f(x) - p_n(x) = R_n(x)$, say, is the *error* introduced when we use $p_n(x)$ as an approximation to $f(x)$ and the following theorem, which is very important in classical mathematical analysis, gives an expression for $R_n(x)$.

Theorem 3.1 (Taylor's theorem) Let $f(x), f'(x), \ldots, f^{(n)}(x)$ exist and be continuous on some interval $[a, b]$, $f^{(n+1)}(x)$ exist for $a < x < b$ and $p_n(x)$ be

† After B. Taylor (1685–1731).

the Taylor polynomial (3.3) for some $x_0 \in [a, b]$. Then, given any $x \in [a, b]$, there exists ξ_x (depending on x) which lies between x_0 and x and is such that

$$R_n(x) = f(x) - p_n(x) = \frac{(x - x_0)^{n+1}}{(n+1)!} f^{(n+1)}(\xi_x). \tag{3.4}$$

Alternatively, if we let $x = x_0 + h$, we may write (3.4) as

$$\begin{aligned} f(x_0 + h) = f(x_0) &+ \frac{h}{1!} f'(x_0) + \frac{h^2}{2!} f''(x_0) + \cdots \\ &+ \frac{h^n}{n!} f^{(n)}(x_0) + \frac{h^{n+1}}{(n+1)!} f^{(n+1)}(x_0 + \theta_h h), \end{aligned} \tag{3.5}$$

where $0 < \theta_h < 1$.

Proof Let $\alpha \in [a, b]$ be the point at which we wish to determine the error. We suppose (without loss of generality) that $\alpha > x_0$. Let

$$g(x) = f(\alpha) - f(x) - \frac{(\alpha - x)}{1!} f'(x) - \cdots - \frac{(\alpha - x)^n}{n!} f^{(n)}(x). \tag{3.6}$$

Then $g'(x)$ exists for $x \in (a, b)$ and

$$g'(x) = -\frac{(\alpha - x)^n}{n!} f^{(n+1)}(x).$$

Now consider the function

$$G(x) = g(x) - \left(\frac{\alpha - x}{\alpha - x_0}\right)^{n+1} g(x_0), \tag{3.7}$$

for which

$$G(x_0) = G(\alpha) = 0.$$

From the differentiability of g and $(\alpha - x)^{n+1}$ it follows that G is differentiable on any subinterval of (a, b) and we may apply Rolle's theorem 2.4 to G on the interval $[x_0, \alpha]$. Thus there exists $\xi_\alpha \in (x_0, \alpha)$ such that

$$G'(\xi_\alpha) = 0,$$

which yields

$$-\frac{(\alpha - \xi_\alpha)^n}{n!} f^{(n+1)}(\xi_\alpha) + (n+1) \frac{(\alpha - \xi_\alpha)^n}{(\alpha - x_0)^{n+1}} g(x_0) = 0,$$

so that, since $\alpha - \xi_\alpha \neq 0$,

$$g(x_0) = \frac{(\alpha - x_0)^{n+1}}{(n+1)!} f^{(n+1)}(\xi_\alpha).$$

Now

$$g(x_0) = f(\alpha) - p_n(\alpha) = R_n(\alpha)$$

and, therefore,

$$R_n(\alpha) = \frac{(\alpha - x_0)^{n+1}}{(n+1)!} f^{(n+1)}(\xi_\alpha),$$

where $\xi_\alpha \in (x_0, \alpha)$. On replacing α by x, we obtain (3.4). □

There are various expressions for the error term R_n other than (3.4). Some of these may be found by making a slightly different choice of $G(x)$ than (3.7), as is seen in Problem 3.3. With $n = 0$, we obtain the mean value theorem and we sometimes call Theorem 3.1 the generalized mean value theorem.

3.3 Convergence of Taylor series

We have seen in the last section how we may construct a polynomial p_n which approximates to a function f near a given point. We now ask whether the error $R_n(x) = f(x) - p_n(x)$ converges to zero as n is increased. Since $p_n(x)$ is the nth partial sum of the *infinite series*

$$f(x_0) + \frac{(x - x_0)}{1!} f'(x_0) + \frac{(x - x_0)^2}{2!} f''(x_0) + \cdots + \frac{(x - x_0)^r}{r!} f^{(r)}(x_0) + \cdots, \quad (3.8)$$

we are equivalently concerned with the convergence of this series to $f(x)$. We call (3.8) the *Taylor series* of f constructed at $x = x_0$ and R_n, the error when the series is terminated after $n + 1$ terms, is called the *remainder*. Sometimes it is possible to investigate the convergence of (3.8) directly, but usually we consider the behaviour of $R_n(x)$ as $n \to \infty$. In Chapter 2, we gave two definitions of convergence of such a sequence of functions, each defined on an interval I. We say that the sequence $(R_n(x))_{n=0}^{\infty}$ converges (pointwise) to zero on I if

$$R_n(x) \to 0 \quad \text{as } n \to \infty, \qquad \text{for all } x \in I \quad (3.9)$$

and $(R_n(x))_{n=0}^{\infty}$ converges uniformly to zero on I if

Corresponding to (3.9) and (3.10), we say respectively that the Taylor series (3.8) is (pointwise) convergent or uniformly convergent to $f(x)$ on I. It is not always easy to determine the behaviour of $R_n(x)$ from (3.4), as ξ_x depends on x and the form of this dependence is not known explicitly.

Example 3.3 We investigate the convergence of the Taylor series of $f(x)=(1+x)^{-1}$ at $x=0$. Clearly

$$f^{(r)}(x)=\frac{(-1)^r r!}{(1+x)^{r+1}}$$

and, therefore, the required Taylor polynomial of nth degree is

$$p_n(x)=1-x+x^2-x^3+\cdots+(-x)^n.$$

On any interval not containing $x=-1$, all the derivatives of f exist and are continuous and from (3.4) the remainder is

$$R_n(x)=\frac{x^{n+1}}{(n+1)!}\,\frac{(-1)^{n+1}(n+1)!}{(1+\xi_x)^{n+2}}=\frac{(-x)^{n+1}}{(1+\xi_x)^{n+2}}. \tag{3.11}$$

We first show that the sequence $(R_n(x))_{n=0}^{\infty}$ is uniformly convergent to zero (and, therefore, that the Taylor series is uniformly convergent) on any interval $[0,b]$ with $0<b<1$. For $0\leqslant x\leqslant b$ we have $0\leqslant \xi_x\leqslant x$ with equality only if $x=0$. Thus

$$\frac{1}{1+\xi_x}\leqslant 1 \tag{3.12}$$

and from (3.11)

$$|R_n(x)|\leqslant x^{n+1}. \tag{3.13}$$

Now $x\leqslant b$ and, therefore,

$$\sup_{x\in[0,b)}|R_n(x)|\leqslant b^{n+1}.$$

From $0<b<1$ it follows that $b^{n+1}\to 0$ as $n\to\infty$ and we have uniform convergence of $(R_n(x))$. In this case we were able to remove ξ_x by using an inequality.

Secondly, we show that $(R_n(x))$ is (pointwise) convergent to zero on $[0,1)$. Again (3.12) and, therefore, (3.13) hold and, for any *given* non-negative $x<1$,

$$|R_n(x)|\leqslant x^{n+1}\to 0 \qquad \text{as } n\to\infty.$$

However, as we have seen before (Example 2.15),

$$\sup_{x\in[0,1)} x^{n+1}=1 \qquad \text{for all } n$$

and, therefore, we cannot prove that the suprema of the $|R_n(x)|$ converge to zero on $[0, 1)$.

In fact, the Taylor series is not uniformly convergent on $[0, 1)$ as can be seen by a more direct approach. For this is a geometric series and

$$p_n(x) = \frac{1-(-x)^{n+1}}{1+x}, \qquad x \neq -1, \tag{3.14}$$

so that

$$R_n(x) = f(x) - p_n(x) = \frac{(-x)^{n+1}}{1+x}. \tag{3.15}$$

Thus

$$\sup_{x \in [0,1)} |R_n(x)| = \tfrac{1}{2}, \qquad \text{for all } n$$

and $(|R_n(x)|)$ does not converge uniformly to zero.

The investigation for all intervals $[a, 0]$ where $-1 < a \leqslant 0$ is not possible using the remainder form (3.4) and the more explicit result (3.15) is required. Alternatively, one of the remainder forms described in Problem 3.3 may be considered. Thus we find that the Taylor series

$$1 - x + x^2 - x^3 + \cdots$$

is uniformly convergent on any interval $[a, b]$ with $-1 < a \leqslant b < 1$ (see Problem 3.6). □

Example 3.4 For $f(x) = \sin x$ the Taylor series at $x = 0$ is (see also Example 3.2)

$$x - \frac{x^3}{3!} + \frac{x^5}{5!} - \frac{x^7}{7!} + \cdots.$$

The remainder is

$$R_n(x) = \begin{cases} \dfrac{x^{n+1}(-1)^{n/2}\cos \xi_x}{(n+1)!}, & \text{for even } n \\[2ex] \dfrac{x^{n+1}(-1)^{(n+1)/2}\sin \xi_x}{(n+1)!}, & \text{for odd } n \end{cases}$$

where ξ_x lies between 0 and x. In both cases

$$|R_n(x)| \leqslant \frac{|x|^{n+1}}{(n+1)!}$$

and on any interval $[a, b]$, with $c = \max\{\,|a|,\,|b|\,\}$,

$$\sup_{x \in [a,b]} |R_n(x)| \leqslant \frac{c^{n+1}}{(n+1)!} \to 0, \qquad \text{as } n \to \infty.$$

Thus $(|R_n(x)|)$ is uniformly convergent to zero (and therefore the Taylor series converges to $f(x) = \sin x$) on any closed interval. This is also true on any open interval as we can always embed such an interval in a larger closed interval. □

If, on some interval $[a, b]$, all the derivatives of f exist and have a common bound M (say), so that

$$\max_{x \in [a,b]} |f^{(n)}(x)| \leqslant M, \qquad \text{for all } n, \tag{3.16}$$

then from (3.4), if $x_0 \in [a, b]$,

$$\sup_{x \in [a,b]} |R_n(x)| \leqslant \frac{(b-a)^{n+1}}{(n+1)!} M.$$

It follows that

$$\max_{x \in [a,b]} |f(x) - p_n(x)| = \sup_{x \in [a,b]} |R_n(x)| \to 0, \qquad \text{as } n \to \infty.$$

Equivalently we can say that, given any $\varepsilon > 0$, there is an N such that

$$\max_{x \in [a,b]} |f(x) - p_n(x)| < \varepsilon, \qquad \text{for all } n \geqslant N.$$

Thus we can choose a polynomial $p_n(x)$ which approximates to f to within any arbitrary accuracy on any interval $[a, b]$, by choosing n sufficiently large, provided f satisfies (3.16). This is a proof of a weak form of Weierstrass's theorem which states that we can approximate to *any continuous function* to within arbitrary accuracy using a polynomial. The condition (3.16) is very severe though it is satisfied by many functions including that of Example 3.4.

3.4 Taylor series in two variables

We extend Theorem 3.1 to functions of two variables.

Theorem 3.2 Suppose that $f(x, y)$ is a function of two real variables such that all the $(n+1)$th order partial derivatives of f exist and are continuous on some circular domain D with centre (x_0, y_0). If $(x_0+h,\ y_0+k)\in D$, then

$$f(x_0+h, y_0+k) = f(x_0, y_0) + \frac{1}{1!}\left(h\frac{\partial}{\partial x} + k\frac{\partial}{\partial y}\right)f(x_0, y_0) + \cdots$$

$$+\frac{1}{n!}\left(h\frac{\partial}{\partial x} + k\frac{\partial}{\partial y}\right)^n f(x_0, y_0)$$

$$+\frac{1}{(n+1)!}\left(h\frac{\partial}{\partial x} + k\frac{\partial}{\partial y}\right)^{n+1} f(x_0+\theta h, y_0+\theta k), \qquad (3.17)$$

where $0<\theta<1$.

Proof Let $x = x_0 + ht$ and $y = y_0 + kt$, where $0 \leqslant t \leqslant 1$. Let

$$g(t) = f(x_0+ht, y_0+kt)$$

when, by the chain rule,

$$g' = \frac{\mathrm{d}g}{\mathrm{d}t} = \frac{\mathrm{d}x}{\mathrm{d}t}\cdot\frac{\partial f}{\partial x} + \frac{\mathrm{d}y}{\mathrm{d}t}\cdot\frac{\partial f}{\partial y} = \left(h\frac{\partial}{\partial x} + k\frac{\partial}{\partial y}\right)f.$$

Similarly

$$g'' = \frac{\mathrm{d}^2 g}{\mathrm{d}t^2} = \left(h\frac{\partial}{\partial x} + k\frac{\partial}{\partial y}\right)\left(h\frac{\partial}{\partial x} + k\frac{\partial}{\partial y}\right)f = \left(h\frac{\partial}{\partial x} + k\frac{\partial}{\partial y}\right)^2 f$$

and

$$g^{(r)} = \frac{\mathrm{d}^r g}{\mathrm{d}t^r} = \left(h\frac{\partial}{\partial x} + k\frac{\partial}{\partial y}\right)^r f, \qquad r = 0, 1, \ldots, n+1.$$

We now construct the Taylor series of $g(t)$ at $t=0$ and apply Theorem 3.1 to obtain

$$g(1) = g(0) + \frac{1g'(0)}{1!} + \frac{1^2 g''(0)}{2!} + \cdots + \frac{1^n g^{(n)}(0)}{n!} + \frac{1^{n+1} g^{(n+1)}(\theta)}{(n+1)!}, \qquad (3.18)$$

where $0<\theta<1$. On substituting for g and its derivatives in (3.18), we obtain (3.17). □

3.5 Power series

In many applications of Taylor's theorem we take $x_0 = 0$. The Taylor series becomes

$$f(0) + \frac{x}{1!}f'(0) + \frac{x^2}{2!}f''(0) + \cdots \qquad (3.19)$$

and is sometimes known as a *Maclaurin* series† in this form. The series (3.19) is a special case of a *power series*, that is, a series of the type

$$a_0 + a_1 x + a_2 x^2 + a_3 x^3 + \cdots,$$

where the coefficients $a_0, a_1, a_2, \ldots$ are real constants, independent of x. As we have seen earlier, such a series may be convergent for only certain values of x. We define the *radius of convergence* of such a series to be the largest positive number r (if it exists) such that the series is convergent for any x with $|x| < r$. Note that this does not imply anything about the convergence or otherwise when $|x| = r$.

Example 3.5 We reconsider the Maclaurin series of $(1+x)^{-1}$ (see Example 3.3),

$$1 - x + x^2 - x^3 + \cdots.$$

The partial sum is (3.14),

$$p_n(x) = \frac{1 - (-x)^{n+1}}{1+x},$$

and this is convergent for $|x| < 1$ and divergent for $|x| > 1$. The radius of convergence is $+1$. □

Example 3.6 The Maclaurin series of the exponential function e^x is

$$1 + \frac{x}{1!} + \frac{x^2}{2!} + \frac{x^3}{3!} + \cdots$$

and this is convergent for any x. We say that the radius of convergence is infinite. □

There are various tests for determining the convergence of a power series and details of these may be found in any good text on advanced calculus. It can also be shown that a power series is uniformly convergent on any interval $[-a, a]$, where $0 < a < r$. We have only considered real power series, but the definitions may easily be extended to such series with complex coefficients and complex argument.

If we use the partial sum of a convergent infinite series as an approximation to the sum of the infinite series, the error thus introduced is called the *truncation error*. In the case of a Taylor series expressed as a power series, this truncation error is the remainder, which takes the form

$$R_n(x_0 + h) = \frac{h^{n+1}}{(n+1)!} f^{(n+1)}(x_0 + \theta h)$$

† Named after the Scots mathematician Colin Maclaurin (1698–1746).

if f satisfies the conditions of Theorem 3.1 on some interval $[a, b]$. If, in addition, for some fixed n we are given that $f^{(n+1)}$ is bounded on (a, b) (for example, this will be true if $f^{(n+1)}$ is continuous for $a \leqslant x \leqslant b$), then we write

$$R_n(x_0+h) = O(h^{n+1})$$

and say that R_n is of order h^{n+1}. By this we mean that there exists M independent of h such that

$$|R_n(x_0+h)| \leqslant Mh^{n+1} \tag{3.20}$$

provided h is sufficiently small. The inequality (3.20) tells us how rapidly $|R_n|$ decreases as $h \to 0$.

Problems

Section 3.2

3.1 Show that the solution of equations (3.2) is

$$a_r = \sum_{j=0}^{n-r} \frac{(-x_0)^j}{j!r!} f^{(r+j)}(x_0), \qquad r = 0, 1, \ldots, n.$$

(*Hint*: see (3.3).)

3.2 Construct the Taylor polynomials of degree 6 at $x = 0$ for the following functions:

$$\cos x, \quad (1-x)^{-1}, \quad (1+x)^{1/2}, \quad \log(1+x), \quad \log(1-x).$$

3.3 Let $f(x)$, $g(x)$, α and x_0 be as in Theorem 3.1. For some positive integer m replace (3.7) by

$$G_m(x) = g(x) - \left(\frac{\alpha - x}{\alpha - x_0}\right)^m g(x_0)$$

and thus show that for some ξ (depending on m and x) between x_0 and x,

$$R_n(x) = \frac{(x-x_0)^m (x-\xi)^{n-m+1}}{m.n!} f^{(n+1)}(\xi).$$

In particular, for $m = 1$, show that

$$R_n(x) = \frac{(x-x_0)(x-\xi)^n}{n!} f^{(n+1)}(\xi) = \frac{h^{n+1}(1-\theta)^n}{n!} f^{(n+1)}(x_0+\theta h), \tag{3.21}$$

where $x = x_0 + h$ and $0 < \theta < 1$.

3.4 If $p_n(x)$ denotes the Taylor polynomial of degree n for e^x constructed at $x = 0$, find the smallest value of n for which

$$\max_{0 \leqslant x \leqslant 1} |\mathrm{e}^x - p_n(x)| \leqslant 10^{-6}.$$

3.5 Show that, if f satisfies the conditions of Theorem 2.7 and h is sufficiently small,

$$f(x_0 + h) = f(x_0) + \tfrac{1}{2} h^2 f''(x_0 + \theta h)$$

and thus prove the theorem.

Section 3.3

3.6 By considering the remainder form (3.21), show that the Taylor series at $x = 0$ of $f(x) = (1 + x)^{-1}$ (see Example 3.3) is uniformly convergent on any interval $[a, 0]$ with $-1 < a < 0$. Combine this result with that of Example 3.3 to show that the series is uniformly convergent on any interval $[a, b]$ with $-1 < a \leqslant b < 1$.

3.7 Show that the Taylor series at $x = 0$ of $f(x) = (1 + x)^{1/2}$ is (pointwise) convergent for $-1 < x < 1$ and uniformly convergent on any interval $[a, b]$ with $-1 < a \leqslant b < 1$. (*Hint*: show from (3.4) that

$$|R_n(x)| = \left| \frac{1.3.5\ldots(2n-1)}{2.4.6\ldots 2n} \frac{x^{n+1}}{2(n+1)(1+\xi)^{n+1/2}} \right|$$
$$< \frac{x^{n+1}}{(1+\xi)^{n+1/2}}, \qquad \text{for } x > 0$$

and thus that $R_n(x) \to 0$ as $n \to \infty$ for $0 \leqslant x < 1$. Similarly show from (3.21) that

$$|R_n(x)| = \left| \frac{1.3.5\ldots(2n-1)}{2.4.6\ldots 2n} \frac{x(x-\xi)^n}{2(1+\xi)^{n+1/2}} \right|$$
$$< \left| \frac{x(x-\xi)^n}{(1+\xi)^{n+1/2}} \right|, \qquad \text{for } x \neq 0$$

so that $R_n(x) \to 0$ as $n \to \infty$ for $-1 < x < 0$.)

3.8 Show that the Taylor series at $x = 0$ of $f(x) = \log(1 + x)$ is (pointwise) convergent for $-1 < x \leqslant 1$. As in Problem 3.7 consider $-1 < x < 0$ and $0 \leqslant x \leqslant 1$ separately. What happens when $x = -1$? (See also Example 2.14.)

Section 3.5

3.9 Show that the Taylor series at $x=0$ of $\log(1+x)$ has radius of convergence $+1$.

3.10 Show that the Taylor series at $x=0$ of $\sin x$ has infinite radius of convergence.

3.11 Show, using Problem 3.7, that

$$(1+x)^{1/2} = 1 + \tfrac{1}{2}x + O(x^2).$$

3.12 If A is a given constant and $h = A/n$ where n is a positive integer, show that

$$(1 + O(h^2))^n = 1 + O(h), \qquad \text{as } n \to \infty$$

and

$$(1 + O(h^3))^n = 1 + O(h^2), \qquad \text{as } n \to \infty.$$

3.13 Show that for $x \geqslant 0$ and any fixed integer $n \geqslant 1$:

$$\begin{aligned} e^x &\geqslant 1 + x, \\ e^x &= 1 + x + O(x^2), \\ e^{nx} &\geqslant (1+x)^n, \\ e^{nx} &= (1+x)^n + O(x^2). \end{aligned}$$

Chapter 4

THE INTERPOLATING POLYNOMIAL

4.1 Linear interpolation

We have seen that the Taylor polynomial is designed to approximate to a given function f very well at one point. There is a much simpler type of polynomial approximation in which the agreement with f is not all focused at one point, but is spread over a number of points. This is the *interpolating polynomial*, which was discovered before the Taylor polynomial.

A simple case is the *linear* interpolating polynomial: given the values of $f(x)$ at two points, say x_0 and x_1, we can write down a first-degree polynomial,

$$p_1(x) = \left(\frac{x - x_1}{x_0 - x_1}\right) f(x_0) + \left(\frac{x - x_0}{x_1 - x_0}\right) f(x_1), \tag{4.1}$$

which takes the same values as $f(x)$ at x_0 and x_1. Geometrically (see Fig. 4.1) it is clear that this is the unique straight line which passes through the two points $(x_0, f(x_0))$ and $(x_1, f(x_1))$. We can easily construct a polynomial of degree greater than one which passes through these two points. However (see Problem 4.3), such a polynomial is not unique.

It will be useful to us later to note that $p_1(x)$ may also be written as

$$p_1(x) = \frac{(x - x_0)f(x_1) - (x - x_1)f(x_0)}{x_1 - x_0} \tag{4.2}$$

or as

$$p_1(x) = f(x_0) + (x - x_0)\left(\frac{f(x_1) - f(x_0)}{x_1 - x_0}\right). \tag{4.3}$$

We will examine (4.3) further and assume that $x_0 < x_1$. From the mean value theorem 2.5, if f is continuous on $[x_0, x_1]$ and is differentiable in (x_0, x_1),

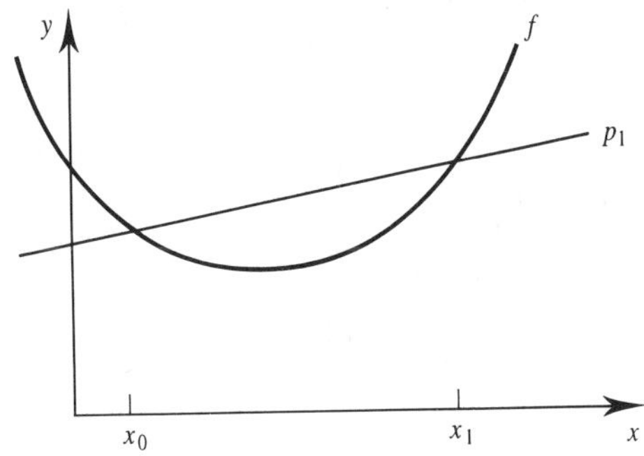

Fig. 4.1 Linear interpolation.

there exists a number $\xi \in (x_0, x_1)$ such that

$$\frac{f(x_1) - f(x_0)}{x_1 - x_0} = f'(\xi). \tag{4.4}$$

Hence (4.3) may be expressed in the form

$$p_1(x) = f(x_0) + (x - x_0)f'(\xi). \tag{4.5}$$

which shows that p_1 approximates to f at $x = x_0$ in a way similar to the approximation by the Taylor polynomial of degree 1. If we keep x_0 fixed and let x_1 tend to x_0, the limit of (4.4) becomes $f'(x_0)$ (assuming continuity of f') and p_1 in (4.5) becomes precisely the Taylor polynomial.

The main purpose of constructing p_1 is to evaluate $p_1(x)$ for a value of x between x_0 and x_1, and to use this as an approximation to $f(x)$; this is called *interpolation*. If x is outside the interval $[x_0, x_1]$ the term *extrapolation* is sometimes used, although we will not normally make this distinction. Since p_1 is a linear function of x, the evaluation of $p_1(x)$ is called *linear interpolation*. The question of how well p_1 approximates to f is one which we naturally ask; we pursue this later using a technique similar to that employed in examining the error of the Taylor polynomial.

Some insight into linear interpolation can be gained by studying tables of standard mathematical functions. Such tables were an indispensable part of the toolkit of the numerical analyst from, say, the seventeenth century to the later part of the twentieth century, when the universal availability of adequate computing power made them obsolete. In these mathematical tables, functions are usually tabulated at equal intervals. A mathematician who constructed such a table had to answer two questions: to how many decimal places and at what interval (between consecutive entries in the table) should the function be evaluated? In practice these questions turned out to be related. The constructor of a mathematical table first decided on the number of decimal places to which the function would be evaluated at the tabulated points. The interval size was then chosen so that linear interpolation between consecutive entries would roughly preserve this level of accuracy. (It is easy to think of this graphically: choose an interval small enough so that the graph of the function is sufficiently close to the graph of a straight line.) As an example, for the function sin x we find that an interval of 0.01 radians is used in the four-figure table, while an interval of 0.001 is used in the six-figure table.

Example 4.1 From a set of four-figure tables of natural logarithms, we find that log 2.1 is 0.7419 and log 2.2 is 0.7885. We use linear interpolation on these two values to estimate log 2.14. Here it is simplest to calculate $p_1(x)$ from (4.3), with $x_0 = 2.1$, $x_1 = 2.2$, $f(x_0) = 0.7419$, $f(x_1) = 0.7885$ and $x = 2.14$. Thus

$$p_1(2.14) = 0.7419 + 0.04\left(\frac{0.0466}{0.1}\right) = 0.76054.$$

For comparison, log $2.14 = 0.76081$ to five decimal places. □

4.2 Polynomial interpolation

We will now generalize most of the results of the last section to the case where we are given the values of the function f at $n+1$ distinct points $x = x_0$, $x_1, \ldots, x_n$. These are not necessarily equally spaced or even arranged in increasing order; the only restriction is that they must be distinct. We want to construct a polynomial p_n which takes the same values as f at the $n+1$ points $x_0, x_1, \ldots, x_n$. We might suspect that $p_n \in P_n$. Certainly this fits in with the case $n = 1$, which we have just discussed, where there are two interpolating points and the interpolating polynomial is of degree 1. We could write

$$p_n(x) = a_0 + a_1 x + \cdots + a_n x^n$$

and, on putting $p_n(x_r) = f(x_r)$, $r = 0, 1, \ldots, n$, we obtain $n+1$ linear equations in the $n+1$ coefficients $a_0, a_1, \ldots, a_n$. Since these equations are difficult to solve unless n is small, we prefer a different approach based on a generalization of (4.1). We write p_n in the form

$$p_n(x) = L_0(x)f(x_0) + L_1(x)f(x_1) + \cdots + L_n(x)f(x_n), \tag{4.6}$$

where each $L_i \in P_n$. The polynomial p_n will have the same values as f at $x = x_0, x_1, \ldots, x_n$ if

$$L_i(x_j) = \delta_{ij}, \qquad 0 \leqslant i \leqslant n. \tag{4.7}$$

In (4.7) we have written δ_{ij} to denote the *Kronecker delta* function, which takes the value 0 when $i \neq j$ and the value 1 when $i = j$. So, for example, L_0 has the value zero at $x = x_1, x_2, \ldots, x_n$ and has the value 1 at $x = x_0$. Therefore, if we put

$$L_0(x) = C(x - x_1)(x - x_2)\cdots(x - x_n), \tag{4.8}$$

where C is a constant, L_0 will indeed be zero at $x = x_1, x_2, \ldots, x_n$ and $L_0 \in P_n$. Putting $x = x_0$ in (4.8), the condition $L_0(x_0) = 1$ gives

$$1 = C(x_0 - x_1)(x_0 - x_2)\cdots(x_0 - x_n), \tag{4.9}$$

which fixes the value of C. On substituting this value in (4.8) we obtain

$$L_0(x) = \frac{(x - x_1)(x - x_2)\ldots(x - x_n)}{(x_0 - x_1)(x_0 - x_2)\ldots(x_0 - x_n)},$$

which may be written more neatly as

$$L_0(x) = \prod_{j=1}^{n}\left(\frac{x - x_j}{x_0 - x_j}\right).$$

Similarly, for a general value of i, $0 \leqslant i \leqslant n$, we obtain

$$L_i(x) = \prod_{\substack{j=0 \\ j \neq i}}^{n}\left(\frac{x - x_j}{x_i - x_j}\right). \tag{4.10}$$

In (4.10) the product is taken over all values of j from 0 to n except for $j = i$. From (4.10) we see immediately that L_i is zero at all values $x = x_0$, $x_1, \ldots, x_n$, except for $x = x_i$ when L_i takes the value 1. This agrees with the requirement (4.7) and the interpolating polynomial can be written in the form

$$p_n(x) = \sum_{i=0}^{n} L_i(x) f(x_i). \tag{4.11}$$

This is known as the Lagrange form of the interpolating polynomial, after the French–Italian mathematician J. L. Lagrange (1736–1813).

Example 4.2 Given the values of $f(x)$ at distinct points $x = x_0$, x_1 and x_2, write out explicitly the interpolating polynomial p_2. We have

$$p_2(x) = \frac{(x - x_1)(x - x_2)}{(x_0 - x_1)(x_0 - x_2)} f(x_0) + \frac{(x - x_0)(x - x_2)}{(x_1 - x_0)(x_1 - x_2)} f(x_1)$$
$$+ \frac{(x - x_0)(x - x_1)}{(x_2 - x_0)(x_2 - x_1)} f(x_2). \tag{4.12}$$

□

Example 4.3 Find the interpolating polynomial for a function f, given that $f(x) = 0$, -3 and 4 when $x = 1$, -1 and 2 respectively. (The reader should draw a graph.) Putting $x_0 = 1$, $x_1 = -1$ and $x_2 = 2$ in (4.12) (these three values could have been assigned in any order), we obtain

$$p_2(x) = \frac{(x+1)(x-2)}{(1+1)(1-2)} \times 0 + \frac{(x-1)(x-2)}{(-1-1)(-1-2)} \times -3$$
$$+\frac{(x-1)(x+1)}{(2-1)(2+1)} \times 4,$$

which simplifies to give $p_2(x) = \frac{1}{6}(5x^2 + 9x - 14)$. As a check on our calculations, it may be verified from this last equation that $p_2(1) = 0$, $p_2(-1) = -3$ and $p_2(2) = 4$. □

The above example invites the question: is the interpolating polynomial unique? With the question put in this way, the answer is obviously no. For example, the polynomial

$$q(x) = \tfrac{1}{6}(5x^2 + 9x - 14) + C(x-1)(x+1)(x-2),$$

with any choice of the constant C, also interpolates the data of Example 4.3. However, the following is true.

Theorem 4.1 Given the values of f at the $n+1$ distinct points $x_0, x_1, \ldots, x_n$, there is a unique polynomial of degree at most n which takes the same value as f at these points.

Proof First, notice the crucial phrase 'of degree at most n' in the statement of the theorem. We already know that there is at least one polynomial $p_n \in P_n$ which interpolates f at $x = x_0, x_1, \ldots, x_n$. This is displayed in (4.11). Now consider any $q_n \in P_n$ such that $q_n(x_i) = f(x_i)$, $0 \leqslant i \leqslant n$. It follows that $p_n - q_n \in P_n$ is zero at the $n+1$ points $x_0, x_1, \ldots, x_n$. But a polynomial of degree at most n has no more than n zeros unless it is zero at every point, that is, it is identically zero. Thus $p_n(x) \equiv q_n(x)$, showing the uniqueness of the interpolating polynomial and completing the proof. □

4.3 Accuracy of interpolation

In this section, we will examine the accuracy with which the interpolating polynomial approximates the function f. First, it is not going to be possible to estimate the size of the error $f - p_n$ from a knowledge of the values of f at $x = x_0, x_1, \ldots, x_n$ alone. Some further information about f is required. To see this, consider the $n+1$ points $(x_0, f(x_0)), (x_1, f(x_1)), \ldots, (x_n, f(x_n))$, where $x_0, x_1, \ldots, x_n$ are distinct. These fix the polynomial p_n, but we are free to draw any curve which passes through these $n+1$ points and let this define the function f. Hence we can arrange for $f(x) - p_n(x)$ to be arbitrarily large at any value of x, except $x = x_0, x_1, \ldots, x_n$, where $f(x) - p_n(x)$ is zero. However, we can estimate the error $f - p_n$ in terms of the $(n+1)$th derivative of f, if this exists.

Theorem 4.2 Let $[a, b]$ be any interval which contains all $n+1$ points $x_0, x_1, \ldots, x_n$. Let $f, f', \ldots, f^{(n)}$ exist and be continuous on $[a, b]$ and let $f^{(n+1)}$ exist for $a < x < b$. Then, given any $x \in [a, b]$, there exists a number ξ_x (depending on x) in (a, b) such that

$$f(x) - p_n(x) = (x - x_0) \cdots (x - x_n) \frac{f^{(n+1)}(\xi_x)}{(n+1)!}. \tag{4.13}$$

Proof The statement of this theorem is similar to that of Taylor's theorem 3.1, whose proof depends on Rolle's theorem 2.4. We use Rolle's theorem again in this proof. First, since $f - p_n$ and $(x - x_0) \cdots (x - x_n)$ have zeros at the $n+1$ points $x_0, x_1, \ldots, x_n$, so also does the function

$$g(x) = f(x) - p_n(x) + \lambda(x - x_0) \cdots (x - x_n), \tag{4.14}$$

where λ is any constant. We now choose λ so as to ensure that g has at least $n+2$ zeros. If we wish to estimate the error at the point $x = \alpha$ in $[a, b]$, we, choose λ in (4.14) so that $g(\alpha) = 0$. This gives

$$0 = f(\alpha) - p_n(\alpha) + \lambda(\alpha - x_0) \cdots (\alpha - x_n).$$

With this choice of λ,

$$g(x) = f(x) - p_n(x) - \frac{(x - x_0) \cdots (x - x_n)}{(\alpha - x_0) \cdots (\alpha - x_n)} \cdot (f(\alpha) - p_n(\alpha)). \tag{4.15}$$

The function g is seen to be zero at not less than $n+2$ points including $x = x_0, x_1, \ldots, x_n$ and $x = \alpha$. On applying the extended Rolle theorem 2.6 to g, we deduce that $g^{(n+1)}$ has at least one zero, say ξ_α in (a, b). Differentiating (4.15) $n+1$ times gives

$$g^{(n+1)}(x) = f^{(n+1)}(x) - \frac{(n+1)!}{(\alpha - x_0) \cdots (\alpha - x_n)} \cdot (f(\alpha) - p_n(\alpha)), \tag{4.16}$$

since, on differentiating $n+1$ times, p_n vanishes and the term x^{n+1} in the product $(x - x_0) \cdots (x - x_n)$ is reduced to $(n+1)!$. Substituting the zero ξ_α in (4.16) we obtain

$$0 = f^{(n+1)}(\xi_\alpha) - \frac{(n+1)!}{(\alpha - x_0) \cdots (\alpha - x_n)} (f(\alpha) - p_n(\alpha)).$$

Lastly, rearranging this last equation and replacing α by x gives

$$f(x) - p_n(x) = (x - x_0) \cdots (x - x_n) \frac{f^{(n+1)}(\xi_x)}{(n+1)!}. \qquad \square$$

To use the error estimate (4.13) we need to estimate $f^{(n+1)}$. Usually we do not know the value of ξ_x in (4.13) and have to work with bounds for the $(n+1)$th derivative, as we show in the following example. In § 4.8, we mention a method for estimating the $(n+1)$th derivative, when the points x_i are equally spaced.

Example 4.4 Use Theorem 4.2 to estimate the error in linear interpolation, with particular reference to Example 4.1, where we estimated $\log 2.14$, given the values of $\log 2.1$ and $\log 2.2$. For linear interpolation of $f(x)$ between $x = x_0$ and x_1, (4.13) yields

$$f(x) - p_1(x) = (x - x_0)(x - x_1) \frac{f''(\xi_x)}{2!}. \tag{4.17}$$

Here $f(x) = \log x$, $f'(x) = 1/x$ and $f''(x) = -1/x^2$. Putting $x = 2.14$, $x_0 = 2.1$ and $x_1 = 2.2$, we obtain from (4.17) an error $-(0.04) \times (-0.06)/(2\xi_x^2)$. Since ξ_x lies between 2.1 and 2.2, the error lies between 0.00024 and 0.00028. $\square$

4.4 The Neville–Aitken algorithm

The interpolating polynomial can be evaluated very elegantly and efficiently by a scheme known as the *Neville–Aitken* algorithm, which we now describe.

Algorithm 4.1 This begins with $x, x_0, \ldots, x_n$ and $f_0, \ldots, f_n$, where f_i denotes $f(x_i)$. It computes $f_{0,n} = p_n(x)$.

for $i := 0$ **to** n

  $f_{i,0} := f_i$

  $y_i := x - x_i$

next i

for $k := 0$ **to** $n - 1$

  for $i := 0$ **to** $n - k - 1$

    $f_{i,k+1} := (y_i f_{i+1,k} - y_{i+k+1} f_{i,k})/(y_i - y_{i+k+1})$ (4.18)

  next i

next k

The heart of the algorithm is (4.18), which repeatedly uses the same pattern of calculation as in the linear case (4.2). □

Each $f_{i,k}$ is a function of x; we could write it as $f_{i,k}(x)$. It can be shown by induction that $f_{i,k}(x)$ is the interpolating polynomial for f at the points $x_i, x_{i+1}, \ldots, x_{i+k}$ so that, in particular,

$$f_{0,n} = f_{0,n}(x) = p_n(x), \tag{4.19}$$

the interpolating polynomial for f at the $n+1$ points $x_0, x_1, \ldots, x_n$. We have to apply the algorithm for every value of x for which we want to evaluate $p_n(x)$.

In order to gain experience of the algorithm, it is helpful to work through a calculation 'by hand', before implementing it on the computer. It is convenient to set out the calculation as in Table 4.1, which illustrates the case where $n=2$. The arrows in Table 4.1 link the sets of four numbers required to calculate $f_{0,1}$ and $f_{1,1}$; the four boxed numbers are used to calculate $f_{0,2}$.

Table 4.1 The Neville–Aitken algorithm

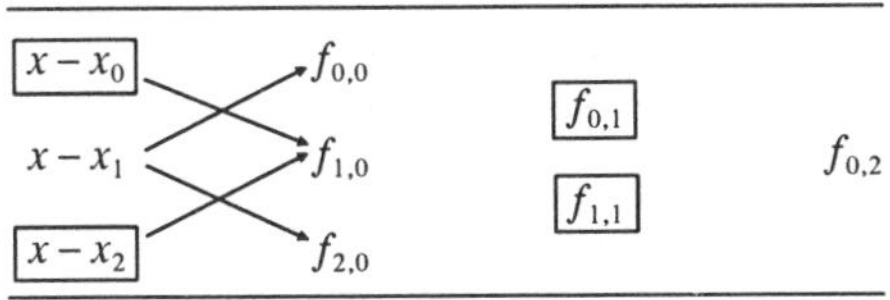

Example 4.5 Suppose we are given the values of the function e^{-x} at the following four points and wish to estimate e^{-x} at $x = 0.2$.

x	0.10	0.15	0.25	0.30
e^{-x}	0.904837	0.860708	0.778801	0.740818

The numbers generated by the algorithm in this case are displayed in Table 4.2. Thus the required interpolated value is 0.818730 to six figures. From (4.13), the interpolation error is

$$E = (0.01)(0.05)(-0.05)(-0.1)e^{-\xi}/24, \tag{4.20}$$

Table 4.2 Application of the Neville–Aitken algorithm

$x - x_i$	e^{-x_i}			
0.10	0.904837			
		0.8165790		
0.05	0.860708		0.8186960	
		0.8197545		0.8187302
−0.05	0.778801		0.8187643	
		0.8167840		
−0.10	0.740818			

where $0.1 < \xi < 0.3$ and so, from the table, $e^{-\xi}$ lies between 0.904837 and 0.740818. From this and (4.20) we see that E satisfies

$$0.77 \times 10^{-6} < E < 0.95 \times 10^{-6}.$$

This shows that the error (neglecting the effects of rounding errors) is almost one unit in the sixth place. □

4.5 Inverse interpolation

Let us write $y = f(x)$. So far we have been interpolating in the x direction. It is possible to consider interpolation in the y direction also. This is referred to as *inverse interpolation*. In this case it is the numbers $f(x_i)$ which we require to be distinct. The roles of x and $f(x)$ are interchanged. Otherwise, the calculation of the interpolating polynomial, say by the Neville–Aitken algorithm, may be carried out as before. The resulting interpolating polynomial may be regarded as a polynomial in y.

Example 4.6 Use inverse interpolation at $x = 0$, $\frac{1}{2}$ and 1 to estimate the only real root of the equation

$$x^3 + x^2 + x - 1 = 0,$$

which lies between $x = 0$ and 1. Secondly, use inverse interpolation at only $x = 0.5$ and 0.6 to estimate this root and estimate the accuracy of the result.

We see that the first derivative of $y = x^3 + x^2 + x - 1$ is always positive, showing that there is exactly one real root of the given equation. At $x = 0$, $\frac{1}{2}$ and 1, y has the values -1, $-\frac{1}{8}$ and 2 respectively. The Neville–Aitken algorithm gives the scheme set out in Table 4.3. We have written $y - y_i$ in the first column, although $y = 0$, to emphasize the pattern of the calculation. From the table, the result for inverse interpolation at the three points is $199/357 = 0.557$ to three decimal places. Secondly, at $x = 0.5$ and 0.6, $y = -0.125$ and 0.176 respectively. Inverse interpolation this time gives, as the estimate of the root, $x = 163/301 = 0.5415$ to four decimal places. We

Table 4.3 Inverse interpolation for Example 4.6

$y - y_i$	x_i		
1	0		
		4/7	
$\frac{1}{8}$	$\frac{1}{2}$		199/347
		9/17	
−2	1		

now want to estimate the accuracy of the last result. Suppose that $x=\phi(y)$ and $y=f(x)$. Assuming that ϕ'' exists on $[-0.125, 0.176]$, the error formula (4.13) gives an error

$$(y-y_0)(y-y_1)\phi''(\eta)/2!. \tag{4.21}$$

In (4.21) η lies in an interval containing y, y_0 and y_1 which, in this case, are the values 0, -0.125 and 0.176. We have

$$\phi''(y)=\frac{\mathrm{d}^2x}{\mathrm{d}y^2}=\frac{\mathrm{d}}{\mathrm{d}y}\left(\frac{\mathrm{d}x}{\mathrm{d}y}\right)=\frac{\mathrm{d}}{\mathrm{d}y}\left(1\bigg/\frac{\mathrm{d}y}{\mathrm{d}x}\right)=\frac{\mathrm{d}}{\mathrm{d}y}\left(\frac{1}{f'(x)}\right),$$

provided $f'(x)\neq 0$. Thus

$$\phi''(y)=\frac{\mathrm{d}}{\mathrm{d}x}\left(\frac{1}{f'(x)}\right)\frac{\mathrm{d}x}{\mathrm{d}y}=-\frac{f''(x)}{[f'(x)]^2}\,\frac{1}{f'(x)},$$

so that

$$\phi''(y)=-\frac{f''(x)}{[f'(x)]^3}.$$

Since $f(x)=x^3+x^2+x-1$, $f'(x)=3x^2+2x+1$, $f''(x)=6x+2$ and so

$$\phi''(y)=-\frac{(6x+2)}{(3x^2+2x+1)^3}.$$

For x between 0.5 and 0.6, $\phi''(y)$ lies between -0.25 and -0.15. From (4.21), it follows that the error lies between 0.001 and 0.003, so the result $x=0.5415$ is certainly correct to two decimal places.

The problem of finding solutions of an algebraic equation is discussed more generally in Chapter 8. □

Uncritical use of inverse interpolation (as also for direct interpolation) can produce misleading results. For example, consider the function $y=x^4$ with $x\geqslant 0$ tabulated at the four points where $y=0$, 1, 16 and 81. It is left to the reader to verify (in Problem 4.20) that cubic inverse interpolation based on these four points produces the approximation $x=-33.4$ at $y=64$, where the correct result is $x=(64)^{1/4}=2\sqrt{2}$. In this case, we are approximating to $x=y^{1/4}$ by a cubic polynomial in y. The error term includes the fourth derivative of $y^{1/4}$ with respect to y, which explains why such a large error can occur, as this derivative is infinite at $y=0$.

4.6 Divided differences

Another approach to the interpolating polynomial is to use divided differences, which will now be discussed. This method is due to Isaac

Newton. As well as being of theoretical interest, it presents the interpolating polynomial in a form which may be readily simplified, when we later choose the interpolating points to be equally spaced.

At the moment, however, we allow the numbers $x_0, x_1, \ldots, x_n$ to be any $n+1$ distinct numbers and attempt to write p_n, the interpolating polynomial for f at these points, as

$$\begin{aligned} p_n(x) = a_0 + (x-x_0)a_1 + (x-x_0)(x-x_1)a_2 + \cdots \\ + (x-x_0)(x-x_1)\cdots(x-x_{n-1})a_n. \end{aligned} \tag{4.22}$$

Substituting $x = x_0, x_1, \ldots, x_n$ in turn into (4.22), we obtain the following simultaneous equations:

$$\begin{aligned} f(x_0) &= a_0 \\ f(x_1) &= a_0 + (x_1 - x_0)a_1 \\ f(x_2) &= a_0 + (x_2 - x_0)a_1 + (x_2 - x_0)(x_2 - x_1)a_2 \\ &\vdots \\ f(x_n) &= a_0 + (x_n - x_0)a_1 + \cdots + (x_n - x_0)\cdots(x_n - x_{n-1})a_n. \end{aligned} \tag{4.23}$$

We see that these equations† determine values for $a_0, \ldots, a_n$ uniquely. The first equation in (4.23) gives a_0 and the second gives a_1, as $(x_1 - x_0) \neq 0$. Knowing a_0 and a_1, the third equation gives a_2, as $(x_2 - x_0)(x_2 - x_1) \neq 0$. Finally, knowing $a_0, a_1, \ldots, a_{n-1}$, the last equation gives a_n, since $(x_n - x_0)\cdots(x_n - x_{n-1}) \neq 0$. Hence $p_n(x)$ can be written, as in (4.22), in a unique way. Moreover, the coefficients which appear in (4.22) enjoy what is called *permanence*. If we add one further point x_{n+1}, distinct from $x_0, \ldots, x_n$, and write the interpolating polynomial constructed at all points $x_0, \ldots, x_{n+1}$ in the form

$$\begin{aligned} p_{n+1}(x) = b_0 + (x-x_0)b_1 + \cdots + (x-x_0)\ldots(x-x_{n-1})b_n \\ + (x-x_0)\cdots(x-x_n)b_{n+1}, \end{aligned} \tag{4.24}$$

we find that $b_0 = a_0$, $b_1 = a_1, \ldots, b_n = a_n$. For, on writing down the linear equations obtained for determining the b_i by substituting $x = x_0, x_1, \ldots, x_{n+1}$ into (4.24) in turn, we have

$$\begin{aligned} f(x_0) &= b_0 \\ f(x_1) &= b_0 + (x_1 - x_0)b_1 \\ &\vdots \\ f(x_n) &= b_0 + (x_n - x_0)b_1 + \cdots + (x_n - x_0)\cdots(x_n - x_{n-1})b_n \\ f(x_{n+1}) &= b_0 + (x_{n+1} - x_0)b_1 + \cdots + (x_{n+1} - x_0)\cdots(x_{n+1} - x_n)b_{n+1}. \end{aligned} \tag{4.25}$$

An examination of the first $n+1$ equations in (4.25) shows that they are the same as the equations (4.23), which justifies the statement that $b_i = a_i$,

† These are triangular equations, the solution of which is discussed in detail in Chapter 9.

$0 \leqslant i \leqslant n$. In fact, we now see that a_0 involves only $f(x_0)$, a_1 involves only $f(x_0)$ and $f(x_1)$, and so on. In general, we see that a_j involves $f(x_0)$, $f(x_1), \ldots, f(x_j)$ only. We now rewrite a_j as

$$a_j = f[x_0, x_1, \ldots, x_j]$$

to emphasize its dependence on these suffixes. It is instructive to compare (4.22) with the Lagrange form (4.6). If we equate coefficients of x^n, we obtain

$$f[x_0, \ldots, x_n] = \sum_{i=0}^{n} \frac{f(x_i)}{\prod\limits_{\substack{j=0 \\ j \neq i}}^{n} (x_i - x_j)}. \tag{4.26}$$

Thus, for example,

$$f[x_0, x_1, x_2] = \frac{f(x_0)}{(x_0 - x_1)(x_0 - x_2)} + \frac{f(x_1)}{(x_1 - x_0)(x_1 - x_2)} + \frac{f(x_2)}{(x_2 - x_0)(x_2 - x_1)}, \tag{4.27}$$

and in general the right side of (4.26) consists of a linear combination of the function values $f(x_i)$. Can we express these coefficients more simply? For example, we might try to find numbers α and β so that

$$f[x_0, x_1, x_2, x_3] = \alpha f[x_0, x_1, x_2] + \beta f[x_1, x_2, x_3], \tag{4.28}$$

for certainly (from (4.26)) each side of this equation is a sum of multiples of $f(x_0)$, $f(x_1)$, $f(x_2)$ and $f(x_3)$. Comparing the term in $f(x_0)$ in (4.26) and (4.28), we need to choose $\alpha = 1/(x_0 - x_3)$, and a comparison of the term in $f(x_3)$ requires $\beta = 1/(x_3 - x_0)$. It may be verified that this choice of α and β agrees also with the terms in $f(x_1)$ and $f(x_2)$, so that

$$f[x_0, x_1, x_2, x_3] = \frac{f[x_1, x_2, x_3] - f[x_0, x_1, x_2]}{x_3 - x_0}. \tag{4.29}$$

This shows why these expressions are called *divided differences*, especially as a similar recurrence relation holds generally. We have

$$f[x_0, \ldots, x_n] = \frac{f[x_1, \ldots, x_n] - f[x_0, \ldots, x_{n-1}]}{x_n - x_0}. \tag{4.30}$$

The last relation may be verified by substituting for each divided difference its representation in the form (4.26). It is helpful to exhibit these divided differences, as in Table 4.4, to remind us of how they are calculated. To

Table 4.4 Calculation of divided differences

x_0	$f[x_0]$			
		$f[x_0, x_1]$		
x_1	$f[x_1]$		$f[x_0, x_1, x_2]$	
		$f[x_1, x_2]$		$f[x_0, x_1, x_2, x_3]$
x_2	$f[x_2]$		$f[x_1, x_2, x_3]$	
		$f[x_2, x_3]$		
x_3	$f[x_3]$			

preserve the pattern, we have written $f[x_i]$ for $f(x_i)$. Thus, for example, we calculate $f[x_2, x_3]$ from

$$f[x_2, x_3] = \frac{f[x_3] - f[x_2]}{x_3 - x_2}$$

and $f[x_1, x_2, x_3]$ from

$$f[x_1, x_2, x_3] = \frac{f[x_2, x_3] - f[x_1, x_2]}{x_3 - x_1}.$$

Applying this notation to (4.22), we write the interpolating polynomial in the Newton form

$$\begin{aligned} p_n(x) &= f[x_0] + (x - x_0)f[x_0, x_1] + \cdots \\ &\quad + (x - x_0) \cdots (x - x_{n-1})f[x_0, \ldots, x_n]. \end{aligned} \tag{4.31}$$

We now give an algorithm for evaluating divided differences.

Algorithm 4.2 This begins with the values $x_0, \ldots, x_n$ and $f_0, \ldots, f_n$, where $f_i = f(x_i)$, $0 \leqslant i \leqslant n$, and computes $a_k = f[x_0, \ldots, x_k]$, $0 \leqslant k \leqslant n$. The general divided difference $f[x_i, \ldots, x_{i+k}]$ is also computed and is denoted by $F_{i,k}$.

$a_0 := f_0$
for $i := 0$ **to** n
  $F_{i,0} := f_i$
next i
for $k := 0$ **to** $n - 1$
  for $i := 0$ **to** $n - k - 1$
    $F_{i,k+1} := (F_{i+1,k} - F_{i,k})/(x_{i+k+1} - x_i)$
  next i
  $a_{k+1} := F_{0,k+1}$
next k □

Table 4.5 Calculation of divided differences for Example 4.7

x	$f(x)$	Divided differences	
1	0		
		$\frac{-3-0}{-1-1}=\frac{3}{2}$	
-1	-3		$\frac{7/3-3/2}{2-1}=\frac{5}{6}$
		$\frac{4-(-3)}{2-(-1)}=\frac{7}{3}$	
2	4		

Example 4.7 By using divided differences, obtain the interpolating polynomial which we sought in Example 4.3, where $f(x)=0$, -3 and 4 at $x=1$, -1 and 2 respectively. The calculation of the divided differences is shown in Table 4.5. We see from this table that $f[x_0, x_1]=\frac{3}{2}$ and $f[x_0, x_1, x_2]=\frac{5}{6}$ so that, from (4.31), the required interpolating polynomial is

$$p_2(x)=0+(x-1)\tfrac{3}{2}+(x-1)(x+1)\tfrac{5}{6}.$$

This simplifies to give

$$p_2(x)=\tfrac{1}{6}(5x^2+9x-14),$$

as we found in Example 4.3 by the Lagrange method. □

We now comment on the relative merits of the divided difference form and the Neville–Aitken algorithm for evaluating the interpolating polynomial. The Neville–Aitken algorithm has the advantage that each number generated by (4.18) is an interpolating polynomial (evaluated at the same point x) for some part of the data. The level of agreement between these numbers provides some empirical evidence about the accuracy of the interpolation (see, for example, Table 4.2). On the other hand, if we wish to evaluate $p_n(x)$ for many values of x, the divided difference form is the more efficient method, since the divided differences $f[x_0, \ldots, x_i]$, being independent of x, only need to be computed once.

4.7 Equally spaced points

We now consider the special case where the interpolating points are equally spaced: we let $x_j=x_0+jh$, $0\leqslant j\leqslant n$, where $h>0$ denotes the equal spacing between the points. Thus the interpolating points are completely determined by only two parameters, x_0 and h, compared with $n+1$ parameters in the general case. We will see that this leads to a simpler form of the interpolating polynomial.

First we note that

$$f[x_i, x_{i+1}] = (f_{i+1} - f_i)/h$$

where we have written f_i as an abbreviation of $f(x_i)$. We also find that

$$f[x_i, x_{i+1}, x_{i+2}] = (f_{i+2} - 2f_{i+1} + f_i)/2h^2.$$

To follow up what happens to higher order divided differences, we introduce the *forward difference operator* Δ. We write

$$\Delta f(x) = f(x+h) - f(x),$$

so that

$$\Delta f_i = f_{i+1} - f_i.$$

This is called a *first difference*. We define *higher order differences* recursively from

$$\Delta^{k+1} f(x) = \Delta(\Delta^k f(x)), \qquad k = 1, 2, \ldots,$$

so that

$$\begin{aligned}\Delta^2 f(x) &= \Delta(\Delta f(x)) = \Delta(f(x+h) - f(x)) \\ &= (f(x+2h) - f(x+h)) - (f(x+h) - f(x)) \\ &= f(x+2h) - 2f(x+h) + f(x)\end{aligned}$$

and

$$\Delta^2 f_i = f_{i+2} - 2f_{i+1} + f_i.$$

This is a *second difference*. For completeness, we also write

$$\Delta^0 f(x) = f(x).$$

An induction argument can be used to verify that

$$f[x_i, x_{i+1}, \ldots, x_{i+k}] = \frac{\Delta^k f_i}{k!h^k}. \tag{4.32}$$

Note that this only holds when the x_j are equally spaced. Table 4.6 shows how the divided difference Table 4.4 simplifies in the equally spaced case.

We now turn to Newton's divided difference formula (4.31). Let us introduce a new variable s, defined by $x = x_0 + sh$. Then

$$(x - x_0)\cdots(x - x_{k-1}) = h^k s(s-1)\cdots(s-k+1).$$

From this and (4.32), the divided difference formula (4.31) becomes

$$p_n(x) = p_n(x_0 + sh) = f_0 + \binom{s}{1}\Delta f_0 + \cdots + \binom{s}{n}\Delta^n f_0, \tag{4.33}$$

Table 4.6 Divided differences with equally spaced points, expressed in terms of forward differences

x_0	f_0			
		$\frac{1}{h}\Delta f_0$		
x_1	f_1		$\frac{1}{2h^2}\Delta^2 f_0$	
		$\frac{1}{h}\Delta f_1$		$\frac{1}{6h^3}\Delta^3 f_0$
x_2	f_2		$\frac{1}{2h^2}\Delta^2 f_1$	
		$\frac{1}{h}\Delta f_2$		
x_3	f_3			

which is called the *forward difference formula*. In (4.33) we have used the binomial coefficient

$$\binom{s}{k} = \frac{s(s-1)\dots(s-k+1)}{k!}$$

for $k = 1, 2, \dots, n$. By convention, when $k = 0$ we write $\binom{s}{0} = 1$. In this notation, the interpolating polynomial plus error term (4.13) can be expressed as

$$f(x_0 + sh) = \sum_{k=0}^{n} \binom{s}{k} \Delta^k f_0 + h^{n+1} \binom{s}{n+1} f^{(n+1)}(\xi_s). \tag{4.34}$$

The forward difference formula is often attributed† to Newton and to his Scots contemporary James Gregory (1638–75). However it was used earlier by the English mathematician Thomas Harriot (1560–1621) and was known very much earlier, at least for small values of n, to the Chinese mathematician Guo Shoujing (1231–1316).

Example 4.8 Given the following table, use the forward difference formula to estimate $\sin x$ at $x = 0.63$ and determine the accuracy of the result.

x	0.6	0.7	0.8	0.9	1.0
$\sin x$	0.564642	0.644218	0.717356	0.783327	0.841471

† Hildebrand (1974, p. 129) makes an interesting comment on the naming of interpolation formulas.

Table 4.7 A table of differences for sin x

x	$\sin x$	Differences			
0.6	0.564642				
		79576			
0.7	0.644218		−6438		
		73138		−729	
0.8	0.717356		−7167		69
		65971		−660	
0.9	0.783327		−7827		
		58144			
1.0	0.841471				

The differences for this data are set out in Table 4.7. For convenience, the decimal point has been omitted from the differences. This is normal practice. In this case, $x_0 = 0.6$ and $h = 0.1$, so in using (4.33) to interpolate at $x = 0.63$, we choose $s = 0.3$ and $n = 4$. We see from Table 4.7 that f_0, Δf_0, $\Delta^2 f_0$, $\Delta^3 f_0$ and $\Delta^4 f_0$ have the values 0.564642, 0.079576, −0.006438, −0.000729 and 0.000069 respectively, which are boxed in the table. With these figures, the estimate for $\sin(0.63)$ provided by (4.33) is 0.589145, to six decimal places. Using (4.34) to estimate the error, the fifth derivative of $\sin x$ is $\cos x$, which never exceeds 1 in modulus. Hence the error is not greater than the modulus of

$$10^{-5}(0.3)(-0.7)(-1.7)(-2.7)(-3.7)/5!,$$

which is smaller than 0.3×10^{-6}. □

In Chapter 13 we require a companion formula to (4.33) which starts at the right-hand end (i.e. at $x = x_n$) of a set of equally spaced data and uses differences constructed from function values to the left of $x = x_n$. To obtain this, we return to the divided difference representation of $p_n(x)$, in (4.31). In this formula, the numbers $x_0, \ldots, x_n$ are distinct, but otherwise arbitrary.

In particular, we could rename $x_0, x_1, \ldots, x_n$ as $x_n, x_{n-1}, \ldots, x_0$ which would give

$$p_n(x) = f[x_n] + (x - x_n)f[x_n, x_{n-1}] + \cdots$$
$$+ (x - x_n)(x - x_{n-1}) \cdots (x - x_1)f[x_n, \ldots, x_0]. \tag{4.35}$$

To simplify this when the x_j are equally spaced, it is convenient to introduce the *backward difference operator* ∇, for which

$$\nabla f_i = f_i - f_{i-1}$$

and $\nabla^{n+1} f_i = \nabla(\nabla^n f_i)$, as for forward differences. Proceeding as we did in deriving the forward difference formula, we obtain from (4.35)

$$p_n(x_n + sh) = \sum_{j=0}^{n} (-1)^j \binom{-s}{j} \nabla^j f_n. \tag{4.36}$$

This is the *backward difference formula.*

It should be emphasized that the forward and backward difference formulas are merely different ways of representing the *same* polynomial. It may also be noted that it is not strictly necessary to have both operators Δ and ∇. For example, ∇f_n, $\nabla^2 f_n$ and $\nabla^3 f_n$ denote the same numbers as Δf_{n-1}, $\Delta^2 f_{n-2}$ and $\Delta^3 f_{n-3}$ respectively. As another illustration, consider Tables 4.8(a) and 4.8(b) whose corresponding entries are numerically the same; only the notation is different. Nevertheless, it is convenient to have these alternative ways of expressing differences. For expressing differences symmetrically, there is a third operator δ, called the *central difference operator*, defined by

$$\delta f(x) = f(x + \tfrac{1}{2}h) - f(x - \tfrac{1}{2}h).$$

Also

$$\begin{aligned}\delta^2 f(x) &= \delta(f(x + \tfrac{1}{2}h) - f(x - \tfrac{1}{2}h)) \\ &= f(x+h) - 2f(x) + f(x-h).\end{aligned}$$

Hildebrand (1974) discusses several interpolating formulas which use central differences. However, we will not pursue this here, since such formulas are primarily of use in highly accurate interpolation of tables, and are rarely required nowadays.

Table 4.8 Forward and backward differences

(a)					(b)				
x_0	f_0				x_0	f_0			
		Δf_0					∇f_1		
x_1	f_1		$\Delta^2 f_0$		x_1	f_1		$\nabla^2 f_2$	
		Δf_1		$\Delta^3 f_0$			∇f_2		$\nabla^3 f_3$
x_2	f_2		$\Delta^2 f_1$		x_2	f_2		$\nabla^2 f_3$	
		Δf_2					∇f_3		
x_3	f_3				x_3	f_3			

We now give algorithms to evaluate forward differences and the forward difference formula.

Algorithm 4.3 This begins with values $f_0, \ldots, f_n$ and computes $a_k = \Delta^k f_0$, $0 \leqslant k \leqslant n$. In the algorithm, $F_{i,k} = \Delta^k f_i$. The $F_{i,k}$ are computed from

$$F_{i,k+1} := F_{i+1,k} - F_{i,k}.$$

Apart from this change in the calculation of the $F_{i,k}$, this algorithm is identical to Algorithm 4.2. □

Having computed the differences $\Delta^k f_0$, we use the following to evaluate $p_n(x) = p_n(x_0 + sh)$ using (4.33).

Algorithm 4.4 This begins with s and $a_k = \Delta^k f_0$, $0 \leq k \leq n$, and computes $p = p_n(x_0 + sh)$ from the forward difference formula (4.33).

$p := a_n$
for $i := 1$ **to** n
  $p := a_{n-i} + (s - n + i)p/(n - i + 1)$
next i □

4.8 Derivatives and differences

One often observes that differences of a function (tabulated at equal intervals) tend to decrease as the order of the differences increases. This appears to be the case in Table 4.7. To explain this, we make use once more of divided differences. Let us write

$$f[x, x_0] = \frac{f(x_0) - f(x)}{x_0 - x},$$

which may be rearranged as

$$f(x) = f(x_0) + (x - x_0)f[x, x_0]. \tag{4.37}$$

Also

$$f[x, x_0, x_1] = \frac{f[x_0, x_1] - f[x, x_0]}{x_1 - x},$$

which gives the equation

$$f[x, x_0] = f[x_0, x_1] + (x - x_1)f[x, x_0, x_1]. \tag{4.38}$$

Substituting for $f[x, x_0]$ from (4.38) into (4.37) gives

$$f(x) = f[x_0] + (x - x_0)f[x_0, x_1] + (x - x_0)(x - x_1)f[x, x_0, x_1],$$

where we have written $f[x_0]$ instead of $f(x_0)$ for the sake of uniformity of notation. This may be extended, by considering $f[x, x_0, x_1, x_2]$, and so on, until we obtain by induction

$$\begin{aligned} f(x) = f[x_0] &+ (x - x_0)f[x_0, x_1] + \cdots \\ &+ (x - x_0)\cdots(x - x_{n-1})f[x_0, \ldots, x_n] \\ &+ (x - x_0)\cdots(x - x_n)f[x, x_0, \ldots, x_n]. \end{aligned} \tag{4.39}$$

Comparing this with (4.31), we see that (4.39), is of the form

$$f(x) = p_n(x) + (x - x_0)\cdots(x - x_n)f[x, x_0, \ldots, x_n] \tag{4.40}$$

and, on comparison with the error formula (4.13), we find that

$$f[x, x_0, \ldots, x_n] = \frac{f^{(n+1)}(\xi_x)}{(n+1)!}, \tag{4.41}$$

where ξ_x is some point in an interval containing $x, x_0, \ldots, x_n$. This last formula holds for any n and any distinct numbers x, $x_0, \ldots, x_n$, provided that the $(n+1)$th derivative of f exists. In particular, omitting the point x_0 in (4.41), we have

$$f[x, x_1, \ldots, x_n] = \frac{f^{(n)}(\eta_x)}{n!}, \tag{4.42}$$

where η_x lies in an interval containing $x, x_1, \ldots, x_n$. Now put $x = x_0$ in (4.42) and let $x_0, \ldots, x_n$ be equally spaced. On using (4.32), we obtain

$$\Delta^n f_0 = h^n f^{(n)}(\eta_0), \tag{4.43}$$

where η_0 lies in the interval $(x_0, x_0 + nh)$. In the same way, we have

$$\Delta^n f_1 = h^n f^{(n)}(\eta_1),$$

where η_1 lies in the interval $(x_1, x_1 + nh)$, and so on. This shows that nth differences behave as h^n times the nth derivative of f. It is now clear why in Table 4.7, where $h = 0.1$, differences are decreasing by approximately a factor of 10, from one column to the next.

As an aside, if in (4.42) we put $x = x_0$ and retain the divided difference notation we have

$$f[x_0, \ldots, x_n] = \frac{f^{(n)}(\eta_0)}{n!}, \tag{4.44}$$

which shows the close connection between the divided difference representation for f, (4.39), and the Taylor polynomial representation. Indeed, allowing $x_0, \ldots, x_n$ to be equally spaced in (4.44) and letting $h \to 0$, we obtain

$$\lim_{h \to 0} f[x_0, \ldots, x_n] = \frac{f^{(n)}(x_0)}{n!} \tag{4.45}$$

and (4.39) becomes the Taylor polynomial for f about x_0, with remainder. What we have given is a proof of Taylor's theorem.

In practice, it is usually difficult to evaluate $f^{(n+1)}(x)$. It is then desirable to use an $(n+1)$th difference in place of $f^{(n+1)}(\xi_x)$ to estimate the error term (4.13) for polynomial interpolation. If in Example 4.4 we are given $\log 2.3 = 0.8329$ we find that, with $h = 0.1$ and $f(x) = \log x$,

Table 4.9 Evaluation of a polynomial by 'building-up' from its differences. Here $p(x) = x^3 - 3x + 5$, so third differences are constant

		Differences		
x	$p(x)$	1st	2nd	3rd
0	5			
		−2		
1	3		6	
		4		6
2	7		12	
		16		6
3	23		18	
		34		6
4	57		24	
		58		
5	115			

$\Delta^2 f(2.1) = -0.0022$. (For x between 2.1 and 2.3, $h^2 f''(x)$ actually lies between -0.00227 and -0.00189.)

The relation (4.43), which connects differences with derivatives, reveals one other property of polynomials. If $p \in P_n$, its nth derivative is constant and higher derivatives are zero. Substituting $p(x)$ into (4.43), we see that the same statement holds for differences so that, for $p \in P_n$, differences higher than the nth must be zero. We can make use of this property to tabulate a polynomial.

Example 4.9 In Table 4.9 the polynomial $p(x) = x^3 - 3x + 5$ is evaluated at $x = 0, 1, 2, 3$ and differenced, to produce all the numbers which appear above the line shown in the table. Further entries are constructed by using the property that third differences are constant. So, below the line, we insert the entries 6, 18, 34, 57 in that order. Next, returning to the constant third differences, we insert the entries 6, 24, 58, 115, and so on, to extend the table. □

4.9 Effect of rounding error

We now consider how the accuracy of the function values $f(x_i)$ affects the accuracy of interpolation. We recall the Lagrange form (4.11),

$$p_n(x) = \sum_{i=0}^{n} L_i(x) f(x_i).$$

Suppose that, instead of exact values $f(x_i)$, we have approximate values $f^*(x_i)$, where

$$|f(x_i) - f^*(x_i)| \leqslant \varepsilon, \tag{4.46}$$

for each i. For example, if the exact values $f(x_i)$ are rounded to four decimal places, we can take $\varepsilon = \frac{1}{2} \times 10^{-4}$. Thus, instead of the true polynomial $p_n(x)$, we evaluate

$$p_n^*(x) = \sum_{i=0}^{n} L_i(x) f^*(x_i). \tag{4.47}$$

(We will ignore the effects of rounding error in evaluating the right side of (4.47), so as not to obscure the main point.) We deduce, using (4.46), that

$$|p_n(x) - p_n^*(x)| \leq \varepsilon \lambda_n(x), \tag{4.48}$$

where

$$\lambda_n(x) = \sum_{i=0}^{n} |L_i(x)|. \tag{4.49}$$

The function $\lambda_n(x)$, which is called a *Lebesgue function*, gives an upper bound for the 'magnification factor' between the errors in the $f(x_i)$ and the error in $p_n(x)$, due to rounding. If we are interpolating within an interval $[a, b]$, we will be interested in the values of λ_n on $[a, b]$. We note that the Lebesgue function $\lambda_n(x)$ depends only on the interpolating points and not on the function f. Suppose the x_i are ordered so that $x_0 < x_1 < \cdots < x_n$. The number

$$\Lambda_n = \max_{a \leq x \leq b} \lambda_n(x) \tag{4.50}$$

is called the *Lebesgue constant* of order n. It is an interesting task to compute Λ_n for different choices of interpolating points. When the x_i are equally spaced on $[-1, 1]$ it may be verified (see Problem 4.33) that $\Lambda_n < 30$ for $1 \leq n \leq 10$, where $[a, b]$ in (4.50) is taken to be $[-1, 1]$. The case $n = 1$ is particularly simple. We find that $\lambda_1(x) = 1$ for $x_0 \leq x \leq x_1$, so that

$$|p_1(x) - p_1^*(x)| \leq \varepsilon$$

for $x_0 \leq x \leq x_1$, whenever (4.46) holds. This can be deduced directly from Fig. 4.1, by drawing above and below p_1, at a distance ε, two straight lines parallel to p_1. We see that p_1^* must lie between these two parallel lines.

4.10 Choice of interpolating points

We now return to the formula (4.13), obtained for the error incurred by using the interpolating polynomial p_n in place of f:

$$f(x) - p_n(x) = (x - x_0) \cdots (x - x_n) \frac{f^{(n+1)}(\xi_x)}{(n+1)!}, \tag{4.51}$$

which is valid provided $f^{(n+1)}$ is continuous. We suppose that $p_n(x)$ is used to approximate to $f(x)$ on some finite interval which, to be definite, we will take to be $[-1, 1]$. Then, if $|f^{(n+1)}(x)|$ is bounded by M_{n+1} on $[-1, 1]$, we have

$$\max_{-1 \leqslant x \leqslant 1} |f(x) - p_n(x)| \leqslant \frac{M_{n+1}}{(n+1)!} \max_{-1 \leqslant x \leqslant 1} |(x - x_0) \cdots (x - x_n)|. \quad (4.52)$$

We naturally ask: is there a choice of interpolating points x_i for which the right side of (4.52) is minimized? This is a very difficult question to answer directly.

The answer is to be found in the study of a sequence of polynomials discovered by the Russian mathematician P. L. Chebyshev (1821–94). By applying an induction argument to the trigonometrical identity

$$\cos(n+1)\theta + \cos(n-1)\theta = 2 \cos n\theta \cos \theta \quad (4.53)$$

we can see that $\cos n\theta$ is a polynomial of degree n in $\cos \theta$ for $n = 0, 1, 2, \ldots$. We therefore put $x = \cos \theta$ and write

$$T_n(x) = \cos(n \cos^{-1} x), \qquad -1 \leqslant x \leqslant 1, \quad (4.54)$$

to denote this polynomial of degree n, which is called a *Chebyshev polynomial*. On rearranging (4.53), we obtain

$$T_{n+1}(x) = 2xT_n(x) - T_{n-1}(x), \qquad n \geqslant 1, \quad (4.55)$$

with $T_0(x) = 1$ and $T_1(x) = x$. This is a recurrence relation which enables us to evaluate the Chebyshev polynomials. By using (4.55), with the above initial values for T_0 and T_1, we derive in turn

$$T_2(x) = 2x^2 - 1, \qquad T_3(x) = 4x^3 - 3x,$$
$$T_4(x) = 8x^4 - 8x^2 + 1, \qquad T_5(x) = 16x^5 - 20x^3 + 5x,$$

to give only the first few. In practice we do not need to display these polynomials in algebraic form, as above. For any given values of n and x we can evaluate $T_n(x)$ by the following algorithm.

Algorithm 4.5

if $n = 0$ **then** $T_n := 1$
else if $n = 1$ **then** $T_n := x$
else $T_0 := 1$, $T_1 := x$
  for $k := 1$ **to** $n - 1$
    $T_{k+1} := 2xT_k - T_{k-1}$
  next k □

It is very easy, using a computer and Algorithm 4.5, to draw graphs of the Chebyshev polynomials. As an example, the graph of $T_4(x)$ is displayed in Fig. 4.2.

These graphs suggest the following three properties of the Chebyshev polynomials:

(i) $|T_n(x)| \leqslant 1$ for $-1 \leqslant x \leqslant 1$,
(ii) T_n has n distinct real zeros in the interior of $[-1, 1]$,
(iii) $|T_n(x)|$ attains its maximum modulus of 1 on $[-1, 1]$ at $n+1$ points, including both endpoints ± 1, and $T_n(x)$ takes the values ± 1 alternately on these points.

We will now verify these properties. First we see that (i) follows immediately from the definition (4.54). Second we note that

$$T_n(x) = 0 \quad \Rightarrow \quad \cos n\theta = 0,$$

whence $n\theta = (2k+1)\pi/2$, with k an integer, and

$$x = \cos\,\theta = \cos\{(2k+1)\pi/(2n)\}.$$

We obtain exactly n distinct values of x on $[-1, 1]$ for which $T_n(x) = 0$ by taking $k = 0,\ 1, \dots, n-1$. Since $T_n \in P_n$, there cannot be more than n such values. Therefore we must have found all the zeros of T_n. Finally,

$$|T_n(x)| = 1 \quad \Rightarrow \quad \cos n\theta = \pm 1$$

whence $n\theta = k\pi$, with k an integer, and

$$x = \cos\,\theta, \qquad \text{with } \theta = k\pi/n.$$

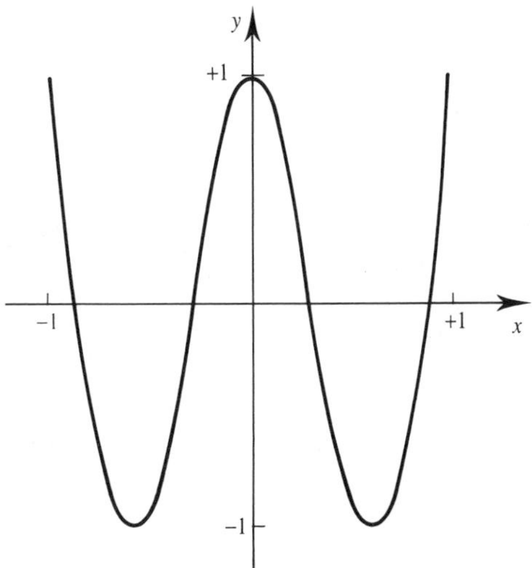

Fig. 4.2 The Chebyshev polynomial $T_4(x)$.

Corresponding to this value of x we have

$$T_n(x) = \cos n\theta, \qquad \text{with } \theta = k\pi/n,$$

so that

$$T_n(x) = \cos k\pi = (-1)^k.$$

This shows that $T_n(x)$ takes the values ± 1 alternately as x takes the values $\cos(k\pi/n)$, $0 \leqslant k \leqslant n$. Note that $k=0$ and n correspond to $x=1$ and -1 respectively. These $n+1$ points where T_n assumes its maximum modulus on $[-1, 1]$ are called the *extreme points* of T_n.

We now state and prove a theorem due to Chebyshev which settles the question posed at the beginning of this section about the choice of interpolating points: the theorem shows that this difficult question has a simple answer.

Theorem 4.3 If $q \in P_n$ has leading coefficient 1 (that is, the coefficient of x^n is $+1$) then†, for $n \geqslant 1$,

$$\|q\|_\infty = \max_{-1 \leqslant x \leqslant 1} |q(x)|$$

is minimized over all such choices of q when $q = T_n/2^{n-1}$.

Proof We suppose that the statement of the theorem is false and seek to establish a contradiction. Thus we assume the existence of a polynomial $r \in P_n$ with leading coefficient 1 such that

$$\|r\|_\infty < 1/2^{n-1}.$$

Let us consider the polynomial

$$r(x) - q(x) = r(x) - \frac{1}{2^{n-1}} T_n(x), \tag{4.56}$$

which belongs to P_{n-1}. On each of the $n+1$ extreme points of T_n, the first term on the right of (4.56), $r(x)$, has modulus less than $1/2^{n-1}$, while the second term, $T_n(x)/2^{n-1}$, takes the values $\pm 1/2^{n-1}$ alternately. Thus the polynomial $r-q$ takes positive and negative values alternately on the $n+1$ extreme points of T_n. This implies that $r - q \in P_{n-1}$ has at least n zeros, which provides the required contradiction. □

Since $(x-x_0)\cdots(x-x_n)$ is a polynomial belonging to P_{n+1}, with leading coefficient 1, we have the following important corollary to Theorem 4.3.

† $\|q\|_\infty$ is called the maximum norm of q (see § 5.1).

Corollary 4.3 The expression

$$\max_{-1 \leqslant x \leqslant 1} |(x-x_0)\cdots(x-x_n)|$$

is minimized, with minimum value $1/2^n$, by choosing $x_0, \ldots, x_n$ as the zeros of the Chebyshev polynomial $T_{n+1}(x)$. □

We now apply this result to the inequality (4.52) which, being derived from the error formula (4.51), requires continuity of $f^{(n+1)}$. If we choose $p_n \in P_n$ as the polynomial which interpolates f on the zeros of T_{n+1}, then

$$\|f - p_n\|_\infty \leqslant \frac{M_{n+1}}{2^n(n+1)!}, \tag{4.57}$$

where M_{n+1} is an upper bound for $|f^{(n+1)}(x)|$ on $[-1, 1]$.

4.11 Examples of Bernstein and Runge

In this section we consider examples which warn us of the limitations of using interpolating polynomials as approximations to functions. First we consider $|x|$ on $[-1, 1]$. Since the first derivative is discontinuous at the origin, the error formula (4.13) is not valid in this case. Nevertheless, since the function $|x|$ is continuous, the reader may believe intuitively that the sequence of interpolating polynomials, whether interpolating at the Chebyshev zeros or at equally spaced points on $[-1, 1]$, will converge uniformly to $|x|$ on $[-1, 1]$. Table 4.10 suggests that such intuition is misleading: for the given range of values of n, we see that the errors grow catastrophically for the equally spaced interpolation points. The behaviour of the errors for interpolation at the Chebyshev zeros is much more satisfactory and indeed these errors diminish with increasing n, although not as rapidly as one might wish. (The entries in Table 4.10 are given to two significant figures.)

Our second example is $1/(1+25x^2)$ on $[-1, 1]$. In this case, all derivatives are continuous on $[-1, 1]$. Table 4.11 presents numerical results for the error of interpolation for this function in the same format as used in Table 4.10 for $|x|$. We note that the results in Table 4.11 are broadly

Table 4.10 Maximum modulus of $f(x) - p_n(x)$ on $[-1, 1]$ for $f(x) = |x|$, for interpolation at equally spaced points and at the Chebyshev zeros

n	2	4	6	8	10	12	14	16	18	20
Equally spaced	0.25	0.15	0.18	0.32	0.66	1.6	4.1	11	32	95
Chebyshev	0.22	0.12	0.087	0.067	0.055	0.046	0.040	0.035	0.031	0.028

Table 4.11 Runge's example: maximum modulus of $f(x)-p_n(x)$ on $[-1,1]$ for $f(x)=1/(1+25x^2)$, where the interpolating points are equally spaced. For comparison, we also give the results for interpolation at the Chebyshev zeros

n	2	4	6	8	10	12	14	16	18	20
Equally spaced	0.65	0.44	0.62	1.0	1.9	3.7	7.2	14	29	60
Chebyshev	0.60	0.40	0.26	0.17	0.11	0.069	0.047	0.033	0.022	0.015

similar to those of Table 4.10, with catastrophic results for the equally spaced case and errors which decrease very slowly in the Chebyshev case. The disastrous behaviour of interpolating polynomials on equally spaced points for the function $1/(1+25x^2)$ on $[-1,1]$ was first demonstrated by the German mathematician C. Runge (1856–1927) near the beginning of the twentieth century. Just a few years later, the corresponding problem for $|x|$, which we have described above, was first discussed by the Russian mathematician S. N. Bernstein (1880–1968).

After these two 'bad' examples, we end this chapter on a more positive note by looking again at (4.57). We deduce that if all the derivatives of a given function f exist and

$$\lim_{n\to\infty}\frac{M_{n+1}}{2^n(n+1)!}=0,$$

where M_{n+1} is the maximum modulus of $f^{(n+1)}$ on $[-1,1]$, then the sequence of polynomials which interpolate f at the Chebyshev zeros converges uniformly to f on $[-1,1]$. An example of such a function is e^x.

Problems

Section 4.1

4.1 If $f(x)=1$ and 5 at $x=0$ and 1 respectively, construct the interpolating polynomial p_1 which matches f at these points.

4.2 Using (4.1) with $x_1=x_0+h$, $x=x_0+sh$ and $f(x)=e^x$, show that

$$p_1(x_0+sh)=[1+s(e^h-1)]e^{x_0}$$

is the polynomial which interpolates e^x at x_0 and x_1.

4.3 Show that for any real number $\lambda\neq 0$ and any positive integers r and s the polynomial

$$q(x)=\lambda(x-x_0)^r(x-x_1)^s+\left(\frac{x-x_1}{x_0-x_1}\right)f(x_0)+\left(\frac{x-x_0}{x_1-x_0}\right)f(x_1),$$

which is of degree $r+s$, passes through the points $(x_0, f(x_0))$ and $(x_1, f(x_1))$.

4.4 Use linear interpolation to estimate cos 50°, given that cos 45° = $1/\sqrt{2}$ (0.7071 to four decimal places) and cos 60° = $\frac{1}{2}$.

4.5 By using interpolation at $x=1$ and 4, find approximations to the function $y=x^{1/2}$ at $x=2$ and $x=3$.

4.6 Write a program for carrying out linear interpolation. Extend the program so that, for given values of x_0 and h, it prints the values

$$|\mathrm{e}^{x_0+sh} - p_1(x_0+sh)|$$

for $s=0.1, 0.2, \ldots, 0.9$, with p_1 as in Problem 4.2. Use this to estimate the maximum error of linear interpolation between successive entries in a table of e^x, assumed to be tabulated at intervals of $h=0.01$ on $[0,1]$.

Section 4.2

4.7 Find the interpolating polynomial for the function f which agrees with the following data: $f(x)=1, -1$ and 1 at $x=-1, 0$ and 1.

4.8 Obtain the interpolating polynomial for the function x^2+1 at $x=0$, 1, 2 and 3.

4.9 Construct a cubic polynomial p such that, at $x=0$ and 1, p takes the same values as f (0 and 1 respectively) and p' has the same values as f' (-3 and 9 respectively). (*Hint*: let $p(x) \equiv ax^3+bx^2+cx+d$, derive four equations in a, b, c, d and solve them. See also § 6.4 on Hermite interpolation.)

Section 4.3

4.10 Estimate the accuracy of the results obtained in Problem 4.4, noting first that each cosine must be expressed in radian measure: for example, cos 50° as $\cos(50\pi/180)$.

4.11 Suppose that a function f is tabulated at equal intervals of length h and that $|f''(x)| \leq M$ throughout the table. Show that the modulus of the error due to the use of linear interpolation between any two adjoining entries in the table cannot be greater than $\frac{1}{8}Mh^2$.

4.12 Verify that your computed results for Problem 4.6 are consistent with the general result in Problem 4.11.

4.13 At what step size h ought the function $\sin x$ to be tabulated so that linear interpolation will produce an error of not more than $\frac{1}{2} \times 10^{-6}$? (Use the result quoted in Problem 4.11.)

Section 4.4

4.14 Verify by induction on k that $f_{i,k}(x)$ is the interpolating polynomial for f at the points $x_i, x_{i+1}, \ldots, x_{i+k}$ (see Algorithm 4.1).

4.15 Set out the Neville–Aitken algorithm (as in Table 4.1) for interpolation of the function $x^{1/2}$ at $x=2$, with $x_0=0$, $x_1=1$ and $x_2=4$. Use the error formula (4.13) to help explain why it is possible for the result of interpolation based on all three points to be less accurate than that of interpolation based on $x=1$ and 4 only.

4.16 Write a computer program to implement Algorithm 4.1 (Neville–Aitken). Use the data of Example 4.5 to test your program.

Section 4.5

4.17 In the following table, y denotes the function $x \sin x - 1$ (to two decimal places). Use inverse interpolation to estimate the smallest positive root of the equation $x \sin x = 1$.

x	1.0	1.1	1.2
y	−0.16	−0.02	0.12

4.18 Write a program to carry out inverse interpolation based on three points. The program should read x_0, x_1, x_2, with $x_0 < x_1 < x_2$, and three numbers $F(x_0)$, $F(x_1)$, $F(x_2)$, with $F(x_0)F(x_2) < 0$. Assuming that $F(x)=0$ has exactly one root in $[x_0, x_2]$, use inverse interpolation to estimate the root. Test your program on the data of Problem 4.17. Also use it to estimate the root of $x - e^{-x} = 0$, taking x_0, x_1 and x_2 as 0.50, 0.55 and 0.60 respectively.

Section 4.6

4.19 It is alleged that a certain cubic polynomial in x matches the following data. Construct the polynomial by using divided differences.

x	−3	−1	0	2	3
y	−9	5	3	11	33

4.20 By using divided differences, construct the cubic polynomial in y which matches the function $x = y^{1/4}$ at $y = 0$, 1, 16 and 81. Evaluate the polynomial at $y = 64$ (see p. 61).

4.21 Let us define

$$f[x_0, \ldots, x_n, x, x] = \lim_{h \to 0} f[x_0, \ldots, x_n, x+h, x].$$

Deduce that

$$f[x_0, \ldots, x_n, x, x] = \frac{\mathrm{d}}{\mathrm{d}x} f[x_0, \ldots, x_n, x].$$

4.22 Write a program to implement Algorithm 4.2 (divided differences). Test it using the data of Table 4.5.

4.23 Extend the above program so that, beginning with x, $x_0, \ldots, x_n$ and $f(x_0), \ldots, f(x_n)$, it evaluates the interpolating polynomial $p_n(x)$.

Section 4.7

4.24 For the following data, calculate the differences, write down the forward and backward difference formulas and show that they give the same polynomial in x.

x	0.0	0.1	0.2	0.3	0.4	0.5
$f(x)$	1.00	1.32	1.68	2.08	2.52	3.00

4.25 In the following table, $S(r)$ refers to the sum of the squares of the first r positive integers. From a table of differences of $S(r)$, form a conjecture that $S(r)$ can be represented exactly by a polynomial of some degree in r. Use the forward difference formula to construct this polynomial in r.

r	1	2	3	4	5	6	7	8
$S(r)$	1	5	14	30	55	91	140	204

4.26 Show that $\Delta(f(x)g(x)) = f(x).\ \Delta g(x) + \Delta f(x).\ g(x+h)$, and note the similarity this result bears to an analogous result for derivatives. Show also that

$$\Delta(f(x)g(x)) = f(x).\ \Delta g(x) + \Delta f(x).\ g(x) + \Delta f(x).\ \Delta g(x).$$

4.27 Construct the difference table for the following data. What is the lowest degree of polynomial which matches the data exactly?

x	0.0	0.1	0.2	0.3	0.4	0.5	0.6	0.7	0.8	0.9	1.0
y	0.000	0.541	1.168	1.887	2.704	3.625	4.656	5.803	7.072	8.469	10.000

4.28 Consider the forward difference formula (4.33) for the case where $f(x) = \mathrm{e}^x$. First show that, for this particular function, $\Delta^k f_0 = u^k f_0$, where $u = \mathrm{e}^h - 1$, and deduce that

$$p_n(x_0 + sh) = \left[1 + \binom{s}{1}u + \binom{s}{2}u^2 + \cdots + \binom{s}{n}u^n\right]\mathrm{e}^{x_0}$$

(Note that the above factor which multiplies e^{x_0} consists of the first $n+1$ terms of the binomial series for $(1+u)^s = \mathrm{e}^{sh}$.)

4.29 Show by induction on k that

$$\Delta^k f_0 = \sum_{i=0}^{k} (-1)^i \binom{k}{i} f_{k-i}.$$

Hint: use $\Delta^{k+1} f_0 = \Delta^k f_1 - \Delta^k f_0$ and Pascal's identity

$$\binom{k}{i} + \binom{k}{i-1} = \binom{k+1}{i}.$$

4.30 For the case where the x_j are equally spaced, write a program to evaluate the interpolating polynomial by the forward difference formula, using Algorithms 4.3 and 4.4. Use the data of Example 4.8 to test your program.

Section 4.8

4.31 Evaluate the polynomial $p(x) = x^2 + x + 7$ for $x = 0, 1, \ldots, 10$ by first evaluating $p(x)$ at $x = -1$, 0 and 1 and then 'building up' from the constant second differences.

Section 4.9

4.32 Consider the Lebesgue function (4.49). Note that, as x varies, $L_i(x)$ changes sign only when x passes through one of the points x_j. Deduce that $\lambda_n(x)$ is a piecewise polynomial; that is, it is a polynomial on each sub-interval $[x_j, x_{j+1}]$. By considering the interpolating polynomial (4.11) for $f = 1$, show that

$$\sum_{i=0}^{n} L_i(x) = 1$$

and deduce that $\lambda_n(x) \geqslant 1$ for all x. Finally, show that $\lambda_n(x_j) = 1, 0 \leqslant j \leqslant n$.

4.33 Write a program to evaluate the Lebesgue function $\lambda_n(x)$ and draw its graph. Estimate

$$\Lambda_n = \max_{-1 \leqslant x \leqslant 1} \lambda_n(x)$$

for $1 \leqslant n \leqslant 10$, when the x_j are equally spaced.

Section 4.10

4.34 Show, by an induction argument based on the recurrence relation (4.55), or otherwise, that

$$T_n(-x) = (-1)^n T_n(x),$$

that is, the Chebyshev polynomials are odd or even functions according to whether n is odd or even respectively.

4.35 Derive the recurrence relation

$$T_{2n+2}(x) = 2(2x^2 - 1)T_{2n}(x) - T_{2n-2}(x)$$

and note that this can be used to compute the even order Chebyshev polynomials recursively, without computing any odd order polynomials. How could the odd order polynomials similarly be computed separately?

4.36 Show that $T_m(T_n(x)) = T_{mn}(x)$.

4.37 Write a program to evaluate $T_n(x)$, using Algorithm 4.5. Hence use the computer to draw the graph of $T_n(x)$ for $-1 \leqslant x \leqslant 1$ and a given choice of n.

4.38 Note how the gradient of $T_n(x)$ near $x = \pm 1$ increases sharply in modulus as n increases. In particular, show that $T_n'(1) = n^2$. Alternatively, note how close the largest zero of T_n is to $x = 1$ by noting that

$$\cos\left(\frac{\pi}{2n}\right) \simeq 1 - \frac{\pi^2}{8n^2}.$$

4.39 By considering the function $x(1-x)$, show that, if n is a positive integer, $r(n-r)/n^2 = (r/n)(1-r/n)$ and

$$0 \leqslant \frac{r(n-r)}{n^2} \leqslant \frac{1}{4}$$

for $0 \leqslant r \leqslant n$. Next show that

$$\max_{0 \leqslant x \leqslant 1} \left| \left(x - \frac{r}{n}\right)\left(x - \frac{n-r}{n}\right) \right| \leqslant \frac{1}{4}$$

and hence that

$$\max_{0 \leqslant x \leqslant 1} \left| x\left(x - \frac{1}{n}\right) \cdots \left(x - \frac{n-1}{n}\right)(x-1) \right| \leqslant \frac{1}{2^{n+1}}.$$

4.40 Use the final result of Problem 4.39 to show that, if p_n denotes the interpolating polynomial constructed for e^x at the $n+1$ equally spaced points $x = r/n$, $0 \leqslant r \leqslant n$, then

$$\max_{0 \leqslant x \leqslant 1} |e^x - p_n(x)| \leqslant \frac{e}{2^{n+1}(n+1)!}.$$

What value of n will ensure that the error of interpolation is less than 10^{-6} in modulus? Compare your result with that of Problem 3.4.

4.41 If $x = \frac{1}{2}(1+t)$, verify that

$$\prod_{i=0}^{n}\left(x - \frac{i}{n}\right) = \frac{1}{2^{n+1}}\prod_{i=0}^{n}\left(t - \frac{2i-n}{n}\right).$$

Deduce from the result of Problem 4.39 that

$$\max_{-1 \leqslant t \leqslant 1} |(t-x_0)\ldots(t-x_n)| \leqslant 1$$

where $x_0 = -1$, $x_n = 1$ and the x_j are equally spaced on $[-1, 1]$.

4.42 In order to be able to evaluate $\sin x$ for any real x, it suffices to be able to evaluate $f(x) = \sin[\pi(1+x)/4]$ on $[-1, 1]$. Show that

$$\max_{-1 \leqslant x \leqslant 1} |f(x) - p_n(x)| \leqslant 2(\pi/8)^{n+1}/(n+1)!,$$

where p_n interpolates f at the Chebyshev zeros. What are the smallest values of n which guarantee accuracy to 2, 4 and 6 decimal places respectively, assuming rounding errors are negligible?

4.43 With f and p_n as in the previous problem, use your computer program for evaluating the interpolating polynomial (Problem 4.23) to estimate the values of

$$\max_{-1 \leqslant x \leqslant 1} |f(x) - p_n(x)|$$

for the first few values of n. How do these compare with the error estimates obtained in the previous problem?

4.44 Show that the change of variable

$$x = \left(\frac{b-a}{2}\right)t + \left(\frac{b+a}{2}\right)$$

maps the interval $-1 \leqslant t \leqslant 1$ onto the interval $a \leqslant x \leqslant b$ and that $x - x_r = \frac{1}{2}(b-a)(t-t_r)$, where x_r and t_r are connected by the above change of variable. Hence show that

$$\min \max_{a \leqslant x \leqslant b} |(x-x_0)\cdots(x-x_n)| = 2\left(\frac{b-a}{4}\right)^{n+1},$$

the minimum being attained when $x_r = \frac{1}{2}(b-a)t_r + \frac{1}{2}(b+a)$ and the t_r are the zeros of T_{n+1}.

4.45 Apply the result of Problem 4.44 to map $[-1, 1]$ onto $[0, 1]$. From this, show that if we construct the interpolating polynomial q_n for e^x at the $n+1$ points $\frac{1}{2}(1+t_r)$, $0 \leqslant r \leqslant n$, where the t_r are the zeros of T_{n+1},

$$\max_{0 \leqslant x \leqslant 1} |e^x - q_n(x)| \leqslant \frac{e}{2^{2n+1}(n+1)!}.$$

Compare this result with that in Problems 3.4 and 4.40.

4.46 Extend your interpolating polynomial program (Problem 4.23) so that it evaluates $f(x)$ and $p_n(x)$, allowing two options: where the x_j are equally spaced and the x_j are the zeros of T_{n+1}. Arrange for the graphs of f and p_n to be drawn for $-1 \leqslant x \leqslant 1$. Test your program on 'good' examples like e^x and the function in Problem 4.42, and on 'bad' examples like $|x|$ and $1/(1+25x^2)$, all on $[-1, 1]$.

Chapter 5

'BEST' APPROXIMATION

5.1 Norms of functions

In earlier chapters we have seen how both the Taylor polynomial and the interpolating polynomial can be used as approximations to a given function f, and in §4.10 we favoured interpolation at the Chebyshev zeros as an approximation to a function $f \in C^{n+1}[-1, 1]$, the class of functions whose $(n+1)$th derivatives are continuous on $[-1, 1]$.

In this chapter we will explore other types of polynomial approximation which are based on minimizing the *norm* of the error function. First we must define the norm of a function, which is a measure of the 'size' of a function. We write $\|f\|$, which we read as 'the norm of f'.

Definition 5.1 The norm of a function belonging to some class of functions C is a mapping from C to the non-negative real numbers which sends $f \in C$ into $\|f\| \geq 0$, subject to the following three properties or axioms.

(i) $\|f\| > 0$ unless f is the zero function, when $\|f\| = 0$.
(ii) For any real number λ and any $f \in C$,

$$\|\lambda f\| = |\lambda| \, \|f\|.$$

(iii) For any $f, g \in C$,

$$\|f + g\| \leq \|f\| + \|g\|. \qquad (5.1) \quad \square$$

Property (iii) is known as the *triangle inequality*. We will be using norms for two classes of functions. One is $C[a, b]$, the class of continuous functions on $[a, b]$. The other is the class of functions defined on a finite set of distinct points $X = \{x_0, x_1, \ldots, x_N\}$ and we will denote this by $C(X)$.

In § 4.10 we have already met an example of a norm:

$$\|f\|_\infty = \max_{a \leq x \leq b} |f(x)|. \qquad (5.2)$$

This is defined on $C[a,b]$ and is called the *maximum, Chebyshev* or *∞-norm.* It is easy to check that all three axioms of Definition 5.1 are satisfied. It is customary to use a suffix, as we have used ∞ in (5.2), to denote a particular norm. This suffix can be omitted if it is clear which norm is meant or if we are making a statement which is valid for any norm.

There is a family of norms, called the *p-norms*, defined by

$$\|f\|_p = \left\{\int_a^b |f(x)|^p \, dx\right\}^{1/p}, \tag{5.3}$$

where $f \in C[a,b]$ and $p \geq 1$. Each value of $p \geq 1$ gives a norm, although the only two finite values of p which are used in practice are 1 and 2. It is easy to verify that (5.3) satisfies axioms (i) and (ii) of Definition 5.1. Axiom (iii) is easily verified for $p=1$ and does not hold for $p<1$, which explains our restriction $p \geq 1$ above. For $p>1$ we can approximate to the integral by sums and show that the verification of axiom (iii) is equivalent to carrying out the same exercise for the discrete p-norm defined by (5.5) below.

As well as the above norms for $C[a,b]$, there are norms defined analogously for $C(X)$. These are the discrete ∞-norm,

$$\|f\|_\infty = \max_{0 \leq i \leq N} |f(x_i)| \tag{5.4}$$

and the discrete p-norm,

$$\|f\|_p = \left\{\sum_{i=0}^{N} |f(x_i)|^p\right\}^{1/p}, \tag{5.5}$$

the latter again being defined for all $p \geq 1$. It can be verified that (5.4) and (5.5) satisfy Definition 5.1 and so are indeed norms. The verification of the triangle inequality, axiom (iii), is difficult for (5.5), except for the important special cases $p=1$ and $p=2$. We will meet these discrete norms again in Chapter 10.

The reader will note that in (5.2) and (5.4) we have used $\|f\|_\infty$ to denote two different norms, one on $C[a,b]$ and one on $C(X)$, and we have similarly used $\|f\|_p$ to denote two different norms. As long as it is clear from the context which class of functions is being discussed, this dual use of notation causes no confusion.

There is a close connection between the maximum norm (5.4) and the p-norms of (5.5), given by

$$\lim_{p\to\infty} \left\{\sum_{i=0}^{N} |f(x_i)|^p\right\}^{1/p} = \max_{0 \leq i \leq N} |f(x_i)|. \tag{5.6}$$

(See Problem 5.5.) This explains the notation $\|f\|_\infty$, and there is an analogous result for the norms (5.2) and (5.3) on $C[a,b]$.

5.2 Best approximations

Suppose we have a function f in C and we have chosen some norm on C, where the class of functions C is either $C[a, b]$ or $C(X)$. This leads us to an interesting class of polynomial approximations to f which are called *best approximations*.

Definition 5.2 We say that $p \in P_n$ is a best approximation to $f \in C$ with respect to a given norm if

$$\|f - p\| = \inf_{q \in P_n} \|f - q\|. \tag{5.7}$$ □

On the right side of (5.7), the infimum is taken over all polynomials $q \in P_n$. Recall that 'infimum' means the greatest lower bound. It is important to realize that a greatest lower bound is not always attained (see Example 2.8). In this case, we want to know whether the infimum is attained by any polynomial $p \in P_n$. If there is such a polynomial p, we will say that the best approximation exists. Anyone who still thinks it is obvious that the best approximation exists should consider another approximation problem: find $\inf |\alpha - \sqrt{2}|$, where the infimum is over all rational numbers α. Obviously the infimum has the value zero, but there is no rational number closest to the irrational number $\sqrt{2}$. This is an example of an infimum not being attained. Besides the question of existence of a best approximating polynomial, we naturally want to know if it is unique, whether it has any characteristic properties and how it can be computed. We should also ask what happens as the degree of the approximating polynomial is increased: can we obtain a sequence of approximating polynomials which converges uniformly to f on $[a, b]$?

In fact, there is always a polynomial p which satisfies (5.7), for any choice of norm, and therefore such a best approximation always exists. For a proof, which relies on the theory of normed linear spaces and so is beyond the scope of our text, see Davis (1976, pp. 137–9). We will consider the question of the uniqueness of best approximations, and the other questions raised above, throughout the rest of this chapter as we study important special cases of best approximations defined by (5.7).

Before getting down to details, we make a simple observation about best approximations for $f \in C(X)$, with $X = \{x_0, x_1, \ldots, x_N\}$. If $n \geq N$, we need only choose $p \in P_n$ as the interpolating polynomial for f on X and then $\|f - p\| = 0$. We therefore exclude this trivial case by seeking best approximations from P_n with $n < N$.

We now give an example on best approximations for a function in $C[0, 1]$ to make the point that, in general, different norms lead to different best approximations.

Example 5.1 The following are best approximations in P_1 for $x^{1/2} \in C[0, 1]$ with respect to the norms given in parentheses.

$$\tfrac{1}{4}(3-\sqrt{3}) + (\sqrt{3}-1)x \qquad \text{(1-norm)}$$
$$\tfrac{4}{15} + \tfrac{4}{5}x \qquad \text{(2-norm)}$$
$$\tfrac{1}{8} + x \qquad (\infty\text{-norm})$$

These can all be found by elementary means and the last two may be derived by methods to be developed later in this chapter. To find the above 1-norm approximation, see Problem 5.32. □

Best approximations with respect to the maximum norm (∞-norm) are called *minimax approximations*, because this best approximation minimizes the maximum error $\| f - q \|_\infty$ over all $q \in P_n$. There is also a special name for best approximations with respect to the 2-norm. These are called *least squares approximations*, for obvious reasons.

Two classes of approximation problems can be distinguished. The first is approximation to a function on a finite interval, for example, the problem of choosing a polynomial which approximates within some given accuracy to e^x on $[0, 1]$. Such a polynomial could be used in a computer for evaluating e^x. For this purpose, a minimax approximation seems the most appropriate to use, amongst all the types of approximations which we have mentioned so far. In practice, as we will see later, it is preferable to use a polynomial which is near to the minimax but is easier to determine.

The second class of problems deals with approximation to a function whose values are given at only a finite number of points. An example is a set of experimental results where the value of some function f has been determined at points $x = x_0, x_1, \ldots, x_N$. If the points are distinct, we could construct the interpolating polynomial which matches f at each point. However, experimental data generally contain errors. It is therefore usually more appropriate to construct a polynomial of lower degree $n < N$ which does not necessarily pass through any of the points, but has the effect of *smoothing out* the errors in the data. We now consider what type of approximation will do this satisfactorily. Suppose we try a minimax approximation and minimize

$$\max_{0 \leqslant i \leqslant N} |f(x_i) - q(x_i)|$$

over all polynomials $q \in P_n$, the set of polynomials of degree at most n. In fact, this can be most unsatisfactory as we now demonstrate by taking an especially bad case. We suppose that in Fig. 5.1 the extreme right-hand point would be in a straight line with the others, but for experimental error. The minimax straight line approximation for this data is that labelled $q_1(x)$ in Fig. 5.1. Notice how the presence of only one inaccurate point has

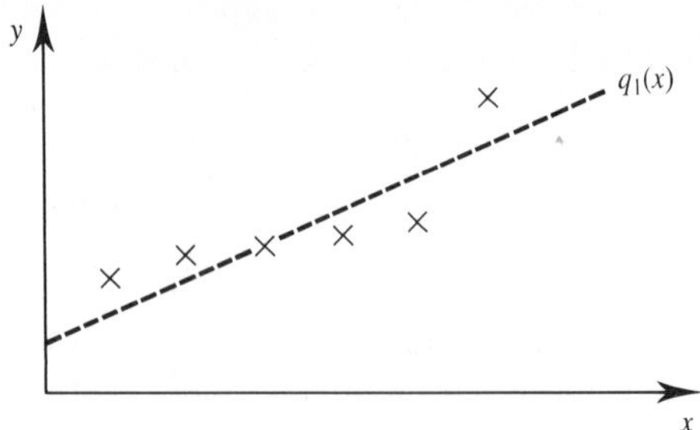

Fig. 5.1 Minimax straight line approximation.

substantially shifted the minimax approximation. Having seen an obvious disadvantage of minimax approximations for a function defined on a finite set of points we naturally turn to a best approximation with respect to one of the p-norms. We will consider in detail least squares approximations, which are the simplest to compute.

5.3 Least squares approximation

In this section, it is expedient to generalize the set of approximating functions. A polynomial of degree n may be thought of as a linear combination of the functions $1, x, x^2, \ldots, x^n$. A key property of these monomials x^r is that no linear combination of them (that is, no polynomial of degree at most n) has more than n zeros, except for the polynomial which is identically zero. It is useful to have a special term to describe functions which have this property.

Definition 5.3 A set of functions $\{\psi_0, \psi_1, \ldots, \psi_n\}$, each continuous on an interval I, is said to be a *Chebyshev set* (or, equivalently, is said to satisfy a *Haar condition*) on I if, for any choice of $a_0, a_1, \ldots, a_n$ *not all zero*, the function

$$\Phi_a(x) = \sum_{r=0}^{n} a_r \psi_r(x) \tag{5.8}$$

has not more than n zeros on I. □

Thus the monomials $\{1, x, \ldots, x^n\}$, for any n, form a Chebyshev set on any interval. Another example of a Chebyshev set is $\{1, \cos x, \sin x, \ldots, \cos kx, \sin kx\}$ on $0 \leq x < 2\pi$, where k is any positive integer.

Suppose we wish to approximate to a function f whose values are known at $x = x_0, \ldots, x_N$ by a function Φ_a defined by (5.8), with $n < N$. Let I be some interval which contains all the x_i. Then the condition that the functions ψ_r form a Chebyshev set on I certainly ensures that the set $\{\psi_0, \ldots, \psi_n\}$ contains no redundant members. For we cannot express any ψ_s as

$$\psi_s(x) = \sum_{\substack{r=0 \\ r \neq s}}^{n} b_r \psi_r(x), \qquad \text{for all } x \in [a, b], \tag{5.9}$$

since this implies that a linear combination of the ψ_r has more than n (in fact, an infinite number) of zeros. A set of functions which satisfy a relationship of the form (5.9) is called *linearly dependent*. It is useful to give a formal definition.

Definition 5.4 A set of functions $\{\psi_0, \ldots, \psi_n\}$ is said to be *linearly independent* on some interval I if

$$\sum_{r=0}^{n} a_r \psi_r(x) = 0, \qquad \text{for all } x \in I$$

only if $a_0 = a_1 = \cdots = a_n = 0$. Otherwise, the functions are said to be linearly dependent. □

The concept of linear independence will be met again in Chapter 9. It follows from Definitions 5.3 and 5.4 that if a set of functions is a Chebyshev set on I, it is also linearly independent on I. The converse does not always hold, as we now show.

Example 5.2 The set $\{1, x, x^3\}$ is *not* a Chebyshev set on $[-1, 1]$, since the function $0.1 + (-1).x + 1.x^3$ has *three* zeros belonging† to $[-1, 1]$. However, these three functions are linearly independent on $[-1, 1]$. □

We now wish to find the least squares approximation to a function f on a finite point set $\{x_0, x_1, \ldots, x_N\}$ by a function of the form Φ_a defined by (5.8). We assume that $\{\psi_0, \ldots, \psi_n\}$, with $n < N$, is a Chebyshev set on some interval $[a, b]$ which contains all the x_i. We require the minimum of

$$E(a_0, \ldots, a_n) = \sum_{i=0}^{N} (f(x_i) - \Phi_a(x_i))^2$$

over all values of $a_0, \ldots, a_n$. A necessary condition for E to have a minimum is $\partial E/\partial a_r = 0$ for $r = 0, 1, \ldots, n$ at the minimum. This gives

$$-2 \sum_{i=0}^{N} [f(x_i) - (a_0\psi_0(x_i) + \cdots + a_n\psi_n(x_i))]\psi_r(x_i) = 0$$

† A simpler example is the set $\{x^3\}$, since x^3 has a triple zero at $x = 0$.

which yields the equations

$$a_0 \sum_{i=0}^{N} \psi_0(x_i)\psi_r(x_i) + \cdots + a_n \sum_{i=0}^{N} \psi_n(x_i)\psi_r(x_i) = \sum_{i=0}^{N} f(x_i)\psi_r(x_i), \qquad (5.10)$$

for $0 \leqslant r \leqslant n$. These $n+1$ linear equations in the $n+1$ unknowns $a_0, \ldots, a_n$ are called the *normal equations*. We state, without proof here,† that these linear equations have a unique solution if there is no set of numbers $b_0, \ldots, b_n$ (except $b_0 = b_1 = \cdots = b_n = 0$) for which

$$\sum_{r=0}^{n} b_r \psi_r(x_i) = 0, \qquad 0 \leqslant i \leqslant N.$$

Since $N > n$ and the ψ_r form a Chebyshev set, this condition is satisfied. It is for this reason that we use functions ψ_r which form a Chebyshev set.

We have seen so far that a *necessary* condition for a minimum of $E(a_0, \ldots, a_n)$ is that the a_r satisfy the normal equations (5.10) and we know that these equations have a unique solution. It still remains to show that the solution of these equations does, in fact, provide the minimum. To see this, let $a_0^*, \ldots, a_n^*$ denote the solution of (5.10). First, we consider the case where there are only two functions ψ_0 and ψ_1. Consider the difference

$$\begin{aligned} E(a_0^* + \delta_0, a_1^* + \delta_1) &- E(a_0^*, a_1^*) \\ &= \sum_{i=0}^{N} [f(x_i) - (a_0^* + \delta_0)\psi_0(x_i) - (a_1^* + \delta_1)\psi_i(x_1)]^2 \\ &\quad - \sum_{i=0}^{N} [f(x_i) - a_0^* \psi_0(x_i) - a_1^* \psi_1(x_i)]^2 \\ &= \sum_{i=0}^{N} [\delta_0 \psi_0(x_i) + \delta_1 \psi_1(x_i)]^2 - 2\delta_0 \sum_{i=0}^{N} \psi_0(x_i)[f(x_i) - a_0^* \psi_0(x_i) - a_1^* \psi_1(x_i)] \\ &\quad - 2\delta_1 \sum_{i=0}^{N} \psi_1(x_i)[f(x_i) - a_0^* \psi_0(x_i) - a_1^* \psi_1(x_i)]. \end{aligned}$$

The last two summations are both zero, since a_0^* and a_1^* satisfy the normal equations (5.10) for the case $n = 1$. Therefore

$$E(a_0^* + \delta_0, a_1^* + \delta_1) - E(a_0^*, a_1^*) = \sum_{i=0}^{N} [\delta_0 \psi_0(x_i) + \delta_1 \psi_1(x_i)]^2. \qquad (5.11)$$

† See Problem 9.39 of Chapter 9.

The right side of (5.11) can only be zero if

$$\delta_0\psi_0(x_i) + \delta_1\psi_1(x_i) = 0$$

for $0 \leq i \leq N$, which cannot be so, unless $\delta_0 = \delta_1 = 0$, since ψ_0 and ψ_1 form a Chebyshev set. Thus the right side of (5.11) is always strictly positive unless $\delta_0 = \delta_1 = 0$, showing that $E(a_0, a_1)$ has indeed a minimum when $a_0 = a_0^*$, $a_1 = a_1^*$. For a general value of $n < N$, we may similarly show that

$$E(a_0^* + \delta_0, \ldots, a_n^* + \delta_n) - E(a_0^*, \ldots, a_n^*) = \sum_{i=0}^{N} [\delta_0\psi_0(x_i) + \cdots + \delta_n\psi_n(x_i)]^2,$$

showing that the minimum value of $E(a_0, \ldots, a_n)$ occurs when $a_0 = a_0^*, \ldots, a_n = a_n^*$.

Example 5.3 Find the least squares straight line approximation for the following data.

x	0.0	0.2	0.4	0.6	0.8
$f(x)$	0.9	1.9	2.8	3.3	4.2

If we take $n = 1$, $\psi_0(x) \equiv 1$ and $\psi_1(x) \equiv x$, the functions Φ_a defined by (5.8) will all be straight lines. To find the least squares straight line we must solve the normal equations (5.10) which in this case are

$$\begin{aligned} a_0(N+1) + a_1 \sum_{i=0}^{N} x_i &= \sum_{i=0}^{N} f(x_i) \\ a_0 \sum_{i=0}^{N} x_i + a_1 \sum_{i=0}^{N} x_i^2 &= \sum_{i=0}^{N} f(x_i)x_i. \end{aligned} \tag{5.12}$$

For the above data, these equations are

$$\begin{aligned} 5a_0 + 2a_1 &= 13.1 \\ 2a_0 + 1.2a_1 &= 6.84 \end{aligned}$$

with solution $a_0 = 1.02$, $a_1 = 4$. The required straight line is therefore $1.02 + 4x$. □

The foregoing method is easily extended to cater for approximation to a function of several variables. To approximate to a function $f(x, y)$ at the points $(x_0, y_0), \ldots, (x_N, y_N)$ by an approximating function of the form $\Sigma_{r=0}^{n} a_r\psi_r(x, y)$, again assuming $N > n$, we minimize

$$\sum_{i=0}^{N} [f(x_i, y_i) - (a_0\psi_0(x_i, y_i) + \cdots + a_n\psi_n(x_i, y_i))]^2.$$

By taking partial derivatives with respect to the a_r, we again obtain a set of linear equations, which are like (5.10) with (x_i, y_i) written in place of (x_i) throughout. For example, if $n=2$, $\psi_0(x, y)=1$, $\psi_1(x, y)=x$, $\psi_2(x, y)=y$, we have the system of three linear equations

$$a_0 \sum_{i=0}^{N} 1 + a_1 \sum_{i=0}^{N} x_i + a_2 \sum_{i=0}^{N} y_i = \sum_{i=0}^{N} f(x_i, y_i)$$

$$a_0 \sum_{i=0}^{N} x_i + a_1 \sum_{i=0}^{N} x_i^2 + a_2 \sum_{i=0}^{N} y_i x_i = \sum_{i=0}^{N} f(x_i, y_i) x_i$$

$$a_0 \sum_{i=0}^{N} y_i + a_1 \sum_{i=0}^{N} x_i y_i + a_2 \sum_{i=0}^{N} y_i^2 = \sum_{i=0}^{N} f(x_i, y_i) y_i$$

to determine a_0, a_1 and a_2.

Although we have discussed the existence and uniqueness of least squares approximations constructed from a Chebyshev set of functions, in practice we tend to use mainly polynomial approximations. It is sometimes quite difficult to decide what degree of approximating polynomial is most appropriate. If we are dealing with experimental results, we can often see that a polynomial of a certain degree appears suitable, judging from the number of large-scale fluctuations in the data and disregarding small-scale fluctuations which may be attributed to experimental error. Thus two large-scale fluctuations suggests a cubic approximation, since a cubic can have two turning values. (Ralston and Rabinowitz, 1978, describe a more rigorous statistical approach to the question of choosing the degree of the approximating polynomial.)

Increasing the degree of the approximating polynomial does not always increase the accuracy of the approximation. If the degree is large, the polynomial may have a large number of maxima and minima. It is then possible for the polynomial to fluctuate considerably more than the data, particularly if this contains irregularities due to errors. For this reason it is often more appropriate to use different low degree polynomials to approximate to different sections of the data. These are called *piecewise approximations*, the most commonly used being spline approximations, which we will discuss in Chapter 6.

5.4 Orthogonal functions

We now consider the problem of finding least squares approximations to a function f on an interval $[a, b]$, rather than on a finite point set. Again, we seek approximations of the form $\sum_{r=0}^{n} a_r \psi_r(x)$ constructed from certain functions ψ_r, $0 \leq r \leq n$. We now assume that the ψ_r are linearly independent on $[a, b]$ and that the ψ_r and f are continuous on $[a, b]$. We wish to minimize

$$\int_a^b \left[f(x) - \sum_{r=0}^{n} a_r \psi_r(x) \right]^2 \mathrm{d}x$$

with respect to $a_0, \dots, a_n$. As before, we set to zero the partial derivatives with respect to each a_r and obtain the (linear) equations

$$a_0 \int_a^b \psi_0(x)\psi_r(x)\,\mathrm{d}x + \cdots + a_n \int_a^b \psi_n(x)\psi_r(x)\,\mathrm{d}x = \int_a^b f(x)\psi_r(x)\,\mathrm{d}x, \quad (5.13)$$

for $0 \leqslant r \leqslant n$. Notice that these equations may be obtained from their counterparts (5.10) in the point set case, on replacing summations by integrals. It can be shown that the equations (5.13) have a unique solution if the ψ_r are linearly independent (see Problem 9.16).

In the particular case where $\psi_r(x) = x^r$ and $[a, b]$ is $[0, 1]$, the equations (5.13) become

$$\begin{aligned} a_0 + \tfrac{1}{2} a_1 + \cdots + \frac{1}{n+1} a_n &= \int_0^1 f(x)\,\mathrm{d}x \\ \tfrac{1}{2} a_0 + \tfrac{1}{3} a_1 + \cdots + \frac{1}{n+2} a_n &= \int_0^1 f(x)x\,\mathrm{d}x \\ \frac{1}{n+1} a_0 + \frac{1}{n+2} a_1 + \cdots + \frac{1}{2n+1} a_n &= \int_0^1 f(x)x^n\,\mathrm{d}x. \end{aligned} \quad (5.14)$$

Unless n is quite small, these equations are of a type for which it is difficult to calculate an accurate solution. Such equations are called *ill-conditioned*, a term which will be made more precise in Chapter 10. As n increases, the accuracy of computed solutions of these linear equations tends to deteriorate rapidly, due to rounding errors (see Example 10.4). Similar behaviour is exhibited by the linear equations which arise in computing least squares polynomial approximations to f on a finite point set. The remedy is to use only low degree approximating polynomials, perhaps $n \leqslant 6$ or so, or to use orthogonal polynomials, which are described in § 5.5.

It is natural to ask whether we can make a more appropriate choice of the functions ψ_r than the monomials x^r. We observe that, if the functions ψ_r satisfy

$$\int_a^b \psi_r(x)\psi_s(x)\,\mathrm{d}x = 0, \qquad r \neq s, \quad (5.15)$$

for all r and s, $0 \leqslant r, s \leqslant n$, then the linear equations (5.13) 'uncouple' to give

$$a_r \int_a^b (\psi_r(x))^2\,\mathrm{d}x = \int_a^b f(x)\psi_r(x)\,\mathrm{d}x, \quad (5.16)$$

for $0 \leqslant r \leqslant n$. Note that both integrals in (5.16) exist, since f and the ψ_r are continuous on $[a, b]$. Further, since the ψ_r are assumed to be linearly independent, the integral on the left of (5.16) is non-zero and therefore a_r exists. This provides the solution immediately.

Definition 5.5 A set of functions $\{\psi_0, \ldots, \psi_n\}$, which does not include the zero function, is said to be orthogonal on $[a, b]$ if the ψ_r satisfy (5.15). □

It can be shown that a set of orthogonal functions is necessarily linearly independent (see Problem 5.14). Before studying particular sets of orthogonal functions, we note the *permanence* property of least squares approximations using orthogonal functions. Suppose we have found the coefficients $a_0, \ldots, a_n$ from (5.16) and wish to *add* a further continuous function ψ_{n+1}, orthogonal to the others, to find the best approximation of the form $\sum_{r=0}^{n+1} a_r\psi_r(x)$. We see that we need only calculate a_{n+1} (from (5.16) with $r = n+1$); the values $a_0, \ldots, a_n$ will be as we have already obtained.

Example 5.4 The functions 1, $\cos x$, $\sin x$, $\cos 2x$, $\sin 2x, \ldots, \cos kx$, $\sin kx$ are orthogonal on $[-\pi, \pi]$. We need to verify that the appropriate integrals are zero (see Problem 5.16). □

With the set of orthogonal functions of Example 5.4 the best approximation to f on $[-\pi, \pi]$ is

$$\tfrac{1}{2} a_0 + \sum_{r=1}^{k} (a_r \cos rx + b_r \sin rx), \tag{5.17}$$

where from (5.16)

$$\begin{aligned}
(\tfrac{1}{2} a_0) \int_{-\pi}^{\pi} \mathrm{d}x &= \int_{-\pi}^{\pi} f(x)\,\mathrm{d}x, \\
a_r \int_{-\pi}^{\pi} \cos^2 rx\,\mathrm{d}x &= \int_{-\pi}^{\pi} f(x) \cos rx\,\mathrm{d}x, \qquad 1 \leqslant r \leqslant k, \\
b_r \int_{-\pi}^{\pi} \sin^2 rx\,\mathrm{d}x &= \int_{-\pi}^{\pi} f(x) \sin rx\,\mathrm{d}x, \qquad 1 \leqslant r \leqslant k.
\end{aligned}$$

We find that the integrals on the left sides of the last two equations have the value π (see Problem 5.16), so that

$$a_r = \frac{1}{\pi}\int_{-\pi}^{\pi} f(x)\cos rx\,\mathrm{d}x, \qquad b_r = \frac{1}{\pi}\int_{-\pi}^{\pi} f(x)\sin rx\,\mathrm{d}x. \tag{5.18}$$

In (5.17), we write $\tfrac{1}{2} a_0$ rather than a_0 so that (5.18) also holds for $r = 0$.

Letting $k \rightarrow \infty$ in (5.17), we obtain an orthogonal series for f which we will refer to as the classical Fourier† series. Some authors use the term 'Fourier series' solely for this particular orthogonal series. The general orthogonal series for f on $[a, b]$ is $\sum_{r=0}^{\infty} a_r \psi_r(x)$, where the ψ_r satisfy the orthogonality property (5.15). To denote the relationship between f and an orthogonal series for f, we will write

$$f(x) \sim \sum_{r=0}^{\infty} a_r \psi_r(x).$$

Note that, in general, we cannot assume that this infinite series converges to the function f nor even that the infinite series has a finite value. From the definition of the least squares approximation we have

$$\left\| f - \sum_{r=0}^{n+1} a_r \psi_r \right\|_2 = \inf \left\| f - \sum_{r=0}^{n+1} b_r \psi_r \right\|_2,$$

where the infimum is taken over all choices of $b_0, \ldots, b_{n+1}$. Since one particular choice of the b_r is $b_r = a_r$ for $0 \leqslant r \leqslant n$ and $b_{n+1} = 0$, it follows that

$$\left\| f - \sum_{r=0}^{n+1} a_r \psi_r \right\|_2 \leqslant \left\| f - \sum_{r=0}^{n} a_r \psi_r \right\|_2.$$

Thus the sequence of numbers $(\| f - \sum_{r=0}^{n} a_r \psi_r \|_2)$ is decreasing. We will be particularly interested in cases where this sequence tends to zero.

Definition 5.6 If $\| f - \sum_{r=0}^{n} a_r \psi_r \|_2 \rightarrow 0$ as $n \rightarrow \infty$, we say that the orthogonal series $\sum_{r=0}^{n} a_r \psi_r$ converges *in the least squares sense* to f on $[a, b]$. □

There is also a stronger form of convergence in which we work with the maximum norm of the error, rather than the least squares norm.

Definition 5.7 If $\| f - \sum_{r=0}^{n} a_r \psi_r \|_\infty \rightarrow 0$ as $n \rightarrow \infty$, we say that the orthogonal series $\sum_{r=0}^{n} a_r \psi_r$ converges *uniformly* to f on $[a, b]$. □

We can verify directly from the definitions of the norms that

$$\left\| f - \sum_{r=0}^{n} a_r \psi_r \right\|_2 \leqslant (b-a)^{1/2} \left\| f - \sum_{r=0}^{n} a_r \psi_r \right\|_\infty. \tag{5.19}$$

This inequality shows that uniform convergence implies convergence in the least squares sense. The following example shows that the converse does not

† Named after the French mathematician J. B. J. Fourier (1768–1830).

hold. This is why we say that uniform convergence is a stronger form of convergence than convergence in the least squares sense.

Example 5.5 Let $f(x) \equiv 0$ on $[0, 1]$ and consider the functions

$$f_n(x) = x^n$$

as approximations to f on $[0, 1]$. We have

$$\int_0^1 [f(x) - f_n(x)]^2 \, dx = \frac{1}{2n+1}$$

and

$$\max_{0 \leq x \leq 1} |f(x) - f_n(x)| = 1.$$

Thus (f_n) converges in the least squares sense to f, but does not converge uniformly to f on $[0, 1]$. □

If f is continuous on $[-\pi, \pi]$, it can be shown that the classical Fourier series converges in the least squares sense to f. Much investigation has been done on the *uniform* convergence of the classical Fourier series. This investigation is concerned with finding conditions on f which ensure that its Fourier series converges uniformly to f on $[-\pi, \pi]$. The assumption that f is continuous on $[-\pi, \pi]$ is not sufficient to guarantee uniform convergence. If, however, f is periodic, with period 2π, and the classical Fourier series converges uniformly to f on $[-\pi, \pi]$ then, due to the periodicity of all the functions involved, the series will converge uniformly to f everywhere. For more information on these convergence questions, see Davis (1976).

Example 5.6 Find the classical Fourier series for the function $f(x) = x$. We have

$$a_r = \frac{1}{\pi} \int_{-\pi}^{\pi} x \cos rx \, dx = 0$$

for all r, since $x \cos rx$ is an *odd* function, and

$$\begin{aligned} b_r &= \frac{1}{\pi} \int_{-\pi}^{\pi} x \sin rx \, dx = \frac{1}{\pi} \int_{-\pi}^{\pi} x \frac{d}{dx} \left(-\frac{1}{r} \cos rx \right) dx \\ &= -\frac{1}{\pi r} [x \cos rx]_{-\pi}^{\pi} + \frac{1}{\pi r} \int_{-\pi}^{\pi} \cos rx \, dx \\ &= (-1)^{r-1} 2/r, \end{aligned}$$

since $\cos r\pi = \cos(-r\pi) = (-1)^r$ and the last integral is zero. Hence

$$x \sim 2 \sum_{r=1}^{\infty} \frac{(-1)^{r-1}}{r} \sin rx. \qquad \square$$

Example 5.7 Find the classical Fourier series for the 'square wave' function defined by

$$f(x) = \begin{cases} -\frac{1}{2}, & -\pi < x < 0 \\ 0, & x = 0 \\ \frac{1}{2}, & 0 < x \leqslant \pi \end{cases}$$

and f periodic elsewhere, with period 2π (see Fig. 5.2). Since f is an odd function, $a_r = 0$ for all r and

$$\begin{aligned} b_r = \frac{1}{\pi} \int_{-\pi}^{\pi} f(x) \sin rx \,\mathrm{d}x &= \frac{2}{\pi} \int_0^{\pi} f(x) \sin rx \,\mathrm{d}x \\ &= \frac{1}{\pi} \int_0^{\pi} \sin rx \,\mathrm{d}x. \end{aligned}$$

We find that

$$b_r = \begin{cases} 0, & r \text{ even} \\ 2/(\pi r), & r \text{ odd} \end{cases}$$

and therefore

$$f(x) \sim \frac{2}{\pi} \sum_{r=0}^{\infty} \frac{\sin(2r+1)x}{2r+1}. \qquad \square$$

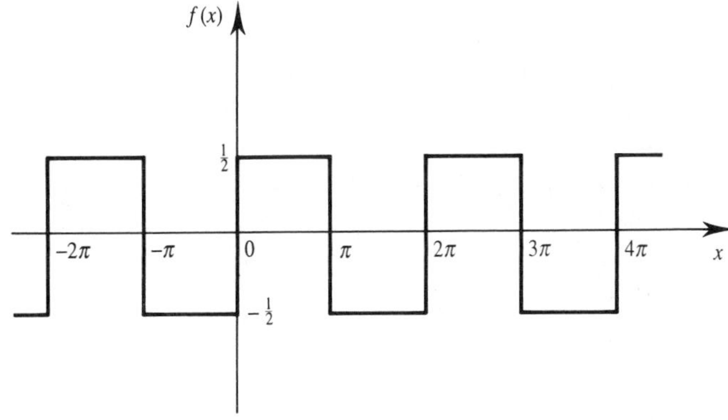

Fig. 5.2 The 'square wave' function.

5.5 Orthogonal polynomials

To give a unified treatment of approximation, it is convenient to generalize the least squares norm by introducing a *weight function.* Let ω denote some function which is integrable, non-negative and not identically zero on a given interval $[a, b]$. Then we define

$$\| f; \omega \|_2 = \left\{ \int_a^b \omega(x)[f(x)]^2 \, \mathrm{d}x \right\}^{1/2}. \tag{5.20}$$

We note that when $\omega(x) \equiv 1$ (5.20) reduces to the 2-norm, defined by (5.3) with $p = 2$, and comparing the two definitions we observe that

$$\| f; \omega \|_2 = \| \omega^{1/2} f \|_2. \tag{5.21}$$

From this it is clear that (5.20) defines a norm. We can determine best approximations with respect to this norm, using Definition 5.2. Given f, $\psi_0, \ldots, \psi_n$ as before, we wish to minimize

$$\int_a^b \omega(x) \left[f(x) - \sum_{r=0}^{n} a_r \psi_r(x) \right]^2 \mathrm{d}x$$

over all choices of $a_0, \ldots, a_n$. Proceeding exactly as before, we find that the introduction of ω does not cause any dramatic changes. The a_r still satisfy linear equations and if we now choose the functions ψ_r so that

$$\int_a^b \omega(x) \psi_r(x) \psi_s(x) \, \mathrm{d}x = 0, \qquad r \neq s, \tag{5.22}$$

the normal equations uncouple, as before, to give

$$a_r \int_a^b \omega(x)(\psi_r(x))^2 \, \mathrm{d}x = \int_a^b \omega(x) f(x) \psi_r(x) \, \mathrm{d}x. \tag{5.23}$$

Note that, when we take $\omega(x) \equiv 1$, (5.23) reduces to (5.16) as we should expect.

Functions which satisfy (5.22) are said to be orthogonal on $[a, b]$ with respect to the weight function ω. The resulting infinite series $\sum_{r=0}^{\infty} a_r \psi_r(x)$ is referred to as an orthogonal series, as before. We now show that we can construct *polynomials* which satisfy (5.22). Since we are now to deal with polynomials, we will write p_r in place of ψ_r. We will construct polynomials p_r, $r = 0, 1, \ldots$, such that $p_r(x)$ is a polynomial of degree r with leading term x^r (with coefficient 1) and

$$\int_a^b \omega(x) p_r(x) p_s(x) \, \mathrm{d}x = 0, \qquad r \neq s. \tag{5.24}$$

This is done recursively. Suppose that, for some integer $k \geqslant 0$, we have constructed polynomials $p_0, p_1, \ldots, p_k$, each with leading coefficient 1, which are mutually orthogonal on $[a, b]$ with respect to ω. (This is easily achieved for $k=0$, where we select $p_0(x) \equiv 1$.) We can verify (Problem 5.14) that $p_0, \ldots, p_k$ are linearly independent and thus any polynomial of degree $k+1$ with leading coefficient 1 can be expressed in the form

$$p_{k+1}(x) = xp_k(x) + \sum_{r=0}^{k} c_r p_r(x). \tag{5.25}$$

We need to choose the c_r so that p_{k+1} is orthogonal to $p_0, p_1, \ldots, p_k$. First we have to cope with the awkward first term on the right of (5.25): we know that the polynomials $p_0, \ldots, p_k$ are orthogonal, but what can be said about $xp_k(x)$? The following lemma comes to the rescue.

Lemma 5.1 If $0 \leqslant s \leqslant k-2$,

$$\int_a^b \omega(x) x p_k(x) p_s(x) \, dx = 0.$$

Proof We may write

$$xp_s(x) = p_{s+1}(x) + \sum_{j=0}^{s} d_j p_j(x), \tag{5.26}$$

for some constants d_j, since $xp_s(x)$ and $p_{s+1}(x)$ are both polynomials of degree $s+1$ with leading coefficient 1 (see Problem 5.23). On multiplying each term of (5.26) by $\omega(x)p_k(x)$ and integrating over $[a, b]$, the proof of the lemma follows from the orthogonality property (5.22). □

By Lemma 5.1, the choice of $c_0 = c_1 = \cdots = c_{k-2} = 0$ in (5.25) makes p_{k+1} orthogonal to $p_0, p_1, \ldots, p_{k-2}$. We now replace c_{k-1} by $-\gamma_k$ and c_k by $-\beta_k$ and write (5.25) as

$$p_{k+1}(x) = (x - \beta_k)p_k(x) - \gamma_k p_{k-1}(x). \tag{5.27}$$

It remains only to choose values of β_k and γ_k to make p_{k+1} orthogonal to p_{k-1} and p_k. From (5.27),

$$\int_a^b \omega(x) p_{k+1}(x) p_s(x) \, dx = \int_a^b \omega(x)(x - \beta_k) p_k(x) p_s(x) \, dx$$
$$-\gamma_k \int_a^b \omega(x) p_{k-1}(x) p_s(x) \, dx \tag{5.28}$$

and we require the integral on the left of (5.28) to be zero for $s = k-1$ and $s = k$. For $s = k-1$, since p_{k-1} and p_k are orthogonal, we obtain

$$\int_a^b \omega(x) x p_k(x) p_{k-1}(x) \, dx - \gamma_k \int_a^b \omega(x)(p_{k-1}(x))^2 \, dx = 0.$$

We can simplify the first of the two integrals above by writing $xp_{k-1}(x)$ as we did with $xp_s(x)$ in (5.26) and obtain, by orthogonality,

$$\gamma_k = \int_a^b \omega(x)(p_k(x))^2\,\mathrm{d}x \Big/ \int_a^b \omega(x)(p_{k-1}(x))^2\,\mathrm{d}x. \tag{5.29}$$

With $s = k$ in (5.28), the required orthogonality of p_{k+1} and p_k immediately yields

$$\beta_k = \int_a^b \omega(x)x(p_k(x))^2\,\mathrm{d}x \Big/ \int_a^b \omega(x)(p_k(x))^2\,\mathrm{d}x. \tag{5.30}$$

It is convenient to define the trivial polynomial $p_{-1}(x) \equiv 0$. This is not, of course, one of the orthogonal polynomials. The introduction of p_{-1}, together with $p_0(x) \equiv 1$, allows us to use (5.27) from $k = 0$ onwards, instead of from $k = 1$. Since $p_{-1} = 0$ the value of γ_0 is irrelevant in (5.27), but to be definite we take $\gamma_0 = 0$. Then (5.27), (5.29) and (5.30) allow us to compute the polynomials $p_1, p_2, \ldots$ recursively. The computation is carried out in the order

$$\beta_0, p_1; \qquad \beta_1, \gamma_1, p_2; \qquad \beta_2, \gamma_2, p_3; \qquad \beta_3, \gamma_3, p_4; \ldots. \tag{5.31}$$

In general, we will need to use numerical integration to evaluate β_k and γ_k. We will return to the practical details of this later.

Given a system of orthogonal polynomials $p_0, p_1, \ldots$, it is clear that the system $\delta_0 p_0$, $\delta_1 p_1, \ldots$ is also orthogonal, where δ_0, $\delta_1, \ldots$ are any non-zero real numbers. These two systems are equivalent in the sense that an orthogonal series in terms of the p_r would be the same as that in terms of the $\delta_r\, p_r$. This is why we have 'normalized' (scaled) the p_r to have leading coefficient 1. Subject to this normalization, there is a unique orthogonal system with respect to a given ω on $[a, b]$.

Example 5.8 The most obvious weight function is $\omega(x) \equiv 1$, and we will take $[a, b]$ as $[-1, 1]$. Then, following the scheme (5.31), we obtain for the first few coefficients and polynomials

$$0, x; \quad 0, \tfrac{1}{3}, x^2 - \tfrac{1}{3}; \quad 0, \tfrac{4}{15}, x^3 - \tfrac{3}{5}x; \quad 0, \tfrac{9}{35}, x^4 - \tfrac{6}{7}x^2 + \tfrac{3}{35}.$$

These are called the *Legendre polynomials*. Traditionally, this name is reserved for the multiples of these polynomials for which $p_r(1) = 1$ for all r. Thus the first few Legendre polynomials proper are 1, x, $(3x^2 - 1)/2$, $(5x^3 - 3x)/2$, $(35x^4 - 30x^2 + 3)/8$. These satisfy the recurrence relation (see Davis, 1976)

$$(k+1)p_{k+1}(x) = (2k+1)x\,p_k(x) - k\,p_{k-1}(x). \tag{5.32}$$ □

Once we have computed the system of orthogonal polynomials p_r, we can find the best approximation, which minimizes (see (5.20))

$$\left\| f - \sum_{r=0}^{n} a_r p_r;\, \omega \right\|_2,$$

by computing the coefficients a_r from (5.23) with ψ_r replaced by p_r. Thus

$$a_r = \int_a^b \omega(x)f(x)p_r(x)\,\mathrm{d}x \bigg/ \int_a^b \omega(x)(p_r(x))^2\,\mathrm{d}x. \tag{5.33}$$

This will require numerical integration, in general.

Example 5.9 Let us find the first three terms of the Legendre series for e^x. We use (5.33) with $\omega(x) \equiv 1$ and $[a, b] = [-1, 1]$, with $p_0 = 1$, $p_1 = x$, $p_2 = x^2 - \frac{1}{3}$ as found in Example 5.8. Using integration by parts, we obtain

$$a_0 = (\mathrm{e} - \mathrm{e}^{-1})/2, \qquad a_1 = 3\mathrm{e}^{-1}, \qquad a_2 = 15(\mathrm{e} - 7\mathrm{e}^{-1})/4.$$

The least squares polynomial for e^x on $[-1, 1]$ is therefore

$$a_0 + a_1 x + a_2(x^2 - \tfrac{1}{3}) \simeq 0.996 + 1.104\,x + 0.537\,x^2$$

which is not so different from the Taylor polynomial. In practice it would be more sensible to use numerical integration to evaluate the a_r. □

In § 4.10 we introduced the Chebyshev polynomials. We now ask if there is a weight function which makes them orthogonal on $[-1, 1]$. For this we require

$$0 = \int_{-1}^{1} \omega(x)T_m(x)T_n(x)\,\mathrm{d}x = \int_0^{\pi} \omega(\cos\theta)\cos m\theta \cos n\theta \sin\theta\,\mathrm{d}\theta,$$

on making the substitution $x = \cos\theta$. The orthogonality property is satisfied if $\omega(\cos\theta)\sin\theta = 1$ since (cf. Problem 5.16)

$$\int_0^{\pi} \cos m\theta \cos n\theta\,\mathrm{d}\theta = 0, \qquad m \neq n.$$

This gives

$$\omega(\cos\theta) = 1/\sin\theta = 1/(1 - \cos^2\theta)^{1/2}$$

and we have shown that the Chebyshev polynomials are orthogonal on $[-1, 1]$ with respect to weight function $(1 - x^2)^{-1/2}$.

The recurrence relation for the Chebyshev polynomials (4.55),

$$T_{k+1}(x) = 2xT_k(x) - T_{k-1}(x),$$

does not fit the general form (5.27) due to the presence of the factor 2 on the right This discrepancy would be righted if each T_r were scaled to have leading coefficient 1.

A third set of orthogonal polynomials which is frequently encountered (although not as commonly as the two sets we have just mentioned) are the *Chebyshev polynomials of the second kind*, defined by

$$U_k(x) = \sin(k + 1)\theta/\sin\theta, \tag{5.34}$$

where $x = \cos\theta$. It is left as an exercise for the reader (in Problem 5.29) to show that these polynomials are orthogonal on $[-1, 1]$ with respect to weight function $(1 - x^2)^{1/2}$. Note that, as defined by (5.34), these polynomials do not have leading coefficient 1. The polynomials U_r play an important role in best approximations with respect to the 1-norm, defined by (5.7) and (5.3) with $p = 1$. We state without proof the following.

Theorem 5.1 If f is continuous on $[-1, 1]$ and $f - p$ has at most $n+1$ zeros in $[-1, 1]$ for every $p \in P_n$, then the best 1-norm approximation from P_n for f on $[-1, 1]$ is simply the interpolating polynomial for f constructed at the zeros of U_{n+1}. □

Thus best 1-norm approximations are surprisingly easy to compute, for functions f which satisfy the conditions of this theorem. This result enables us to justify that the best 1-norm approximation for $x^{1/2}$ on $[0, 1]$ is that given in Example 5.1 (see Problem 5.32). We will not pursue 1-norm approximations any further. (See Rivlin, 1981.)

Of all the orthogonal polynomials, those which concern us most here are the polynomials T_r. The orthogonal series based on these, called the *Chebyshev series*, is of considerable theoretical and practical interest. It is usual to write the series as

$$\sum_{r=0}^{\infty}{}' a_r T_r(x), \tag{5.35}$$

where $\sum'$ denotes a summation whose first term is halved. To find the coefficients we use (5.33) with $\omega(x) = (1 - x^2)^{-1/2}$, $[a, b] = [-1, 1]$ and $p_r = T_r$. Because we have $\frac{1}{2}a_0$ as the first term in (5.35) we need to replace a_0 by $\frac{1}{2}a_0$ in (5.33) when $r = 0$. Putting $x = \cos\theta$, the denominator on the right of (5.33) is

$$\int_{-1}^{1} (1 - x^2)^{-1/2} (T_r(x))^2 \, dx = \int_0^{\pi} \cos^2 r\theta \, d\theta, \tag{5.36}$$

which has the value $\pi/2$ for $r > 0$ and the value π for $r = 0$. Thus from (5.33)

$$a_r = \frac{2}{\pi} \int_{-1}^{1} (1 - x^2)^{-1/2} f(x) T_r(x) \, dx. \tag{5.37}$$

The $\sum'$ notation allows us to use the one formula (5.37) to stand for both $r = 0$ and $r > 0$, despite the two different values of (5.36). On making the substitution $x = \cos\theta$ in (5.37), we obtain

$$a_r = \frac{2}{\pi} \int_0^{\pi} f(\cos\theta) \cos r\theta \, d\theta, \tag{5.38}$$

which reminds us of the classical Fourier coefficients. Denoting the classical Fourier coefficients of $f(\cos\theta)$ by a_r^* and b_r^*, we have from (5.18) that

$$a_r^* = \frac{1}{\pi}\int_{-\pi}^{\pi} f(\cos\theta)\cos r\theta\,\mathrm{d}\theta, \qquad b_r^* = \frac{1}{\pi}\int_{-\pi}^{\pi} f(\cos\theta)\sin r\theta\,\mathrm{d}\theta.$$

Since the above two integrands are respectively even and odd functions on $[-\pi,\pi]$, we have $b_r^*=0$ and

$$a_r^* = \frac{2}{\pi}\int_0^{\pi} f(\cos\theta)\cos r\theta\,\mathrm{d}\theta = a_r.$$

Thus the Chebyshev series for $f(x)$ is just the classical Fourier series for $f(\cos\theta)$ and therefore we may apply results concerning convergence of the latter series. For example (see Davis, 1976), if f' is piecewise continuous on $[-1,1]$, the Chebyshev series converges uniformly to f on $[-1,1]$.

Example 5.10 Find the Chebyshev series for $f(x)=((1+x)/2)^{1/2}$. Since, with $x=\cos\theta$, $((1+x)/2)^{1/2}=\cos\frac{1}{2}\theta$, we obtain

$$a_r = \frac{2}{\pi}\int_0^{\pi}\cos\tfrac{1}{2}\theta\cos r\theta\,\mathrm{d}\theta.$$

From (2.8) we have

$$2\cos\tfrac{1}{2}\theta\cos r\theta = \cos(r+\tfrac{1}{2})\theta + \cos(r-\tfrac{1}{2})\theta$$

and so obtain

$$a_r = (-1)^{r-1}\frac{4}{\pi(4r^2-1)}$$

for all r. This gives the Chebyshev series

$$\left(\frac{1+x}{2}\right)^{1/2} \sim \frac{2}{\pi} + \frac{4}{\pi}\sum_{r=1}^{\infty}(-1)^{r-1}T_r(x)/(4r^2-1). \qquad \square$$

The above account of weighted orthogonal expansions was derived from the weighted 2-norm defined by (5.20). We can analogously define

$$\|f;\omega\|_2 = \left\{\sum_{i=0}^{N}\omega_i(f(x_i))^2\right\}^{1/2} \tag{5.39}$$

for a function defined on a point set $\{x_0,\ldots,x_N\}$, given a finite set of positive weights $\{\omega_0,\ldots,\omega_N\}$. We can verify that (5.39) defines a norm, in the same way as we did for (5.20). Given a function f and linearly independent functions $\psi_0, \ldots, \psi_n$ we can define a best approximation (Definition 5.2) with respect to the norm (5.39) and we can repeat the discussion presented at the beginning of this section. We can likewise construct a system of orthogonal polynomials induced by this norm. These satisfy

$$\sum_{i=0}^{N} \omega_i p_r(x_i) p_s(x_i) = 0, \qquad r \neq s,$$

which is analogous to (5.24). If we believe that, for the given f, the data at some x_i are more reliable than at other points, we can choose values for the weights ω_i accordingly. Otherwise, we can set each $\omega_i = 1$ for equal weighting. As for approximation on an interval, we find that the orthogonal polynomials satisfy the same form of recurrence relation (see (5.27),

$$p_{k+1}(x) = (x - \beta_k) p_k(x) - \gamma_k p_{k-1}(x). \tag{5.40}$$

In this case, we compute β_k and γ_k from

$$\beta_k = \sum_{i=0}^{N} \omega_i x_i (p_k(x_i))^2 \Big/ \sum_{i=0}^{N} \omega_i (p_k(x_i))^2, \tag{5.41}$$

$$\gamma_k = \sum_{i=0}^{N} \omega_i (p_k(x_i))^2 \Big/ \sum_{i=0}^{N} \omega_i (p_{k-1}(x_i))^2, \tag{5.42}$$

which are analogous to (5.30) and (5.29) respectively. If the denominator in (5.41) is zero, β_k is not defined. This can only happen if $p_k(x_i) = 0$, $i = 0, \ldots, N$, and therefore only if $k \geq N + 1$, since p_k is a polynomial of degree k. The least squares polynomial for f of degree $\leq n$ is

$$\sum_{r=0}^{n} a_r p_r(x), \tag{5.43}$$

where

$$a_r = \sum_{i=0}^{N} \omega_i f(x_i) p_r(x_i) \Big/ \sum_{i=0}^{N} \omega_i (p_r(x_i))^2, \tag{5.44}$$

which is analogous to (5.33). If in (5.43) we choose $n = N$, the least squares polynomial must coincide with the interpolating polynomial for f constructed at $x = x_0, \ldots, x_N$ and, therefore, we terminate the orthogonal series for f after $N + 1$ terms. Thus the orthogonal series for a function on a finite point set is a *finite* series, which is evaluated in the same way as described above for the evaluation of weighted orthogonal series on $[a, b]$. In fact, a weighted orthogonal series on $[a, b]$ will usually be computed as if it were an orthogonal series on a finite point set, since the integrals in (5.29), (5.30) and (5.33) will be estimated by numerical integration. If we use an integration rule with all weights positive (see Chapter 7), the integrals become sums exactly as in (5.41), (5.42) and (5.44) above.

The determination of least squares approximations using orthogonal polynomials avoids the ill-conditioning, referred to in §5.4, which is encountered in the direct method of solving the normal equations.

Example 5.11 Construct least squares approximations of degrees 1, 2 and 3 for the following data.

x_i	-2	-1	0	1	2
f_i	-1	-1	0	1	1

We take every $\omega_i=1$, $p_{-1}(x)\equiv 0$, $p_0(x)\equiv 1$, $\gamma_0=0$ and find that $\beta_0=0$. Thus $p_1(x)=x$, $\beta_1=0$, $\gamma_1=2$; $p_2(x)=x^2-2$, $\beta_2=0$, $\gamma_2=\frac{7}{5}$; $p_3(x)=x^3-\frac{17}{5}x$. Hence $a_0=0$, $a_1=\frac{3}{5}$, $a_2=0$, $a_3=-\frac{1}{6}$, so that the best approximation of degrees 1 and 2 is $\frac{3}{5}x$; that of degree 3 is $(7x-x^3)/6$, which interpolates f at all five points. □

The above example was so simple that we were able to write down the polynomials. In the algorithm which follows, the polynomials are evaluated in a purely arithmetical way without being displayed in algebraic form. In the first instance, the algorithm evaluates the least squares approximation defined by (5.43) and (5.44). It can also be used to calculate the least squares approximation with respect to ω on $[a,b]$, whose coefficients are given by (5.33). For example, if $\omega(x)\equiv 1$ on $[-1,1]$, we can evaluate the Legendre series by taking equally spaced points

$$x_i=-1+2i/N, \qquad 0\leqslant i\leqslant N,$$

and choosing weights

$$\omega_0=\omega_N=\tfrac{1}{2}, \qquad \omega_1=\cdots=\omega_{N-1}=1,$$

for some suitably large N. Then the integrals (5.29), (5.30) and (5.33) will be estimated by the trapezoidal rule (see Chapter 7).

Algorithm 5.1 The input consists of $x_0,\ldots,x_N$, $\omega_0,\ldots,\omega_N$ and $f_i=f(x_i)$, $0\leqslant i\leqslant N$, with $n<N$.

for $i:=0$ **to** N
  $p_{-1,i}:=0$, $p_{0,i}:=1$
next i
$\gamma_0:=0$

```
compute β_0 and a_0 from (5.41) and (5.44)
for r:=1 to n
    for i:=0 to N
        compute p_{r,i} = p_r(x_i) from (5.40)
    next i
    if r<n compute β_r and γ_r from (5.41) and (5.42)
    compute a_r from(5.44)
next r
```

For any required value of x, we can now find $p_r(x)$, $0 \leqslant r \leqslant n$, using the recurrence relation (5.40) and hence compute $\sum_{r=0}^{n} a_r p_r(x)$. □

Note that we do not compute β_r and γ_r as above in the case of the Legendre series, since the recurrence relation is known explicitly (see(5.32)). Likewise, for the Chebyshev series on $[-1, 1]$, we should not use Algorithm 5.1 as it stands, since we again know the coefficients β_r and γ_r of the recurrence relation. We will return to the evaluation of Chebyshev series later, when the following property of Chebyshev polynomials will prove useful.

Example 5.12 Let x_j denote $\cos(\pi j/N)$. The point set $\{x_0, \ldots, x_N\}$ consists of the extreme points (that is, points of maximum modulus) of T_N. The Chebyshev polynomials T_r, for $0 \leqslant r \leqslant N$, are orthogonal on this point set with respect to weights $\{\frac{1}{2}, 1, 1, \ldots, 1, 1, \frac{1}{2}\}$. We can show (see Problem 5.37) that

$$\sum_{j=0}^{N}{}'' T_r(x_j)T_s(x_j) = 0, \qquad r \neq s, \tag{5.45}$$

where Σ'' denotes a summation whose first and last terms are halved. We now use our results on orthogonal series to obtain least squares approximations for a function f for the given sets of points and weights. To obtain polynomials with leading coefficient 1, we write $p_r = T_r/2^{r-1}$ (for $r \geqslant 1$) in (5.43) and define $\alpha_r = a_r/2^{r-1}$, where a_r is given by the appropriate form of (5.44). On checking the coefficients a_r for the three cases $r=0$, N and $0<r<N$ and using (5.45) we find that the least squares polynomial of degree $n<N$ on the given point set is

$$\sum_{r=0}^{n}{}' \alpha_r T_r(x), \tag{5.46}$$

where

$$\alpha_r = \frac{2}{N} \sum_{j=0}^{N}{}'' f(x_j)T_r(x_j). \tag{5.47}$$

If we take $n = N$, the least squares polynomial, which coincides with the interpolating polynomial for f at the x_j, is given by

$$\sum_{r=0}^{N}{}'' \alpha_r T_r(x). \tag{5.48}$$ □

5.6 Minimax approximation

As we have just seen, the Chebyshev series for a given function f is the solution of a certain least squares problem. We shall see later that, surprisingly, this series is closely related to the best polynomials in the minimax sense.

If f is continuous on $[a, b]$, we write

$$E_n(f) = \min_{p \in P_n} \| f - p \|_\infty . \tag{5.49}$$

We found the least squares problem easy to solve because it required the minimization of a differentiable function of several variables, for which we have the standard technique of setting partial derivatives to zero. It is also helpful that the resulting normal equations are linear. However, the solution of the minimax problem (5.49) is much more difficult and we cannot solve it by direct assault! It is helpful to look first at the following examples.

Example 5.13 If we replace n by $n+1$ in Theorem 4.3 we see that the minimax approximation from P_n for x^{n+1} on $[-1, 1]$ is the polynomial p such that

$$x^{n+1} - p(x) = T_{n+1}(x)/2^n. \tag{5.50}$$ □

Example 5.14 Find the minimax straight line approximation $a + bx$ to the function $x^{1/2}$ on $[\frac{1}{4}, 1]$. It is convenient to draw a graph of $x^{1/2}$ on $[\frac{1}{4}, 1]$. On the graph, we can visualize a straight line segment $y = a + bx$ which we move around in different positions, seeking

$$E = \min_{a,b} \max_{1/4 \leqslant x \leqslant 1} | x^{1/2} - (a + bx) |,$$

the minimum being over all a and b. It is obvious geometrically that, as shown in Fig. 5.3, the best straight line is that for which the maximum error occurs at $x = \frac{1}{4}$, $x = 1$ and at some interior point, say $x = \xi$. We therefore have

$$a + b.\tfrac{1}{4} - (\tfrac{1}{4})^{1/2} = E \tag{5.51}$$

$$a + b\xi - \xi^{1/2} = -E \tag{5.52}$$

$$a + b.1 - 1 = E. \tag{5.53}$$

So far, to find four unknowns a, b, E and ξ, we have only three equations. From Fig. 5.3 it is clear that the error $y = x^{1/2} - (a + bx)$ has a turning value at $x = \xi$ and so has zero derivative there. This gives, as a fourth equation,

$$\tfrac{1}{2}\xi^{-1/2} - b = 0. \tag{5.54}$$

On subtracting (5.51) from (5.53) we obtain $b = \frac{2}{3}$ and so, from (5.54), $\xi = \frac{9}{16}$. Adding (5.52) and (5.53) yields $a = \frac{17}{48}$ and finally we obtain, say from (5.53), $E = \frac{1}{48}$. Thus the best straight line is $y = \frac{17}{48} + \frac{2}{3}x$, which approximates $x^{1/2}$ on $[\frac{1}{4}, 1]$ with error in modulus not greater than $\frac{1}{48}$. As an aside, we point out that this approximation may be used to estimate the square root of any positive number. We need to multiply it by an appropriate power of 4 to put it into the range $[\frac{1}{4}, 1]$. See Problem 5.41. □

The latter example also shows us how minimax approximations from P_1 behave for all functions f whose second derivative has constant sign. (Draw a graph.) In such a case we can move the straight line by raising, lowering or rotating it around the graph of f until the maximum error occurs at the two end points and at some interior point, with the maximum error oscillating in sign over these three points. A definition is helpful here.

Definition 5.8 A continuous function E is said to *equioscillate* on n points of $[a, b]$ if there exist n points x_i, with $a \leqslant x_1 < x_2 < \cdots < x_n \leqslant b$, such that

$$|E(x_i)| = \max_{a \leqslant x \leqslant b} |E(x)|, \qquad i = 1, \ldots, n,$$

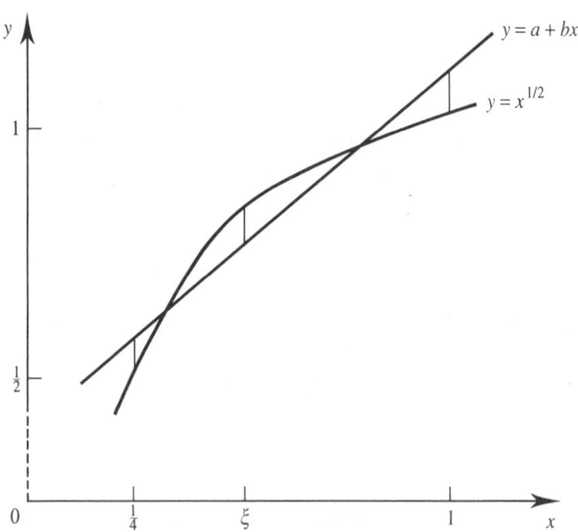

Fig. 5.3 Minimax straight line approximation for $x^{1/2}$ on $[\frac{1}{4}, 1]$.

and

$$E(x_i) = -E(x_{i+1}), \qquad i = 1, \ldots, n-1. \qquad \square$$

Thus for the minimax approximation from P_n in Example 5.13 the error function equioscillates on $n+2$ points and the same is true with $n=1$ in Example 5.14. The equioscillation of the error on $n+2$ points or more turns out to be the property which characterizes minimax approximation from P_n for any $f \in C[a, b]$.

Theorem 5.2 Let f be continuous on $[a, b]$ and suppose that, for a certain polynomial $p \in P_n$, $f-p$ equioscillates on $n+2$ points of $[a, b]$. It follows that p is the minimax approximation from P_n for f on $[a, b]$.

Proof If p is not the minimax polynomial, we can 'adjust' p by subtracting some non-zero $q \in P_n$ so as to reduce the maximum error. Thus q must have the same sign as $f-p$ on its $n+2$ equioscillation points. This entails that $q \in P_n$ oscillates in sign on $n+2$ points, which is impossible as q has at most n zeros. □

Example 5.15 Find the minimax polynomial from P_2 for $|x|$ on $[-1, 1]$. We might guess this to be of the form $a + bx^2$, since $|x|$ is even. Thus on $[0, 1]$ we require a minimax approximation for x of the form $a + bx^2$. This is equivalent to finding a minimax approximation for $x^{1/2}$ from P_1 on $[-1, 1]$ which, using the method of Example 5.14, is found to be $\frac{1}{8} + x$ (see also Example 5.1 and Problem 5.38). So we conjecture that the required minimax approximation for $|x|$ on $[-1, 1]$ is $x^2 + \frac{1}{8}$. Since the error function $|x| - (x^2 + \frac{1}{8})$ equioscillates on the points $-1, -\frac{1}{2}, 0, \frac{1}{2}, 1$, Theorem 5.2 shows that $x^2 + \frac{1}{8}$ is the minimax approximation for $|x|$ on $[-1, 1]$ from P_3, and so also from P_2. □

There is a converse of Theorem 5.2 which justifies our claim above that equioscillation of the error function on $n+2$ points characterizes minimax approximation.

Theorem 5.3 If f is continuous on $[a, b]$ and p denotes a minimax approximation $\in P_n$ to f on $[a, b]$, then $f-p$ equioscillates on at least $n+2$ points of $[a, b]$.

Proof Write $E(x) = f(x) - p(x)$ and suppose that E equioscillates on $k < n+2$ points. Then (see Problem 5.45) we can construct $q \in P_n$ which takes the same sign as E on all points where $\|E\|_\infty$ is attained and is scaled so that $\|q\|_\infty = 1$. We denote by S_- the set of points where $E(x)q(x) \leqslant 0$

and S_+ the set where $E(x)q(x)>0$. Thus $S_+ \cup S_- = [a, b]$ and S_+ includes all points where $\| E \|_\infty$ is attained. Let

$$d = \max_{x \in S_-} |E(x)| < \|E\|_\infty. \tag{5.55}$$

We choose any positive θ and define ξ as any point in $[a, b]$ for which

$$|E(\xi) - \theta q(\xi)| = \| E - \theta q \|_\infty. \tag{5.56}$$

If $\xi \in S_-$, we obtain from (5.56) that

$$\| E - \theta q \|_\infty = |E(\xi)| + \theta |q(\xi)|$$

and so

$$\| E - \theta q \|_\infty \leqslant d + \theta. \tag{5.57}$$

If $\xi \in S_+$, we deduce from (5.56) that

$$\| E - \theta q \|_\infty < \max\{ |E(\xi)|, \theta |q(\xi)| \}$$

so that

$$\| E - \theta q \|_\infty < \max\{ \| E \|_\infty, \theta\}. \tag{5.58}$$

By (5.55), we can now restict θ so that

$$0 < \theta < \| E \|_\infty - d$$

and the two possibilities (5.57) and (5.58) both yield

$$\| E - \theta q \|_\infty < \| E \|_\infty$$

which contradicts the statement that p is the minimax approximation. □

Corollary 5.3 If f is continuous on $[a, b]$, the minimax $p \in P_n$ is an interpolating polynomial on certain $n+1$ points of $[a, b]$. (This interpolating property is shared also by the weighted least squares approximations. See Problem 5.25.) □

Knowing the equioscillation property, it is not difficult to show the uniqueness of the minimax polynomial.

Theorem 5.4 If f is continuous on $[a, b]$, there is a unique minimax polynomial approximation of degree at most n for f on $[a, b]$.

Proof This proof is essentially that of Theorem 5.2. We argue that if $p \in P_n$ is not the unique minimax polynomial, we can subtract some non-zero $q \in P_n$ from p so that $p - q$ is also a minimax polynomial. Since

$$|f(x) - (p(x) - q(x))| = |(f(x) - p(x)) + q(x)|$$

cannot exceed $\|f - p\|_\infty$ at each of the $n+2$ equioscillation points of $f - p$, it follows that $q \in P_n$ is alternately $\geqslant$ and $\leqslant$ on these points. This is impossible (see Problems 2.3 and 2.4). □

5.7 Chebyshev series

Having studied the equioscillation property, we can now appreciate the relevance of Chebyshev series. Intuitively, if the Chebyshev series $\sum_{r=0}^{\prime\infty} a_r T_r(x)$ converges rapidly to $f(x)$, the error in the partial sum,

$$f(x) - \sum_{r=0}^{n}{}' a_r T_r(x) = \sum_{r=n+1}^{\infty} a_r T_r(x)$$

will be dominated by the leading term $a_{n+1} T_{n+1}(x)$, which equioscillates on $n+2$ points. (We assume here that $a_{n+1} \neq 0$, although if $a_{n+1} = 0$, the first non-zero term in the error will still equioscillate on $n+2$ points.) From Theorem 5.2 we see that if the error were *exactly* $a_{n+1} T_{n+1}(x)$, then the partial sum

$$s_n(x) = \sum_{r=0}^{n}{}' a_r T_r(x)$$

would be precisely the minimax approximation for $f(x)$. If the series converges rapidly, so that $a_{n+2}, a_{n+3}, \ldots$ are small compared with a_{n+1}, we can expect s_n to be close to the minimax approximation.

There are algorithms for computing minimax approximations. However, these algorithms are not easy to use and hence the emphasis on Chebyshev series, which are much easier to handle.

Example 5.16 Consider the function $x^{1/2}$ on $[\frac{1}{4}, 1]$, for which we found the minimax straight line in Example 5.14. We make a linear change of variable, so that $x^{1/2}$ for $\frac{1}{4} \leqslant x \leqslant 1$ becomes $(\frac{3}{8}u + \frac{5}{8})^{1/2}$ for $-1 \leqslant u \leqslant 1$. To four decimal places, the first few Chebyshev coefficients a_r for this function are

r	0	1	2	3	4	5
a_r	1.5420	0.2465	−0.0202	0.0033	−0.0007	0.0002

The remaining coefficients are zero, to four decimal places. Knowing these coefficients, we construct partial sums of the Chebyshev series. In Table 5.1 these are compared with the minimax polynomials of the same degree. To make it easy to compare the polynomials themselves, we have expressed the Chebyshev partial sums explicitly in powers of u. □

Table 5.1 Comparison of Chebyshev series and minimax approximations of degree 0, 1, 2 for the function $(\frac{3}{8}u+\frac{5}{8})^{1/2}$ on $[-1, 1]$, which is $x^{1/2}$ on $[\frac{1}{4}, 1]$

Degree of polynomial	Minimax polynomial	Maximum error	Chebyshev partial sum	Maximum error	$\lvert a_{n+1}\rvert$
0	0.75	0.25	0.7710	0.2710	0.2465
1	$0.7708+0.2500u$	0.0208	$0.7710+0.2465u$	0.0245	0.0202
2	$0.7920+0.2465u$ $-0.0420u^2$	0.0035	$0.7912+0.2465u$ $-0.0404u^2$	0.0043	0.0033

The Chebyshev coefficients, as we saw in (5.38), are given by

$$a_r=\frac{2}{\pi}\int_0^{\pi} f(\cos\theta)\cos r\theta\,\mathrm{d}\theta. \tag{5.59}$$

It is not always possible to evaluate these integrals exactly, as we were able to do in Example 5.10. We have to find some method for approximating to these coefficients. The most obvious approach is to approximate to the integral in (5.59) by some numerical integration rule. On using the trapezoidal rule with N equal sub-intervals (see Chapter 7), we obtain from (5.59), as an approximation for a_r,

$$\alpha_r=\frac{2}{N}\sum_{j=0}^{N}{}'' f(x_j)T_r(x_j), \tag{5.60}$$

where $x_j=\cos(\pi j/N)$. In Example 5.12 we saw that these same numbers α_r are coefficients in a finite orthogonal series for f defined on the point set $\{x_0, \ldots, x_N\}$, whose members are the extreme points of T_N. This series has partial sums

$$\sum_{r=0}^{n}{}' \alpha_r T_r(x) \tag{5.61}$$

which we can now think of as approximations to the partial sums of the Chebyshev series. Of course each α_r depends on N; we expect that, in general, as N is increased in (5.60), we will obtain closer approximations α_r to the Chebyshev coefficient a_r, given by (5.59).

We now derive an explicit relation between the α_r and the a_r. First we extend the orthogonality relation (5.45) to polynomials T_r with $r>N$ by noting that

$$T_{2kN\pm r}(x_j)=\cos\left(\frac{(2kN\pm r)\pi j}{N}\right)=\cos\left(2k\pi j\pm\frac{r\pi j}{N}\right).$$

From the periodicity of the cosine function, we have

$$T_{2kN\pm r}(x_j) = T_r(x_j). \tag{5.62}$$

We also extend (5.45) to give, with $r = s$ (see Problem 5.37),

$$\sum_{j=0}^{N}{}'' (T_r(x_j))^2 = \begin{cases} \frac{1}{2}N, & 0 < r < N \\ N, & r = 0 \text{ or } N. \end{cases} \tag{5.63}$$

Let us now assume that the Chebyshev series for f converges uniformly to f on $[-1, 1]$. Then it follows from (5.60) that

$$\begin{aligned} \alpha_r &= \frac{2}{N} \sum_{j=0}^{N}{}'' \left(\sum_{s=0}^{\infty}{}' a_s T_s(x_j) \right) T_r(x_j) \\ &= \frac{2}{N} \sum_{s=0}^{\infty}{}' a_s \sum_{j=0}^{N}{}'' T_s(x_j) T_r(x_j). \end{aligned} \tag{5.64}$$

From (5.45), (5.62) and (5.63), if $r \neq 0$ or N, the only summations over j which are non-zero in (5.64) are those for which $s = r$, $2N - r$, $2N + r$, $4N - r$, $4N + r$, and so on. So, if $r \neq 0$ or N,

$$\alpha_r = a_r + \sum_{k=1}^{\infty} (a_{2kN-r} + a_{2kN+r}). \tag{5.65}$$

We find that (5.65) holds for $r = 0$ and N also, on examining these cases separately. If we keep r fixed and take N large enough, we expect from (5.65) that α_r will approximate closely to a_r.

We now consider an alternative method for approximating to the Chebyshev coefficients a_r, which is based on another orthogonality property of the Chebyshev polynomials. This time, the orthogonality is on the *zeros* of T_N, instead of on the extreme points as discussed above. We have (see Problem 5.47) that for $0 \leqslant r, s \leqslant N$,

$$\sum_{j=1}^{N} T_r(x_j^*) T_s(x_j^*) = \begin{cases} 0, & r \neq s \quad \text{or} \quad r = s = N \\ N/2, & r = s \neq 0 \quad \text{or} \quad N \\ N, & r = s = 0 \end{cases} \tag{5.66}$$

where the x_j^* are the zeros of T_N. Following the same procedure as in Example 5.12, we find that for $n \leqslant N - 1$

$$\sum_{r=0}^{n}{}' \alpha_r^* T_r(x) \tag{5.67}$$

with

$$\alpha_r^* = \frac{2}{N} \sum_{j=1}^{N} f(x_j^*) T_r(x_j^*), \tag{5.68}$$

is the least squares approximation to f on the point set $\{x_1^*, \ldots, x_N^*\}$. The weights are all equal in this case. Thus the polynomial (5.67) is another approximation, easily computed, to the partial Chebyshev series. If we put $n = N$ in (5.67) the least squares approximation must coincide with the interpolating polynomial for f constructed at the x_j^*, that is, the zeros of $T_N(x)$, which we discussed at the end of Chapter 4. Since then, we have in this chapter followed a full circle from interpolation via least squares and minimax approximations back to interpolation again.

We now establish two relations between α_r^*, and a_r as we did for α_r and a_r. First, (5.68) may be regarded as an approximation to (5.59) obtained by using a quadrature rule. In this case (see Chapter 7), it is the *midpoint rule*. Second, we replace $f(x_j^*)$ in (5.68) by its Chebyshev series (again assumed to be uniformly convergent) and extend the orthogonality relation (5.66) by the relation

$$T_{2kN \pm r}(x_j^*) = (-1)^k T_r(x_j^*). \tag{5.69}$$

Thus we obtain

$$\alpha_r^* = a_r + \sum_{k=1}^{\infty} (-1)^k (a_{2kN-r} + a_{2kN+r}) \tag{5.70}$$

for $0 \leqslant r < N$.

The above analysis suggests a simple, reliable method for estimating Chebyshev coefficients a_r, for $0 \leqslant r \leqslant n$. First we select a value of $N > n$ and compute α_r and α_r^* $(0 \leqslant r \leqslant n)$ from (5.60) and (5.68). If $|\alpha_r - \alpha_r^*|$ is sufficiently small, we use $(\alpha_r + \alpha_r^*)/2$ to approximate to a_r. Otherwise, we increase N and re-calculate α_r and α_r^*. Finally, we observe from (5.65) and (5.70) that, if the Chebyshev coefficients tend to zero rapidly, we have

$$a_r \simeq \alpha_r - (a_{2N-r} + a_{2N+r})$$

and

$$a_r \simeq \alpha_r^* + (a_{2N-r} + a_{2N+r}).$$

Thus we expect that a_r will usually lie between α_r and α_r^* and will be more closely estimated by $(\alpha_r + \alpha_r^*)/2$.

5.8 Economization of power series

Suppose we wish to approximate to the sine function with error not greater than 0.5×10^{-3}. It will be sufficient to approximate to $\sin(\pi x/2)$ on $[0, 1]$. Forgetting all knowledge of interpolating polynomials, minimax approximations and Chebyshev series, we might use the Taylor polynomial

$$p(x) = \left(\frac{\pi x}{2}\right) - \frac{1}{3!}\left(\frac{\pi x}{2}\right)^3 + \frac{1}{5!}\left(\frac{\pi x}{2}\right)^5 - \frac{1}{7!}\left(\frac{\pi x}{2}\right)^7. \tag{5.71}$$

We may verify that $\| \sin(\pi x/2) - p \|_\infty = 0.157 \times 10^{-3}$, where the norm is on $[-1, 1]$, and the first three terms of p do not give the required accuracy. We now express the powers of x which appear in (5.71) in terms of the Chebyshev polynomials:

$$\begin{aligned} x &= T_1(x) \\ x^3 &= (3T_1(x) + T_3(x))/4 \\ x^5 &= (10T_1(x) + 5T_3(x) + T_5(x))/16 \\ x^7 &= (35T_1(x) + 21T_3(x) + 7T_5(x) + T_7(x))/64. \end{aligned}$$

Substituting these into (5.71) and collecting like terms, we obtain

$$\begin{aligned} p(x) = 1.133571\,T_1(x) - 0.138123\,T_3(x) \\ + 0.004469\,T_5(x) - 0.000073\,T_7(x), \end{aligned} \tag{5.72}$$

where we have rounded the coefficients to six decimal places. Since $\| T_7 \|_\infty = 1$ we may delete the last term in (5.72) to give a polynomial $q \in P_5$ which approximates $\sin(\pi x/2)$ on $[-1, 1]$ with error not greater than $(0.157 + 0.073) \times 10^{-3}$, which is still within the required accuracy. In fact, by direct calculation we find that, with

$$q(x) = 1.133571\,T_1(x) - 0.138123\,T_3(x) + 0.004469\,T_5(x),$$

$\| \sin(\pi x/2) - q \|_\infty = 0.146 \times 10^{-3}$ so that in this case $q \in P_5$ is an even better approximation than $p \in P_7$. This simple process is known as *economization* of power series. It is sometimes possible to neglect further terms besides the last one in the transformed series of Chebyshev polynomials.

We can also turn an infinite power series into a Chebyshev series, as follows. Write

$$x = \cos\theta = \tfrac{1}{2}(\mathrm{e}^{i\theta} + \mathrm{e}^{-i\theta}),$$

where $i^2 = -1$. Then $x^r = (\mathrm{e}^{i\theta} + \mathrm{e}^{-i\theta})^r/2^r$ and we apply the binomial expansion. For $0 \leqslant s \leqslant \frac{1}{2}r$ the coefficients of $\mathrm{e}^{i(r-2s)\theta}$ and $\mathrm{e}^{-i(r-2s)\theta}$ are equal. (Look at the first and last terms, the second and second last terms, and so on.) Since

$$T_{r-2s}(x) = \tfrac{1}{2}(\mathrm{e}^{i(r-2s)\theta} + \mathrm{e}^{-i(r-2s)\theta}),$$

x^r can be expressed in terms of T_r, T_{r-2}, and so on (see Problem 5.50). Thus, if

$$f(x) = \sum_{r=0}^{\infty} c_r x^r, \tag{5.73}$$

and the series converges uniformly to f on $[-1,1]$, we can express each x^r in terms of Chebyshev polynomials and collect the terms in $T_0, T_1, \ldots,$ to give the Chebyshev series

$$f(x) = {\sum_{r=0}^{\infty}}' a_r T_r(x), \tag{5.74}$$

where

$$a_r = \frac{1}{2^{r-1}} \sum_{s=0}^{\infty} \frac{c_{r+2s}}{2^{2s}} \binom{r+2s}{s}. \tag{5.75}$$

5.9 The Remez algorithms

We now return to the construction of minimax approximations and discuss a class of algorithms based on one first given by the Ukrainian mathematician E. Ya. Remez (1896–1975) in the 1930s.

Algorithm 5.2 (Remez) This computes iteratively the minimax polynomial $p \in P_n$ for $f \in C[a,b]$. Each time we carry out Steps 2 and 3 below, we compute a polynomial $p \in P_n$ and a set X of $n+2$ points. The sequence of such polynomials p converges to the minimax polynomial $p^* \in P_n$ and the sequence of sets X converges to a set X^* on which $f - p^*$ equioscillates.

Step 1 Choose a set $X = \{x_1, \ldots, x_{n+2}\}$.

Step 2 Solve the system of linear equations

$$f(x_i) - p(x_i) = (-1)^i e, \qquad 1 \leqslant i \leqslant n+2, \tag{5.76}$$

to determine a real number e and a polynomial $p \in P_n$.

Step 3 Change the set X in a suitable way (see below) and, unless some 'stopping criterion' has been attained, go to Step 2. □

When $[a,b]$ is $[-1,1]$, a common choice for the initial set X in Step 1 is the set of extreme points of T_{n+1}. (For this initial set, Step 2 gives the minimax polynomial immediately when $f(x) \equiv x^{n+1}$.) In the general case (see Problem 5.52) the linear equations in Step 2 always have a unique solution.

To carry out Step 3, we first need to estimate $\|f-p\|_\infty$, where $p \in P_n$ is the polynomial which has just been computed in Step 2. If $\|f-p\|_\infty$ is sufficiently close to $|e|$ then, by Theorem 5.2, p is nearly the minimax polynomial and we terminate the iterative process. Otherwise, let ξ be a point such that $|f(\xi) - p(\xi)| = \|f-p\|_\infty$. We include ξ in the point set X and delete one of the existing points in such a way that $f-p$ still alternates in sign on the $n+2$ points of X. If ξ happens to lie between a and x_1 we discard x_1 or x_{n+2}, and similarly if ξ happens to lie between x_{n+2} and b we discard x_1 or x_{n+2}. Otherwise, ξ lies between two consecutive x_i and we discard one of these appropriately.

We have described the simplest version of the Remez algorithms, where we change only one point at a time. In another version, which converges more rapidly, we change X by including not only a point where $\|f-p\|_\infty$ is attained, but we include $n+2$ local maxima or minima and delete existing points so that the new set X still consists of $n+2$ points on which $f-p$ alternates in sign.

Example 5.17 We will find the minimax $p \in P_2$ for $f(x) = (\frac{3}{8}x + \frac{5}{8})^{1/2}$ on $[-1, 1]$, which was quoted in Example 5.16. We choose $X = \{-1, -0.5, 0.5, 1\}$ initially, the extreme points of T_3, and solve the linear equations (5.76) to give

$$p(x) = 0.79188 + 0.24665x - 0.04188x^2, \qquad |e| = 0.00335, \quad (5.77)$$

to five decimal places. On evaluating $f-p$ at intervals of 0.01, we find that there are local maxima and minima at -1, -0.61, 0.40 and 1 and $f-p$ alternates in sign on these points, which we take as the new set X. Solving (5.76) again gives

$$p(x) = 0.79196 + 0.24650x - 0.04196x^2, \qquad |e| = 0.00350. \quad (5.78)$$

We find that $\|f-p\|_\infty = 0.00350$ to five decimal places and therefore we accept this p as the minimax polynomial.

Let us also carry out the simpler version of the Remez algorithm, where we change only one point at a time. We take the same initial X as above and obtain (5.77). Since, again working at intervals of 0.01, we find that $\|f-p\|_\infty$ is attained at approximately -0.61, we change X to $\{-1, -0.61, 0.5, 1\}$ and solve (5.76) to give

$$p(x) = 0.79209 + 0.24655x - 0.04209x^2, \qquad |e| = 0.00345.$$

For this p, $\|f-p\|_\infty = 0.00363$, attained at $x = 0.40$. So we take the next X as $\{-1, -0.61, 0.40, 1\}$ and solve (5.76) to give the minimax p as found in (5.78). □

We now state a theorem due to the Belgian mathematician C. de La Vallée-Poussin (1866–1962) which allows us to determine lower and upper bounds for the minimax error

$$E_n(f) = \min_{p \in P_n} \|f-p\|_\infty$$

at each stage of the Remez algorithm.

Theorem 5.5 If $f-p$ alternates in sign on $n+2$ points x_i, with $a \leqslant x_1 < \cdots < x_{n+2} \leqslant b$, where $f \in C[a, b]$ and $p \in P_n$, then

$$\min_i |f(x_i) - p(x_i)| \leqslant E_n(f) \leqslant \|f-p\|_\infty. \tag{5.79}$$

Proof The right-hand inequality is a consequence of the definition of $E_n(f)$. To pursue the left-hand inequality in (5.79), we write $p^* \in P_n$ to denote the minimax polynomial and consider

$$(f(x_i) - p(x_i)) - (f(x_i) - p^*(x_i)) = p^*(x_i) - p(x_i). \tag{5.80}$$

If the left inequality of (5.79) is false,

$$|f(x_i) - p^*(x_i)| \leqslant E_n(f) < |f(x_i) - p(x_i)|$$

for all i and the sign on the left of (5.80) will be that of $f(x_i) - p(x_i)$, which alternates. Thus $p^* - p \in P_n$ alternates in sign on $n+2$ points, which is impossible. □

Corollary 5.5 At each stage of the Remez algorithm,

$$|e| \leqslant E_n(f) \leqslant \|f-p\|_\infty, \tag{5.81}$$

where p and e are obtained from the solution of the linear equations (5.76). □

As we carry out the Remez algorithm, the sequence of numbers $|e|$ increases and the sequence $\|f-p\|_\infty$ decreases. In principle, both sequences converge to the common limit $E_n(f)$. Although in practice this limit is not attained because $f-p$ is evaluated only on a finite set of points, and we carry out only a finite number of iterations, the inequalities (5.79) allow us to monitor the progress of the algorithm after each iteration.

5.10 Further results on minimax approximation

The interpolatory property of minimax approximations (see Corollary 5.3) can be used to derive error estimates when f is sufficiently smooth.

Theorem 5.6 If $f \in C^{n+1}[-1, 1]$, the minimax error satisfies

$$E_n(f) = \frac{|f^{(n+1)}(\xi)|}{2^n(n+1)!}, \tag{5.82}$$

where ξ is some point in $(-1, 1)$.

Proof First we have

$$E_n(f) \leqslant \|f-p_n\|_\infty \leqslant \frac{1}{2^n(n+1)!} \max_{-1 \leqslant x \leqslant 1} |f^{(n+1)}(x)|, \tag{5.83}$$

where p_n is the interpolating polynomial for f on the zeros of T_{n+1}. The left-hand inequality above follows from the definition of $E_n(f)$ and the right inequality (see (4.57)) was obtained in § 4.10.

Let $p \in P_n$ denote the minimax polynomial for f on $[-1, 1]$. From Corollary 5.3, p interpolates f on certain $n+1$ points of $[-1, 1]$, say $x_0, \ldots, x_n$, and we can write

$$f(x) - p(x) = (x - x_0) \ldots (x - x_n) \frac{f^{(n+1)}(\xi_x)}{(n+1)!}.$$

It follows that

$$E_n(f) = \| f - p \|_\infty \geqslant \frac{1}{(n+1)!} \| (x - x_0) \ldots (x - x_n) \|_\infty \min_{-1 \leqslant x \leqslant 1} | f^{(n+1)}(x) |.$$

From Corollary 4.3,

$$\| (x - x_0) \ldots (x - x_n) \|_\infty \geqslant \frac{1}{2^n}$$

and therefore

$$E_n(f) \geqslant \frac{1}{2^n (n+1)!} \min_{-1 \leqslant x \leqslant 1} | f^{(n+1)}(x) |. \tag{5.84}$$

The theorem follows from (5.83), (5.84) and the continuity of $f^{(n+1)}$. □

The following corollary is verified in Problem 5.54.

Corollary 5.6 If $f \in C^{n+1}[a, b]$,

$$E_n(f) = \frac{2}{(n+1)!} \left(\frac{b-a}{4} \right)^{n+1} | f^{(n+1)}(\xi) |, \tag{5.85}$$

for some $\xi \in (a, b)$. □

As we remarked in Chapter 3, the main justification for using polynomials to approximate to a function f which is continuous on a finite interval $[a, b]$ is Weierstrass's theorem, which we restate more rigorously here.

Theorem 5.7 (Weierstrass's theorem) If f is continuous on $[a, b]$ then, given any $\varepsilon > 0$, there exists a polynomial p such that $| f(x) - p(x) | < \varepsilon$ for all $x \in [a, b]$. □

There are several quite different proofs of this fundamental theorem. One proof, published by S. N. Bernstein in 1912, uses the following polynomials.

Definition 5.9 The nth Bernstein polynomial for f on $[0, 1]$ is

$$B_n(f;x) = \sum_{j=0}^{n} f\left(\frac{j}{n}\right)\binom{n}{j} x^j(1-x)^{n-j}. \tag{5.86}$$ □

By making a linear change of variable, we may construct similar polynomials on any finite interval $[a, b]$. It can be shown rigorously that, if f is continuous on $[0, 1]$, the sequence $(B_n(f;\, x))$ converges uniformly to $f(x)$ on $[0, 1]$. In addition, derivatives of the Bernstein polynomials converge to derivatives of f (if these exist). Also, if f is convex (Definition 2.6), so is each Bernstein polynomial. Unfortunately, such good behaviour has to be paid for: the Bernstein polynomials converge *very slowly* to f, and thus are of little direct use for approximating to f. However, the uniform convergence of the Bernstein polynomials to f gives us a constructive proof of Weierstrass's theorem and shows us that the minimax polynomials must also converge uniformly to f.

The Bernstein polynomials are most efficiently computed as follows.

Algorithm 5.3 This process, called the *de Casteljau algorithm*, begins with n and x and the numbers $f(i/n)$, $0 \leqslant i \leqslant n$, and computes $f_0^{[n]} = B_n(f;x)$.

for $i := 0$ **to** n
  $f_i^{[0]} := f(i/n)$
next i
for $m := 1$ **to** n
  for $i := 0$ **to** $n - m$
    $f_i^{[m]} := x f_{i+1}^{[m-1]} + (1-x) f_i^{[m-1]}$
  next i
next m □

The following theorem justifies the algorithm.

Theorem 5.8 For $0 \leqslant m \leqslant n$ and $0 \leqslant i \leqslant n - m$,

$$f_i^{[m]} = \sum_{r=0}^{m} f\left(\frac{i+r}{n}\right)\binom{m}{r} x^r(1-x)^{m-r}.$$

The proof is by induction on m. □

Corollary 5.8 $f_0^{[n]} = B_n(f;x)$. □

We will find it useful to have the following alternative expression for $f_i^{[m]}$.

Theorem 5.9 For $0 \leqslant m \leqslant n$ and $0 \leqslant i \leqslant n-m$,

$$f_i^{[m]} = \sum_{r=0}^{m} \binom{m}{r} x^r \Delta^r f_i,$$

where $f_i = f(i/n)$.

The proof is again by induction on m. □

Corollary 5.9

$$B_n(f;x) = \sum_{r=0}^{n} \binom{n}{r} x^r \Delta^r f_0. \qquad (5.87)$$ □

Example 5.18 Let us use (5.87) to evaluate $B_n(x^2;x)$. For $f(x) = x^2$, we have $f_0 = f(0) = 0$, $\Delta f_0 = f(1/n) - f(0) = 1/n^2$ and $\Delta^2 f_0 = (2/n)^2 - 2(1/n)^2 + 0 = 2/n^2$. The third and higher differences are zero (see § 4.8) and therefore from (5.87)

$$B_n(x^2;x) = \binom{n}{1} x \frac{1}{n^2} + \binom{n}{2} x^2 \frac{2}{n^2}.$$

Thus

$$B_n(x^2;x) = x^2 + x(1-x)/n.$$ □

This illustrates the point made above that $B_n(f;x)$ converges very slowly to f. Indeed, for any $f \in C^2[0,1]$, the error tends to zero like $1/n$, as in the above example.

Problems

Section 5.1

5.1 Verify that $\|f\|_\infty$ in (5.2) defines a norm and that (5.4) also defines a norm.

5.2 Verify that the p-norm in (5.3) satisfies the norm axioms (i) and (ii) for all p and that axiom (iii) holds for $p = 1$. Repeat this exercise for the norm defined by (5.5).

5.3 Show that, for the p-norm (5.3) with $p = 2$, norm axiom (iii) is equivalent to

$$\left(\int_a^b f(x)g(x)\,\mathrm{d}x\right)^2 \leqslant \int_a^b (f(x))^2\,\mathrm{d}x \int_a^b (g(x))^2\,\mathrm{d}x.$$

5.4 Prove that, for the p-norm (5.5) with $p=2$, norm axiom (iii) is equivalent to

$$\sum_{i=0}^{N} f(x_i)g(x_i) \leqslant \|f\|_2\|g\|_2$$

and, by squaring both sides, verify the above inequality. (*Hint*: begin by squaring both sides of (5.1).)

5.5 Show that

$$\|f\|_\infty \leqslant \|f\|_p \leqslant (N+1)^{1/p}\|f\|_\infty,$$

where the above two norms are defined by (5.4) and (5.5). By taking the logarithm, or otherwise, show that $(N+1)^{1/p} \to 1$ and hence $\|f\|_p \to \|f\|_\infty$ as $p \to \infty$.

Section 5.2

5.6 If $p \in P_n$ is a best approximation for f with respect to a given norm, and $q \in P_n$, deduce from (5.7) that $p+q$ is a best approximation for $f+q$.

Section 5.3

5.7 Find the least squares straight line for $f(x)=x^{1/2}$ on $[0, 1]$.

5.8 Find the least squares straight line approximation for the following data:

x	0.0	0.1	0.2	0.3	0.4	0.5	0.6
$f(x)$	2.9	2.8	2.7	2.3	2.1	2.1	1.7

5.9 Show that, if the functions $\psi_0, \ldots, \psi_n$ form a Chebyshev set on some interval $[a, b]$, they are also linearly independent on $[a, b]$.

5.10 A set of functions $\{\psi_0, \ldots, \psi_n\}$ is said to be linearly dependent on a point set $\{x_0, \ldots, x_N\}$ if there exist numbers $a_0, \ldots, a_n$, not all zero, such that

$$\sum_{r=0}^{n} a_r\psi_r(x_i) = 0, \qquad i=0, 1, \ldots, N.$$

Show that the functions x, x^3, x^5 are linearly dependent on the point set $\{-2, -1, 0, 1, 2\}$.

5.11 If a_1 and a_2 are not both zero, find a number α so that

$$a_0 + a_1 \cos x + a_2 \sin x = a_0 + (a_1^2 + a_2^2)^{1/2} \cos(x-\alpha).$$

Then argue that the right side of the equation can be zero at no more than two points of $[0, 2\pi)$ and so $\{1, \cos x, \sin x\}$ is a Chebyshev set on $[0, 2\pi)$. (*Hint*: use the identity $\cos(x-\alpha) = \cos x \cos \alpha + \sin x \sin \alpha$.)

5.12 For the data of Problem 5.8, make a linear transformation so that $\sum x_i = \sum x_i^3 = 0$. Hence find the least squares quadratic polynomial approximation for this data.

5.13 Find the least squares planar approximation, that is, of the form $z = a_0 + a_1 x + a_2 y$, for the following five points (x, y, z): $(0, 0, 1.1)$, $(0, 1, 1.9)$, $(1, 0, 2.2)$, $(0, -1, 0.0)$, $(-1, 0, 0.1)$.

Section 5.4

5.14 Show that a set of orthogonal functions is necessarily linearly independent. (*Hint*: show that

$$\sum_{r=0}^{n} a_r^2 \int_a^b [\psi_r(x)]^2 \, dx = \int_a^b \left(\sum_{r=0}^{n} a_r \psi_r(x) \right)^2 dx = 0,$$

if $\sum_{r=0}^{n} a_r \psi_r = 0$.)

5.15 If the ψ_r satisfy the orthogonality conditions (5.15) and the orthogonal coefficients a_r are given by (5.16), show that

$$\int_a^b \left[f(x) - \sum_{r=0}^{n} a_r \psi_r(x) \right]^2 dx = \int_a^b [f(x)]^2 \, dx - \sum_{r=0}^{n} a_r^2 \left(\int_a^b [\psi_r(x)]^2 \, dx \right).$$

5.16 By using the well-known trigonometrical identities, including (2.7) and (2.8), show that the set of functions $\{1, \cos x, \sin x, \ldots, \cos kx, \sin kx\}$ is orthogonal on $[-\pi, \pi]$ and verify the formulas (5.18) for the Fourier coefficients.

5.17 Obtain the classical Fourier series for $f(x) = |x|$ on $[-\pi, \pi]$.

5.18 Find the classical Fourier series for $f(x) = x^2$ on $[-\pi, \pi]$.

5.19 If $S_k(x)$ denotes the partial Fourier series (5.17), obtain as a special case of the result of Problem 5.15 that

$$\int_{-\pi}^{\pi} [f(x) - S_k(x)]^2 \, dx = \int_{-\pi}^{\pi} [f(x)]^2 \, dx - \pi \left(\tfrac{1}{2} a_0^2 + \sum_{r=1}^{k} (a_r^2 + b_r^2) \right).$$

5.20 For the 'square wave' function of Example 5.7, deduce from the result of the previous problem that

$$\lim_{k \to \infty} \int_{-\pi}^{\pi} [f(x) - S_k(x)]^2 \, dx = \frac{\pi}{2} - \frac{4}{\pi} \left(\frac{1}{1^2} + \frac{1}{3^2} + \frac{1}{5^2} + \cdots \right)$$

and show that this Fourier series converges to f in the least squares sense. (*Hint*: you may assume that $\sum_{r=1}^{\infty} (1/r^2) = \pi^2/6$.)

5.21 Write a program which evaluates the partial Fourier series

$$\frac{1}{2} a_0 + \sum_{r=1}^{n} (a_r \cos rx + b_r \sin rx)$$

for a given x, f and n, computing the coefficients by a simple integration rule (see Chapter 7). Your program should ask whether the function f is odd, even or neither. If the function is odd, each a_r should immediately be set to zero and if it is even each b_r should be set to zero.

Section 5.5

5.22 Consider the polynomials which are orthogonal on $[-1, 1]$ with respect to weight function ω. If ω is an even function, use an induction argument on (5.27) and (5.30) to show that $\beta_k = 0$ for all k and p_k is even or odd according as k is even or odd. Further, deduce from (5.33) that $a_r = 0$ for odd r if f is even, and $a_r = 0$ for even r if f is an odd function.

5.23 The set of polynomials $\{p_0, \ldots, p_n\}$ with $p_r \in P_r$ is orthogonal with respect to a weight function ω on $[a, b]$. Show that, given any $q \in P_n$, there is a set of coefficients $\{a_0, \ldots, a_n\}$ such that

$$q(x) = \sum_{r=0}^{n} a_r p_r(x).$$

Obtain an expression for the a_r. (*Hint*: the weighted least squares polynomial approximation to q out of P_n is unique and must be q.)

5.24 Suppose the orthogonal polynomial p_n defined in Problem 5.23 has exactly $k \leqslant n$ zeros on $[a, b]$, at $x_1, \ldots, x_k$, and write $q(x) = (x - x_1) \cdots (x - x_k)$. Why is $k \geqslant 1$? Then

$$\int_a^b \omega(x) p_n(x) q(x)\, \mathrm{d}x \neq 0.$$

By applying the result of the last problem, deduce that the orthogonal polynomial p_n has exactly n zeros on $[a, b]$.

5.25 Following the method of the previous problem, show that the error

$$f(x) - \sum_{r=0}^{n} a_r p_r(x)$$

has at least $n+2$ changes of sign on $[a, b]$, where the set $\{p_0, \ldots, p_n\}$ is as defined in Problem 5.23. (*Hint*: recall that

$$\int_a^b \omega(x)\left(f(x) - \sum_{r=0}^{n} a_r p_r(x)\right) p_s(x)\, dx = 0$$

for $0 \leqslant s \leqslant n$.)

5.26 With p_r as in the previous problem (and having leading coefficient 1), let us write

$$x^{n+1} - p_{n+1}(x) = \sum_{r=0}^{n} b_r p_r(x).$$

Multiply each term above by $\omega(x)p_s(x)$, integrate over $[a, b]$ and use the orthogonality property to show that $x^{n+1} - p_{n+1}(x)$ is the least squares approximation with respect to $\omega(x)$ for x^{n+1} on $[a, b]$.

5.27 By combining the results of Problems 5.24 and 5.26, show that

$$\int_a^b \omega(x)(x - x_0)^2 \cdots (x - x_n)^2\, dx$$

is minimized over all choices of $x_0, \ldots, x_n$ by choosing $(x - x_0)\cdots(x - x_n) = p_{n+1}(x)$, where p_{n+1} is the orthogonal polynomial defined in the previous problems.

5.28 Suppose that the polynomials p_r are orthogonal on $[a, b]$ with respect to weight function ω. By making the change of variable $u = (2x - b - a)/(b - a)$, find polynomials which are orthogonal on $[-1, 1]$. What is the weight function in this case?

5.29 Show that the Chebyshev polynomials of the second kind U_r, defined by (5.34), are orthogonal on $[-1, 1]$ with respect to weight function $(1 - x^2)^{1/2}$.

5.30 From (5.34) deduce that the orthogonal polynomials U_r satisfy the same recurrence relation as the T_r.

5.31 Show that the zeros of U_n are $x = \cos(j\pi/(n+1))$, $1 \leqslant j \leqslant n$, and show from the recurrence formula that U_n has leading term $2^n x^n$.

5.32 Verify that the conditions of Theorem 5.1 apply to the function $f(x) = x^{1/2}$ on $[0, 1]$. Hence verify that the best 1-norm straight line approximation for $x^{1/2}$ on $[0, 1]$ is that given in Example 5.1.

5.33 Find the Chebyshev series for $f(x) = \cos^{-1} x$.

5.34 Derive the Chebyshev series for $f(x) = (1 - x^2)^{1/2}$.

5.35 By calculating the coefficients β_k, γ_k which are used in the recurrence formula (5.40), construct polynomials of degree r, $0 \leqslant r \leqslant 3$, which are orthogonal on the point set $\{-2, -1, 1, 2\}$.

5.36 Write a program to compute the Legendre series, based on the adaption of Algorithm 5.1 described in the text.

5.37 To verify (5.45), one may proceed as follows. Express each $T_r(x_j)$ as $\cos(\pi rj/N)$ and apply the trigonometrical identity (2.8) to give the left side of (5.45) as a *sum* of cosines. Find the sum by using the identity

$$\cos k\theta = \frac{\sin(k+\frac{1}{2})\theta - \sin(k-\frac{1}{2})\theta}{2\sin\frac{1}{2}\theta}.$$

Verify (5.63) similarly.

Section 5.6

5.38 Obtain the minimax first degree polynomial for $f(x) = x^{1/2}$ on $[0, 1]$.

5.39 Obtain the minimax first degree polynomial for $f(x) = 1/(1+x)$ on $[0, 1]$.

5.40 Let m and M denote the minimum and maximum values of a continuous function f on $[a, b]$. Show that the minimax polynomial of degree zero for f on $[a, b]$ is $\frac{1}{2}(m+M)$ and find the maximum error.

5.41 Given any $a>0$, show that there is a unique integer k such that $\frac{1}{4} \leqslant 4^k a < 1$. If $x = 4^k a$, we saw in Example 5.14 that, for $\frac{1}{4} \leqslant x \leqslant 1$, $x^{1/2} \simeq \frac{17}{48} + \frac{2}{3}x$. Deduce that $a^{1/2} \simeq \frac{17}{48}2^{-k} + \frac{2}{3}2^k a$ and find the maximum error incurred by using this approximation. Thus estimate $\sqrt{2}$, $\sqrt{3}$ and $\sqrt{10}$ and estimate the error in each case.

5.42 Show that the minimax polynomial from P_2 for $1/(3x+5)$ on $[-1, 1]$ is $(9 - 8x + 6x^2)/48$. (*Hint*: first show that the error function has turning values at $\frac{1}{3}$ and $-\frac{2}{3}$.)

5.43 Suppose f is even on $[-a, a]$ and that $p^* \in P_n$ is the minimax approximation for f. By considering the $n+2$ equioscillation points, deduce that $p^*(-x)$ is the minimax polynomial for $f(-x)$. Since $f(-x) \equiv f(x)$ and the minimax polynomial is unique, deduce that $p^*(-x) \equiv p^*(x)$.

5.44 Adapt the argument used in the previous problem to show that the minimax approximation for an odd function on $[-a, a]$ is itself odd.

5.45 Let $E = f - p$ equioscillate on exactly $k < n+2$ points on $[a, b]$, where $f \in C[a, b]$. Then there are points x_i such that $a < x_1 < \cdots < x_{k-1} < b$, $E(x_i) = 0$, $1 \leqslant i \leqslant k-1$, and there is one equioscillation point in each of the intervals $[a, x_1)$, $(x_1, x_2), \ldots, (x_{k-2}, x_{k-1})$, $(x_{k-1}, b]$. Show that we can choose a value of the constant C so that $q(x) = C(x - x_1) \cdots (x - x_{k-1})$ satisfies the requirements of the q used in the proof of Theorem 5.3.

5.46 Let $p \in P_n$ denote the minimax approximation for $f \in C^{n+1}[a, b]$. Suppose there are k *interior* points of $[a, b]$ where $f - p$ equioscillates. By Theorem 5.3, $k \geq n$. Apply the extended Rolle theorem 2.6 to $g = f' - p'$ to show that $f^{(n+1)}$ has at least $k - n$ zeros. Deduce that if $f^{(n+1)}$ has constant sign there are exactly n interior equioscillation points plus two at the endpoints a and b.

Section 5.7

5.47 Verify (5.66) by following the method of Problem 5.37.

5.48 Write a program to evaluate the partial Chebyshev series $\sum\nolimits'^{n}_{r=0} a_r T_r(x)$. For each r choose a value of N and compute α_r and α_r^* from (5.60) and (5.68). If $|\alpha_r - \alpha_r^*|$ is sufficiently small, let $a_r = \frac{1}{2}(\alpha_r + \alpha_r^*)$ otherwise increase N and repeat the process. (Take account of the final part of Problem 5.22 to avoid the unnecessary calculation of zero coefficients when f is an even or odd function.)

Section 5.8

5.49 Show that, to approximate to e^x on $[-1, 1]$ to an accuracy within 10^{-6} by a Taylor polynomial constructed at $x = 0$, we require the polynomial of degree $n = 9$ or more (cf. Problem 3.4). Verify that if $n = 9$ the maximum error is less than 0.75×10^{-6}. Given that

$x^9 = T_9(x)/256 +$ (terms involving lower *odd order* Chebyshev polynomials),

$x^8 = T_8(x)/128 +$ (terms involving lower *even order* Chebyshev polynomials),

find how many terms of the above Taylor polynomial can be 'economized' so that the maximum error remains less than 10^{-6}.

5.50 Show that when r is odd

$$x^r = \frac{1}{2^{r-1}} \sum_{0 \leq s < r/2} \binom{r}{s} T_{r-2s}(x)$$

and when r is even we need to add the further term $(1/2^r)\binom{r}{r/2}$.

5.51 Deduce from (5.73) and (5.75) that the Chebyshev coefficients for e^x are given by

$$a_r = \frac{1}{2^{r-1}} \sum_{s=0}^{\infty} \frac{1}{2^{2s}(r+s)!s!}.$$

Section 5.9

5.52 Consider the matrix of the linear equations (5.76) and assume that the matrix is singular. Then its columns would be linearly dependent and there would exist numbers $c_0, \ldots, c_{n+1}$ not all zero such that

$$c_0 + c_1 x_i + \cdots + c_n x_i^n = (-1)^i c_{n+1}, \qquad 1 \leqslant i \leqslant n+2.$$

Obtain a contradiction, considering separately the cases $c_{n+1} = 0$ and $c_{n+1} \neq 0$ and deduce that the linear equations have a unique solution.

5.53 Find the minimax $p \in P_2$ for e^x on $[-1, 1]$, using the Remez algorithm. Compare this with the corresponding partial Chebyshev series, whose coefficients are given in Problem 5.51.

Section 5.10

5.54 Let $p \in P_n$ be the minimax approximation for $f \in C^{n+1}[a, b]$. Write $x = (b-a)t/2 + (b+a)/2$ and let $f(x) = g(t)$, $p(x) = q(t)$. Deduce from Theorem 5.3 that $q \in P_n$ is the minimax approximation for g on $[-1, 1]$. Verify Corollary 5.6 by applying Theorem 5.6 to g.

5.55 Apply the methods of Theorem 5.6 and the result of Problem 5.25 to show that for $f \in C^{n+1}[a, b]$ there is a $\xi \in [-1, 1]$ such that

$$\inf_{p \in P_n} \| f - q \|_2 = C_n \frac{| f^{(n+1)}(\xi) |}{(n+1)!},$$

where

$$C_n = \inf_{x_i} \| (x - x_0) \cdots (x - x_n) \|_2.$$

5.56 Write a program to evaluate $B_n(f; x)$ using the de Casteljau algorithm. Hence draw graphs of $B_n(f; x)$ for the function $|x|$ on $[-1, 1]$, for various values of n. (Note that $|x|$ on $[-1, 1]$ is replaced by $|2x - 1|$ on $[0, 1]$.)

5.57 Verify Theorems 5.8 and 5.9 by induction on m and thus verify (5.87).

5.58 Deduce directly from (5.86) and the binomial expansion that

$$B_n(e^x; x) = (1 + (e^{1/n} - 1)x)^n$$

and expand the latter expression in powers of x to show agreement with (5.87).

Chapter **6**

SPLINES AND OTHER APPROXIMATIONS

6.1 Introduction

We begin with a gentle warning that some of the material in this chapter is a little harder. For example, we have left the details of the proof of Theorem 6.1 to be filled in by the reader, who may wish to defer this task to a second reading.

In the previous three chapters we have discussed various types of polynomial approximations for a continuous function f defined on $[a, b]$. The emphasis on polynomials is justified by Weierstrass's theorem 5.7 which assures us that by choosing a polynomial of sufficiently high degree we can approximate as closely as we wish to the given function. However there are cases where the convergence of polynomial approximants is very slow (cf. § 4.11) and this provides the motivation for developing other types of approximations. We now consider one of the simplest of these.

Let us partition the interval $[a, b]$ into n sub-intervals $[t_{i-1}, t_i]$, $1 \leqslant i \leqslant n$, where

$$a = t_0 < t_1 < \cdots < t_n = b.$$

We refer to the t_i as the *knots*. We approximate to a given function f in $C[a, b]$ by the polygonal arc $S_n(x)$ which is formed by connecting consecutive pairs of points (t_i, y_i), $0 \leqslant i \leqslant n$, with straight line segments. As a special case we may choose $y_i = f(t_i)$, and then S_n interpolates f at the knots t_i. In this case, by choosing n sufficiently large and ensuring that each interval $[t_{i-1}, t_i]$ is sufficiently small (for example, by taking equal sub-intervals) the continuity of f ensures that S_n is as close to f as we wish.

Example 6.1 Let us construct the polygonal arc S_n for $x^{1/2}$ on $[0, 1]$, where $t_i = i/n$ and $y_i = t_i^{1/2}$. Thus S_n interpolates $x^{1/2}$ at the knots, which are equally spaced. It may be verified (see Problems 6.1 and 6.2) that

$$\max_{0 \leqslant x \leqslant 1} |x^{1/2} - S_n(x)| = \frac{1}{4\sqrt{n}},$$

the maximum error being attained on the first sub-interval, $[0, 1/n]$. It follows that the sequence of approximating functions $(S_n(x))$ converges uniformly to $x^{1/2}$ on $[0, 1]$. □

By taking some care with the choice of knots and not insisting that S_n interpolates the function, we can sometimes obtain a much better approximation, as achieved in the following example.

Example 6.2 Again consider $x^{1/2}$ on $[0, 1]$ and define S_n as the polygonal arc connecting the points (t_i, y_i), $0 \leqslant i \leqslant n$, where

$$t_i = \left(\frac{i(i+1)}{n(n+1)}\right)^2, \qquad y_i = t_i^{1/2} + \frac{1}{4n(n+1)}.$$

In this case it is clear that $S_n(x)$ does *not* interpolate $x^{1/2}$ at the knots t_i. We may verify that $S_n(x)$ restricted to $[t_{i-1}, t_i]$ is the minimax straight line for $x^{1/2}$ on that interval and that

$$\max_{0 \leqslant x \leqslant 1} |x^{1/2} - S_n(x)| = \frac{1}{4n(n+1)}. \tag{6.1}$$

The maximum error is attained at each of the knots t_i, $0 \leqslant i \leqslant n$, and also at one point $\xi_i = (i^2/n(n+1))^2$ in the interior of each sub-interval. Figure 6.1 shows the graphs of $x^{1/2}$ and S_n for the case $n = 2$. □

The polygonal arc functions S_n defined above are called *first degree splines*, a special case of the general spline which we now define.

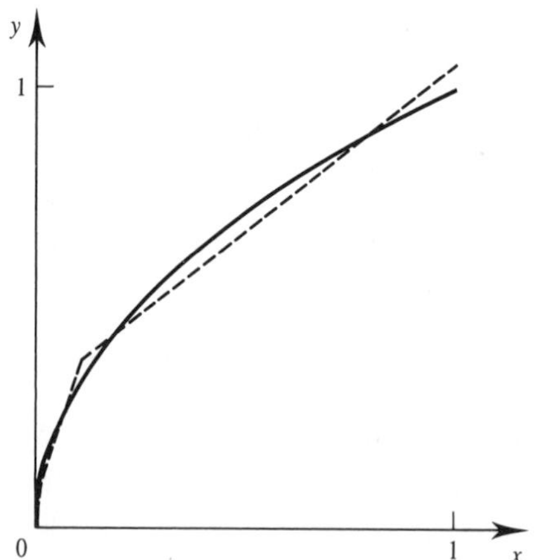

Fig. 6.1 Graph of $x^{1/2}$ and the linear spline $S_2(x)$.

Definition 6.1 Let $[a,b]$ be partitioned into sub-intervals $[t_{i-1}, t_i]$, $1 \leqslant i \leqslant n$, where $a = t_0 < t_1 < \cdots < t_n = b$. A *spline* S_n of degree k in $[a,b]$ satisfies the following two properties:

(i) S_n restricted to $[t_{i-1}, t_i]$ is a polynomial of degree at most $k \geqslant 1$,
(ii) $S_n \in C^{k-1}[a,b]$. □

Thus, for a given partition of $[a,b]$, the spline consists of n polynomial segments with the appropriate continuity condition at each of the interior knots, which gives the spline a certain degree of *smoothness*. The first degree spline discussed above (satisfying Definition 6.1 with $k=1$) is continuous but may have discontinuities in its first derivative at the knots. In many practical applications a greater degree of smoothness is usually desirable and we may choose $k=2$ above to give a *quadratic spline*, which has a continuous first derivative, or choose $k=3$ to give a *cubic spline* with continuous first and second derivatives. It is rare to use a spline of higher order than the cubic spline, which is the one most commonly used. For the general spline of degree k we require $k+1$ conditions to determine the polynomial of degree k on each interval, less k conditions at each interior knot (to satisfy property (ii) of Definition 6.1). Thus, to determine a spline of degree k we require

$$n(k+1) - (n-1)k = n + k$$

conditions.

Example 6.3 Let us define the function C by

$$C(x) = \begin{cases} \frac{2}{3} - \frac{1}{2}x^2(2 - |x|), & 0 \leqslant |x| \leqslant 1 \\ \frac{1}{6}(2 - |x|)^3, & 1 < |x| \leqslant 2 \\ 0, & |x| > 2 \end{cases} \tag{6.2}$$

It is easily verified that C is a cubic spline with respect to the knots $-2, -1, 0, 1, 2$ and that C has a continuous second derivative for all real x. The reader is encouraged to sketch the graph of C. This is a cubic B-spline, to be defined in the next section (see Fig. 6.5). □

We are familiar with the fact that any polynomial $p \in P_n$ can be expressed as a linear combination of $n+1$ fixed linearly independent polynomials, collectively referred to as a *basis*. For example, we can take $\{1, x, \ldots, x^n\}$ as a basis for P_n. We will now obtain a basis for first degree splines. It is helpful to extend the sequence of $n+1$ knots $t_0, \ldots, t_n$ so that they become part of an infinite sequence of knots

$$\ldots, t_{-2}, t_{-1}, t_0, \ldots, t_n, t_{n+1}, t_{n+2}, \ldots$$

where $t_{-m} \to -\infty$ and $t_m \to \infty$ as $m \to \infty$. We then define, for $-\infty < i < \infty$,

$$B_i^1(x) = \begin{cases} 0, & x \leqslant t_i \\ \dfrac{x - t_i}{t_{i+1} - t_i}, & t_i < x \leqslant t_{i+1} \\ \dfrac{t_{i+2} - x}{t_{i+2} - t_{i+1}}, & t_{i+1} < x \leqslant t_{i+2} \\ 0, & t_{i+2} < x. \end{cases} \tag{6.3}$$

(See Fig. 6.2.) Note that $B_i^1(t_{i+1}) = 1$ and $B_i^1(x) = 0$ at all of the other knots. It is easily verified (see Problem 6.5) that the functions B_i^1 are linearly independent and are a basis for the first degree splines. The expression of the spline S_n in terms of the B_i^1 is remarkably simple. We have

$$S_n(x) = \sum_{i=0}^{n} y_i B_{i-1}^1(x), \quad t_0 \leqslant x \leqslant t_n. \tag{6.4}$$

From (6.4) we see immediately that $S_n(t_i) = y_i$, as required, and it remains only to verify that the right side of (6.4) reduces to a straight line segment on each sub-interval $[t_{i-1}, t_i]$ (see Problem 6.4).

In the next section we will define the B-splines of degree k, written B_i^k, which generalize the first degree splines B_i^1 defined above, and form a basis for splines of degree k. To aid our discussion of this later, we conclude this section by finding an alternative basis for the splines of degree k.

We define the *truncated power* of degree $k \geqslant 0$ as a function of the form

$$(x - t)_+^k = \begin{cases} (x - t)^k, & x \geqslant t \\ 0, & x < t \end{cases} \tag{6.5}$$

where t is some fixed real number. Note that this is a spline of degree k with a single knot at $x = t$. We now show that we can build up the representation of a spline, one sub-interval at a time, in terms of a polynomial and truncated powers. First, on $[t_0, t_1]$ we can write a spline of degree k as

$$S_n(x) = \sum_{i=0}^{k} c_i x^i. \tag{6.6}$$

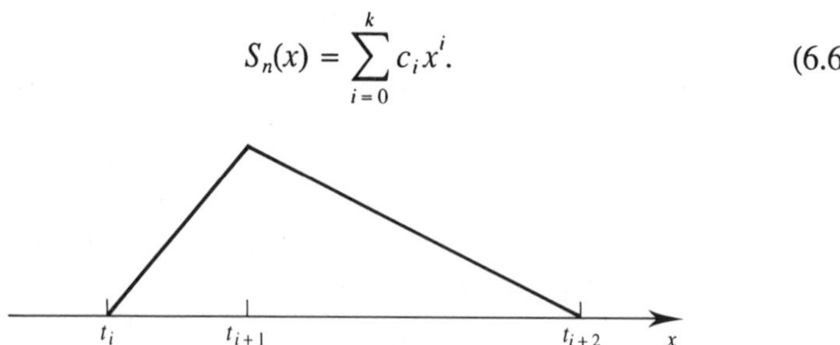

Fig. 6.2 Graph of a linear B-spline $B_i^1(x)$.

We now seek to add a suitable function to the right side of (6.6) to extend the representation of S_n from $[t_0, t_1]$ to $[t_0, t_2]$. Note that any polynomial in P_k can be written in the form

$$\sum_{i=0}^{k} a_i(x - t_1)^i$$

and if we adapt this to

$$\sum_{i=0}^{k} a_i(x - t_1)^i_+ \tag{6.7}$$

we obtain a function (no longer a polynomial, but a spline of degree k) which is zero for $x < t_1$, and so can be added to the right side of (6.6) without disturbing the representation of S_n on $[t_0, t_1]$. However, in order to satisfy the smoothness requirement for the spline S_n at t_1, we must not disturb the continuity of either S_n or its first $k-1$ derivatives at t_1. Thus in (6.7) we must take $a_0 = a_1 = \cdots = a_{k-1} = 0$ and, on renaming a_k as d_1, we find we can write

$$S_n(x) = \sum_{i=0}^{k} c_i x^i + d_1(x - t_1)^k_+, \tag{6.8}$$

and this is valid for the interval $[t_0, t_2]$. Thus we need to choose $c_0, \ldots, c_k$ to give the appropriate polynomial to define S_n on $[t_0, t_1]$ and choose only one further coefficient, d_1, so that the right side of (6.8) gives also the correct polynomial which represents S_n on $[t_1, t_2]$. We continue in this way, adding a suitable truncated power to give the correct polynomial on the next sub-interval while not changing the representation of S_n in earlier sub-intervals. Finally, we obtain

$$S_n(x) = \sum_{i=0}^{k} c_i x^i + \sum_{j=1}^{n-1} d_j(x - t_j)^k_+, \tag{6.9}$$

which is valid on the whole of $[t_0, t_n]$. We note that the number of coefficients c_i and d_j is $n + k$, which agrees with the number of conditions noted above to determine a spline of degree k on an interval with $n-1$ interior knots. We emphasize that the representation of S_n in the form (6.9) is only of theoretical interest. In order to evaluate a spline we will prefer to express it in terms of B-splines, which we will discuss in the next section.

Example 6.4 Let us express the cubic spline of Example 6.3 in the form of truncated powers. We obtain, for $x \geqslant -2$,

$$S(x) = \tfrac{1}{6}(x+2)^3 - \tfrac{2}{3}(x+1)^3_+ + (x)^3_+ - \tfrac{2}{3}(x-1)^3_+ + \tfrac{1}{6}(x-2)^3_+.$$

This coincides with the function $C(x)$ of Example 6.3 for $-2 \leqslant x < \infty$. If we replace the first term $\frac{1}{6}(x+2)^3$ by the truncated power $\frac{1}{6}(x+2)_+^3$ we obtain an expression which agrees with $C(x)$ for all x. □

6.2 B-splines

As we have seen, the first degree splines B_i^1 form a basis for all first degree splines (with respect to a given set of knots). We now show that the B_i^1 can themselves be expressed in terms of still simpler functions. Let us define

$$B_i^0(x) = \begin{cases} 1, & t_i < x \leqslant t_{i+1} \\ 0, & \text{otherwise.} \end{cases} \tag{6.10}$$

Note that B_i^0 is not continuous at t_i and t_{i+1} (see Fig. 6.3). It is easily verified from (6.3) and (6.10) that

$$B_i^1(x) = \left(\frac{x - t_i}{t_{i+1} - t_i}\right) B_i^0(x) + \left(\frac{t_{i+2} - x}{t_{i+2} - t_{i+1}}\right) B_{i+1}^0(x). \tag{6.11}$$

The B_i^0 are piecewise constant functions. They are clearly a basis for the set of all piecewise constant functions. Such a function which takes the value y_i on the interval $t_i < x \leqslant t_{i+1}$, for $0 \leqslant i \leqslant n-1$, and the value zero elsewhere can be expressed as

$$\sum_{i=0}^{n-1} y_i B_i^0(x).$$

The functions B_i^0 and B_i^1 are called B-splines of degree 0 and 1 respectively. We now introduce functions B_i^2, B_i^3, ..., defined recursively by

$$B_i^k(x) = \left(\frac{x - t_i}{t_{i+k} - t_i}\right) B_i^{k-1}(x) + \left(\frac{t_{i+k+1} - x}{t_{i+k+1} - t_{i+1}}\right) B_{i+1}^{k-1}(x) \tag{6.12}$$

for each i and for $k = 1, 2, \ldots$. Note that for $k = 1$ (6.12) reduces to (6.11). Graphs of B_i^2 and B_i^3 with equally-spaced knots are given in Figs. 6.4 and 6.5

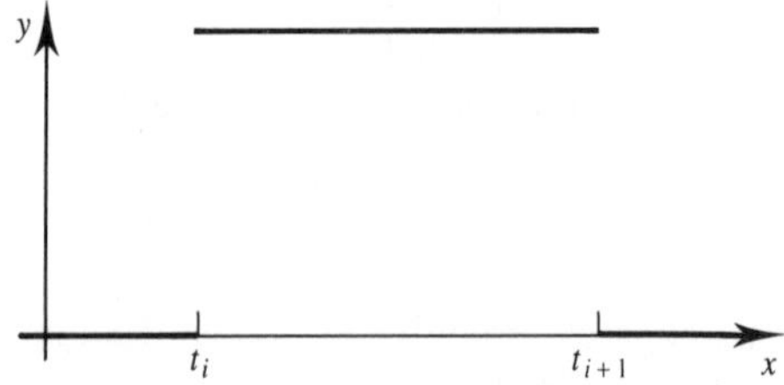

Fig. 6.3 Graph of the simplest B-spline $B_i^0(x)$.

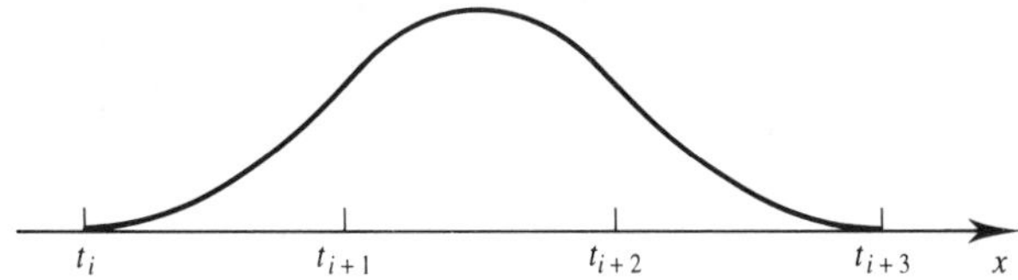

Fig. 6.4 Graph of a quadratic B-spline $B_i^2(x)$ with equally spaced knots.

respectively. The function B_i^k is called a B-spline of degree k. We need to justify that it is indeed a spline. First, by using induction on k in (6.12) it is clear that B_i^k is a piecewise polynomial of degree k, that is, on each interval $[t_{j-1}, t_j]$, B_i^k is a polynomial of degree k. An induction argument on (6.12) also reveals how the *support* of B_i^k grows with k. (The support of a function f is the smallest closed set containing all x such that $f(x) \neq 0$.) For $k \geqslant 0$ the support of B_i^k is the interval $[t_i, t_{i+k+1}]$, that is, it consists of $k+1$ of the subintervals created by our set of knots.

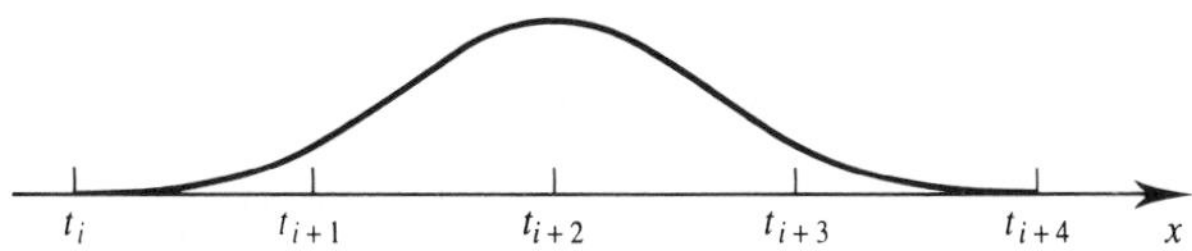

Fig. 6.5 Graph of a cubic B-spline $B_i^3(x)$ with equally spaced knots.

It is not immediately obvious from (6.12) that B_i^k satisfies the *smoothness* condition necessary for a spline of degree k, namely that $B_i^k \in C^{k-1}(-\infty, \infty)$. However, this smoothness condition can be deduced from the following result.

Theorem 6.1 For $k > 1$

$$\frac{\mathrm{d}}{\mathrm{d}x} B_i^k(x) = \left(\frac{k}{t_{i+k} - t_i}\right) B_i^{k-1}(x) - \left(\frac{k}{t_{i+k+1} - t_{i+1}}\right) B_{i+1}^{k-1}(x) \qquad (6.13)$$

for all x. For $k = 1$, (6.13) holds for all x except at $x = t_i$, t_{i+1} and t_{i+2}, where the derivative of B_i^1 is not defined.

Proof We first prove the weaker result that (6.13) holds for all x excluding the knots t_j. It is easily verified that (6.13) holds for $k = 1$ except at the knots t_i, t_{i+1} and t_{i+2}. Next we assume that (6.13) holds for some fixed value of k and all i. Differentiate both sides of (6.12) with k replaced by $k+1$ and replace the derivatives of B_i^k and B_{i+1}^k, using (6.13). Combine the terms in B_i^{k-1}, B_{i+1}^{k-1} and B_{i+2}^{k-1}, using (6.12), to give terms involving B_i^k and B_{i+1}^k. (Note that this involves splitting the term B_{i+1}^{k-1} into two parts.)

Thus verify that (6.13) holds when k is replaced by $k+1$ and so complete the proof of Theorem 6.1 by induction (see Problem 6.7).

We now pay careful attention to what happens at the knots. By induction on (6.12) it is obvious that, for $k \geqslant 1$, B_i^k is continuous for all x. Thus for $k \geqslant 2$ we see that the right side of (6.13) is continuous for all x. Also, we already know that (6.13) holds for all x except at the knots. It follows from the continuity of the right side that, for $k \geqslant 2$, (6.13) holds for all x. □

Corollary 6.1 For $k \geqslant 1$, $B_i^k \in C^{k-1}(-\infty, \infty)$.

Proof This follows from (6.13) by induction on k. □

We will require the following identity in the proof of our next theorem:

$$t - x = \left(\frac{t - t_{i+k+1}}{t_i - t_{i+k+1}}\right)(t_i - x) + \left(\frac{t - t_i}{t_{i+k+1} - t_i}\right)(t_{i+k+1} - x). \tag{6.14}$$

This is easily verified directly or we may observe that the right side of (6.14) is the linear interpolating polynomial for $t - x$ (regarded as a function of t) at $t = t_i$ and t_{i+k+1}.

Theorem 6.2 (Marsden's identity) For any fixed $k = 0, 1, \ldots,$

$$(t - x)^k = \sum_{i=-\infty}^{\infty} (t - t_{i+1}) \cdots (t - t_{i+k}) B_i^k(x) \tag{6.15}$$

where, for $k = 0$, the empty product $(t - t_{i+1}) \cdots (t - t_{i+k})$ is replaced by 1.

Proof For $k = 0$, (6.15) gives

$$1 = \sum_{i=-\infty}^{\infty} B_i^0(x) \tag{6.16}$$

which follows immediately from the definition of B_i^0. The proof is completed by induction on k. Let us suppose that (6.15) holds for some $k \geqslant 0$. We multiply both sides of (6.15) by $t - x$ and, in each term in the summation on the right side, we express $t - x$ as in (6.14). This gives two summations on the right which we combine into one, using (6.12) with k replaced by $k+1$. This gives (6.15) with k replaced by $k+1$. □

Corollary 6.2 Every monomial x^r, $0 \leqslant r \leqslant k$, can be written as a sum of multiples of the B_i^k, with k fixed.

Proof Compare coefficients of powers of t^{k-r} on both sides of (6.15). □

As a special case, comparing coefficients of t^k in (6.15) yields

$$1 = \sum_{i=-\infty}^{\infty} B_i^k(x) \tag{6.17}$$

which generalizes (6.16). Note that Corollary 6.2 implies that any element in P_k can be expressed in terms of the B_i^k with k fixed. Now a polynomial in P_k is a special case of a spline of degree k on our given system of knots. Not only elements of P_k, but in fact *all* splines of degree k can be expressed as a linear combination of the B_i^k with k fixed. We will content ourselves with justifying this in the next section for the special case of equally spaced knots.

Example 6.5 Let us take $k = 2$ in Marsden's identity (6.15) to give

$$t^2 - 2xt + x^2 = \sum_{i=-\infty}^{\infty} (t^2 - (t_{i+1} + t_{i+2})t + t_{i+1}t_{i+2})B_i^2(x)$$

and we may compare coefficients of t^2, t and 1 to give expressions for 1, x and x^2 respectively. Thus any element of P_2 can be written as

$$a_0 + a_1x + a_2x^2 = \sum_{i=-\infty}^{\infty} (a_0 + \tfrac{1}{2}a_1(t_{i+1} + t_{i+2}) + a_2t_{i+1}t_{i+2})B_i^2(x). \qquad \square$$

6.3 Equally spaced knots

B-splines become simpler when the knots are equally spaced. Let us choose $t_i = i$, since any equally spaced system of knots can be obtained from these by making a linear change of variable. The recurrence relation (6.12) then becomes

$$B_i^k(x) = \frac{1}{k}(x - i)B_i^{k-1}(x) + \frac{1}{k}(i + k + 1 - x)B_{i+1}^{k-1}(x). \tag{6.18}$$

These are called *uniform* B-splines. We find that for any fixed value of k the functions B_i^k, as i varies, are all 'copies' of each other. Specifically, we have

$$B_i^k(x) = B_{i+1}^k(x+1) \tag{6.19}$$

and we say that B_{i+1}^k is a *translate* of B_i^k. Also, for $k > 0$, each B_i^k is symmetric about the centre of its interval of support $[i, i + k + 1]$, that is,

$$B_i^k(x) = B_i^k(2i + k + 1 - x). \tag{6.20}$$

Both (6.19) and (6.20) may be proved by induction on (6.18).

Example 6.6 Using (6.10) with $t_i = i$ and (6.18) we obtain the uniform quadratic B-spline

$$B_i^2(x) = \begin{cases} \frac{1}{2}(x-i)^2, & i < x \leq i+1 \\ \frac{3}{4} - (x - (i + \frac{3}{2}))^2, & i+1 < x \leq i+2 \\ \frac{1}{2}(i+3-x)^2, & i+2 < x \leq i+3 \\ 0, & \text{otherwise.} \end{cases}$$

The corresponding cubic spline $B_i^3(x) = C(x - i - 2)$, where C is the spline defined in Example 6.3 above. □

We will require the following lemma.

Lemma 6.1 For $k = 0, 1, \ldots,$

$$\Delta^k(f_i g_i) = \sum_{r=0}^{k} \binom{k}{r} \Delta^{k-r} f_{i+r} \Delta^r g_i. \tag{6.21}$$

Proof We call (6.21) the Leibniz formula, after the analogous result for the kth derivative of a product. The proof of (6.21) is by induction on k, making use of the relation

$$\Delta^r g_{i+1} = \Delta^{r+1} g_i + \Delta^r g_i$$

and Pascal's identity

$$\binom{k}{r-1} + \binom{k}{r} = \binom{k+1}{r}.$$

□

As an application of Lemma 6.1 we take $f_i = (i-x)_+^m$, $g_i = i - x$ and we note that $f_i g_i = (i-x)_+^{m+1}$ for $m \geqslant 0$. Since second and higher differences of $i - x$ are zero it follows from (6.21) that

$$\Delta^k (i-x)_+^{m+1} = (i-x) \cdot \Delta^k (i-x)_+^m + k\Delta^{k-1}(i+1-x)_+^m. \tag{6.22}$$

We now present a beautiful result due to I. J. Schoenberg (1903–90), who carried out much of the pioneering work on splines. This shows the simple relation between the uniform B-splines and the truncated powers.

Theorem 6.3 For any $k \geqslant 0$ and all i,

$$B_i^k(x) = \frac{1}{k!} \Delta^{k+1}(i-x)_+^k. \tag{6.23}$$

Proof We emphasize that in (6.23), as in (6.22), the forward difference operates on i and not on x. The verification of (6.23) is by induction on k. First we show that (6.23) holds for $k = 0$. Then we assume (6.23) holds for some $k - 1 \geqslant 0$ and use this in the right side of (6.18). The first term of (6.18) is then a multiple of $(x - i) \cdot \Delta^k (i-x)_+^{k-1}$ which we rewrite using (6.22) with $m = k - 1$. The second term is a multiple of

$$(i+k+1-x)\Delta^k(i+1-x)_+^{k-1} = k\Delta^k(i+1-x)_+^{k-1} + (i+1-x)\Delta^k(i+1-x)_+^{k-1}. \tag{6.24}$$

We also apply (6.22), with i replaced by $i+1$ and $m=k-1$, to the latter term in (6.24). On simplifying the transformed right side of (6.18) we obtain (6.23). □

There is also an inverse result to (6.23), in which a truncated power can be expressed in terms of uniform B-splines.

Theorem 6.4 For any $k \geqslant 0$ and all i,

$$(i-x)_+^k = k! \sum_{r=0}^{\infty} \binom{k+r}{r} B_{i-k-1-r}^k(x). \tag{6.25}$$

Proof We substitute (6.23) in the right side of (6.25) and expand the forward differences (see Problem 4.29) to give

$$\sum_{r=0}^{\infty} \sum_{s=0}^{k+1} (-1)^s \binom{k+r}{r} \binom{k+1}{s} (i-r-s-x)_+^k.$$

We write $r+s=m$ and express the above as

$$\sum_{m=0}^{\infty} \left(\sum_{r+s=m} (-1)^s \binom{k+r}{r} \binom{k+1}{s} \right) (i-m-x)_+^k.$$

The inner summation above, in which $r+s=m$ and also $r \geqslant 0$ and $0 \leqslant s \leqslant k+1$, is seen to be the coefficient of x^m in

$$1 = (1-x)^{-(k+1)}(1-x)^{k+1}.$$

Thus the inner summation has the value 1 for $m=0$ and zero for $m>0$ and hence the double summation reduces to $(i-x)_+^k$. □

Example 6.7 For a cubic truncated power, (6.25) gives

$$(i-x)_+^3 = \sum_{r=0}^{\infty} (r+1)(r+2)(r+3) B_{i-4-r}^3(x).$$

Note how the growth of the coefficients of B_{i-4-r}^3 follows the cubic nature of the function $(i-x)_+^3$. □

When the knots are not equally spaced, a truncated power can still be expressed in terms of B-splines, and vice versa. In this case the finite difference on the right of (6.23) must be replaced by a divided difference. In view of (6.9) and the identity

$$(t_i - x)^k = (t_i - x)_+^k + (-1)^k (x - t_i)_+^k \tag{6.26}$$

(see Problem 6.12) the above results show that any spline of degree k can be expressed as a linear combination of the B_i^k, with k fixed.

To compute a spline of degree k on the knots $0, 1, \ldots, n$, we need to use the B-splines $B^k_{-k}, B^k_{-k+1}, \ldots, B^k_{n-1}$, for these are the only B-splines of degree k which are non-zero on $[0, n]$. Note that there are $n+k$ of these, in keeping with our earlier observation that $n+k$ conditions need to be satisfied in determining a spline of degree k. For example, a cubic spline on $[0, n]$ is of the form

$$S(x) = \sum_{r=-3}^{n-1} a_r B^3_r(x), \quad 0 \leqslant x \leqslant n. \tag{6.27}$$

From (6.19) we have, due to the equal spacing of the knots, that

$$B^3_r(x) = B^3_{r-1}(x-1) = \cdots = B^3_{-2}(x-r-2).$$

Let us write $C(x)$ to denote $B^3_{-2}(x)$, the uniform cubic B-spline whose interval of support is $[-2, 2]$.

Then we obtain from (6.27) that

$$S(x) = \sum_{r=-3}^{n-1} a_r C(x-r-2), \qquad 0 \leqslant x \leqslant n. \tag{6.28}$$

We have already given the function C explicitly in Example 6.3, from which we note that

$$C(0) = \tfrac{2}{3}, \qquad C(\pm 1) = \tfrac{1}{6}$$

and $C(x)$ is zero at all other knots. Also

$$C'(-1) = \tfrac{1}{2}, \qquad C'(1) = -\tfrac{1}{2}$$

and $C'(x)$ is zero at all other knots. Thus from (6.28) we obtain

$$S(i) = \tfrac{1}{6}(a_{i-3} + 4a_{i-2} + a_{i-1}) \tag{6.29}$$

and

$$S'(i) = \tfrac{1}{2}(a_{i-1} - a_{i-3}), \tag{6.30}$$

both valid for $0 \leqslant i \leqslant n$.

To determine the $n+3$ values $a_{-3}, a_{-2}, \ldots, a_{n-1}$, thus determining the spline S given by (6.28), we need $n+3$ appropriate conditions. Now suppose that S is an *interpolating spline* which interpolates a given function f at the knots $x = 0, 1, \ldots, n$. We still need two further conditions, and obtain these by supposing that S' matches f' at the endpoints $x = 0$ and $x = n$. Thus

$$S'(0) = f'(0) \quad \text{and} \quad S'(n) = f'(n).$$

(This is not the only possibility. One alternative, which yields what is called a *natural spline*, is to choose $S''(0) = S''(n) = 0$.) Then from (6.29) and

(6.30) we can write down a system of $n+3$ linear equations to determine the $n+3$ unknowns $a_{-3}, a_{-2}, \ldots, a_{n-1}$. These are

$$f(i) = \tfrac{1}{6}(a_{i-3} + 4a_{i-2} + a_{i-1}), \qquad 0 \leqslant i \leqslant n \tag{6.31}$$

$$f'(0) = \tfrac{1}{2}(a_{-1} - a_{-3}) \tag{6.32}$$

$$f'(n) = \tfrac{1}{2}(a_{n-1} - a_{n-3}). \tag{6.33}$$

(If the values of f' are not available, they may be replaced by a finite difference.) It is convenient to eliminate a_{-3} between (6.32) and (6.31) with $i=0$ to give

$$f(0) + \tfrac{1}{3}f'(0) = \tfrac{1}{6}(4a_{-2} + 2a_{-1}) \tag{6.34}$$

and eliminate a_{n-1} between (6.33) and (6.31) with $i=n$ to give

$$f(n) - \tfrac{1}{3}f'(n) = \tfrac{1}{6}(2a_{n-3} + 4a_{n-2}). \tag{6.35}$$

We now solve the system of linear equations consisting of (6.31) for $1 \leqslant i \leqslant n-1$ together with (6.34) and (6.35). This can be written in the form

$$\mathbf{Ma} = \mathbf{b}, \tag{6.36}$$

where

$$\mathbf{a}^{\mathrm{T}} = [a_{-2}, a_{-1}, \ldots, a_{n-2}], \tag{6.37}$$

$$\mathbf{b}^{\mathrm{T}} = [6f(0) + 2f'(0), 6f(1), \ldots, 6f(n-1), 6f(n) - 2f'(n)] \tag{6.38}$$

and

$$\mathbf{M} = \begin{bmatrix} 4 & 2 & & & & \\ 1 & 4 & 1 & & & \\ & 1 & 4 & 1 & & \mathbf{0} \\ & & \ddots & \ddots & \ddots & \\ & \mathbf{0} & & 1 & 4 & 1 \\ & & & & 2 & 4 \end{bmatrix}.$$

The solution of these equations yields $a_{-2}, a_{-1}, \ldots, a_{n-2}$ and the remaining coefficients a_{-3} and an a_{n-1} are obtained from (6.32) and (6.33) respectively, thus determining the spline (6.28). The above system of $n+1$ linear equations is called *tridiagonal*. It is also strictly diagonally dominant, meaning that in every row of the matrix the modulus of the element on the diagonal (4 for the matrix $\mathbf{M}$ above) exceeds the sum of the moduli of the off-diagonal elements. As we will see in Chapters 9 and 10, a diagonally dominant system has a unique solution and a tridiagonal system is easily solved numerically.

The following example illustrates the above procedure for finding a cubic interpolating spline and also makes it clear that our choice of knots at integer points imposes no real restrictions.

Example 6.8 Let us apply the above method to the function $e^{x/n}$ on $[0, n]$ for different values of n and so obtain cubic splines for e^x on $[0, 1]$. For $n = 1$ the system (6.36) has only two equations which we solve to give a_{-2} and a_{-1}. We then obtain a_{-3} and a_0 from (6.32) and (6.33), to give

$$S(x) = \tfrac{2}{3}(-5 + 2e)C(x+1) + \tfrac{1}{3}(8 - 2e)C(x) + \tfrac{2}{3}(-2 + 2e)C(x-1) + \tfrac{1}{3}(8 + 4e)C(x-2)$$

as an approximation for e^x on $[0, 1]$. Since (see Example 6.3) $C(\tfrac{1}{2}) = C(-\tfrac{1}{2}) = \tfrac{23}{48}$ and $C(\tfrac{3}{2}) = C(-\tfrac{3}{2}) = \tfrac{1}{48}$ we obtain $S(\tfrac{1}{2}) = \tfrac{1}{8}(5 + 3e) \simeq 1.644$ as an approximation for $e^{1/2} \simeq 1.649$. For a general value of n we obtain the coefficients a_j numerically by solving the tridiagonal system (6.36) and evaluate the spline using (6.28) and (6.2) for each required value of x. In Table 6.1 we list the maximum errors of these interpolating cubic splines for e^x on $[0, 1]$. The coefficients for these splines mimic the growth of the exponential function. The coefficients for $n = 1$, given above, are approximately 0.2910, 0.8545, 2.2910 and 6.2910. For $n = 5$ the coefficients are approximately 0.8133, 0.9934, 1.2133, 1.4819, 1.8100, 2.2108, 2.7002 and 3.2981. The ratios of consecutive pairs of the latter coefficients are very close to $e^{0.2}$. □

For a *quadratic* spline, it is not appropriate to interpolate at the knots, $0, 1, \ldots, n$, for then the typical equation is

$$f(i) = \tfrac{1}{2}(a_{i-2} + a_{i-1})$$

and the linear system is not strictly diagonally dominant. Further, in providing the additional condition required to make up the necessary $n + 2$ conditions for a quadratic spline, we may even create a system of linear equations which is singular. Instead we interpolate at each of the midpoints of the n sub-intervals between the knots and also interpolate at both endpoints $x = 0$ and n. Intuitively this seems more satisfactory, not only because we are observing symmetry between the endpoints but, more important, because by interpolating at the midpoints we are picking up the maximum values of the quadratic B-splines (see Example 6.6). Thus we express the quadratic spline in the form

$$S(x) = \sum_{r=-2}^{n-1} a_r B_r^2(x). \tag{6.39}$$

Table 6.1 Maximum errors of interpolating cubic splines for e^x with $n + 1$ equally spaced knots on $[0, 1]$

n	1	2	3	4	5
Maximum modulus of error	4.3×10^{-3}	4.0×10^{-4}	8.1×10^{-5}	2.6×10^{-5}	1.1×10^{-5}

Applying the above interpolatory conditions we find that, for $n>1$, a_{-1}, $a_0, \ldots, a_{n-2}$ satisfy the tridiagonal system $\mathbf{Ma}=\mathbf{b}$, where

$$\mathbf{M} = \begin{bmatrix} 5 & 1 & & & & \\ 1 & 6 & 1 & & & \\ & 1 & 6 & 1 & & \mathbf{0} \\ & & \ddots & \ddots & \ddots & \\ & \mathbf{0} & & 1 & 6 & 1 \\ & & & & 1 & 5 \end{bmatrix} \tag{6.40}$$

$$\mathbf{a}^{\mathrm{T}} = [a_{-1}, a_0, \ldots, a_{n-2}]$$

and

$$\mathbf{b}^{\mathrm{T}} = [8f(\tfrac{1}{2}) - 2f(0), 8f(\tfrac{3}{2}), \ldots, 8f(n-\tfrac{3}{2}), 8f(n-\tfrac{1}{2}) - 2f(n)]. \tag{6.41}$$

Then a_{-2} and a_{n-1} are computed from

$$a_{-2} = 2f(0) - a_{-1}, \qquad a_{n-1} = 2f(n) - a_{n-2}.$$

In checking the details, the reader needs to be aware (Example 6.6) that

$$B^2_{-2}(-\tfrac{3}{2}) = B^2_{-2}(\tfrac{1}{2}) = \tfrac{1}{8}, \quad B^2_{-2}(-1) = B^2_{-2}(0) = \tfrac{1}{2}, \quad B^2_{-2}(-\tfrac{1}{2}) = \tfrac{3}{4}. \tag{6.42}$$

6.4 Hermite interpolation

Suppose we are given the values of a function f and its first derivative f' at $n+1$ points

$$a = x_0 < x_1 < \cdots < x_n = b.$$

Since we have $2n+2$ conditions it seems possible that we can construct a polynomial $p_{2n+1} \in P_{2n+1}$ such that

$$p_{2n+1}(x_i) = f(x_i) \quad \text{and} \quad p'_{2n+1}(x_i) = f'(x_i), \qquad 0 \leqslant i \leqslant n. \tag{6.43}$$

One way of obtaining p_{2n+1} is to construct the polynomial which interpolates f on the $2n+2$ points $x_0, \ldots, x_n, x_0+h, \ldots, x_n+h$ and then let h tend to zero. We write the interpolating polynomial based on the above $2n+2$ points in the Lagrange form

$$\sum_{i=0}^{n} (\alpha_i(x;h)f(x_i) + \beta_i(x;h)f(x_i+h)), \tag{6.44}$$

where the Lagrange coefficients $\boldsymbol{\alpha}_i$ and β_i depend on h as well as on x. It is helpful to rearrange (6.44) in the form

$$\sum_{i=0}^{n} \left((\alpha_i(x;h) + \beta_i(x;h)) f(x_i) + h\beta_i(x;h) \left(\frac{f(x_i+h) - f(x_i)}{h} \right) \right). \tag{6.45}$$

On letting h tend to zero we obtain

$$p_{2n+1}(x) = \sum_{i=0}^{n} (u_i(x) f(x_i) + v_i(x) f'(x_i)), \tag{6.46}$$

where

$$u_i(x) = (1 - 2L_i'(x_i)(x - x_i))(L_i(x))^2, \tag{6.47}$$

$$v_i(x) = (x - x_i)(L_i(x))^2 \tag{6.48}$$

and

$$L_i(x) = \prod_{\substack{j=0 \\ j \neq i}}^{n} \left(\frac{x - x_j}{x_i - x_j} \right)$$

is the usual Lagrange coefficient. We call p_{2n+1} the *Hermite interpolating polynomial* for f on the given $n+1$ points x_i. Note that u_i, v_i and hence p_{2n+1} belong to P_{2n+1}. The derivation of (6.48) by writing

$$v_i(x) = \lim_{h \to 0} h\beta_i(x; h)$$

in (6.45) is straightforward, although that of (6.47) from

$$u_i(x) = \lim_{h \to 0} (\alpha_i(x; h) + \beta_i(x; h))$$

is a little more complicated. However, once one has obtained (6.47) and (6.48) it is easy to verify that (6.46) gives the required polynomial, by checking that

$$u_i(x_j) = \delta_{ij}, \qquad u_i'(x_j) = 0,$$
$$v_i(x_j) = 0, \qquad v_i'(x_j) = \delta_{ij},$$

for all i, j. Above, δ_{ij} is the Kronecker delta function, which has the value 1 for $i = j$ and zero otherwise.

Example 6.9 For $\sin \pi x$ with interpolating points at 0, $\frac{1}{2}$ and 1 we find that (6.46) simplifies to give

$$p_5(x) = (16 - 4\pi)x^2(1 - x)^2 + \pi x(1 - x).$$

Since both $\sin \pi x$ and $p_5(x)$ are symmetric about $x = \frac{1}{2}$, we tabulate these for values of x in $[0, \frac{1}{2}]$ only in Table 6.2. □

From (4.13) we know that the error of interpolation at the points $x_0, \ldots, x_n, x_0 + h, \ldots, x_n + h$ of a function f whose $(2n+2)$th derivative is continuous is

$$(x - x_0) \cdots (x - x_n)(x - x_0 - h) \cdots (x - x_n - h) \frac{f^{(2n+2)}(\xi_x)}{(2n+2)!}$$

Table. 6.2 Comparison of $\sin \pi x$ with the Hermite interpolating polynomial constructed at 0, $\frac{1}{2}$ and 1

x	0	0.1	0.2	0.3	0.4	0.5
$\sin \pi x$	0	0.3090	0.5878	0.8090	0.9511	1
$p_5(x)$	0	0.3106	0.5906	0.8112	0.9518	1

where ξ_x depends on h as well as on x, f and the x_j. Since p_{2n+1} was obtained by taking the limit of the above interpolating polynomial as h tends to zero, we deduce that for Hermite interpolation

$$f(x) - p_{2n+1}(x) = (x - x_0)^2 \cdots (x - x_n)^2 \frac{f^{(2n+2)}(\eta_x)}{(2n+2)!} \tag{6.49}$$

for some $\eta_x \in (a, b)$.

The form of the error (6.49) prompts us to ask whether we can expect more favourable results when using Hermite interpolation by making a particular choice of interpolating points x_j. Let us assume the interval $[a, b]$ is $[-1, 1]$, making a linear change of variable if necessary. Then the above question is easily answered: if we can select the interpolating points we should choose the zeros of the Chebyshev polynomial T_{n+1} since such x_j minimize $\| (x - x_0) \cdots (x - x_n) \|_\infty$. Then by (6.49) the maximum error on $[-1, 1]$ satisfies

$$\| f - p_{2n+1} \|_\infty \leqslant \frac{1}{4^n} \frac{\| f^{(2n+2)} \|_\infty}{(2n+2)!}, \tag{6.50}$$

valid only for Hermite interpolation at the Chebyshev zeros. In this case p_{2n+1} is given by (6.46) with

$$u_i(x) = \left(\frac{T_{n+1}(x)}{n+1} \right)^2 \frac{(1 - x_i x)}{(x - x_i)^2} \tag{6.51}$$

and

$$v_i(x) = \left(\frac{T_{n+1}(x)}{n+1} \right)^2 \left(\frac{1 - x_i^2}{x - x_i} \right), \tag{6.52}$$

where $x_i = \cos[(2n - 2i + 1)\pi/(2n+2)]$, $0 \leqslant i \leqslant n$, denote the zeros of T_{n+1} arranged in order on $[-1, 1]$ (see Problems 6.18 and 6.19).

Example 6.10 If we interpolate at $x = -1$, 0 and 1 and know that $| f^{(6)}(x) | \leqslant M$ on $[-1, 1]$ we find from (6.49) that

$$\| f - p_5 \|_\infty \leqslant \frac{M}{6!} \| x^2(x-1)^2(x+1)^2 \|_\infty = \frac{M}{4860},$$

the norm of $x(x-1)(x+1)$ being attained at $x=\pm 1/\sqrt{3}$. For comparison, if we interpolate at $x=0$ and $\pm\sqrt{3}/2$, the zeros of T_3, (6.50) shows that $\| f-p_5 \|_\infty \leqslant M/11520$. □

Instead of constructing *one* polynomial as above which interpolates a function and its derivative at certain points, we can construct separate cubic polynomials which interpolate the function and its derivative at the endpoints of each sub-interval. Thus we take

$$a=t_0<t_1<\cdots<t_N=b$$

and obtain a unique function H such that

$$H(t_i)=f(t_i), \qquad H'(t_i)=f'(t_i), \qquad 0\leqslant i\leqslant N,$$

and H is a cubic polynomial on each sub-interval. This is called *piecewise cubic Hermite* approximation. Note that H is not in general a cubic spline since H is only in $C^1[a,b]$ and not in $C^2[a,b]$.

Let the piecewise cubic $H(x)$ coincide with the cubic polynomial $H_i(x)$ on $[t_i, t_{i+1}]$. Although $H_i(x)$ is simply (6.46) with $n=1$ and x_0 and x_1 renamed as t_i and t_{i+1}, we will derive H_i by using the divided difference form of the interpolating polynomial (see (4.31)) with interpolating points t_i, t_i+h, t_{i+1} and $t_{i+1}+h$ and letting h tend to zero. We obtain, for $t_i\leqslant x\leqslant t_{i+1}$,

$$H_i(x)=a_0+a_1(x-t_i)+a_2(x-t_i)^2+a_3(x-t_i)^2(x-t_{i+1}), \tag{6.53}$$

where

$$\begin{aligned} &a_0=f(t_i), \qquad a_1=f'(t_i),\\ &a_2=(f[t_i, t_{i+1}]-f'(t_i))/(t_{i+1}-t_i),\\ &a_3=(f'(t_{i+1})-2f[t_i, t_{i+1}]+f'(t_i))/(t_{i+1}-t_i)^2. \end{aligned}$$

Note that the coefficients a_0, a_1, a_2 and a_3 need be evaluated only once for each sub-interval $[t_i, t_{i+1}]$ and then $H_i(x)$ can most efficiently be evaluated for values of x in $[t_i, t_{i+1}]$ by nested multiplication, that is by expressing

$$H_i(x)=((a_3(x-t_{i+1})+a_2)(x-t_i)+a_1)(x-t_i)+a_0.$$

Recalling that H_i denotes the function H restricted to $[t_i, t_{i+1}]$ we have

$$\|f-H\|_\infty=\max_i \max_{t_i\leqslant x\leqslant t_{i+1}} |f(x)-H_i(x)|$$

and by using (6.49) with $n=1$ and x_0, x_1 replaced by t_i, t_{i+1} we deduce that

$$\|f-H\|_\infty\leqslant \max_i \frac{(t_{i+1}-t_i)^4}{384} M_i, \tag{6.54}$$

Table 6.3 Comparison of the maximum error of piecewise cubic Hermite interpolation with the error bound for $\sin \pi x$ on $[0, 1]$

N	Maximum modulus of error	Error bound $(\pi/N)^4/384$
1	0.2146	0.2537
2	0.0108	0.0159
3	0.0031	0.0031
4	0.0010	0.0010
5	0.0004	0.0004

where

$$M_i = \max_{t_i \leqslant x \leqslant t_{i+1}} |f^{(4)}(x)|.$$

For the commonly used special case where the t_i are equally spaced on $[a, b]$, that is $t_i = a + (b-a)i/N$, then (6.54) becomes

$$\|f - H\|_\infty \leqslant \frac{1}{384}\left(\frac{b-a}{N}\right)^4 \|f^{(4)}\|_\infty. \tag{6.55}$$

Example 6.11 Let us examine the accuracy of piecewise cubic Hermite interpolation for $\sin \pi x$ on $[0, 1]$. In Table 6.3 we compare the error of maximum modulus with the error bound on the right of (6.55), for various values of N. □

6.5 Padé and rational approximation

Suppose that for $x \in (-R, R)$ a function f has a Taylor series about $x = 0$,

$$f(x) = c_0 + c_1 x + c_2 x^2 + \cdots, \tag{6.56}$$

where $c_j = f^{(j)}(0)/j!$. For given integers $m, n \geqslant 0$ let us denote polynomials p_m, q_n by

$$p_m(x) = a_0 + a_1 x + \cdots + a_m x^m$$

and

$$q_n(x) = b_0 + b_1 x + \cdots + b_n x^n,$$

where we will take $b_0 \neq 0$. We are familiar with using a polynomial (the Taylor polynomial) to match the values of a function f and its derivatives at a given point. Here we explore the use of a rational function p_m/q_n to match

f and its derivatives at the origin. Since multiplying p_m and q_n by the same constant does not alter the value of p_m/q_n we will 'normalize' p_m and q_n by choosing $b_0 = 1$.

We write

$$-y = b_1 x + \cdots + b_n x^n$$

so that $q_n(x) = 1 - y$ and

$$1/q_n(x) = 1 + y + y^2 + \cdots.$$

This expansion is valid for $|y| < 1$. Thus we can express p_m/q_n as an infinite series in powers of x provided $|y| < 1$. In what follows we restrict x so that

$$|x| < R \quad \text{and} \quad |b_1 x + \cdots + b_n x^n| < 1, \tag{6.57}$$

giving a sufficient condition for both f and p_m/q_n to have expansions in power series. Since, with $b_0 = 1$, p_m and q_n have $m + n + 1$ coefficients to be determined we can hope to construct a rational function p_m/q_n to match f and its first $m + n$ derivatives at $x = 0$. We therefore seek values of the coefficients a_j and b_j so that

$$f(x) - \frac{p_m(x)}{q_n(x)} = \sum_{j=m+n+1}^{\infty} d_j x^j. \tag{6.58}$$

We note from (6.57) that $q_n(x) \neq 0$ for the values of x we are considering. On multiplying (6.58) throughout by $q_n(x)$ we obtain

$$\left(\sum_{j=0}^{\infty} c_j x^j\right)\left(\sum_{j=0}^{n} b_j x^j\right) - \sum_{j=0}^{m} a_j x^j = \sum_{j=m+n+1}^{\infty} d'_j x^j. \tag{6.59}$$

(The relation between the d'_j and the d_j is obvious but is irrelevant since we are not interested in these coefficients.) Thus in (6.59) we seek values of the a_j and b_j so that the coefficients of x^j are zero for $j = 0, \ldots, m + n$. This gives the equations

$$\sum_{k=0}^{j} c_{j-k} b_k - a_j = 0, \qquad 0 \leqslant j \leqslant m + n, \tag{6.60}$$

where $b_0 = 1$, $b_j = 0$ for $j > n$ and $a_j = 0$ for $j > m$. We may write these equations in matrix form

$$\mathbf{A}\begin{bmatrix} a_0 \\ \vdots \\ a_m \\ b_1 \\ \vdots \\ b_n \end{bmatrix} = \begin{bmatrix} -c_0 \\ \vdots \\ \vdots \\ \vdots \\ \vdots \\ -c_{m+n} \end{bmatrix}. \tag{6.61}$$

The matrix is

$$\mathbf{A} = \begin{bmatrix} -1 & 0 & 0 & 0 & \cdots & 0 & 0 & 0 & 0 & \cdots & 0 \\ 0 & -1 & 0 & 0 & \cdots & 0 & c_0 & 0 & 0 & \cdots & 0 \\ 0 & 0 & -1 & 0 & \cdots & 0 & c_1 & c_0 & 0 & \cdots & 0 \\ 0 & 0 & 0 & -1 & \cdots & 0 & c_2 & c_1 & c_0 & \cdots & 0 \\ \cdot & & & \cdot & & \cdot & \cdot & & & & \cdot \\ 0 & 0 & 0 & 0 & \cdots & -1 & c_{m-1} & \cdot & \cdot & \cdot & c_{m-n} \\ 0 & 0 & \cdot & \cdot & \cdot & 0 & c_m & \cdot & \cdot & \cdot & c_{m-n+1} \\ \vdots & & & & & \vdots & \vdots & & & & \vdots \\ 0 & 0 & \cdot & \cdot & \cdot & 0 & c_{m+n-1} & \cdot & \cdot & \cdot & c_m \end{bmatrix} \tag{6.62}$$

where $c_j = 0$ for $j < 0$. (We have allowed the possibility of zero values of c_j in (6.62) to avoid having to write down two versions of $\mathbf{A}$, corresponding to $m \geqslant n$ and $m < n$. A similar point applies to the a_j and b_j in (6.60).)

On examining $\mathbf{A}$ further we see (Problem 6.24) that the equations (6.61) have a unique solution if and only if the matrix

$$\mathbf{C} = \begin{bmatrix} c_m & \cdots & c_{m-n+1} \\ \vdots & & \vdots \\ c_{m+n-1} & \cdots & c_m \end{bmatrix}$$

is non-singular, since $\det \mathbf{A} = (-1)^{m+1} \det \mathbf{C}$. Thus if $\mathbf{C}$ is non-singular we solve

$$\mathbf{C}\begin{bmatrix} b_1 \\ \vdots \\ b_n \end{bmatrix} = \begin{bmatrix} -c_{m+1} \\ \vdots \\ -c_{m+n} \end{bmatrix} \tag{6.63}$$

to determine the b_j uniquely. The a_j are then obtained immediately from (6.60) with $j = 0, \ldots, m$ only. The resulting rational function

$$R_{m,n}(x) = \frac{a_0 + a_1 x + \cdots + a_m x^m}{1 + b_1 x + \cdots + b_n x^n} \tag{6.64}$$

is called the (m, n) *Padé* approximation for f.

Example 6.12 Find Padé approximations for e^x. When $n = 0$ we obtain simply the Taylor polynomial and when $m = 0$, we obtain $1/(1 + b_1 x + \cdots + b_n x^n)$, where the denominator is the Taylor polynomial for e^{-x}. Neither of these possibilities gives anything new. We will choose $m = n$ as a 'mean' between these two extremes, although $m = n \pm 1$ is commonly favoured. For $m = n = 2$ the equations (6.63) are

$$\begin{pmatrix} \frac{1}{2} & 1 \\ \frac{1}{6} & \frac{1}{2} \end{pmatrix} \begin{pmatrix} b_1 \\ b_2 \end{pmatrix} = \begin{pmatrix} -\frac{1}{6} \\ -\frac{1}{24} \end{pmatrix}$$

with solution $b_1 = -\frac{1}{2}$, $b_2 = \frac{1}{12}$. From (6.60) we then obtain $a_0 = 1$, $a_1 = \frac{1}{2}$, $a_2 = \frac{1}{12}$ to give

$$R_{2,2}(x) = \frac{1 + \frac{1}{2}x + \frac{1}{12}x^2}{1 - \frac{1}{2}x + \frac{1}{12}x^2}.$$

We similarly obtain

$$R_{1,1}(x) = \frac{1 + \frac{1}{2}x}{1 - \frac{1}{2}x}$$

and

$$R_{3,3}(x) = \frac{1 + \frac{1}{2}x + \frac{1}{10}x^2 + \frac{1}{120}x^3}{1 - \frac{1}{2}x + \frac{1}{10}x^2 - \frac{1}{120}x^3}.$$

In the Padé approximation $R_{m,m}$ for e^x, due to the fact that e^x satisfies $f(-x) = 1/f(x)$ the coefficients satisfy $b_j = (-1)^j a_j$, as seen above (see also Problem 6.26). □

An obvious disadvantage of Padé approximations is that as we increase m and n the earlier coefficients may change, as for the coefficients of x^2 in Example 6.12. Another negative feature of Padé approximations is that an error estimate is not readily available. On the positive side, Padé approximations can sometimes give good approximations to a function even at points where the series (6.56) diverges (see Problem 6.31).

We might expect the errors of Padé approximations to grow with $|x|$. We illustrate this in Table 6.4 where we have chosen to display the *relative errors* $|e^x - R_{m,m}(x)|/e^x$. Note that $R_{1,1}$ is not defined at $x = 2$.

An efficient way of evaluating a rational function is to express it as a continued fraction. This is analogous to writing a rational number as a continued fraction, for example†

$$\frac{355}{113} = 3 + \cfrac{1}{7 + \cfrac{1}{16}}.$$

Let $p_0 \in P_{n_0}$, $p_1 \in P_{n_1}$, where $n_0 \geq n_1$. If p_1 divides exactly into p_0 to give a

† The fraction $\frac{355}{113}$ is the very accurate approximation of π obtained by the Chinese mathematician Zu Chongzhi (429–500).

Table 6.4 Relative errors $| e^x - R_{m,m}(x) | / e^x$

x	$m=1$	$m=2$	$m=3$
−2	1.000000	0.055580	0.001479
−1	0.093906	0.001472	0.000010
1	0.103638	0.001470	0.000010
2	–	0.052653	0.001481

polynomial α_0 we simply write

$$\frac{p_0(x)}{p_1(x)} = \alpha_0(x).$$

Otherwise we divide p_1 into p_0 to produce a quotient polynomial plus a remainder p_2 of degree lower than that of p_1. Thus

$$\frac{p_0(x)}{p_1(x)} = \alpha_0(x) + \frac{p_2(x)}{p_1(x)} = \alpha_0(x) + \frac{1}{p_1(x)/p_2(x)}, \tag{6.65}$$

where $\alpha_0 \in P_{n_0 - n_1}$ and $p_2 \in P_{n_2}$ with $n_2 < n_1$. We note that the quotient α_0 and the remainder p_2 are unique. Similarly we obtain

$$\frac{p_1(x)}{p_2(x)} = \alpha_1(x) + \frac{1}{p_2(x)/p_3(x)} \tag{6.66}$$

unless p_2 divides exactly into p_1, when we have only α_1 on the right of (6.66). Note that $\alpha_1 \in P_{n_1 - n_2}$. We continue this process, until finally we obtain

$$\frac{p_0(x)}{p_1(x)} = \alpha_0(x) + \cfrac{1}{\alpha_1(x) + \cfrac{1}{\cdots + \cfrac{1}{\alpha_k(x)}}}, \tag{6.67}$$

where $\alpha_k(x) = p_k(x)/p_{k+1}(x)$ and $p_{k+1}(x)$ divides exactly into $p_k(x)$. The expression on the right of (6.67) is called a *continued fraction*. Since it looks so cumbersome, we usually write it in the more concise form

$$\frac{p_0(x)}{p_1(x)} = \alpha_0(x) + \frac{1}{\alpha_1(x)+}\;\frac{1}{\alpha_2(x)+}\cdots\frac{1}{\alpha_k(x)}. \tag{6.68}$$

We can modify (6.67) by introducing real numbers β_j chosen so that α_j has leading coefficient 1 for $1 \leqslant j \leqslant k$. This saves up to one multiplication in the evaluation of each α_j, $1 \leq j \leq k$, and gives

$$\frac{p_0(x)}{p_1(x)} = \alpha_0(x) + \frac{\beta_1}{\alpha_1(x)+}\ \frac{\beta_2}{\alpha_2(x)+} \cdots \frac{\beta_k}{\alpha_k(x)}. \tag{6.69}$$

This process is called the *Euclidean algorithm*, which we state more formally below. The final polynomial p_{k+1} is the polynomial of largest degree which divides p_0 and p_1 exactly (see Problem 6.29).

Algorithm 6.1 (Euclidean) Given $p_0 \in P_{n_0}$, $p_1 \in P_{n_1}$, with $n_0 \geq n_1$, compute polynomials α_{j-1} of largest degree (and leading coefficient 1 for $j \geq 2$), polynomials p_{j+1} and real numbers β_j such that

$$p_{j-1}(x) = \alpha_{j-1}(x)p_j(x) + \beta_j p_{j+1}(x) \tag{6.70}$$

for $j = 1, 2, \ldots$ until $\beta_j = 0$. □

To implement the algorithm we represent each polynomial $p_j \in P_{n_j}$ by a vector $\mathbf{v}_j$ with $n_j + 1$ elements and by using (6.70) we compute $\mathbf{v}_{j+1}$, together with β_j and the vector corresponding to α_{j-1}, by carrying out suitable elementary operations on $\mathbf{v}_{j-1}$ and $\mathbf{v}_j$. We omit the details. Thus we obtain the continued fraction (6.69) for p_0/p_1. Since the degree of p_j is n_j, α_j is of degree $n_j - n_{j+1}$. If we evaluate the α_j by nested multiplication, we see that the evaluation of (6.69) requires at most $n_0 - n_{k+1} + k$ additions, $n_0 - n_{k-1} - k$ multiplications and exactly k divisions.

Example 6.13 We apply the Euclidean algorithm 6.1 to obtain a continued fraction of the form (6.69) for the Padé approximant $R_{3,3}$ obtained for e^x in Example 6.12. We obtain

$$R_{3,3}(x) = -1 + \frac{-24}{x-12+}\ \frac{50}{x+}\ \frac{10}{x}. \tag{6.71}$$

This continued fraction is not defined at $x = 0$, nor at $x \simeq 4.644$, the only real zero of the denominator of $R_{3,3}$. □

In (6.71) we note that $R_{3,3}(x) \rightarrow -1$ as $x \rightarrow \infty$ and this illustrates the point that the Euclidean algorithm 6.1 develops a continued fraction 'about infinity'. We can also develop a continued fraction about the origin by reversing the order of the coefficients in p_0 and p_1 before applying the algorithm to the resulting vectors of coefficients. This is equivalent to applying the algorithm to $t^{n_0}p_0(1/t)$ and $t^{n_1}p_1(1/t)$. Then we need to replace t by $1/x$ and eliminate negative powers of x. Applying this modified algorithm to $R_{3,3}$ above, we obtain

$$R_{3,3}(x) = 1 + \frac{x}{1-\frac{1}{2}x+}\ \frac{\frac{1}{12}x^2}{1+\frac{1}{60}x^2}. \tag{6.72}$$

In this representation the 'beginning' of $R_{3,3}$ is

$$1+\frac{x}{1-\frac{1}{2}x}=\frac{1+\frac{1}{2}x}{1-\frac{1}{2}x}=R_{2,2}(x),$$

as given in Example 6.12.

A further modification proves fruitful. Let us consider again the quotient $p_0(x)/p_1(x)$, but now assume that the polynomials p_0 and p_1 are of the same degree, say n, and $p_0(0)=p_1(0)=1$. Then if

$$q_0(t)=t^n p_0(1/t), \qquad q_1(t)=t^n p_1(1/t),$$

we see that q_0 and q_1 are both polynomials of degree n and have leading coefficient 1. Unless the polynomials q_0 and q_1 are identical, we now compute further polynomials $q_2, q_3, \ldots$ from

$$q_{j-1}(t)=t^{m_{j-1}}q_j(t)+\beta_j q_{j+1}(t) \tag{6.73}$$

for $j=1,2,\ldots$ until $\beta_j=0$, where the integer m_{j-1} is chosen so that $t^{m_{j-1}}q_j(t)$ has the same degree as $q_{j-1}(t)$ and β_j is a real number chosen so that the polynomial q_{j+1} has leading coefficient 1. Note that it follows from the definition of q_0 and q_1 that $m_0=0$ and the degree of q_2 is $n-1$ or less. Observe, also, the similarity (and the difference!) between (6.73) and the Euclidean algorithm. From (6.73) we can now construct the continued fraction which begins

$$\frac{q_0(t)}{q_1(t)}=1+\frac{\beta_1}{t^{m_1}+}\,\frac{\beta_2}{t^{m_2}+}\cdots. \tag{6.74}$$

For example, we have

$$\frac{t^2+2t+1}{t^2+t+2}=1+\frac{1}{t+}\,\frac{2}{1+}\,\frac{-2}{t+}\,\frac{1}{1}. \tag{6.75}$$

This corresponds to the case where $p_0(x)=1+2x+x^2$ and $p_1(x)=1+x+2x^2$. If we now write $t=1/x$ in (6.75) and eliminate negative powers of x we obtain

$$\frac{p_0(x)}{p_1(x)}=\frac{1+2x+x^2}{1+x+2x^2}=1+\frac{x}{1+}\,\frac{2x}{1+}\,\frac{-2x}{1+}\,\frac{x}{1}.$$

This illustrates the (commonly encountered) case where, in (6.74), the m_j are alternately 1 and 0, beginning with $m_1=1$ and $m_2=0$. In this case we see that (6.74) gives the continued fraction which begins

$$\frac{p_0(x)}{p_1(x)}=\frac{q_0(1/x)}{q_1(1/x)}=1+\frac{\beta_1 x}{1+}\,\frac{\beta_2 x}{1+}\cdots. \tag{6.76}$$

(We will not discuss further the more general case where the m_j are not alternately 1 and 0.) If we apply the process embodied in (6.76) to the Padé approximant $R_{m,m}$ for e^x for m large enough, we obtain $\beta_1 = 1$ and the next β_j are $-\frac{1}{2}, \frac{1}{6}, -\frac{1}{6}, \frac{1}{10}, -\frac{1}{10}, \frac{1}{14}, -\frac{1}{14}, \ldots$. For example, in this latter form, we have

$$R_{3,3}(x) = 1 + \frac{x}{1+}\,\frac{-\frac{1}{2}x}{1+}\,\frac{\frac{1}{6}x}{1+}\,\frac{-\frac{1}{6}x}{1+}\,\frac{\frac{1}{10}x}{1+}\,\frac{-\frac{1}{10}x}{1}. \tag{6.77}$$

The popularity of least squares approximations by *polynomials* is due to the fact that they can be computed by solving linear equations. This simplicity is lost if we replace polynomials by rational functions, which explains why rational least squares approximations are not commonly studied. However, there is a well-developed theory of minimax approximations by rational functions, which includes an equioscillation property similar to that which holds for minimax polynomial approximations.

Problems

Section 6.1

6.1 Let $f \in C^2[0, 1]$ and let S_n denote the first degree spline which interpolates f at the knots $t_i = i/n$. If $|f''(x)|$ is monotonic decreasing on $[0, 1]$, deduce from the error of interpolation (4.13) that

$$\max_{0 \leqslant x \leqslant 1} |f(x) - S_n(x)| = \max_{0 \leqslant x \leqslant 1/n} |f(x) - S_n(x)|.$$

6.2 Verify the result in Example 6.1 by applying the result of Problem 6.1.

6.3 Repeat the analysis of Example 6.2 for the function $1/x$ on $[1, 2]$, that is, find points

$$1 = t_0 < t_1 < \cdots < t_n = 2$$

and a first degree spline S_n so that

$$\max_{1 \leqslant x \leqslant 2} |1/x - S_n(x)|$$

is minimized. Verify that we need to choose $t_i = (1 - (1 - (1/\sqrt{2}))i/n)^{-2}$ and that the minimax error is $\frac{1}{2}(1 - 1/\sqrt{2})^2/n^2$.

6.4 Show that, on $[t_{i-1}, t_i]$, (6.4) is just the straight line joining the points (t_{i-1}, y_{i-1}) and (t_i, y_i), by first showing that only two terms in (6.4) are non-zero in this interval.

6.5 Observe that at the knot t_{i+1} every $B_j^1(x)$ is zero except for $j = i$ and deduce that the B_i^1 are linearly independent. Then (6.4) shows that the B_i^1 form a basis.

6.6 Verify that

$$(x-t_i)_+ = \sum_{r=i}^{\infty} (t_{r+1}-t_i)B_r^1(x).$$

(*Hint*: check that both sides are zero for $x \leqslant t_i$ and that they agree on the interval $[t_s, t_{s+1}]$ for every $s \geqslant i$.)

Section 6.2

6.7 Complete the details of the proof of Theorem 6.1. Note that at some stage it is recommended to split the term in B_{i+1}^{k-1} into two parts: one part is to be combined with the term in B_i^{k-1} and the other with the term in B_{i+2}^{k-1}.

6.8 Show that for any fixed value of x the right side of (6.15) has at most $k+1$ non-zero terms.

6.9 Verify that

$$B_i^1(x) = (t_{i+2}-t_i)f_x[t_i, t_{i+1}, t_{i+2}]$$

where the divided difference is of the function of t

$$f_x(t) = (x-t)_+.$$

(*Hint*: expand the divided difference in the symmetric form (4.27) and verify that both sides agree on each interval $[t_s, t_{s+1}]$.)

Section 6.3

6.10 Verify (6.19) and (6.20).

6.11 Verify directly that the expression for B_i^2 given in Example 6.6 satisfies the required smoothness conditions at the knots.

6.12 Show that $(t_i-x)^k = (t_i-x)_+^k + (-1)^k\,(x-t_i)_+^k$ by checking that the splines on each side of the equation are equal for $x \leqslant t_i$ and for $x > t_i$.

6.13 To emphasize the symmetry of the uniform B-splines, let us write

$$C^{2k-1}(x) = B_{-k}^{2k-1}(x) \quad \text{and} \quad C^{2k}(x) = B_{-k}^{2k}\,(x+\tfrac{1}{2}).$$

Show by induction that

$$C^k(x) = \frac{1}{2k}\left((k+1+2x)C^{k-1}(x+\tfrac{1}{2}) + (k+1-2x)C^{k-1}(x-\tfrac{1}{2})\right),$$

with $C^0(x) = 1$ for $-\tfrac{1}{2} \leqslant x < \tfrac{1}{2}$ and zero otherwise.

6.14 Use induction on the above relation connecting C^k and C^{k-1} to show directly that each C^k is an even function.

6.15 Derive the linear equations which must be solved to obtain the quadratic spline $S(x)$ such that $S(i)=f(i)$, $0 \leqslant i \leqslant n$, and $S'(0)=f'(0)$. Verify that this system of equations is non-singular but is not strictly diagonally dominant.

Section 6.4

6.16 Show that there is a unique $p \in P_{2n+1}$ such that p and p' take given values at $n+1$ distinct values of x.

6.17 Verify that, as defined by (6.47) and (6.48), u_i and v_i and their first derivatives are all zero at $x_0, \ldots, x_n$ except that $u_i(x_i) = v_i'(x_i) = 1$.

6.18 To obtain (6.52) from (6.48) when the x_i are the zeros of T_{n+1}, use (4.54) to show that

$$L_i(x) = \frac{T_{n+1}(x)}{(x-x_i)T'_{n+1}(x_i)} = (-1)^i \frac{T_{n+1}(x)\sqrt{1-x_i^2}}{(n+1)(x-x_i)}.$$

6.19 To obtain (6.51) from (6.47) for the Chebyshev case, first show that

$$L_i'(x_i) = \frac{1}{2}\frac{T''_{n+1}(x_i)}{T'_{n+1}(x_i)} = \frac{1}{2}\frac{x_i}{1-x_i^2}.$$

6.20 Show that $f[t_i, t_i+h] \rightarrow f'(t_i)$ as $h \rightarrow 0$. Similarly consider the limits of $f[t_i, t_i+h, t_{i+1}]$ and $f[t_i, t_i+h, t_{i+1}, t_{i+1}+h]$ as h tends to zero and so verify (6.53).

6.21 Show that the cubic Hermite approximation on the single interval $[0,1]$ can be expressed in the symmetric form

$$\begin{aligned} H(x) = {} & \tfrac{1}{2}(f(0)+f(1)) + (x-\tfrac{1}{2})f[0,1] + \tfrac{1}{2}x(x-1)(f'(1)-f'(0)) \\ & + x(x-\tfrac{1}{2})(x-1)(f'(1)-2f[0,1]+f'(0)). \end{aligned}$$

(Although non-symmetric, the nested form of (6.53) is preferable to the above for reasons of computational efficiency.)

6.22 For which functions would it be advantageous to use piecewise cubic Hermite approximation with *unequally* spaced interpolating points?

6.23 Write a computer program to implement piecewise cubic Hermite approximation with equally spaced interpolating points on $[a,b]$. The program should be provided with procedures to calculate f and f'.

Section 6.5

6.24 For the matrices **A** and **C** defined in and following (6.62), show by expanding det **A** by its first row, and so on, that det $\mathbf{A} = (-1)^{m+1}$ det **C**, showing that **A** is non-singular if and only if **C** is non-singular.

6.25 Obtain the Padé approximations $R_{2,1}$, $R_{1,2}$, $R_{3,2}$ and $R_{2,3}$ for e^x.

6.26 The Padé approximation $R_{m,m}$ for e^x satisfies

$$q_m(x)e^x - p_m(x) = d'_{m+n+1}x^{m+n+1} + \cdots.$$

Multiply both sides above by $e^{-x}/p_m(x)$ and then replace x by $-x$ to give

$$q_m(-x)/p_m(-x) - e^x = d''_{m+n+1}x^{m+n+1} + \cdots.$$

Deduce from the uniqueness of $R_{m,m}(x)$ that

$$p_m(-x) = q_m(x)$$

so that $b_j = (-1)^j a_j$.

6.27 For $-1 < x \leqslant 1$, $\tan^{-1}x = x - \frac{1}{3}x^3 + \frac{1}{5}x^5 - \cdots$. If $f(x) = (\tan^{-1}x^{1/2})/x^{1/2}$ obtain the $R_{1,1}$ and $R_{2,2}$ Padé approximations for $f(x)$ and thus the $R_{3,2}$ and $R_{5,4}$ approximations for $\tan^{-1}x$. Hence estimate π by using the relation $\tan^{-1}(1/\sqrt{3}) = \pi/6$.

6.28 In Example 6.12 we obtained the $R_{2,2}$ Padé approximation for e^x. Express $R_{2,2}$ as a continued fraction in the form (6.69).

6.29 Use the fact that in (6.70) any polynomial which exactly divides p_j and p_{j+1} must also exactly divide p_{j-1} to show that the final polynomial produced by the Euclidean algorithm is the polynomial of highest degree which divides both p_0 and p_1.

6.30 As an application of the Euclidean algorithm, devise a procedure to test whether a given polynomial has a multiple zero. (*Hint*: consider the derivative.)

6.31 Find the Padé approximations $R_{1,1}$ and $R_{2,2}$ for the function $(1/x)\log(1+x)$ and express each of these as a continued fraction of the form described in the material leading up to (6.76). (Note how (6.76) has to be modified in the case of $R_{2,2}$.) Investigate how well these rational functions approximate to $(1/x)\log(1+x)$, both for $-1 < x \leqslant 1$ and for $x > 1$ where the Taylor series diverges.

Chapter **7**

NUMERICAL INTEGRATION AND DIFFERENTIATION

7.1 Numerical integration

Given a function f defined on a finite interval $[a, b]$, we wish to evaluate the definite integral

$$\int_a^b f(x)\,\mathrm{d}x, \tag{7.1}$$

assuming that f is integrable (see §2.4). If $f(x) \geqslant 0$ on $[a, b]$ the integral (7.1) has the value of the area bounded by the curve $y = f(x)$, the x-axis and the ordinates $x = a$ and $x = b$. The fundamental theorem of the calculus (Theorem 2.11) shows that integration is the inverse process to differentiation. If we can find a function F such that $F' = f$, then we can evaluate the integral using the relation

$$\int_a^b f(x)\,\mathrm{d}x = F(b) - F(a).$$

Sometimes considerable skill is required to obtain F, perhaps by making a change of variable or integrating by parts. Even if F can be found, it may still be more convenient to use a numerical method to estimate (7.1) if the evaluation of F requires a great deal of computation. If we cannot find F or if f is known only for certain values of x, we use a numerical method for evaluating (7.1).

An obvious approach is to replace the integrand f in (7.1) by an approximating polynomial and integrate the polynomial.

Example 7.1 To evaluate

$$\int_0^1 \mathrm{e}^{x^2}\,\mathrm{d}x$$

we can replace the integrand by its Taylor series constructed at $x=0$,

$$e^{x^2}=1+x^2+\frac{x^4}{2!}+\frac{x^6}{3!}+\cdots=\sum_{n=0}^{\infty}\frac{(x^2)^n}{n!}.$$

Since this series converges uniformly to e^{x^2} on any finite interval, it may be integrated term-by-term (Theorem 2.16) to give

$$\int_0^1 e^{x^2}\,dx=\sum_{n=0}^{\infty}\frac{1}{(2n+1)n!}.$$

This series converges very rapidly. For example, to estimate the integral to six decimal places, we require only nine terms of the series. □

Despite its success in Example 7.1, the Taylor polynomial is not suitable in general since it requires the evaluation of derivatives of f. We will prefer to use an *interpolating* polynomial. First we use an interpolating polynomial constructed at equally spaced points $x_r=x_0+rh$, $0\leqslant r\leqslant n$. If $f^{(n+1)}$ is continuous on $[x_0, x_n]$, we have from (4.34) that

$$f(x_0+sh)=f_0+\binom{s}{1}\Delta f_0+\cdots+\binom{s}{n}\Delta^n f_0+h^{n+1}\binom{s}{n+1}f^{(n+1)}(\xi_s), \tag{7.2}$$

where $\xi_s\in(x_0, x_n)$. (We need to replace $[x_0, x_n]$ by an appropriate larger interval if s is outside the interval $0\leqslant s\leqslant n$.) We can construct different integration rules by choosing different values of n in (7.2). Thus, with $n=0$,

$$f(x_0+sh)=f_0+h\binom{s}{1}f'(\xi_s).$$

Integrating $f(x)$ over $[x_0, x_1]$, that is, integrating with respect to s over $[0,1]$ we obtain

$$\int_{x_0}^{x_1} f(x)\,dx=hf_0+h^2\int_0^1 sf'(\xi_s)\,ds. \tag{7.3}$$

We replaced dx by $h.ds$, as $x=x_0+sh$. In the integrand on the right of (7.3), s does not change sign on $[0,1]$ and $f'(\xi_s)$ is a continuous function of s. Thus by the mean value theorem for integrals, Theorem 2.13, there is a number $s=\bar{s}$ with $\xi_{\bar{s}}\in(x_0, x_1)$ such that

$$\int_{x_0}^{x_1} f(x)\,dx=hf_0+h^2f'(\xi_{\bar{s}})\int_0^1 s\,ds.$$

Writing ξ in place of $\xi_{\bar{s}}$, we have

$$\int_{x_0}^{x_1} f(x)\,dx=hf_0+\tfrac{1}{2}h^2f'(\xi). \tag{7.4}$$

The first term on the right of (7.4) gives the integration rule (also called a *quadrature rule*); the second term is the error term. This is called the

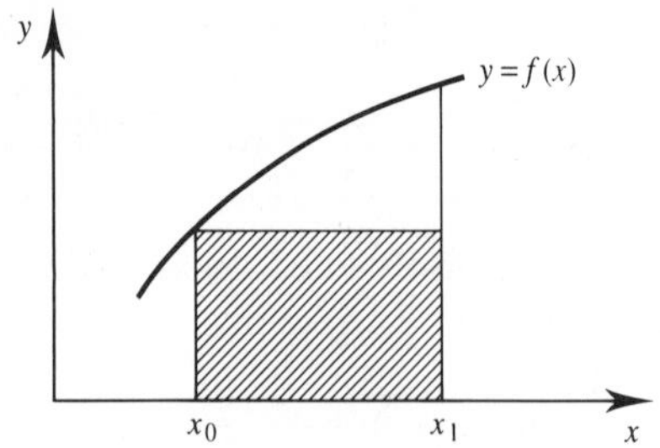

Fig. 7.1 The regular quadrature rule.

rectangular quadrature rule, since the integral (the area under the curve $y = f(x)$ in Fig. 7.1) is approximated by the rectangle of width $h = x_1 - x_0$ and height f_0.

Putting $n = 1$ and integrating $f(x)$ over $[x_0, x_1]$, we similarly obtain

$$\int_{x_0}^{x_1} f(x)\,\mathrm{d}x = hf_0 + \tfrac{1}{2}h\,\Delta f_0 + h^3 f''(\xi)\int_0^1 \binom{s}{2}\mathrm{d}s,$$

where f'' is assumed continuous on $[x_0, x_1]$. We have again been able to apply Theorem 2.13, since $\binom{s}{2}$ does not change sign on $[0, 1]$. This gives

$$\int_{x_0}^{x_1} f(x)\,\mathrm{d}x = \frac{h}{2}(f_0 + f_1) - \frac{h^3}{12}f''(\xi), \tag{7.5}$$

where $\xi \in (x_0, x_1)$ and is usually distinct from the ξ appearing in (7.4). This is the *trapezoidal rule* plus error term. The integral is approximated by $\frac{1}{2}h(f_0 + f_1)$, the area of the trapezium which appears shaded in Fig. 7.2.

The trapezoidal rule is usually applied in a *composite form*. To estimate the integral of f over $[a, b]$, we divide $[a, b]$ into N sub-intervals of equal length $h = (b - a)/N$. The endpoints of the sub-intervals are $x_i = a + ih$, $i = 0, 1, \ldots, N$, so that $x_0 = a$ and $x_N = b$. We now apply the trapezoidal rule to each sub-interval $[x_{i-1}, x_i]$, $i = 1, 2, \ldots, N$ (see Fig. 7.3). Thus we have from (7.5), on distinguishing the numbers ξ occurring in the error terms,

$$\int_{x_0}^{x_N} f(x)\,\mathrm{d}x = \frac{h}{2}(f_0 + f_1) - \frac{h^3}{12}f''(\xi_1) + \cdots + \frac{h}{2}(f_{N-1} + f_N) - \frac{h^3}{12}f''(\xi_N).$$

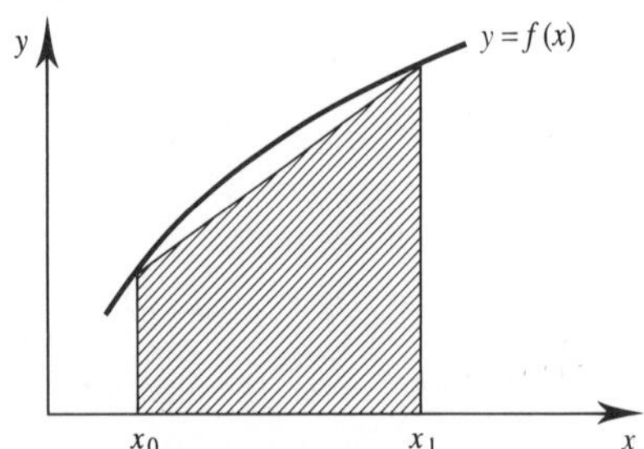

Fig. 7.2 The trapezoidal rule.

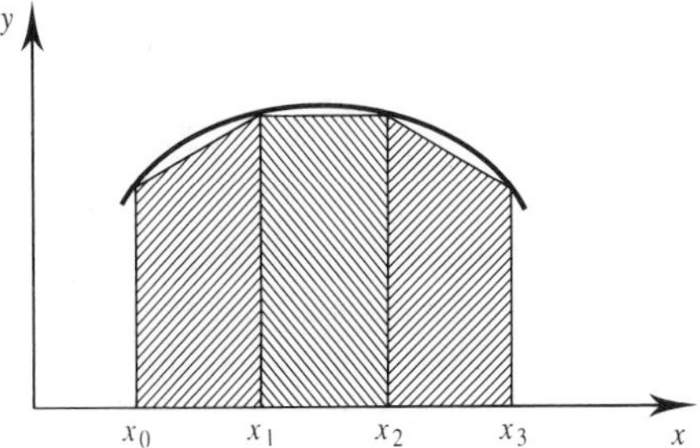

Fig. 7.3 Composite form of the trapezoidal rule.

Maintaining the assumption that f'' is continuous on $[a, b]$, we can combine the error terms (see Theorem 2.3) to give

$$\int_a^b f(x)\,\mathrm{d}x = \frac{h}{2}(f_0 + 2f_1 + 2f_2 + \cdots + 2f_{N-1} + f_N) - \frac{Nh^3}{12} f''(\xi), \quad (7.6)$$

for some $\xi \in (a, b)$. This may be rewritten as

$$\int_a^b f(x)\,\mathrm{d}x = h(f_0 + f_1 + \cdots + f_{N-1}) + \frac{h}{2}(f_N - f_0) - (x_N - x_0)\frac{h^2}{12} f''(\xi), \quad (7.7)$$

since $x_N - x_0 = Nh$. Incidentally, we note from (7.7) that the composite trapezoidal rule consists of the composite rectangular rule plus the *correction term* $h(f_N - f_0)/2$. We note that the error in (7.7) is of order h^2 and that the rule is exact for every polynomial belonging to P_1. (The second derivative in the error term is zero for such polynomials.)

Now suppose that, due to rounding, the f_i are in error by at most $\frac{1}{2} \times 10^{-k}$. Then we see from (7.6) that the error in the trapezoidal rule due to rounding is not greater than

$$\tfrac{1}{2}h(1 + 2 + 2 + \cdots + 2 + 1) \times \tfrac{1}{2} \times 10^{-k} = Nh \times \tfrac{1}{2} \times 10^{-k} = \tfrac{1}{2}(b - a)10^{-k}.$$

Thus, rounding errors do not seriously affect the accuracy of the quadrature rule. This is generally true of numerical integration (see Problem 7.8), unlike numerical differentiation, as we shall see later.

There is a generalization of the composite trapezoidal rule called *Gregory's formula:*

$$\int_{x_0}^{x_N} f(x)\,\mathrm{d}x \simeq h(f_0 + \cdots + f_N) + h\sum_{j=0}^{m} c_{j+1}(\nabla^j f_N + (-1)^j \Delta^j f_0), \quad (7.8)$$

where

$$c_j = (-1)^j \int_{-1}^{0} \binom{-s}{j} \, \mathrm{d}s. \tag{7.9}$$

If $m \leqslant N$, the rule uses values $f(x_i)$ with every $x_i \in [x_0, x_N]$. With $m=0$, since $c_1 = -\frac{1}{2}$, (7.8) is simply the trapezoidal rule. (The next few values of the c_j are $-\frac{1}{12}$, $-\frac{1}{24}$, $-\frac{19}{720}$. We see from (7.9) that $c_j < 0$ for all j.) We state without proof that if $m=2k$ or $m=2k+1$, then the rule (7.8) has error of order h^{m+2} and integrates exactly every polynomial belonging to P_{2k+1}. For example if $m=2$ and $N \geqslant 2$, we have the integration rule

$$\int_{x_0}^{x_N} f(x)\,\mathrm{d}x \simeq h(f_0 + \cdots + f_N) - \frac{5h}{8}(f_0 + f_N) + \frac{h}{6}(f_1 + f_{N-1}) - \frac{h}{24}(f_2 + f_{N-2}), \tag{7.10}$$

which is exact if $f \in P_3$. For the derivation of Gregory's formula, see Ralston and Rabinowitz (1978).

Returning to (7.2), we now choose $n=3$ and integrate $f(x)$ over $[x_0, x_2]$. This time the error term contains the binomial coefficient $\binom{s}{4}$ which *does* change sign over $[x_0, x_2]$. However, it can be shown that, in this case, the error term can still be written in the form

$$h^5 f^{(4)}(\xi) \int_0^2 \binom{s}{4} \mathrm{d}s = -\frac{h^5}{90} f^{(4)}(\xi),$$

as if Theorem 2.13 were applicable. We obtain

$$\int_{x_0}^{x_2} f(x)\,\mathrm{d}x = \frac{h}{3}(f_0 + 4f_1 + f_2) - \frac{h^5}{90} f^{(4)}(\xi), \tag{7.11}$$

where $\xi \in (x_0, x_2)$. This is called *Simpson's rule*. Like the trapezoidal rule, Simpson's rule is usually applied in composite form. Combining the error terms as we did for the trapezoidal rule, we obtain

$$\int_{x_0}^{x_{2N}} f(x)\,\mathrm{d}x = \frac{h}{3}(f_0 + 4f_1 + 2f_2 + 4f_3 + \cdots + 2f_{2N-2} + 4f_{2N-1} + f_{2N}) - (x_{2N} - x_0)\frac{h^4}{180} f^{(4)}(\xi), \tag{7.12}$$

where $\xi \in (x_0, x_{2N})$. Note that the composite form of Simpson's rule requires an even number of sub-intervals. For a rule of comparable accuracy which permits an odd number of sub-intervals, we have the Gregory formula (7.10). Note that both (7.10) and (7.12) are exact if the integrand $f \in P_3$.

To verify (7.11) we may let $x_1 = 0$ without loss of generality. We write p_2 to denote the interpolating polynomial for f constructed at $x = -h$, 0 and h (i.e. x_0, x_1 and x_2). Then from (4.40) we have

$$f(x) = p_2(x) + (x+h)x(x-h)f[-h, 0, h, x]$$

and integrate to give

$$\int_{-h}^{h} f(x)\,\mathrm{d}x = \frac{h}{3}(f(-h) + 4f(0) + f(h)) + E(f),$$

say, where the error term is

$$E(f) = \int_{-h}^{h} f[-h, 0, h, x]\, x(x^2 - h^2)\,\mathrm{d}x.$$

Integration by parts yields

$$E(f) = -\int_{-h}^{h} \frac{\mathrm{d}}{\mathrm{d}x} f[-h, 0, h, x]\, \tfrac{1}{4}(x^2 - h^2)^2\,\mathrm{d}x, \tag{7.13}$$

since $\frac{1}{4}(x^2 - h^2)^2$ is zero at $x = \pm h$. From Problem 4.21 and (4.41) we may write

$$\frac{\mathrm{d}}{\mathrm{d}x} f[-h, 0, h, x] = f[-h, 0, h, x, x] = f^{(4)}(\xi_x)/4!$$

where $\xi_x \in (-h, h)$, assuming that $-h \leqslant x \leqslant h$. Finally, since $(x^2 - h^2)^2 \geqslant 0$, we may apply Theorem 2.13 to (7.13) to obtain

$$E(f) = -\frac{h^5}{90} f^{(4)}(\xi),$$

thus justifying (7.11).

The basic trapezoidal and Simpson rules (7.5) and (7.11) are special cases of the form

$$\int_{x_0}^{x_n} f(x)\,\mathrm{d}x \simeq \sum_{i=0}^{n} w_i f(x_0 + ih) \tag{7.14}$$

where the w_i are chosen so that the rule is exact if $f \in P_n$. These are called the *closed Newton–Cotes formulas*. For $n = 3$ we have $w_0 = w_3 = 3h/8$ and $w_1 = w_2 = 9h/8$. For *even* values of n the rules turn out to be exact for $f \in P_{n+1}$ (as for Simpson's rule, with $n = 2$). In any integration rule, such as (7.14), the numbers w_i are called the *weights* since the rule consists of a weighted sum of certain function values. There are also *open Newton–Cotes formulas*, which are of the form

$$\int_{x_0}^{x_n} f(x)\,dx \simeq \sum_{i=1}^{n-1} w_i f(x_0 + ih).$$

The terms 'closed' and 'open' refer to whether the endpoints x_0 and x_n are or are not included in the rule.

To obtain the simplest open Newton–Cotes rule, we integrate (7.2), with $n = 1$, over $[x_0 - h, x_0 + h]$. This gives

$$\int_{x_0-h}^{x_0+h} f(x)\,dx = 2hf_0 + h^3 \int_{-1}^{1} \binom{s}{2} f''(\xi_s)\,ds.$$

As in the derivation of Simpson's rule, the error term here is troublesome, since $\binom{s}{2}$ changes sign on $[-1, 1]$ and Theorem 2.13 is inapplicable. However, again it is possible to write the error in the form

$$h^3 f''(\xi) \int_{-1}^{1} \binom{s}{2} ds = \tfrac{1}{3} h^3 f''(\xi),$$

as we will see shortly. Applying this rule to f on the interval $[x_0, x_1]$, we obtain

$$\int_{x_0}^{x_1} f(x)\,dx = hf(x_0 + \tfrac{1}{2}h) + \frac{h^3}{24} f''(\xi), \tag{7.15}$$

where $\xi \in (x_0, x_1)$. This is called the *midpoint rule*. At first sight this rule appears to be more economical than the trapezoidal rule on $[x_0, x_1]$, since both have accuracy of order h^3 but (7.15) requires only one evaluation of f instead of two evaluations for the trapezoidal rule. This is misleading, since the composite form of (7.15) is

$$\int_{x_0}^{x_N} f(x)\,dx = h \sum_{r=0}^{N-1} f(x_r + \tfrac{1}{2}h) + (x_N - x_0)\frac{h^2}{24} f''(\xi), \tag{7.16}$$

where $\xi \in (x_0, x_N)$. This requires N evaluations of f, which is only one fewer than the $N + 1$ evaluations required by the composite trapezoidal rule (7.7). Note the opposite signs taken by the error terms in (7.7) and (7.16). If f'' has constant sign on $[x_0, x_N]$ the value of the integral must therefore lie *between* the estimates provided by the midpoint and trapezoidal rules.

Let $T(h)$ and $M(h)$ denote the approximations to $\int_a^b f(x)\,dx$ obtained by using respectively the trapezoidal and midpoint rules with sub-intervals of width h. If f'' is constant, we can eliminate the error terms between (7.7) and (7.16) and obtain the quadrature rule

$$\int_a^b f(x)\,\mathrm{d}x \simeq (2M(h)+T(h))/3. \tag{7.17}$$

From (7.7) and (7.16), we see that this rule involves values of $f(x)$ at the $2N+1$ equally spaced points $x = x_0 + \frac{1}{2}rh$, $r = 0, 1, \ldots, 2N$. Indeed,

$$(2M(h)+T(h))/3 = S(\tfrac{1}{2}h) \tag{7.18}$$

where $S(\frac{1}{2}h)$ is the result of applying Simpson's rule with sub-intervals of width $\frac{1}{2}h$. Thus the elimination of the h^2 error term in this way gives Simpson's rule.

Example 7.2 In Table 7.1 are listed the results of approximating to $\int_0^1 (1+x)^{-1}\,\mathrm{d}x$ by the trapezoidal, midpoint and Simpson rules, for different numbers (N) of sub-intervals. As the second derivative of $1/(1+x)$ is positive on $[0, 1]$, the numbers in columns 2 and 3 of the table provide bounds for the integral, whose value is $\log 2 = 0.6931$, to four decimal places. □

Truncation errors for integration rules may also be estimated using Taylor series. For example, if $F'(x) = f(x)$ and we write $u = x_0 + \frac{1}{2}h$, we have

$$\begin{aligned}\int_{x_0}^{x_1} f(x)\,\mathrm{d}x &= F(u+\tfrac{1}{2}h) - F(u-\tfrac{1}{2}h)\\ &= F(u) + \tfrac{1}{2}hF'(u) + \tfrac{1}{8}h^2F''(u) + \tfrac{1}{48}h^3F^{(3)}(\xi_1)\\ &\quad - [F(u) - \tfrac{1}{2}hF'(u) + \tfrac{1}{8}h^2F''(u) - \tfrac{1}{48}h^3F^{(3)}(\xi_0)],\end{aligned} \tag{7.19}$$

where ξ_0 and $\xi_1 \in (x_0, x_1)$. Using Theorem 2.3 to combine the two error terms, we see that (7.19) yields the midpoint rule plus error, as given by (7.15). Alternatively, we can find the error term by a method similar to that used for Simpson's rule (see Problem 7.9).

Each quadrature rule studied so far consists of a weighted sum of values $f(x_i)$, where the points x_i are *within* the range of integration. It is not always

Table 7.1 Approximations to $\int_0^1 (1+x)^{-1}\,\mathrm{d}x$

N	Trapezoidal rule	Midpoint rule	Simpson's rule
1	0.7500	0.6667	–
2	0.7083	0.6857	0.6944
3	0.7000	0.6898	–
4	0.6970	0.6912	0.6933

possible to use such rules. For example, suppose we wish to integrate a function f over $[x_n, x_{n+1}]$, where $x_n = x_0 + nh$, and the only known values of f are $f(x_i)$, $0 \leqslant i \leqslant n$. This type of problem arises in the numerical solution of ordinary differential equations (see Chapter 13) and it is convenient to use the backward difference form of the interpolating polynomial, constructed at $x = x_n, x_{n-1}, \ldots, x_{n-m}$, where $m \leqslant n$. From (4.13) and (4.36), we have for any $s > 0$

$$f(x_n + sh) = \sum_{j=0}^{m} (-1)^j \binom{-s}{j} \nabla^j f_n + (-1)^{m+1} h^{m+1} \binom{-s}{m+1} f^{(m+1)}(\xi_s). \quad (7.20)$$

In (7.20), $\xi_s \in (x_{n-m}, x_n + sh)$ and $f^{(m+1)}$ is assumed continuous on this interval. We now integrate (7.20) over the interval $x_n \leqslant x \leqslant x_{n+1}$. Because of the change of variable $x = x_n + sh$, we must replace $\mathrm{d}x$ by $h\mathrm{d}s$ to give

$$\int_{x_n}^{x_{n+1}} f(x)\,\mathrm{d}x = h \sum_{j=0}^{m} (-1)^j \nabla^j f_n \int_0^1 \binom{-s}{j} \mathrm{d}s$$
$$+ (-1)^{m+1} h^{m+2} \int_0^1 \binom{-s}{m+1} f^{(m+1)}(\xi_s)\,\mathrm{d}s. \quad (7.21)$$

Again, Theorem 2.13 is applicable and we obtain

$$\int_{x_n}^{x_{n+1}} f(x)\,\mathrm{d}x = h \sum_{j=0}^{m} b_j \nabla^j f_n + h^{m+2} b_{m+1} f^{(m+1)}(\xi_n), \quad (7.22)$$

where $\xi_n \in (x_{n-m}, x_{n+1})$ and

$$b_j = (-1)^j \int_0^1 \binom{-s}{j} \mathrm{d}s.$$

It is sometimes appropriate to make use of $f(x_{n+1})$ also and construct the interpolating polynomial at $x = x_{n+1}, x_n, \ldots, x_{n-m}$. We have

$$f(x_{n+1} + sh) = \sum_{j=0}^{m+1} (-1)^j \binom{-s}{j} \nabla^j f_{n+1} + (-1)^{m+2} h^{m+2} \binom{-s}{m+2} f^{(m+2)}(\xi_s'), \quad (7.23)$$

which is just (7.20) with n and m increased by 1. In this case, $x_n \leqslant x \leqslant x_{n+1}$ corresponds to $-1 \leqslant s \leqslant 0$. We apply Theorem 2.12 and obtain

$$\int_{x_n}^{x_{n+1}} f(x)\,\mathrm{d}x = h \sum_{j=0}^{m+1} c_j \nabla^j f_{n+1} + h^{m+3} c_{m+2} f^{(m+2)}(\xi_n'), \quad (7.24)$$

where $\xi_n' \in (x_{n-m}, x_{n+1})$ and the c_j (see (7.9)) are the same numbers as occur in Gregory's formula.

7.2 Romberg integration

We have already seen in (7.18) how Simpson's rule can be expressed as a linear combination of the midpoint and trapezoidal rules. By comparing $T(h)$ and $T(\frac{1}{2}h)$ we can similarly discover the relation

$$S(\tfrac{1}{2}h) = (4T(\tfrac{1}{2}h) - T(h))/3. \tag{7.25}$$

We can develop this further by using the Euler–Maclaurin formula: if all the derivatives of f exist,

$$\int_a^b f(x)\,\mathrm{d}x - T(h) = h^2E_2 + h^4E_4 + h^6E_6 + \cdots, \tag{7.26}$$

where

$$E_{2r} = -\frac{B_{2r}}{(2r)!}\,(f^{(2r-1)}(b) - f^{(2r-1)}(a))$$

and the coefficients B_{2r} are the *Bernoulli numbers*, defined by

$$\frac{x}{\mathrm{e}^x - 1} = \sum_{r=0}^{\infty} B_r x^r/r!\,.$$

See Ralston and Rabinowitz (1978) for the derivation of (7.26). (Note the similarity between (7.26) and Gregory's formula (7.8). The first correction term in the latter yields the trapezoidal rule and the remaining terms consist of finite differences in place of the *derivatives* in (7.26).)

We note that in (7.26) h must be of the form $(b-a)/N$, where N is a positive integer. If we replace h by $h/2$ in (7.26), thus doubling the number of sub-intervals, we obtain

$$\int_a^b f(x)\,\mathrm{d}x - T(h/2) = \left(\frac{h}{2}\right)^2 E_2 + \left(\frac{h}{2}\right)^4 E_4 + \left(\frac{h}{2}\right)^6 E_6 + \cdots. \tag{7.27}$$

We now *eliminate* the term in h^2 between (7.26) and (7.27), that is, we multiply (7.27) by 4, subtract (7.26) and divide by 3 to give

$$\int_a^b f(x)\,\mathrm{d}x - T^{(1)}(h) = h^4E_4^{(1)} + h^6E_6^{(1)} + \cdots, \tag{7.28}$$

where

$$T^{(1)}(h) = \frac{4T(h/2) - T(h)}{3} \tag{7.29}$$

and $E_4^{(1)}$, $E_6^{(1)}, \ldots$ are multiples of E_4, E_6, $\ldots$. The important point is that $T^{(1)}(h)$ gives, at least for sufficiently small values of h, an approximation to the integral which is more accurate than either $T(h)$ or $T(h/2)$. (Of course, we have already seen that $T^{(1)}(h) = S(\frac{1}{2}h)$.) Also note that we do not need to know the values of the coefficients E_{2r} in (7.26) in order to obtain (7.29).

This process of eliminating the leading power of h in the error term is an example of *extrapolation to the limit*, also known as *Richardson extrapolation*. We can take this process a stage further, eliminating the term in h^4 between $T^{(1)}(h)$ and $T^{(1)}(h/2)$ to give

$$\int_a^b f(x)\,dx - T^{(2)}(h) = h^6 E_6^{(2)} + h^8 E_8^{(2)} + \cdots,$$

where

$$T^{(2)}(h) = \frac{16T^{(1)}(h/2) - T^{(1)}(h)}{15}. \tag{7.30}$$

The numbers $E_6^{(2)}$, $E_8^{(2)}, \ldots$ are multiples of $E_6^{(1)}, E_8^{(1)}, \ldots$ and are thus multiples of E_6, E_8, ... More generally, we find that

$$\int_a^b f(x)\,dx - T^{(k)}(h) = h^{2(k+1)} E_{2(k+1)}^{(k)} + \cdots, \tag{7.31}$$

where

$$T^{(k)}(h) = \frac{4^k T^{(k-1)}(h/2) - T^{(k-1)}(h)}{4^k - 1}. \tag{7.32}$$

Note that, in order to compute $T^{(k)}(h)$, we first need to compute the trapezoidal approximations $T(h)$, $T(h/2)$, $T(h/4)$, ..., $T(h/2^k)$. This particular application of repeated extrapolation is known as *Romberg integration*, named after W. Romberg who discussed the method in a paper published in 1955.

Table 7.2 The Romberg table

$T(h)$	$T^{(1)}(h)$	$T^{(2)}(h)$	$T^{(3)}(h)$
$T(h/2)$	$T^{(1)}(h/2)$	$T^{(2)}(h/2)$	
$T(h/4)$	$T^{(1)}(h/4)$		
$T(h/8)$			

It may be helpful to set out the extrapolated values as in Table 7.2, where we illustrate the case $k=3$. In the Romberg table let $R(i,0)$ denote $T(h/2^i)$ and, for $j=1,2,\ldots$, let $R(i,j)$ denote $T^{(j)}(h/2^i)$. The $R(i,j)$ may be computed by the following algorithm.

Algorithm 7.1 This approximates to the integral of f over $[a, b]$ and begins with $h = (b - a)/N$ and $h_k = h/2^k$, for some positive integers N and k, and the values of f at $x = a + ih_k$ for $0 \leqslant i \leqslant N.2^k$.

> **for** $i := 0$ **to** k
> $R(i, 0) := T(h/2^i)$
> **next** i
>
> **for** $j := 1$ **to** k
> $\rho := 1/(1 - 4^{-j})$
>   **for** $i := 0$ **to** $k - j$
> $R(i, j) := \rho\, R(i + 1, j - 1) + (1 - \rho)\, R(i, j - 1)$
>   **next** i
> **next** j □

Example 7.3 Table 7.3 gives the results of applying Algorithm 7.1 to the integral

$$\int_0^1 \frac{dx}{1 + x}$$

with $N = 1$ and $k = 3$. For comparison, the result to six decimal places is 0.693147. Simpson's rule with eight sub-intervals gives 0.693155, as in the second column of the table, which is consistent with (7.25). Thus Romberg integration has estimated the integral with an error not greater than 1.5×10^{-6}, using only nine function evaluations. From the error term of (7.7), we can estimate how many evaluations are required to achieve such accuracy, using the trapezoidal rule. Since in this case $\frac{1}{4} \leqslant f''(x) \leqslant 2$, for $0 \leqslant x \leqslant 1$, we see that over 140 evaluations are needed. □

In the above example, the final number in the Romberg table (0.693148) is about 1000 times as accurate as the most accurate trapezoidal approximation in the first column of the table. However, it is important to remember that the justification of Romberg's method lies in the existence of the error series (7.26), which depends on the existence of the appropriate derivatives of the integrand f. If one of the derivatives of f does not exist at some point in the range of integration, the results of applying the formula can be misleading, as the next example illustrates.

Table 7.3 Results of applying Algorithm 7.1 with $N = 1$ and $k = 3$ to $\int_0^1 (1 + x)^{-1}\, dx$ (Example 7.3)

0.750000	0.694444	0.693175	0.693148
0.708333	0.693254	0.693148	
0.697024	0.693155		
0.694122			

Table 7.4 Erroneous Romberg table for $\int_0^{0.8} x^{1/2}\,\mathrm{d}x$

0.35777				
	0.45657			
0.43187		0.47066		
	0.46978		0.47484	
0.46030		0.47477		0.47626
	0.47446		0.47625	
0.47092		0.47623		
	0.47612			
0.47482				

Example 7.4 We try repeated extrapolation to the limit for the integral

$$\int_0^{0.8} x^{1/2}\,\mathrm{d}x.$$

The integrand is not differentiable at $x=0$. Table 7.4 shows results taking $h=0.8/2^n$, for $0 \leqslant n \leqslant 4$. The two boxed entries are quite inaccurate, as the result should be 0.47703 to five decimal places. □

In looking at a table of extrapolated results, we should look for signs of convergence as we progress down each column. Looking at just the last entries of the columns is misleading, as we found in Example 7.4.

7.3 Gaussian integration

We now consider quadrature formulas of the form

$$\int_a^b f(x)\,\mathrm{d}x \simeq \sum_{i=0}^{n} w_i f(x_i), \tag{7.33}$$

where the points are not necessarily equally spaced, as hitherto in this chapter. We again replace f by p_n, the interpolating polynomial constructed at $x=x_0, \ldots, x_n$. We write p_n in the Lagrange form (4.6),

$$p_n(x) = L_0(x)\,f(x_0) + \cdots + L_n(x)\,f(x_n)$$

and integrate p_n over $[a, b]$ to give the right side of (7.33). We thus obtain the weights,

$$w_i = \int_a^b L_i(x)\,\mathrm{d}x, \qquad 0 \leqslant i \leqslant n. \tag{7.34}$$

An explicit expression for $L_i(x)$ is given in (4.10). If we take the x_i as equally spaced on $[a, b]$, we have an alternative derivation of the Newton–Cotes rules discussed in § 7.1. We now study the error of the quadrature rule (7.33) for arbitrarily spaced $x_i \in [a, b]$. Integrating the error of the interpolating polynomial (4.13), we obtain

$$\int_a^b f(x)\,dx - \sum_{i=0}^{n} w_i f(x_i) = \frac{1}{(n+1)!}\int_a^b \pi_{n+1}(x) f^{(n+1)}(\xi_x)\,dx, \qquad (7.35)$$

where $f^{(n+1)}$ is assumed to exist on $[a, b]$ and

$$\pi_{n+1}(x) = (x - x_0)\cdots(x - x_n). \qquad (7.36)$$

Thus the integration rule is exact if $f \in P_n$, since then $f^{(n+1)}(x) \equiv 0$ and the right side of (7.35) is zero.

We can regard (7.33) in another way. First we note that if the rule (7.33) is exact for functions f and g, it is also exact for any function $\alpha f + \beta g$, where α and β are arbitrary real numbers. For

$$\begin{aligned}\int_a^b (\alpha f(x) + \beta g(x))\,dx &= \alpha\int_a^b f(x)\,dx + \beta\int_a^b g(x)\,dx\\ &= \sum_{i=0}^{n} w_i(\alpha f(x_i) + \beta g(x_i)).\end{aligned}$$

Hence, the rule (7.33) is exact for $f \in P_n$ if, and only if, it is exact for the monomials $f(x) = 1, x, \dots, x^n$. If (7.33) is to be exact for $f(x) = x^j$, we require

$$\int_a^b x^j\,dx = \sum_{i=0}^{n} w_i x_i^j. \qquad (7.37)$$

The left side of (7.37) is known. This suggests that, by taking $j = 0, 1, \dots, 2n+1$, we set up $2n+2$ equations to solve for the $2n+2$ unknowns w_i and x_i, $i = 0, 1, \dots, n$. If these equations have a solution, the resulting integration rule will be exact for $f \in P_{2n+1}$. We will not pursue the direct solution of these equations, although this is not as difficult as might at first appear (see Davis and Rabinowitz, 1984).

Instead, we return to (7.35). It is convenient to use the divided difference form of the error $f - p_n$, as in (4.40),

$$f(x) - p_n(x) = \pi_{n+1}(x) f[x, x_0, \dots, x_n].$$

Thus we have

$$\int_a^b f(x) - \sum_{i=0}^{n} w_i f(x_i) = \int_a^b \pi_{n+1}(x) f[x, x_0, \dots, x_n]\,dx. \qquad (7.38)$$

We now prove the following result.

Lemma 7.1 If $f[x, x_0, \dots, x_k]$ is a polynomial (in x) of degree $m>0$, then $f[x, x_0, \dots, x_{k+1}]$ is a polynomial of degree $m-1$.

Proof From (4.30),

$$f[x, x_0, \dots, x_{k+1}] = \frac{f[x_0, \dots, x_{k+1}] - f[x, x_0, \dots, x_k]}{x_{k+1} - x}. \tag{7.39}$$

Now $x - x_{k+1}$ is a factor of the numerator on the right of (7.39), since

$$f[x_0, \dots, x_{k+1}] - f[x_{k+1}, x_0, \dots, x_k] = 0.$$

This follows from the fact that the order of the arguments in a divided difference is irrelevant, which is apparent from the symmetric form of divided differences, (4.26). Since the numerator on the right of (7.39) is a polynomial of degree m and $x - x_{k+1}$ is a factor, it follows that $f[x, x_0, \dots, x_{k+1}]$ is a polynomial of degree $m-1$. □

Returning to (7.38), if $f \in P_{2n+1}$, an induction argument using Lemma 7.1 shows that $f[x, x_0, \dots, x_n] \in P_n$. Suppose that $\{p_0, p_1, \dots, p_{n+1}\}$ is a set of orthogonal polynomials on $[a, b]$, that is,

$$\int_a^b p_r(x) p_s(x)\, \mathrm{d}x \quad \begin{cases} = 0, & r \neq s \\ \neq 0, & r = s \end{cases}$$

and $p_r \in P_r$, $r = 0, 1, \dots, n+1$. For some set of real numbers $\alpha_0, \alpha_1, \dots, \alpha_n$.

$$f[x, x_0, \dots, x_n] = \alpha_0 p_0(x) + \alpha_1 p_1(x) + \cdots + \alpha_n p_n(x)$$

(see Problem 5.23) and, from the orthogonality relation, the right side of (7.38) will be zero if $\pi_{n+1}(x) = \alpha p_{n+1}(x)$ for some real number $\alpha \neq 0$.

On $[-1, 1]$, these orthogonal polynomials p_n are the Legendre polynomials (see Example 5.8). Thus, if the x_i are chosen as the zeros of the Legendre polynomial of degree $n+1$ and the weights w_i are determined from (7.34) with $[-1, 1]$ for $[a, b]$, the integration rule

$$\int_{-1}^{1} f(x)\, \mathrm{d}x \simeq \sum_{i=0}^{n} w_i f(x_i) \tag{7.40}$$

is exact for $f \in P_{2n+1}$. These are called *Gaussian integration rules*, after K. F. Gauss.

The first few Legendre polynomials (see Example 5.8) are 1, x, $(3x^2-1)/2$, $(5x^3-3x)/2$. With $n=0$ in (7.40) we have a rule with one point, $x=0$. This gives

$$\int_{-1}^{1} f(x)\, \mathrm{d}x \simeq 2f(0) \tag{7.41}$$

which is exact for $f \in P_1$. This is the midpoint rule, which we derived earlier as (7.15). With $n=1$, the points are the zeros of $(3x^2-1)/2$, which are $x = \pm 1/\sqrt{3}$. The weights w_0 and w_1 are easily found from the requirement that (7.40), with $n=1$, must integrate the monomials 1 and x exactly. Thus we obtain the Gaussian rule

$$\int_{-1}^{1} f(x)\,\mathrm{d}x \simeq f(-1/\sqrt{3}) + f(1/\sqrt{3}), \tag{7.42}$$

which is exact for $f \in P_3$ and so is comparable with Simpson's rule. With $n=2$, the points are the zeros of $(5x^3-3x)/2$, which are $x=0, \pm\sqrt{\frac{3}{5}}$. Since the rule (7.40) with $n=2$ has to integrate 1, x and x^2 exactly, we have:

$$\begin{aligned} 2 &= w_0 + w_1 + w_2 \\ 0 &= -\sqrt{\tfrac{3}{5}}\,w_0 + \sqrt{\tfrac{3}{5}}\,w_2 \\ \tfrac{2}{3} &= \tfrac{3}{5}w_0 + \tfrac{3}{5}w_2. \end{aligned}$$

The integration rule is therefore

$$\int_{-1}^{1} f(x)\,\mathrm{d}x \simeq \left(5f\left(-\sqrt{\tfrac{3}{5}}\right) + 8f(0) + 5f\left(\sqrt{\tfrac{3}{5}}\right)\right)/9, \tag{7.43}$$

which is exact for $f \in P_5$.

Gaussian rules are also used in composite forms, similar to the trapezoidal, midpoint and Simpson rules. It can be shown that if $f^{(2n+2)}$ is continuous on $[-1,1]$, the error of the $(n+1)$-point Gaussian rule (7.40) is of the form $d_{n+1}f^{(2n+2)}(\xi)$, where $\xi \in (-1,1)$ and

$$d_n = \frac{2^{2n+1}(n!)^4}{(2n+1)[(2n)!]^3}, \tag{7.44}$$

which decreases rapidly as n increases. To apply a Gaussian rule to the integral

$$\int_{a}^{a+h} g(t)\,\mathrm{d}t, \tag{7.45}$$

we make the change of variable $t = a + \frac{1}{2}h(x+1)$, which maps $a \leqslant t \leqslant a+h$ onto $-1 \leqslant x \leqslant 1$. We have

$$g(t) = g(a + \tfrac{1}{2}h(x+1)) = f(x),$$

say, and on differentiating k times, we have

$$\frac{\mathrm{d}^k}{\mathrm{d}x^k} f(x) = \left(\frac{h}{2}\right)^k \frac{\mathrm{d}^k}{\mathrm{d}t^k} g(t).$$

Thus, the error in using the Gaussian $(n+1)$-point rule to estimate (7.45) is

$$\left(\frac{h}{2}\right)^{2n+2} d_{n+1} g^{(2n+2)}(\eta), \tag{7.46}$$

where $\eta \in (a, a+h)$. The high power of $h/2$ in (7.46) makes the error very small for small values of h.

Example 7.5 To illustrate the accuracy of Gaussian rules, we apply the 3-point rule to

$$\int_0^1 \frac{\mathrm{d}x}{1+x}$$

to give the result 0.693122. The same rule applied in composite form on $[0, \frac{1}{2}]$ and $[\frac{1}{2}, 1]$, that is with six function evaluations, gives 0.693146. The last error is not greater than 1.5×10^{-6}, since the correct result is $\log 2 =$ 0.693147 to six decimal places. The accuracy is thus comparable with that achieved by Romberg integration, with nine function evaluations, in Example 7.3. □

There also exist quadrature formulas of the form

$$\int_a^b \omega(x) f(x)\,\mathrm{d}x \simeq \sum_{i=0}^{n} w_i f(x_i), \tag{7.47}$$

where ω is some fixed weight function. Weights w_i and points x_i can be found so that (7.47) is exact for all $f \in P_{2n+1}$. A similar argument to that used for the derivation of the classical Gaussian rules shows that the x_i in (7.47) must be chosen as the zeros of the polynomial of degree $n+1$ belonging to the sequence of polynomials which are orthogonal on $[a, b]$ with respect to the weight function ω. The resulting formulas (7.47) are all referred to as *Gaussian rules*. In particular, if $[a, b]$ is taken as $[-1, 1]$ and $\omega(x) = (1-x^2)^{-1/2}$, the orthogonal polynomials are the Chebyshev polynomials. (The resulting formulas are also called the *Chebyshev quadrature rules*.) In this case, the interpolating polynomial is

$$p_n(x) = \sum_{i=0}^{n} L_i(x) f(x_i),$$

where the x_i are the zeros of T_{n+1} and (see Problem 7.14) we may write

$$L_i(x) = \frac{T_{n+1}(x)}{(x-x_i)T'_{n+1}(x_i)}.$$

Thus from (7.47) the weights are given by

$$w_i = \int_{-1}^{1} \frac{(1-x^2)^{-1/2} T_{n+1}(x)}{(x-x_i)T'_{n+1}(x_i)}\,\mathrm{d}x. \tag{7.48}$$

We now use the formula

$$\tfrac{1}{2}(T_{n+1}(x)T_n(y) - T_{n+1}(y)T_n(x)) = (x-y)\sum_{r=0}^{n}{}' \, T_r(x)T_r(y), \qquad (7.49)$$

which is derived in Problem 7.15. (This is a special case of the Christoffel–Darboux formula, which holds for any set of orthogonal polynomials. See Davis (1976). Putting $y = x_i$ in (7.49) we have $T_{n+1}(y) = 0$ and we divide both sides of (7.49) by $\frac{1}{2}(x - x_i)T_n(x_i)T'_{n+1}(x_i)$ to give for (7.48)

$$w_i = \frac{2}{T_n(x_i)T'_{n+1}(x_i)} \sum_{r=0}^{n}{}' \, T_r(x_i) \int_{-1}^{1} (1-x^2)^{-1/2} T_r(x)\,\mathrm{d}x.$$

From the orthogonality of $T_0 = 1$ and T_r, $r \neq 0$, only the first term of the summation is non-zero, so that

$$w_i = \frac{\pi}{T_n(x_i)T'_{n+1}(x_i)}. \qquad (7.50)$$

Putting $x = \cos\theta$, since $T_{n+1}(x) = \cos(n+1)\theta$ we obtain

$$T_n(x)T'_{n+1}(x) = (n+1)\sin(n+1)\theta\cos n\theta/\sin\theta. \qquad (7.51)$$

We now write $x_i = \cos\theta_i$ and use the identity

$$\cos n\theta = \cos((n+1)\theta - \theta) = \cos(n+1)\theta\cos\theta + \sin(n+1)\theta\sin\theta.$$

Since $\cos(n+1)\theta_i = 0$, we have

$$\cos n\theta_i = \sin(n+1)\theta_i \sin\theta_i$$

and, from (7.51),

$$\begin{aligned} T_n(x_i)T'_{n+1}(x_i) &= (n+1)\sin^2(n+1)\theta_i \\ &= (n+1)(1-\cos^2(n+1)\theta_i) = n+1. \end{aligned}$$

Thus from (7.50) we have

$$w_i = \pi/(n+1)$$

and therefore, for the Chebyshev rules, all the weights are equal.

Note that the Gaussian rules with weight functions may be used to integrate $\int_a^b g(x)\,\mathrm{d}x$ by writing

$$g(x) = \omega(x)[g(x)/\omega(x)]$$

and applying the rule (7.47) to the function $f(x) = g(x)/\omega(x)$.

An obvious disadvantage of Gaussian rules is that their weights and points are a little awkward to handle: we need to compute them with appropriate accuracy and store them in the computer. Romberg integration, whose weights and points are very simple, is used more often than Gaussian rules.

7.4 Indefinite integrals

Suppose we wish to evaluate the *indefinite integral*

$$F(x) = \int_a^x f(t)\,\mathrm{d}t \tag{7.52}$$

for $a < x \leq b$, rather than the definite integral (7.1). We can choose an appropriate step size h and approximate to the right side of (7.52) for $x = a + h,\ a + 2h, \ldots$ in turn. Each integral can be estimated by some quadrature rule, for example the trapezoidal rule. Intermediate values of $F(x)$ can be estimated by using interpolation.

Another method is to replace f by some approximating polynomial p and approximate to F by integrating p. If $|f(x) - p(x)| < \varepsilon$ for $a \leq x \leq b$, then

$$\left| F(x) - \int_a^x p(t)\,\mathrm{d}t \right| < \int_a^x \varepsilon\,\mathrm{d}x \leq (b-a)\varepsilon$$

for $a \leq x \leq b$. Taking $[a, b]$ as $[-1, 1]$, a suitable choice of p (see § 5.7) is the modified Chebyshev series

$$\sum_{r=0}^{n}{}' \alpha_r T_r(x), \tag{7.53}$$

where

$$\alpha_r = \frac{2}{N} \sum_{j=0}^{N}{}'' f(x_j) T_r(x_j) \tag{7.54}$$

and $x_j = \cos(\pi j/N)$, with $N > n$. To integrate (7.53) we need to integrate $T_r(x)$, whose indefinite integral is clearly a polynomial of degree $r + 1$ and is thus expressible as a sum of multiples of $T_0, T_1, \ldots, T_{r+1}$. In fact, we find that the indefinite integral is quite simple for, putting $t = \cos\theta$, we have

$$\begin{aligned}
\int T_r(t)\,\mathrm{d}t &= -\int \cos r\theta \sin\theta\,\mathrm{d}\theta \\
&= \frac{1}{2}\int [\sin(r-1)\theta - \sin(r+1)\theta]\,\mathrm{d}\theta \\
&= \frac{1}{2}\left(\frac{\cos(r+1)\theta}{r+1} - \frac{\cos(r-1)\theta}{r-1}\right) + C
\end{aligned}$$

if $r \neq 1$, C being an arbitrary constant. Choosing C in agreement with the condition that $T_r(-1)=(-1)^r$, we deduce that

$$\int_{-1}^{x} T_r(t)\,\mathrm{d}t = \frac{1}{2}\left(\frac{T_{r+1}(x)}{r+1} - \frac{T_{r-1}(x)}{r-1}\right) + \frac{(-1)^{r+1}}{r^2-1}, \tag{7.55}$$

where $r>1$. We also find that

$$\int_{-1}^{x} T_1(t)\,\mathrm{d}t = \tfrac{1}{4}T_2(x) - \tfrac{1}{4},$$

$$\int_{-1}^{x} T_0(t)\,\mathrm{d}t = T_1(x) + 1.$$

Integrating (7.53) we have

$$\int_{-1}^{x}\left(\sum_{r=0}^{n}{}' \alpha_r T_r(t)\right)\mathrm{d}t = \sum_{j=0}^{n+1}\beta_j T_j(x),$$

where, from (7.55) and the relations following it, we find that

$$\beta_j = \frac{1}{2j}(\alpha_{j-1} - \alpha_{j+1}), \qquad 1 \leqslant j \leqslant n-1, \tag{7.56}$$

and

$$\beta_0 = \tfrac{1}{2}\alpha_0 - \tfrac{1}{4}\alpha_1 + \sum_{r=2}^{n}(-1)^{r+1}\alpha_r/(r^2-1).$$

If we define $\alpha_{n+1}=\alpha_{n+2}=0$, (7.56) holds for $j=n$ and $n+1$.

7.5 Improper integrals

Hitherto we have restricted our attention to so-called proper integrals, whose range of integration is finite and whose integrands are bounded. We now consider the following two problems.

(i) Estimate $\int_a^b f(x)\,\mathrm{d}x$, where f has a singularity at $x=b$, but is defined on any interval $[a, b-\varepsilon]$, where $0<\varepsilon<b-a$. We suppose that

$$\lim_{\varepsilon\to 0}\int_a^{b-\varepsilon} f(x)\,\mathrm{d}x$$

exists, write $\int_a^b f(x)\,\mathrm{d}x$ to denote this limit and say that the integral converges. An example of this type of integral is $\int_0^1 (1-x)^{-1/2}\,\mathrm{d}x$.

(ii) Estimate $\int_a^\infty f(x)\,\mathrm{d}x$, where f is defined on any finite interval $[a,b]$ and

$$\lim_{b\to\infty}\int_a^b f(x)\,\mathrm{d}x$$

exists. We write $\int_a^\infty f(x)\,dx$ to denote the value of this limit and say that the integral converges. As an example, we have $\int_1^\infty x^{-2}\,dx$ which, like our previous example, can be evaluated without recourse to numerical methods.

A technique which is sometimes useful in both cases (i) and (ii) (and also for proper integrals) is to make a suitable change of variable. Thus the substitution $t = (1-x)^{1/2}$ changes the improper integral (case (i) above) $\int_0^1 (1-x)^{-1/2} \sin x\,dx$ into the proper integral $2\int_0^1 \sin(1-t^2)\,dt$, which may be estimated by using a standard quadrature rule.

Another useful technique is that of 'subtracting out' the singularity. Consider the improper integral

$$I = \int_0^1 x^{-p} \cos x\,dx, \qquad 0 < p < 1,$$

with a singularity at $x = 0$. We can approximate to $\cos x$ by the first term of its Taylor series about $x = 0$, to give

$$I = \int_0^1 x^{-p}\,dx + \int_0^1 x^{-p}(\cos x - 1)\,dx.$$

The first integral has the value $(1-p)^{-1}$. A numerical method may be used for the second integral, whose integrand has no singularity, since $\cos x - 1$ behaves like $-\frac{1}{2}x^2$ at $x = 0$. However, $x^{-p}(\cos x - 1)$ behaves like $-\frac{1}{2}x^{2-p}$, whose second and higher derivatives are singular at $x = 0$. We therefore could not guarantee that numerical integration would give very accurate results and could subtract out a little more. For example, we have

$$I = \int_0^1 x^{-p} p_4(x)\,dx + \int_0^1 x^{-p}(\cos x - p_4(x))\,dx, \tag{7.57}$$

where

$$p_4(x) = 1 - \frac{x^2}{2!} + \frac{x^4}{4!}.$$

At $x = 0$, the second integrand in (7.57) is zero and its first five derivatives exist on $[0, 1]$. We could therefore safely use Simpson's rule (whose error term involves the fourth derivative) to estimate the second integral. The first integral in (7.57) may be evaluated explicitly.

Infinite integrals can be estimated by truncating the interval at a suitable point. We replace $\int_a^\infty f(x)\,dx$ by $\int_a^b f(x)\,dx$, where b is chosen large enough to give a sufficiently good approximation. We then use a quadrature rule to estimate the latter integral. For example,

$$\left| \int_b^\infty \frac{\sin^2 x}{1 + e^x}\,dx \right| < \int_b^\infty e^{-x}\,dx = e^{-b}$$

and $e^{-b} < \frac{1}{2} \times 10^{-6}$ for $b \geq 15$. Thus we can estimate $\int_0^\infty \sin^2 x/(1+e^x)\,dx$ correct to six decimal places by calculating a sufficiently accurate approximation to the integral over the range $[0, 15]$.

It is sometimes convenient to use Gaussian rules to estimate improper integrals. Thus to evaluate $\int_a^b \omega(x)f(x)\,dx$, where f is 'well-behaved' on $[a, b]$, but $\omega(x) \geq 0$ has a singularity, we could use a Gaussian rule based on the polynomials which are orthogonal with respect to ω on $[a, b]$. The obvious disadvantage is the need to compute the required points and weights. For infinite integrals we have Gaussian rules of the form

$$\int_0^\infty \omega(x)f(x)\,dx \simeq \sum_{i=0}^{n} w_i f(x_i) \tag{7.58}$$

and

$$\int_{-\infty}^\infty \omega(x)f(x)\,dx \simeq \sum_{i=0}^{n} w_i f(x_i) \tag{7.59}$$

which are exact for $f \in P_{2n+1}$. The choice of $\omega(x) = \exp(-x)$ in (7.58) gives the *Gauss-Laguerre rules*. With $\omega(x) = \exp(-x^2)$ in (7.59), we obtain the *Gauss–Hermite rules*. The names of Laguerre and Hermite are given to the respective sets of orthogonal polynomials associated with these formulas. For further information on these two formulas, see Davis and Rabinowitz (1984).

7.6 Multiple integrals

It is obvious that we cannot easily give a comprehensive treatment of numerical integration over domains of more than one dimension. For in one dimension we have essentially only one domain of integration, the finite interval $[a, b]$, with the possibility that we may need to take a limit as $a \rightarrow -\infty$ or as $b \rightarrow \infty$. Yet in two or more dimensions the number of possible domains is infinite. Here we shall discuss only two-dimensional integrals over a rectangular region. By means of a linear change of variable, the rectangle may be transformed into a square. Now, using any one-dimensional quadrature rule, we can write

$$\begin{aligned}
\int_a^b \left(\int_a^b f(x, y)\,dx \right) dy &\simeq \int_a^b \left(\sum_{i=0}^{N} w_i f(x_i, y) \right) dy \\
&= \sum_{i=0}^{N} w_i \int_a^b f(x_i, y)\,dy \\
&\simeq \sum_{i=0}^{N} w_i \left(\sum_{j=0}^{N} w_j f(x_i, y_j) \right),
\end{aligned}$$

where, in fact, $x_i = y_i$. The numbers x_i and w_i are the points and weights of the one-dimensional quadrature rule. Thus we have the two-dimensional rule

$$\int_a^b \int_a^b f(x, y)\, dx\, dy \simeq \sum_{i,j=0}^{N} w_i w_j f(x_i, y_j). \tag{7.60}$$

This is called a *product integration* rule. It generalizes obviously to higher dimensional multiple integrals over rectangular regions. For a product composite Simpson rule in two dimensions with, say, four sub-intervals in each direction, the pattern of relative weights 1, 4, 2, 4, 1 is 'multiplied' to give the array in Table 7.5. The sum of the numbers in the table is $(1+4+2+4+1)^2 = 144$. Since the rule has to integrate the function $f(x, y) \equiv 1$ exactly over the region with area $(b-a)^2$, each number in the table must be multiplied by $(b-a)^2/144$ to give the appropriate weight $w_i w_j$ for (7.60).

It is easy to derive error estimates for product rules, given an error estimate for the one-dimensional rule on which the product formula is based. We will obtain an error bound for (7.60) based on Simpson's rule, with step-size h. Let R denote the square $a \leqslant x \leqslant b$, $a \leqslant y \leqslant b$ and write

$$\int_a^b f(x, y)\, dx = \sum_{i=0}^{N} w_i f(x_i, y) - (b-a)\frac{h^4}{180} f_{4x}(\xi_y, y), \tag{7.61}$$

using (7.12). We have written f_{4x} for $\partial^4 f/\partial x^4$, assumed continuous on R. Then

$$\int_a^b \left(\int_a^b f(x, y)\, dx \right) dy = \sum_{i=0}^{N} w_i \int_a^b f(x_i, y)\, dy + E, \tag{7.62}$$

where

$$E = -(b-a)\frac{h^4}{180} \int_a^b f_{4x}(\xi_y, y)\, dy.$$

Table 7.5 Relative weights of a product Simpson rule

1	4	2	4	1
4	16	8	16	4
2	8	4	8	2
4	16	8	16	4
1	4	2	4	1

Replacing $\int_a^b f(x_i, y)\,dy$, for each i, by Simpson's rule with error term, we obtain the total error in the product rule,

$$-(b-a)\,\frac{h^4}{180}\sum_{i=0}^{N} w_i f_{4y}(x_i, \eta_i) + E, \tag{7.63}$$

assuming that $\partial^4 f/\partial y^4$ is continuous on R. If the moduli of f_{4x} and f_{4y} are bounded by M_x and M_y respectively, we deduce from (7.63) that the modulus of the error of the product rule is not greater than

$$(b-a)^2\,\frac{h^4}{180}\,(M_x + M_y). \tag{7.64}$$

7.7 Numerical differentiation

We will consider methods for approximating to f', given the values of $f(x)$ at certain points. If p is some polynomial approximation to f, we might use p' as an approximation to f'. However, we need to be careful: the maximum modulus of $f'(x) - p'(x)$ on a given interval $[a, b]$ can be much larger than the maximum modulus of $f(x) - p(x)$, as the following example shows.

Example 7.6 Suppose that $f(x) - p(x) = 10^{-2}T_n(x)$. Putting $x = \cos\theta$ and recalling that $T_n(x) = \cos n\theta$ we obtain

$$T_n'(x) = n\,\frac{\sin n\theta}{\sin\theta} = n^2\left(\frac{\sin n\theta}{n\theta}\right)\left(\frac{\theta}{\sin\theta}\right).$$

Since (see § 2.2)

$$\lim_{\theta\to 0}\frac{\sin\theta}{\theta} = 1,$$

we find that $T_n'(1) = n^2$. It can be shown that this is the maximum modulus of T_n' on $[-1, 1]$. If $n = 10$, say, the maximum modulus of $f - p$ is 10^{-2} on $[-1, 1]$, whereas that of $f' - p'$ is unity. □

As an approximation to f, we now take the polynomial p_n which interpolates f at distinct points $x_0, \ldots, x_n$. The error formula (4.13) is

$$f(x) - p_n(x) = \pi_{n+1}(x)\,\frac{f^{(n+1)}(\xi_x)}{(n+1)!}, \tag{7.65}$$

where

$$\pi_{n+1}(x) = (x - x_0)\cdots(x - x_n). \tag{7.66}$$

This is valid (see Theorem 4.2) if $f^{(n+1)}$ exists on some interval $[a, b]$ which contains $x, x_0, \ldots, x_n$. The number ξ_x (depending on x) belongs to (a, b). Differentiating (7.65) we obtain

$$f'(x) - p_n'(x) = \pi_{n+1}'(x)\,\frac{f^{(n+1)}(\xi_x)}{(n+1)!} + \frac{\pi_{n+1}(x)}{(n+1)!}\,\frac{\mathrm{d}}{\mathrm{d}x}\,f^{(n+1)}(\xi_x). \quad (7.67)$$

In general, we can say nothing further about the second term on the right of (7.67). We cannot perform the differentiation *with respect to* x of $f^{(n+1)}(\xi_x)$, since ξ_x is an unknown function of x. Thus, for general values of x, (7.67) is useless for determining the accuracy with which p_n' approximates to f'. However, if we restrict x to one of the values $x_0, \ldots, x_n$, then $\pi_{n+1}(x) = 0$ and the unknown second term on the right of (7.67) becomes zero. Putting $x = x_r$, say, we have

$$f'(x_r) - p_n'(x_r) = \pi_{n+1}'(x_r)\,\frac{f^{(n+1)}(\xi_r)}{(n+1)!}, \quad (7.68)$$

where we have written ξ_r for the value of ξ_x when $x = x_r$. For equally spaced x_r, we write p_n in the forward difference form (4.33):

$$p_n(x) = p_n(x_0 + sh) = f_0 + \binom{s}{1}\Delta f_0 + \binom{s}{2}\Delta^2 f_0 + \cdots + \binom{s}{n}\Delta^n f_0. \quad (7.69)$$

From $x = x_0 + sh$, we have $\mathrm{d}x/\mathrm{d}s = h$ and therefore

$$p_n'(x) = \frac{\mathrm{d}s}{\mathrm{d}x}\,\frac{\mathrm{d}}{\mathrm{d}s}\,p_n(x_0 + sh) = \frac{1}{h}\,\frac{\mathrm{d}}{\mathrm{d}s}\,p_n(x_0 + sh).$$

Thus

$$p_n'(x) = \frac{1}{h}\left[\Delta f_0 + \tfrac{1}{2}(2s-1)\,\Delta^2 f_0 + \cdots + \frac{\mathrm{d}}{\mathrm{d}s}\binom{s}{n}.\Delta^n f_0\right]. \quad (7.70)$$

To calculate $\pi_{n+1}'(x_r)$, we write

$$\pi_{n+1}(x) = (x - x_r)\prod_{j \neq r}(x - x_j) \quad (7.71)$$

and obtain π_{n+1}' by differentiating the product on the right of (7.71). We have

$$\pi_{n+1}'(x) = \left(\frac{\mathrm{d}}{\mathrm{d}x}(x - x_r)\right)\prod_{j \neq r}(x - x_j) + (x - x_r)\,\frac{\mathrm{d}}{\mathrm{d}x}\prod_{j \neq r}(x - x_j).$$

On putting $x = x_r$, the second term becomes zero and we obtain

$$\pi_{n+1}'(x_r) = \prod_{j \neq r}(x_r - x_j) = (-1)^{n-r} h^n r!(n-r)!, \quad (7.72)$$

since $x_r - x_j = (r-j)h$. We therefore now write (7.68) as

$$f'(x_r) - p_n'(x_r) = (-1)^{n-r} h^n \frac{r!(n-r)!}{(n+1)!} f^{(n+1)}(\xi_r). \tag{7.73}$$

We now give some differentiation rules obtained from (7.70), with error estimates provided by (7.73), for different values of n and r.

Putting $n = 1$ in (7.70), we obtain

$$p_1'(x) = \frac{1}{h} \Delta f_0 = \frac{f_1 - f_0}{h}$$

and with $r = 0$ in (7.73) we have

$$f'(x_0) = \frac{f_1 - f_0}{h} - \tfrac{1}{2} h f''(\xi_0), \tag{7.74}$$

where $\xi_0 \in (x_0, x_1)$. If we use $(f_1 - f_0)/h$ as an approximation to $f'(x_0)$, (7.74) provides an estimate for the error.

We now choose $n = 2$ in (7.70) and choose $r = 1$, since we are then obtaining the derivative at x_1, which is the centre of the interpolating points x_0, x_1 and x_2. If $x = x_1$ then $s = 1$ and (7.70) becomes

$$p_2'(x_1) = \frac{1}{h} [\Delta f_0 + \tfrac{1}{2} \Delta^2 f_0] = \frac{f_2 - f_0}{2h}. \tag{7.75}$$

This is an approximation to $f'(x_1)$ with an error (from (7.73) with $n = 2$, $r = 1$)

$$f'(x_1) - \frac{f_2 - f_0}{2h} = -\tfrac{1}{6} h^2 f^{(3)}(\xi_1). \tag{7.76}$$

Both differentiation rules obtained above,

$$f'(x) \simeq \frac{f(x+h) - f(x)}{h} \tag{7.77}$$

and

$$f'(x) \simeq \frac{f(x+h) - f(x-h)}{2h} \tag{7.78}$$

are in common use. The former is the more obvious rule, since $f'(x)$ is defined as the limit of the right side of (7.77), as $h \to 0$. However, the midpoint rule (7.78), which uses symmetrically placed interpolating points, is usually preferred because the error (7.76) is of order h^2. For some applications, it is convenient to use the approximation

$$f'(x) \simeq \frac{f(x)-f(x-h)}{h} \tag{7.79}$$

(see Problem 7.27). Note that all three differentiation rules have a simple geometrical interpretation: the right sides of (7.77), (7.78) and (7.79) are the gradients of the chords BC, AC and AB in Fig. 7.4.

Higher derivatives of f are approximated by higher derivatives of the interpolating polynomial p_n. If we differentiate (7.65) k times ($k>1$), we again face the insoluble problem of finding derivatives of $f^{(n+1)}(\xi_x)$ with respect to x. If $k>1$ we cannot even estimate $f^{(k)}(x)-p_n^{(k)}(x)$ at tabulated points $x=x_r$. At the end of this section we describe another way of estimating this error. Meanwhile, we write

$$p_n''(x) = \frac{1}{h^2}\frac{\mathrm{d}^2}{\mathrm{d}s^2}p_n(x_0+sh)$$

and obtain from (7.70)

$$p_n''(x) = \frac{1}{h^2}\left[\Delta^2 f_0 + \frac{\mathrm{d}^2}{\mathrm{d}s^2}\binom{s}{3}\Delta^3 f_0 + \cdots + \frac{\mathrm{d}^2}{\mathrm{d}s^2}\binom{s}{n}\Delta^n f_0\right], \tag{7.80}$$

where $x=x_0+sh$. To obtain a non-zero value for $p_n''(x)$, we must take $n \geqslant 2$. Putting $n=2$ in (7.80), we find that

$$p_2''(x) = \frac{1}{h^2}\Delta^2 f_0,$$

which is a constant. We take this as an approximation to $f''(x)$ at $x=x_1$, the central interpolating point, since we find that at this point the error is of

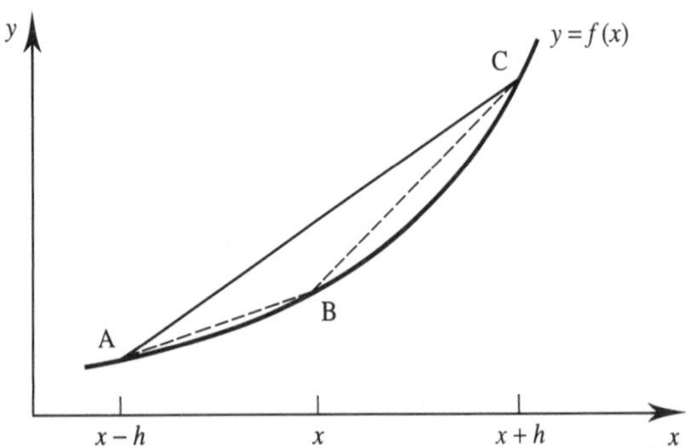

Fig. 7.4 Compare the gradients of the chords BC, AC and AB with $f'(x)$, the gradient of the tangent to $y=f(x)$ at B.

largest order in h. Thus

$$f''(x_1) \simeq \frac{f_2 - 2f_1 + f_0}{h^2}. \tag{7.81}$$

This is the usual approximation to the second derivative. More generally, for any $k \leqslant n$, we obtain

$$p_n^{(k)}(x) = \frac{1}{h^k}\left[\frac{\mathrm{d}^k}{\mathrm{d}s^k}\binom{s}{k}\Delta^k f_0 + \cdots + \frac{\mathrm{d}^k}{\mathrm{d}s^k}\binom{s}{n}\Delta^n f_0\right]. \tag{7.82}$$

If we choose $n = k$ (for k fixed), there is only one term on the right of (7.82). Since

$$\frac{\mathrm{d}^k}{\mathrm{d}s^k}[s(s-1)\cdots(s-k+1)] = \frac{\mathrm{d}^k}{\mathrm{d}s^k}[s^k] = k!,$$

we obtain from (7.82) the approximation (cf. (4.43))

$$f^{(k)}(x) \simeq \frac{\Delta^k f_0}{h^k}. \tag{7.83}$$

We normally use (7.83) to approximate to $f^{(k)}(x)$ for x lying at the centre of the interval $[x_0, x_k]$. If $k = 2m$, we use the approximation

$$f^{(2m)}(x_m) \simeq \frac{\Delta^{2m} f_0}{h^{2m}}. \tag{7.84}$$

If $k = 2m - 1$, the interval $[x_0, x_{2m-1}]$ has no central tabulated point. In this case, generalizing the procedure adapted for estimating the first derivative, we use (7.82) with $n = k + 1 = 2m$. We find that

$$f^{(2m-1)}(x_m) \simeq \frac{1}{h^{2m-1}}(\Delta^{2m-1} f_0 + \tfrac{1}{2}\Delta^{2m} f_0). \tag{7.85}$$

(See Problem 7.28 for the details.) Since

$$\Delta^{2m} f_0 = \Delta(\Delta^{2m-1} f_0) = \Delta^{2m-1} f_1 - \Delta^{2m-1} f_0,$$

we obtain

$$f^{(2m-1)}(x_m) \simeq \frac{\Delta^{2m-1} f_0 + \Delta^{2m-1} f_1}{2h^{2m-1}}, \tag{7.86}$$

which generalizes the result for f' $(m = 1)$ given by (7.78). From (7.86) and (7.84), we obtain the following approximations for the third and fourth derivatives:

$$f^{(3)}(x_2) \simeq \frac{1}{2h^3}(f_4 - 2f_3 + 2f_1 - f_0),$$

$$f^{(4)}(x_2) \simeq \frac{1}{h^4}(f_4 - 4f_3 + 6f_2 - 4f_1 + f_0).$$

Obviously we can derive other differentiation rules by retaining more terms of (7.82), that is, choosing a larger value of n for a given value of k. See, for example, Problem 7.29.

We can find error estimates for all the above rules by the use of Taylor series. As an example, we find an error estimate for the rule (7.81) for approximating to $f''(x_1)$. We assume that $f^{(4)}$ is continuous on $[x_0, x_2]$. Writing $x_0 = x_1 - h$ and $x_2 = x_1 + h$, we have

$$f(x_0) = f_1 - hf'_1 + \frac{h^2}{2!}f''_1 - \frac{h^3}{3!}f^{(3)}_1 + \frac{h^4}{4!}f^{(4)}(\xi_0), \tag{7.87}$$

$$f(x_2) = f_1 + hf'_1 + \frac{h^2}{2!}f''_1 + \frac{h^3}{3!}f^{(3)}_1 + \frac{h^4}{4!}f^{(4)}(\xi_2), \tag{7.88}$$

where $\xi_0 \in (x_0, x_1)$ and $\xi_2 \in (x_1, x_2)$. By Theorem 2.3, there exists a number $\xi \in (x_0, x_2)$ such that

$$f^{(4)}(\xi_0) + f^{(4)}(\xi_2) = 2f^{(4)}(\xi).$$

Adding (7.87) and (7.88), we find that

$$f_0 + f_2 = 2f_1 + h^2 f''_1 + \tfrac{1}{12}h^4 f^{(4)}(\xi).$$

This gives, as an error estimate for (7.81),

$$f''(x_1) = \frac{f_2 - 2f_1 + f_0}{h^2} - \frac{h^2}{12}f^{(4)}(\xi), \tag{7.89}$$

where $\xi \in (x_0, x_2)$.

7.8 Effect of errors

All numerical differentiation formulas are sensitive to rounding errors in the function values f_i. To illustrate this, we consider the relation for f' in (7.76),

$$f'(x_1) = \frac{f_2 - f_0}{2h} + E_T, \tag{7.90}$$

where

$$E_T = -\tfrac{1}{6}h^2 f^{(3)}(\xi_1) \tag{7.91}$$

denotes the truncation error. Usually we do not have the exact values f_0 and f_2, but have approximations f_0^* and f_2^*. If

$$|f_i - f_i^*| \leqslant \tfrac{1}{2}10^{-k}, \qquad i=0,2, \tag{7.92}$$

then

$$f'(x_1) = (f_2^* - f_0^*)/2h + E_R + E_T, \tag{7.93}$$

where

$$E_R = \frac{f_2 - f_0}{2h} - \frac{f_2^* - f_0^*}{2h}$$

denotes the rounding error. From (7.92) we obtain

$$|E_R| \leqslant \frac{1}{2h}(|f_0 - f_0^*| + |f_2 - f_2^*|) \leqslant \frac{1}{2h}10^{-k}. \tag{7.94}$$

If $|f^{(3)}(x)| \leqslant M_3$ for $x \in [x_0, x_2]$, we deduce from (7.93), (7.91) and (7.94) that the *total error* in evaluating f', due to truncation and rounding, is not greater than

$$|E_R| + |E_T| \leqslant \frac{1}{2h}10^{-k} + \frac{1}{6}h^2 M_3.$$

We write

$$E(h) = \frac{1}{2h}10^{-k} + \frac{1}{6}h^2 M_3. \tag{7.95}$$

As h decreases, the first term on the right of (7.95) increases and the second term decreases. To find a value of h which minimizes $E(h)$, we calculate

$$E'(h) = -\frac{1}{2h^2}10^{-k} + \frac{1}{3}hM_3.$$

We see that $E'(h) = 0$ if $h^3 = 3.10^{-k}/(2M_3)$. Thus $E(h)$ is a minimum when

$$h = \left(\frac{3}{2M_3 10^k}\right)^{1/3} = h_{\min}, \tag{7.96}$$

say. Substituting $h = h_{\min}$ into (7.95) we find that

$$E(h_{\min}) = \frac{1}{2}\left(\frac{3}{2\times 10^k}\right)^{2/3} M_3^{1/3}. \tag{7.97}$$

For example, if the data is accurate to six decimal places and M_3 is approximately unity, we expect from (7.97) that the differentiation rule will be accurate to about four decimal places.

Differentiation rules based on higher order interpolation are even more sensitive to rounding errors. If a formula is of the form

$$\frac{1}{h}\sum_{i=0}^{m} \alpha_i f_i,$$

the maximum error due to rounding is $\Sigma|\alpha_i|10^{-k}/(2h)$ and $\Sigma|\alpha_i|$ becomes large as we increase the order of the interpolation. For example, the rule in Problem 7.29 has a truncation error $O(h^4)$ and the maximum error due to rounding is $3\times 10^{-k}/(4h)$, which is larger than (7.94).

Example 7.7 To illustrate the effect of different choices of h on the accuracy of the differentiation rule (7.90), we list the moduli of the errors of approximating to the derivative of e^x at $x=1$, assuming that values of e^x correct to four decimal places are used. In this case, $k=4$, M_3 is approximately e and, from (7.96), $h_{\min}\simeq 0.04$. The results in Table 7.6 support this choice of h. □

Table 7.6 Accuracy of approximations to the derivative of e^x at $x=1$, using (7.90) with different choices of h

h	0.002	0.004	0.01	0.02	0.04	0.1	0.2	0.4
Modulus of error	0.0183	0.0067	0.0017	0.0008	0.0005	0.0047	0.0182	0.0731

Extrapolation techniques can be applied to differentiation rules, although caution is needed since differentiation rules are highly sensitive to rounding error. Extrapolation will only reduce the truncation error and the over-all error will continue to be limited by the size of the rounding error. For example, by extending the Taylor expansions in (7.87) and (7.88) we see that the truncation error in (7.90),

$$E_T = f'(x_1) - \frac{f_2 - f_0}{2h},$$

can be written as a series in powers of h^2, provided the relevant derivatives of f exist, and so we can apply repeated extrapolation exactly as in Romberg's method (see § 7.2).

Example 7.8 The numbers in the first column of Table 7.7 are the results

Table 7.7 Extrapolation to the limit (Example 7.8)

3.0176		
	2.7160	
2.7914		2.7183
	2.7182	
2.7365		

of applying the differentiation rule $(f_2 - f_0)/2h$ for e^x at $x = 1$ with $h = 0.8$, 0.4 and 0.2, where f_0 and f_2 are given to four decimal places. The remaining numbers of Table 7.7 are the result of repeated extrapolation. Given the expected effects of rounding, the last number 2.7183 is fortuitously accurate to four decimal places. □

Problems

Section 7.1

7.1 Verify that Simpson's rule is exact for all $f \in P_3$ by verifying it holds exactly for $f(x) = 1,\ x,\ x^2$ and x^3. (*Hint*: apply the rule to the interval $[-1, 1]$.)

7.2 Derive an integration rule

$$\int_{x_0}^{x_3} f(x)\,dx \simeq h(a_0 f_0 + a_1 f_1 + a_2 f_2 + a_3 f_3)$$

which is exact for $f \in P_3$. (*Hint*: set the midpoint $x_0 + 3h/2$ to zero, put $h = 2$ and make the rule exact for $f(x) = 1,\ x,\ x^2$ and x^3.)

7.3 Obtain the open Newton–Cotes formula of the form

$$\int_{x_0}^{x_3} f(x)\,dx \simeq h(b_1 f_1 + b_2 f_2).$$

7.4 Show that the integration rule

$$\int_{x_0}^{x_1} f(x)\,dx \simeq \frac{h}{2}(f(x_0) + f(x_1)) - \frac{h^2}{12}(f'(x_1) - f'(x_0))$$

is exact for $f \in P_3$.

7.5 Deduce from the result of Problem 7.4 that

$$\int_{x_0}^{x_N} f(x)\,dx \simeq T(h) - \frac{h^2}{12}(f'(x_N) - f'(x_0)),$$

where $T(h)$ denotes the result of applying the composite trapezoidal rule, is exact for $f \in P_3$. (This is called the *trapezoidal rule with end correction* and is obtained by truncating the right side of the Euler–Maclaurin formula (7.26) after one correction term.)

7.6 Verify that the special case of Gregory's formula given by (7.10) is exact for all $f \in P_3$. (*Hint*: put $x_0 = 0$, $h = 1$ and use the identities $\Sigma_{r=1}^{N} r = \frac{1}{2}N(N+1)$; $\Sigma_{r=1}^{N} r^2 = \frac{1}{6}N(N+1)(2N+1)$; $\Sigma_{r=1}^{N} r^3 = \frac{1}{4}N^2(N+1)^2$.)

7.7 Verify (7.18), that $(2M(h) + T(h))/3 = S(\frac{1}{2}h)$.

7.8 Consider the quadrature formula

$$\int_a^b f(x) \approx \sum_{i=0}^{n} w_i f(x_i),$$

where each $w_i > 0$ and the rule is exact for the function $f(x) \equiv 1$. If the $f(x_i)$ are in error by at most $\frac{1}{2}10^{-k}$, show that the error in the quadrature formula due to this is not greater than $\frac{1}{2}(b-a)10^{-k}$.

7.9 Derive the error of the mid-point rule by following the method used in the text for obtaining the error of Simpson's rule. (*Hint*: write $f(x) = f(0) + x f[0, x]$ and integrate over the interval $[-h, h]$, replacing x by $\frac{1}{2}\mathrm{d}/\mathrm{d}x(x^2 - h^2)$.)

Section 7.2

7.10 Verify that one step of Romberg integration, that is the elimination of the h^2 error term between $T(h)$ and $T(\frac{1}{2}h)$, is equivalent to the composite Simpson's rule.

7.11 Write a program to implement Romberg integration and test it on

$$\int_0^1 (1 + \mathrm{e}^x)^{-1}\, \mathrm{d}x.$$

7.12 Assuming that f is sufficiently differentiable, use Taylor series to show that the 'half-way' interpolation formula

$$f(x_0) \simeq \tfrac{1}{2}(f(x_0 + \tfrac{1}{2}h) + f(x_0 - \tfrac{1}{2}h)) = H(x_0; h)$$

has an error of a type which makes extrapolation to the limit valid.

Section 7.3

7.13 Verify directly that the 2-point Gaussian quadrature rule (7.42) is exact for all $f \in P_3$ and that the 3-point Gaussian quadrature rule (7.43) is exact for all $f \in P_5$.

7.14 With π_{n+1} defined by (7.36), show that

$$L_r(x) = \frac{(x-x_0)\cdots(x-x_{r-1})(x-x_{r+1})\cdots(x-x_n)}{(x_r-x_0)\cdots(x_r-x_{r-1})(x_r-x_{r+1})\cdots(x_r-x_n)}$$
$$= \frac{\pi_{n+1}(x)}{(x-x_r)\pi'_{n+1}(x_r)}.$$

(*Hint*: use (7.72).)

7.15 If $\chi_n(x,y) = T_{n+1}(x)T_n(y) - T_{n+1}(y)T_n(x)$, where T_n denotes the Chebyshev polynomial, show that

$$\chi_n(x,y) = 2(x-y)T_n(x)T_n(y) + \chi_{n-1}(x,y)$$

for $n \geq 1$ and deduce that

$$\tfrac{1}{2}\chi_n(x,y) = (x-y)\sum_{r=0}^{n}{}' T_r(x)T_r(y).$$

7.16 For a given function f, the Chebyshev coefficient a_r (see (5.37)) is

$$a_r = \frac{2}{\pi}\int_{-1}^{1}(1-x^2)^{-1/2}f(x)T_r(x)\,dx.$$

Show that, if the N-point Chebyshev quadrature formula is used to estimate the integral, we obtain the approximation $a_r \simeq \alpha_r^*$, where α_r^* is defined by (5.68).

7.17 The points of the 4 point Gaussian rule are $\pm[(15-2\sqrt{30})/35]^{1/2}$, each with weight $\frac{1}{2}+\sqrt{30}/36$, and $\pm[(15+2\sqrt{30})/35]^{1/2}$, each with weight $\frac{1}{2}-\sqrt{30}/36$. Test numerically that, to within the limits imposed by rounding error, this rule is exact for 1, x^2, x^4 and x^6.

7.18 Write and test a program to apply the composite 2 and 3-point Gaussian rules. (*Hint*: note the change of variable given below (7.45).)

Section 7.4

7.19 Obtain a quadratic polynomial approximation for

$$F(x) = \int_0^x e^{-t^2}\,dt$$

by integrating the interpolating polynomial for e^{-t^2} constructed at $t=0$ and 0.1. Estimate the error in the resulting approximation to F on $[0, 0.1]$.

7.20 One method of obtaining a polynomial approximation to

$$F(x) = \int_a^x f(t)\, \mathrm{d}t$$

is to use the Taylor polynomial approximation of degree $n-1$ to $f(t)$ at $t=a$. Show that the resulting approximation to $F(x)$ is the Taylor polynomial of degree n for $F(x)$ at $x=a$.

Section 7.5

7.21 Deduce from the inequality $\mathrm{e}^{-x^2} < \mathrm{e}^{-Mx}$ for $x > M$ that

$$\int_M^\infty \mathrm{e}^{-x^2}\, \mathrm{d}x < \mathrm{e}^{-M^2}/M.$$

Find a value of M so that this last number is smaller than 10^{-7} and hence estimate $\int_0^\infty \mathrm{e}^{-x^2}\, \mathrm{d}x$ to six decimal places.

Section 7.6

7.22 Obtain an error estimate for the product trapezoidal rule analogous to the result (7.64) obtained for the product Simpson rule.

7.23 Use the product Simpson rule to estimate

$$\int_0^1 \int_0^1 (1 + x^2 + y^2)^{-1}\, \mathrm{d}x\, \mathrm{d}y.$$

Section 7.7

7.24 Write down the forward difference formula up to the second differences and differentiate to obtain the approximation

$$f'(x_0) \simeq (-3f_0 + 4f_1 - f_2)/2h.$$

7.25 Obtain the error term for the approximation in Problem 7.24.

7.26 Show that a differentiation rule of the form

$$f'(x_0) \simeq \alpha_0 f_0 + \alpha_1 f_1 + \alpha_2 f_2$$

is exact for all $f \in P_2$ if, and only if, it is exact for $f(x) = 1$, x and x^2. Hence find values of α_0, α_1 and α_2 so that the rule is exact for $f \in P_2$.

7.27 Use the backward difference formula with error term,

$$f(x_1 + sh) = f_1 + s\nabla f_1 + h^2 \binom{-s}{2} f''(\xi_s),$$

to obtain

$$f'(x_1) = \frac{f_1 - f_0}{h} + \tfrac{1}{2}hf''(\xi).$$

7.28 Show that the first two leading terms of $s(s-1)\cdots(s-k)$, regarded as a polynomial in s, are $s^{k+1} - \frac{1}{2}k(k+1)s^k$. Deduce that

$$\frac{\mathrm{d}^k}{\mathrm{d}s^k}\binom{s}{k+1} = s - \tfrac{1}{2}k.$$

Put $k = 2m-1$, $n = 2m$ and $s = m$ in (7.82), and so obtain the differentiation rule (7.86).

7.29 Put $n = 4$ in (7.70) and then put $s = 2$ to obtain

$$f'(x_2) \simeq \frac{1}{12h}(f_0 - 8f_1 + 8f_3 - f_4),$$

which is exact for $f \in P_4$.

7.30 By differentiating the best first degree polynomial, in the least squares sense, for f at $x = x_0,\ x_0 \pm h,\ x_0 \pm 2h$, obtain the differentiation rule

$$f'(x_0) \simeq [2f(x_0+2h) + f(x_0+h) - f(x_0-h) - 2f(x_0-2h)]/10h.$$

(*Hint*: it is sufficient to consider the case where $x_0 = 0$.)

7.31 If we repeat Problem 7.30, but construct the least squares straight line at $x = x_0,\ x_0 \pm h$ only, what approximation is obtained for $f'(x_0)$?

Section 7.8

7.32 Consider the effect of rounding errors and the truncation error in the rule (7.89) for estimating f''. If $|f^{(4)}(x)| \leqslant M_4$ and the f_i are in error by at most $\frac{1}{2} \times 10^{-k}$, show that a choice of h near to

$$h = \left(\frac{24 \times 10^{-k}}{M_4}\right)^{1/4}$$

may be expected to minimize the total error.

7.33 Suppose that $|f^{(4)}(x)| \leqslant 1$ for the function f tabulated below. From these tabulated values, estimate $f''(1.8)$ using the rule (7.89) with all three possible choices of step size h. Given that $f(x) = \sin x$, see how the best value of h (that is $h = 0.1, 0.2$ or 0.3) compares with that predicted in the last problem.

x	1.5	1.6	1.7	1.8	1.9	2.0	2.1
$f(x)$	0.9975	0.9996	0.9917	0.9738	0.9463	0.9093	0.8632

Chapter **8**

SOLUTION OF ALGEBRAIC EQUATIONS OF ONE VARIABLE

8.1 Introduction

In this chapter, we discuss numerical methods for solving equations of the form

$$f(x)=0, \tag{8.1}$$

where both x and $f(x)$ are real. Real values of x for which (8.1) holds are called *roots* of the equation. (We shall not consider any numerical methods for finding complex roots.) Examples of such equations are

$$x^2-5x+2=0, \tag{8.2}$$

$$2^x-5x+2=0. \tag{8.3}$$

The familiar quadratic equation (8.2) appears easier to solve than (8.3) since we can express its roots in the form $x=\frac{5}{2}\pm((\frac{5}{2})^2-2)^{1/2}$ and we can compute the square root by Newton's method, which we will discuss in §8.6. For a general equation (8.1) we need to evaluate $f(x)$ at enough points so that we can sketch the curve $y=f(x)$ with sufficient accuracy to *locate* all the roots of (8.1). Once we have located the roots approximately, we set about *refining* them. In locating the roots it sometimes helps to rearrange the equation first. For instance, if we rewrite (8.3) as $2^x=5x-2$, we see that the required roots are the values of x where the graphs of $y=2^x$ and $y=5x-2$ intersect. From Fig. 8.1 it follows that (8.3) has exactly one root (denoted by α) between 0 and 1.

We shall consider only generally applicable methods of refining the roots. Thus we shall not discuss any of the several methods available for solving *polynomial equations* (see Ralston and Rabinowitz, 1978). We first describe some elementary methods which have the following property in common: beginning with an interval I_0 which contains at least one root of (8.1), we construct a sequence of intervals I_n such that $I_{n+1}\subset I_n$ and that each I_n contains at least one root of (8.1). This will be called a *bracketing*

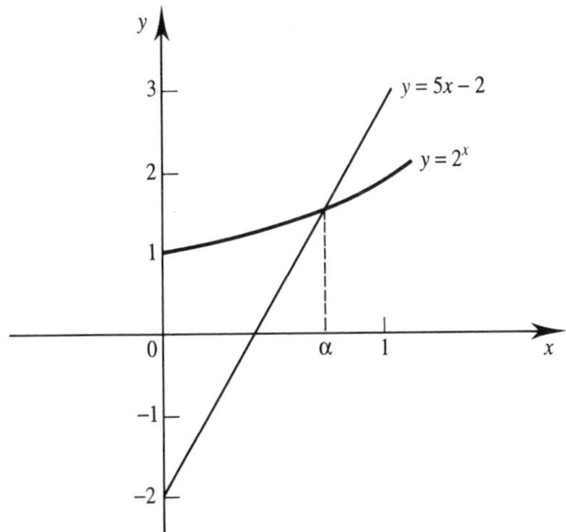

Fig. 8.1 Locating the root of $2^x = 5x - 2$.

method: a root is bracketed by the endpoints of each interval I_n. Obviously, we would like the length of the interval I_n to tend to zero as $n \to \infty$ so that in principle we may compute the root as accurately as we wish.

8.2 The bisection method

Suppose that f is continuous on some interval $[x_0, x_1]$ and that $f(x_0)$ and $f(x_1)$ have opposite signs, as in Fig. 8.2. There is at least one root of $f(x) = 0$ in the interval $[x_0, x_1]$, which we shall denote by I_0. We now bisect I_0, writing $x_2 = (x_0 + x_1)/2$, and let I_1 denote the sub-interval $[x_0, x_2]$ or $[x_2, x_1]$ at whose endpoints f takes opposite signs. Similarly I_1 is bisected to give an interval I_2 (half the width of I_1) at whose endpoints f still has opposite signs, and so on (see Fig. 8.2). This bracketing method is called the *bisection* method.

Algorithm 8.1 (bisection method) We begin with $x_0 < x_1$, and $y_0 = f(x_0)$, $y_1 = f(x_1)$ with $y_0 y_1 < 0$.

repeat
  $x_2 := (x_0 + x_1)/2$ (8.4)
  $y_2 := f(x_2)$
  if $y_1 y_2 < 0$ **then** $x_0 := x_2 \quad y_0 := y_2$
    else $x_1 := x_2 \quad y_1 := y_2$
until $x_1 - x_0 \leqslant 10^{-6}$ □

The algorithm terminates when the root is pinned down in an interval of at most 10^{-6}. We may select the midpoint of this last interval to

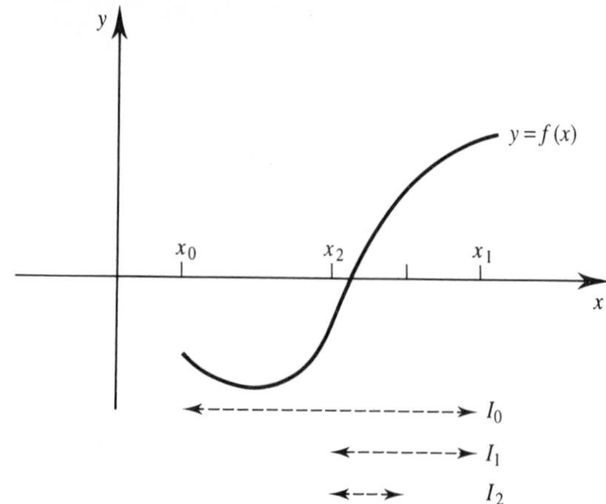

Fig. 8.2 The bisection method.

approximate the root, with an error not greater than $\frac{1}{2} \times 10^{-6}$. (The number 10^{-6} which we have chosen in our stopping criterion, can be adjusted. However, we need to avoid making this number so small that the operation of the algorithm is confused by rounding error.) Advantages of the bisection method are its simplicity and the fact that we may predict, in advance, how many iterations of the main 'loop' of calculations will be required since the interval is halved during each iteration. For example, if initially $x_1 - x_0 = 1$, the number of iterations or bisections necessary will be 20, since $2^{-20} < 10^{-6} < 2^{-19}$.

Example 8.1 For equation (8.3), $f(0) = 3$, $f(1) = -1$ and Algorithm 8.1 can be used on the interval $[0, 1]$. With 20 iterations we find that the root lies between $x_0 = 0.7322435$ and $x_1 = 0.7322445$ (to seven decimal places). □

8.3 Interpolation methods

The bisection method uses the fact that $f(x_0)$ and $f(x_1)$ have opposite signs but takes no account of their *magnitudes*. This suggests replacing f by its interpolating polynomial p_1 (see (4.1)) constructed at x_0 and x_1 and then solving $p_1(x) = 0$. Since $f(x_0)f(x_1) < 0$ we obtain a unique zero of p_1 defined by

$$x_2 = \frac{x_0 f(x_1) - x_1 f(x_0)}{f(x_1) - f(x_0)}. \tag{8.5}$$

Alternatively we see from the similar triangles AA′C and BB′C of Fig. 8.3 that

$$\frac{\mathrm{AA'}}{\mathrm{A'C}} = \frac{\mathrm{BB'}}{\mathrm{CB'}}$$

and (8.5) follows from this. If we then modify Algorithm 8.1, replacing (8.4) by

$$x_2 := (x_0 y_1 - x_1 y_0)/(y_1 - y_0), \tag{8.6}$$

we obtain a method called the *rule of false position* or *regula falsi* (from the Latin).

If f'' exists and has constant sign in the vicinity of the root α, one of the two points x_0, x_1 remains unchanged throughout the operation of the *regula falsi* algorithm. Therefore $x_1 - x_0$ does not tend to zero but has limit $\alpha - x_0$ or $x_1 - \alpha$ (see Problem 8.4). We then need to modify the stopping criterion, ending when $|y_2|$, rather than $x_1 - x_0$, is small. Even so, without further modification this method generally converges rather slowly and we do not recommend it in practice.

There exists a method for solving equations which is sometimes confused with the *regula falsi* method. This is the *secant* method, in which we begin with any two numbers $x_0 < x_1$ and calculate

$$x_{r+1} = \frac{x_{r-1}f(x_r) - x_r f(x_{r-1})}{f(x_r) - f(x_{r-1})} = x_r - \frac{f(x_r)(x_r - x_{r-1})}{f(x_r) - f(x_{r-1})}, \qquad r = 1, 2, \ldots, \tag{8.7}$$

assuming that each denominator $f(x_r) - f(x_{r-1})$ is non-zero. Note that, unlike the *regula falsi* method, no attention is paid here to the signs of the numbers $f(x_r)$ and, therefore, it is *not* a bracketing method. The second

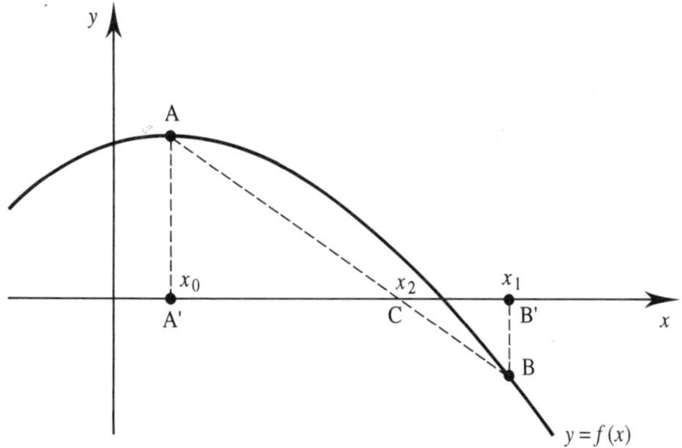

Fig. 8.3 The *regula falsi* method.

formula in (8.7) is preferable to the first as it expresses x_{r+1} in terms of x_r minus a correction term. We now give a formal algorithm for the secant method, followed by an example.

Algorithm 8.2 (secant method) This begins with x_0, x_1 and $y_0 = f(x_0)$, $y_1 = f(x_1)$.

$r := 0$
repeat
$r := r + 1$

$$x_{r+1} := x_r - \frac{y_r(x_r - x_{r-1})}{y_r - y_{r-1}}$$

$y_{r+1} := f(x_{r+1})$
until $|\, y_{r+1} \,| \leqslant 10^{-6}$ □

Example 8.2 Let us apply the secant method to equation (8.3) with $x_0 = 0$ and $x_1 = 1$, so that $y_0 = 3$ and $y_1 = -1$. Algorithm 8.2 gives 0.75, 0.731700, 0.732245 and 0.732244 as the next iterates, and then the stopping criterion is reached. □

Suppose $y = f(x)$ and that the inverse function, $x = \phi(y)$, exists. Then, in particular, $0 = f(\alpha)$ corresponds to $\alpha = \phi(0)$. Each step of the secant method, as we have already seen in Example 4.6, may be regarded as inverse interpolation at two points x_0 and x_1. We replace $\phi(y)$ by the linear interpolating polynomial $p_1(y)$ constructed at y_0 and y_1. To estimate the accuracy attained at any stage by the *regula falsi* method, we consider the error formula (from (4.13)):

$$\phi(y) - p_1(y) = (y - y_0)(y - y_1)\phi''(\eta_y)/2!. \tag{8.8}$$

This is valid on any interval $c \leqslant y \leqslant d$ containing y_0 and y_1 and on which ϕ'' exists; η_y is some point of that interval. We now assume that $|\phi''(y)| \leqslant M_2$ for $y \in [c, d]$. Putting $y = 0$ in (8.8) we obtain

$$|\alpha - x_2| \leqslant \tfrac{1}{2} M_2 y_0 y_1, \tag{8.9}$$

as by construction $p_1(0) = x_2$. We can estimate M_2 from the relation

$$\phi''(y) = -f''(x)/[f'(x)]^3, \tag{8.10}$$

assuming $f'(x) \neq 0$ (see Example 4.6, where (8.10) is derived).

Example 8.3 Let us pursue (8.10) with $f(x) = 2^x - 5x + 2$ so that $f'(x) = 2^x \log 2 - 5$ and $f''(x) = 2^x (\log 2)^2$. Taking $x_0 = 0.75$, $x_1 = 0.731700$ and $x_2 = 0.732245$ (three consecutive iterates obtained in Example 8.2 above) we find that $|\phi''(y)| < 0.0144$ for $0.73 \leqslant x \leqslant 0.75$ and obtain from (8.9) that the error in the iterate 0.732245 is bounded by 1.03×10^{-6}. □

An obvious extension of the secant method is to use three points at a time instead of two. Suppose we begin with two approximations, x_0 and x_1, to a root of $f(x)=0$ and that the secant method is used to compute a third approximation x_2. Instead of discarding x_0 or x_1, we may construct the unique (quadratic) interpolating polynomial p_2 for f at all three points. Suppose that $p_2(x)=0$ has two real roots, as in Fig. 8.4, and that x_3 denotes the root nearer to x_2. We then repeat the above procedure with x_1, x_2 and x_3 in place of x_0, x_1 and x_2, and so on. This is called *Muller's* method.

To avoid the possibility of the polynomial equation $p_2(x)=0$ having complex roots, we may adopt a variant of the above method in which the root is bracketed at any stage. This time we begin with numbers $x_0 < x_1$ such that $f(x_0)f(x_1)<0$ and select x_2 as some number in $[x_0, x_1]$. We could select x_2 as the midpoint of $[x_0, x_1]$ or as the number obtained by one iteration of the secant method. The polynomial equation $p_2(x)=0$ must then have real roots. Only *one* of these, denoted by x_3, will be inside $[x_0, x_1]$. Next, we replace either x_0 or x_1 by x_2, exactly as in the *regula falsi* method, and replace x_2 by x_3. For example, in Fig. 8.4 we replace x_0 by x_2 and x_2 by x_3. This completes one iteration of the method. We repeat the process with the new set $\{x_0, x_1, x_2\}$ until our stopping criterion is satisfied, as mentioned above.

It is necessary to add one further set of instructions to this algorithm. If $|x_1 - x_0|$ approaches zero, the effect of rounding error can perturb the coefficients of the polynomial p_2 sufficiently to create complex roots instead of the predicted real roots. If this occurs, we can arrange to drop the point x_2 and continue the search for the root by applying the bisection or *regula falsi* method.

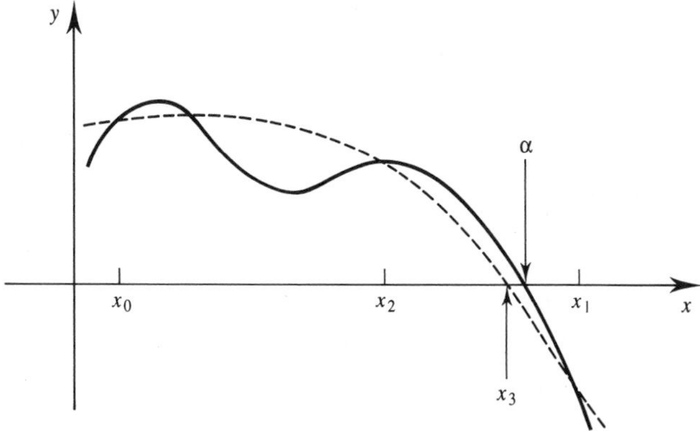

Fig. 8.4 Muller's method. The dotted line is $p_2(x)$.

Example 8.4 For the equation $2^x - 5x + 2 = 0$ with $x_0 = 0$ and $x_1 = 1$ the secant method (see Example 8.2) gives $x_2 = 0.75$. Two applications of Muller's method gives the iterates 0.732166 and 0.732244.

8.4 One-point iterative methods

Hitherto we have used at least two values of x in order to predict a 'better' approximation to the root, at any stage. We naturally ask whether we can devise methods which use only *one* value of x at any time. These will be called *one-point* methods. To find a root α of $f(x) = 0$, we must therefore construct a sequence (x_r) which satisfies two criteria:

(i) the sequence (x_r) converges to α,
(ii) x_{r+1} depends directly only on its predecessor x_r.

From (ii), we need to find some function g, so that the sequence (x_r) may be computed from

$$x_{r+1} = g(x_r), \qquad r = 0, 1, \ldots, \tag{8.11}$$

given an initial value x_0. If we happened to choose $x_0 = \alpha$, we would also want (8.11) to give $x_1 = \alpha$ and therefore g satisfies the property $\alpha = g(\alpha)$. Thus, to obtain a one-point method for finding a root α of the equation $f(x) = 0$, we rearrange the equation in a form $x = g(x)$, with $\alpha = g(\alpha)$, so that (8.11) yields a sequence (x_r) which converges to α.

Example 8.5 Consider the equation $x = \mathrm{e}^{-x}$, which is already of the form $x = g(x)$. The graphs of $y = x$ and $y = \mathrm{e}^{-x}$ intersect at only one point $x = \alpha$ (between 0 and 1), which is therefore the only root of $x = \mathrm{e}^{-x}$. Following (8.11), let us choose $x_0 = 0.5$ and compute $x_{r+1} = \mathrm{e}^{-x_r}$, $r = 0, 1, \ldots$. Rounding the iterates to three decimal places at each stage, we obtain the numbers displayed in Table 8.1. These numbers suggest that the process is converging, though slowly, and that the root α is 0.567, to three decimal places. Assuming that this is true, the last row of Table 8.1 suggests further that the errors $|x_r - \alpha|$ tend monotonically to zero. □

To investigate the behaviour of errors in the general case, we return to (8.11) and subtract from it the equation $\alpha = g(\alpha)$. Thus

$$x_{r+1} - \alpha = g(x_r) - g(\alpha) = (x_r - \alpha)g'(\xi_r), \tag{8.12}$$

with ξ_r some point between x_r and α. This last step follows from the mean value theorem 2.5, assuming that g' exists on an interval containing x_r and α. It is easy to see how to impose conditions on g to ensure convergence of the one-point iterative method. For, if

$$|g'(x)| \leqslant L < 1, \qquad \text{for all } x, \tag{8.13}$$

Table 8.1 Solution of $x = e^{-x}$ by iterating $x_{r+1} = e^{-x_r}$

r	0	1	2	3	4	5	6	7	8	9
x_r	0.5	0.607	0.545	0.580	0.560	0.571	0.565	0.568	0.567	0.567
$x_r - \alpha$	−0.067	0.040	−0.022	0.013	−0.007	0.004	−0.002	0.001	0.000	0.000

we deduce from (8.12) that

$$|x_{r+1} - \alpha| \leqslant L|x_r - \alpha|, \tag{8.14}$$

so that the errors are diminishing. Replacing r by $r-1$ in (8.14), we obtain

$$|x_r - \alpha| \leqslant L|x_{r-1} - \alpha|, \qquad r \geqslant 1,$$

and, on applying this inequality repeatedly,

$$|x_r - \alpha| \leqslant L^r|x_0 - \alpha|, \qquad r \geqslant 0. \tag{8.15}$$

Since $0 \leqslant L < 1$, $L^r \to 0$ as $r \to \infty$. From (8.15), $|x_r - \alpha| \to 0$ as $r \to \infty$ and the sequence (x_r) converges to the root α. In fact, the above condition on g' allows us to show also the *existence* of the root α. The proof requires the following lemma.

Lemma 8.1 If f has continuous first derivative f' with

$$f'(x) \geqslant M > 0, \qquad \text{for all real } x$$

then the equation $f(x) = 0$ has a unique root and this root lies between $x = 0$ and $x = -f(0)/M$.

Proof First we assume that $f(0) > 0$ when by the mean value theorem 2.5 there is a point ξ such that

$$\begin{aligned} f(-f(0)/M) - f(0) &= f'(\xi)(-f(0)/M - 0) \\ &= -f'(\xi)f(0)/M \\ &\leqslant -f(0) \end{aligned}$$

as $-f'(\xi) \leqslant -M$. Thus

$$f(-f(0)/M) \leqslant 0,$$

showing that there must be a root between $x = 0$ and $x = -f(0)/M$. Figure 8.5 gives a graphical representation of this argument. The slope of f must be positive and larger than that of the line AB, which has slope M. Similarly we can deal with the case $f(0) < 0$; finally $f(0) = 0$ automatically gives a root.

It remains to show that the root is unique and this follows from Rolle's theorem 2.4. Two distinct roots would imply the existence of a point η at which $f'(\eta) = 0$ which contradicts the required property of f'. □

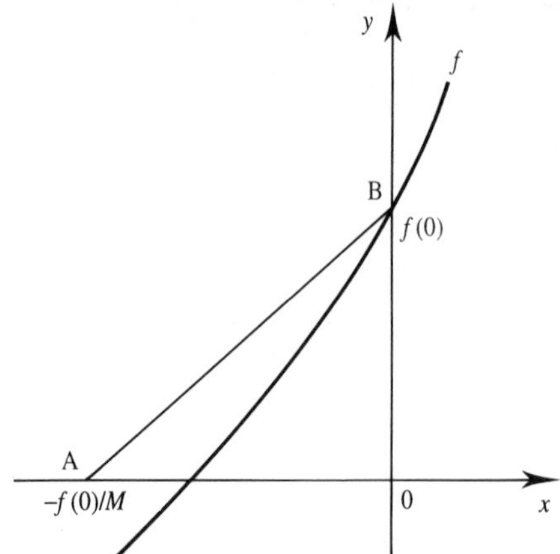

Fig. 8.5 Representation of Lemma 8.1.

Theorem 8.1 If g' exists, is continuous and $|g'(x)| \leqslant L < 1$ for all x, then the equation $x = g(x)$ has a unique root, say α. Further, given any x_0, the sequence (x_r) defined by

$$x_{r+1} = g(x_r), \qquad r = 0, 1, \ldots, \tag{8.16}$$

converges to α.

Proof Since

$$\frac{\mathrm{d}}{\mathrm{d}x}(x - g(x)) = 1 - g'(x) \geqslant 1 - L > 0, \tag{8.17}$$

the function $x - g(x)$ satisfies the conditions of the lemma and the existence and uniqueness of α follows. We have already shown that the sequence (x_r) converges to α. □

It is not sufficient to have $f'(x) > 0$ in Lemma 8.1 and, similarly, $|g'(x)| < 1$ in Theorem 8.1. For example, the function $f(x) = \mathrm{e}^x$ has $f'(x) > 0$ everywhere, but there are no zeros of f. In fact the condition $|g'(x)| \leqslant L < 1$ for *all* values of x is rather restrictive. For example, this condition is not satisfied by the function $g(x) = \mathrm{e}^{-x}$ in Example 8.5. We shall return to this point in § 8.7. Meanwhile we note that the one-point method has a simple graphical interpretation, as depicted in Fig. 8.6. In the figure we start with the iterate x_0 and determine the point A_0 which has coordinates $(x_0, g(x_0))$. We now draw the line A_0A_1 parallel to the x-axis. Since A_1 lies on the line $y = x$ and $x_1 = g(x_0)$, A_1 has coordinates (x_1, x_1). Next we compute the position of A_2 which has coordinates $(x_1, g(x_1))$. We continue along the arrowed lines. In Fig. 8.6, where $|g'(x)| < 1$, the

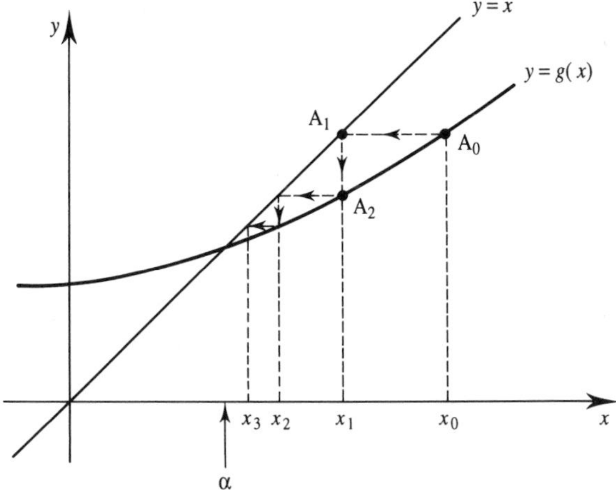

Fig. 8.6 A graphical interpretation of the one-point method.

geometrical construction suggests that the sequence (x_r) is converging to α. The reader is urged to use the above geometrical approach to 'prove' non-convergence in a case where $|g'(x)| > 1$.

8.5 Faster convergence

As we have seen, to solve an equation $f(x) = 0$ by a one-point method, we first replace $f(x) = 0$ by an equivalent equation $x = g(x)$ and then (from (8.16)) compute a sequence (x_r). We have to find a function g which will yield a convergent sequence. Obviously we ought to aim for rapid convergence, if possible. Assuming that $x_r \neq \alpha$ for all r, we obtain

$$\lim_{r\to\infty} \frac{x_{r+1} - \alpha}{x_r - \alpha} = \lim_{r\to\infty} \frac{g(x_r) - g(\alpha)}{x_r - \alpha} = g'(\alpha), \tag{8.18}$$

from the definition of the derivative g', which is assumed to exist. Thus the ratio of successive errors tends to the value $g'(\alpha)$. This explains the behaviour of the errors $x_r - \alpha$ in Table 8.1 for the equation $x = e^{-x}$, where $g'(\alpha) \simeq -0.6$. Since $(0.6)^4 \simeq 0.1$, we expect to gain only about one decimal place of accuracy every four iterations. This is in agreement with Table 8.1.

Now, given an equation

$$x = g(x) \tag{8.19}$$

and any constant $\lambda \neq -1$, we may add λx to each side of the equation and divide by $1 + \lambda$. This gives a family of equations

$$x = \left(\frac{\lambda}{1+\lambda}\right)x + \left(\frac{1}{1+\lambda}\right)g(x), \tag{8.20}$$

each of the form

$$x = G(x)$$

and having the same roots as (8.19). Bearing in mind (8.18), we choose λ so as to make $|G'(x)|$ small (and certainly less than unity) at a root α. From (8.20),

$$G'(x) = \left(\frac{1}{1+\lambda}\right)(\lambda + g'(x)), \tag{8.21}$$

so we choose a value of λ near to $-g'(\alpha)$.

Example 8.6 For the equation $x = \mathrm{e}^{-x}$, with $g'(\alpha) \simeq -0.6$, we modify the equation as in (8.20) with $\lambda = 0.6$. This yields the iterative method

$$x_{r+1} = (3x_r + 5\mathrm{e}^{-x_r})/8.$$

As in Example 8.5 we choose $x_0 = 0.5$ but we now obtain the iterates (rounded to six decimal places) shown in Table 8.2. Even x_1 is nearly as accurate as the x_9 of Table 8.1 and x_3 here is correct to six decimal places. □

The above procedure for improving the rate of convergence requires an estimate of g'. We examine a method for accelerating convergence which does not require explicit knowledge of the derivative. From (8.12), if $g'(x)$ were constant, we could deduce that

$$\frac{x_{r+1} - \alpha}{x_r - \alpha} = \frac{x_r - \alpha}{x_{r-1} - \alpha}, \tag{8.22}$$

with both sides of (8.22) taking the value $g'(x)$ (which is constant). We could solve (8.22) to determine the root α. Given a function g for which g' is not constant, we may still expect that the number x^*_{r+1} satisfying the equation

$$\frac{x_{r+1} - x^*_{r+1}}{x_r - x^*_{r+1}} = \frac{x_r - x^*_{r+1}}{x_{r-1} - x^*_{r+1}}$$

Table 8.2 Faster convergence to the root of $x = \mathrm{e}^{-x}$

r	0	1	2	3
x_r	0.5	0.566582	0.567132	0.567143

will be *close* to α. Thus

$$x^*_{r+1} = \frac{x_{r+1}x_{r-1} - x_r^2}{x_{r+1} - 2x_r + x_{r-1}}, \tag{8.23}$$

assuming the denominator is non-zero. The right side of (8.23) may be rewritten as x_{r+1} plus a 'correction',

$$x^*_{r+1} = x_{r+1} - \frac{(x_{r+1} - x_r)^2}{x_{r+1} - 2x_r + x_{r-1}}. \tag{8.24}$$

In terms of the backward difference operator ∇, introduced in Chapter 4, we have

$$x^*_{r+1} = x_{r+1} - \frac{(\nabla x_{r+1})^2}{\nabla^2 x_{r+1}}.$$

The calculation of a sequence (x_r^*) in this way from a given sequence (x_r) is called Aitken's delta-squared process or *Aitken acceleration*, after A. C. Aitken (1895–1967), who first suggested it in 1926.

In passing we note a connection between Aitken acceleration and the secant method. If we apply the secant method to the equation $f(x) = 0$ where $f(x) = g(x) - x$, beginning with x_{r-1} and x_r, we obtain as the next iterate (see (8.7))

$$x_r - \frac{f(x_r)(x_r - x_{r-1})}{f(x_r) - f(x_{r-1})}. \tag{8.25}$$

If we now write

$$f(x_r) = g(x_r) - x_r = x_{r+1} - x_r$$

and similarly replace $f(x_{r-1})$ by $x_r - x_{r-1}$, then (8.25) becomes

$$x_r - \frac{(x_{r+1} - x_r)(x_r - x_{r-1})}{x_{r+1} - 2x_r + x_{r-1}}$$

which may be simplified to give the right side of (8.23) or (8.24).

Example 8.7 To show how rapidly the Aitken sequence (x_r^*) can converge compared with the original sequence (x_r), we apply the method to the sequence discussed in Example 8.5. The members of the sequence (x_r) are given to five decimal figures (to do justice to the accuracy of the Aitken process) instead of the three figures quoted in Table 8.1. The results are shown in Table 8.3. Note from (8.24) that x_r^* is defined only for $r \geqslant 2$.

Often a more satisfactory way of using Aitken acceleration is to choose x_0 and calculate x_1 and x_2, as usual, from $x_{r+1} = g(x_r)$, $r = 0$ and 1. Then use

Table 8.3 Aiken's delta-squared process

r	0	1	2	3	4	5
x_r	0.5	0.60653	0.54524	0.57970	0.56007	0.57117
x_r^*	–	–	0.56762	0.56730	0.56719	0.56716

the acceleration technique on x_0, x_1 and x_2 (that is, (8.24) with $r = 1$) to find x_2^*. We use x_2^* as a *new* value for x_0, calculate new values $x_1 = g(x_0)$, $x_2 = g(x_1)$ and accelerate again, and so on. Thus, for the equation $x = e^{-x}$ with $x_0 = 0.5$ initially, we obtain:

$$\left.\begin{aligned} x_0 &= 0.50000 \\ x_1 &= 0.60653 \\ x_2 &= 0.54524 \end{aligned}\right\} \qquad x_2^* = 0.56762$$

$$\left.\begin{aligned} x_0 &= 0.56762 \\ x_1 &= 0.56687 \\ x_2 &= 0.56730 \end{aligned}\right\} \qquad x_2^* = 0.56714$$

So, by calculating only two sets of x_0, x_1 and x_2, we find the root correct to five decimal places. This is very satisfactory, considering the slowness of convergence of the original iterative process. Sometimes Aitken acceleration will not work, especially if g' fluctuates a large amount and x_0 is a long way from the root. □

8.6 Higher order processes

Let (x_r) denote a sequence defined by $x_{r+1} = g(x_r)$, for some choice of x_0, which converges to a root α of $x = g(x)$. We define

$$e_r = x_r - \alpha, \qquad r = 0, 1, \ldots, \tag{8.26}$$

to denote the error at any stage. Thus $x_{r+1} = g(x_r)$ may be written as

$$\alpha + e_{r+1} = g(\alpha + e_r).$$

If $g^{(k)}$ exists, we may replace $g(\alpha + e_r)$, by its Taylor polynomial with remainder, to give

$$\alpha + e_{r+1} = g(\alpha) + e_r g'(\alpha) + \cdots + \frac{e_r^{k-1}}{(k-1)!} g^{(k-1)}(\alpha) + \frac{e_r^k}{k!} g^{(k)}(\xi_r),$$

where ξ_r lies between α and $\alpha + e_r$. Since $\alpha = g(\alpha)$, we find that the relation between successive errors e_r and e_{r+1} has the form

$$e_{r+1} = e_r g'(\alpha) + \frac{e_r^2}{2!} g''(\alpha) + \cdots + \frac{e_r^{k-1}}{(k-1)!} g^{(k-1)}(\alpha) + \frac{e_r^k}{k!} g^{(k)}(\xi_r). \quad (8.27)$$

If $g'(\alpha) \neq 0$ and e_r is 'quite small', we may neglect terms in e_r^2, e_r^3 and so on, and see that

$$e_{r+1} \simeq e_r g'(\alpha),$$

which is merely an imprecise way of saying what we knew before from (8.12). However, we see from (8.27) that if $g'(\alpha) = 0$ but $g''(\alpha) \neq 0$, e_{r+1} behaves like a multiple of e_r^2 and the process is called *second order*. More generally, if for some fixed $k \geqslant 2$,

$$g'(\alpha) = g''(\alpha) = \cdots = g^{(k-1)}(\alpha) = 0 \quad (8.28)$$

and

$$g^{(k)}(\alpha) \neq 0,$$

then, from (8.27)

$$e_{r+1} = e_r^k g^{(k)}(\xi_r)/k! \quad (8.29)$$

and the process is said to be *kth order*. If $g'(\alpha) \neq 0$, the process is called *first order*. If $g^{(k)}$ is continuous, we see from (8.29) that

$$\lim_{r \to \infty} e_{r+1}/e_r^k = g^{(k)}(\alpha)/k! \quad (8.30)$$

This discussion may appear to be rather academic, as it may seem unlikely that the conditions (8.28) will often be fulfilled in practice, for any $k \geqslant 2$. On the contrary, however, we can easily construct second and higher order processes, since $f(x) = 0$ may be rearranged in an infinite number of ways in the form $x = g(x)$. For a given root α, different choices of g lead to processes of possibly different orders.

To derive a second order process, we may return to (8.20) where we constructed a family of equations with the same roots as $x = g(x)$. These equations are of the form $x = G(x)$, with

$$G(x) = \left(\frac{\lambda}{1+\lambda}\right)x + \left(\frac{1}{1+\lambda}\right)g(x)$$

with λ a constant, and, therefore,

$$G'(x) = \left(\frac{\lambda}{1+\lambda}\right)(\lambda + g'(x)).$$

We can make $G'(\alpha) = 0$ by choosing $\lambda = -g'(\alpha)$. To relate this to an arbitrary equation $f(x) = 0$ with a root α, we replace $g(x)$ by $x - f(x)$ so that $\lambda = f'(\alpha) - 1$. Then $x = G(x)$ becomes

$$x = x - \frac{f(x)}{f'(\alpha)},$$

suggesting the second order iterative process

$$x_{r+1} = x_r - \frac{f(x_r)}{f'(\alpha)}. \tag{8.31}$$

This is of no practical use, since we cannot compute $f'(\alpha)$ without knowing α, which is what we are trying to find. However, let us replace $f'(\alpha)$ by $f'(x_r)$ in (8.31). This gives the method

$$x_{r+1} = x_r - \frac{f(x_r)}{f'(x_r)}, \tag{8.32}$$

which is of the form $x_{r+1} = g(x_r)$, with

$$g(x) = x - \frac{f(x)}{f'(x)}. \tag{8.33}$$

Thus $\alpha = g(\alpha)$ (since $f(\alpha) = 0$) and, on differentiating (8.33),

$$g'(x) = \frac{f(x)f''(x)}{[f'(x)]^2}, \tag{8.34}$$

provided $f'(x) \neq 0$. Thus, provided $f'(\alpha) \neq 0$, we obtain $g'(\alpha) = 0$, showing that the method (8.32) is at least second order. Differentiating (8.34) and putting $x = \alpha$, we find that

$$g''(\alpha) = f''(\alpha)/f'(\alpha),$$

and thus the method (8.32) is exactly second order if $f''(\alpha) \neq 0$. This second order process, called *Newton's method*, has a simple geometrical interpretation. The tangent to the curve $y = f(x)$ at $x = x_r$ cuts the x-axis at $x = x_{r+1}$. From Fig. 8.7 we see that

$$f'(x_r) = \frac{f(x_r)}{x_r - x_{r+1}},$$

which is in agreement with (8.32). Note that we require the condition $f'(x_r) \neq 0$ so that the tangent will cut the x-axis. This is in accordance with the restriction $f'(x) \neq 0$ made above.

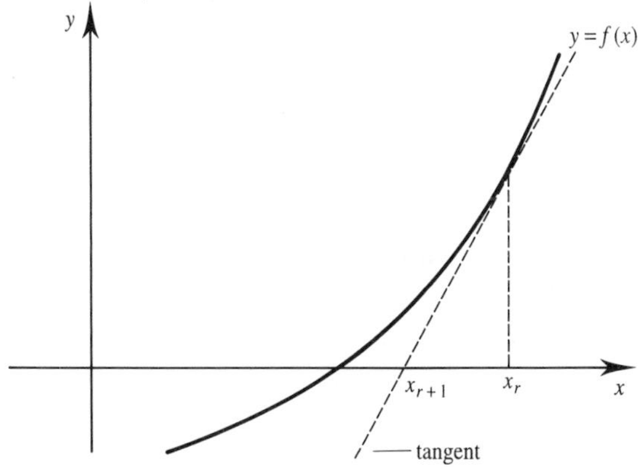

Fig. 8.7 Newton's method.

Example 8.8 For the equation $xe^x - 1 = 0$ (which we have used in earlier examples in the form $x = e^{-x}$),

$$f(x) = xe^x - 1, \qquad f'(x) = e^x(1 + x),$$

so that, for Newton's method,

$$g(x) = x - \frac{xe^x - 1}{e^x(1 + x)} = x - \frac{x - e^{-x}}{1 + x}.$$

Newton's method is

$$x_{r+1} = x_r - \frac{x_r - e^{-x_r}}{1 + x_r}.$$

With $x_0 = 0.5$, the next three iterates are given to eight decimal places in Table 8.4, x_3 being the exact solution correct to all eight digits. □

To investigate the *convergence* of Newton's method to a root α of $f(x) = 0$, we consider (8.29) with $k = 2$. Successive errors behave as

$$e_{r+1} = \tfrac{1}{2} e_r^2 g''(\xi_r), \tag{8.35}$$

where ξ_r lies between α and x_r and g'' may be found by differentiating (8.34).

We assert, without formal proof, that provided $f'(x) \neq 0$ and x_0 is chosen

Table 8.4 Newton's method

r	1	2	3
x_r	0.57102044	0.56715557	0.56714329

sufficiently close to α, we may deduce from (8.35) that the sequence (e_r) converges to zero. Thus Newton's method converges. We also state a more useful result on convergence.

Theorem 8.2 If there exists an interval $[a, b]$ such that

(i) $f(a)$ and $f(b)$ have opposite signs,
(ii) f'' does not change sign on $[a, b]$,
(iii) the tangents to the curve $y = f(x)$ at both a and b cut the x-axis within $[a, b]$,

then $f(x) = 0$ has a unique root α in $[a, b]$ and Newton's method converges to α, for *any* $x_0 \in [a, b]$.

The proof is omitted. Condition (i) guarantees there is at least one root $\alpha \in [a, b]$ and (ii) means that f is convex or concave on $[a, b]$, which allows only one root of $f(x) = 0$ in $[a, b]$. Condition (iii) together with (ii) ensures that, if $x_r \in [a, b]$, the next iterate x_{r+1}, produced by Newton's method, is also in $[a, b]$. □

The best-known application of Newton's method is the *square root* algorithm. The square root of a positive number c is the positive root of the equation $x^2 - c = 0$. In this case, $f(x) = x^2 - c$ and $f'(x) = 2x$. So Newton's method is

$$x_{r+1} = x_r - \frac{x_r^2 - c}{2x_r}, \qquad x_r \neq 0.$$

This simplifies to give

$$x_{r+1} = \frac{1}{2}\left(x_r + \frac{c}{x_r}\right). \tag{8.36}$$

All three conditions of Theorem 8.2 hold for the function $x^2 - c$ on an interval $[a, b]$ such that $0 < a < c^{1/2}$ and $b > \frac{1}{2}(a + c/a)$. The inequality for b ensures that both conditions (i) and (iii) hold (see Problem 8.18). Thus the square root algorithm converges for any $x_0 > 0$. For a more elementary proof of the convergence of the square root method (8.36) see Problem 8.19. Notice that it is easy to demonstrate the expected second order convergence of the sequence generated by (8.36). From

$$x_{r+1} - c^{1/2} = \frac{1}{2}\left(x_r + \frac{c}{x_r}\right) - c^{1/2} = \frac{1}{2x_r}(x_r^2 - 2c^{1/2}x_r + c),$$

we obtain

$$x_{r+1} - c^{1/2} = \frac{1}{2x_r}(x_r - c^{1/2})^2. \tag{8.37}$$

Note the squaring of the error $x_r - c^{1/2}$, associated with second order convergence.

Given any $c>0$, there exists an integer n such that $\frac{1}{4} \leqslant 4^n c < 1$ and thus $\frac{1}{2} \leqslant 2^n c^{1/2} < 1$. To calculate $c^{1/2}$, it therefore suffices to consider c in the range $\frac{1}{4} \leqslant c < 1$ and we choose

$$x_0 = \tfrac{2}{3}c + \tfrac{17}{48},$$

which is the minimax approximation to $c^{1/2}$ on $[\frac{1}{4}, 1]$. We have (see Example 5.14)

$$|x_0 - c^{1/2}| \leqslant \tfrac{1}{48}, \tag{8.38}$$

for this choice of initial iterate x_0 and after very few iterations we obtain highly accurate approximations to $c^{1/2}$ (Problem 8.21).

We cannot always use Newton's method as it is not always feasible to evaluate f'. An obvious way to modify (8.32) is to replace $f'(x_r)$ by the divided difference approximation, $(f(x_r) - f(x_{r-1}))/(x_r - x_{r-1})$. Thus (8.32) becomes

$$x_{r+1} = x_r - \frac{f(x_r)(x_r - x_{r-1})}{f(x_r) - f(x_{r-1})},$$

which we have already met as the secant method.

In Problem 8.24 an alternative method of obtaining Newton's method is described and in Problem 8.25 an extension of Newton's method to a third order process is given.

8.7 The contraction mapping theorem

Recall Theorem 8.1, which states that if g' is continuous and $|g'(x)| \leqslant L < 1$ for all values of x, then the iterative process

$$x_{r+1} = g(x_r), \qquad x_0 \text{ arbitrary},$$

converges to the unique root of $x = g(x)$. The conditions of Theorem 8.1 are not satisfied by the function $g(x) = e^{-x}$, yet convergence appeared to take place in that case (Example 8.5). It is not necessary to require $|g'(x)| \leqslant L < 1$ for *all* values of x. We will make use of a Lipschitz condition (Definition 2.12) and will require:

Definition 8.1 A function g is said to be a *contraction mapping* on an interval $[a, b]$ if

(i) $x \in [a, b] \Rightarrow g(x) \in [a, b]$, (8.39)

(ii) g satisfies a Lipschitz condition on $[a, b]$ with a Lipschitz constant $L<1$. Thus

$$x, y \in [a, b] \quad \Rightarrow \quad |g(x) - g(y)| \leqslant L|x - y|,$$

where $L<1$.

Condition (i) is sometimes referred to as the *closure condition.* □

We can now give a less restrictive result than Theorem 8.1.

Theorem 8.3 If there is an interval $[a, b]$ on which g is a contraction mapping, then

(i) the equation $x = g(x)$ has a unique root (say α) in $[a, b]$,
(ii) for any $x_0 \in [a, b]$, the sequence defined by

$$x_{r+1} = g(x_r), \qquad r = 0, 1, \ldots, \tag{8.40}$$

converges to α.

Proof For any $x \in [a, b]$, $g(x) \in [a, b]$. In particular, for $x = a$,

$$a - g(a) \leqslant 0. \tag{8.41}$$

and, for $x = b$,

$$b - g(b) \geqslant 0. \tag{8.42}$$

The Lipschitz condition implies that g is continuous on $[a, b]$. Thus $x - g(x)$ is continuous on $[a, b]$ and, from the inequalities (8.41) and (8.42), we deduce that the equation $x - g(x) = 0$ has at least one root belonging to $[a, b]$. Let us assume that there is more than one root and let α and β ($a \leqslant \alpha, \beta \leqslant b$) denote two distinct roots. Therefore

$$\alpha = g(\alpha), \qquad \beta = g(\beta)$$

and, subtracting these equations,

$$\alpha - \beta = g(\alpha) - g(\beta).$$

From the contraction mapping property, we deduce that

$$|\alpha - \beta| = |g(\alpha) - g(\beta)| \leqslant L|\alpha - \beta|,$$

with $0 \leqslant L < 1$. This entails $|\alpha - \beta| < |\alpha - \beta|$, which is impossible. We deduce that there is only one root, say α, belonging to $[a, b]$.

Note that, since g is a contraction mapping on $[a, b]$, $x_r \in [a, b] \Rightarrow g(x_r) \in [a, b]$, which shows that $x_{r+1} \in [a, b]$. By induction, every $x_r \in [a, b]$ if $x_0 \in [a, b]$, From the Lipschitz condition, it follows that

$$|x_{r+1} - \alpha| = |g(x_r) - g(\alpha)| \leqslant L|x_r - \alpha|.$$

Hence, as in the proof of Theorem 8.1, we deduce that

$$|x_r - \alpha| \leqslant L^r |x_0 - \alpha| \tag{8.43}$$

and, since $0 \leqslant L < 1$, we conclude that the sequence (x_r) converges to α. □

Looking back over the proof, we see that it is not *essential* to use a Lipschitz condition. If we prefer, we could use the mean value theorem and require instead that $|g'(x)| \leqslant L < 1$ on $[a, b]$. See Chapter 2, p. 23. It is, however, informative to use the Lipschitz condition to prepare for the extension of Theorem 8.3 to deal with iterative methods for a *system* of algebraic equations, which we discuss in Chapters 10 and 12.

From (8.43), we can derive an *a priori* estimate of the error at any stage. Since both x_0 and α belong to $[a, b]$, we obtain

$$|x_r - \alpha| \leqslant L^r(b - a). \tag{8.44}$$

See also Problem 8.28.

Example 8.9 We apply Theorem 8.3 to the equation $x = e^{-x}$ of Example 8.5. For $g(x) = e^{-x}$, $g(0.4) = 0.67$ and $g(0.7) = 0.50$, to two decimal places. Since g is also a decreasing function,

$$x \in [0.4, 0.7] \quad \Rightarrow \quad g(x) \in [0.4, 0.7].$$

We have $g'(x) = -e^{-x}$ and, therefore, $|g'(x)| < 0.7$ on $[0.4, 0.7]$. Thus g is a contraction mapping with $L = 0.7$ and Theorem 8.3 shows that the process $x_{r+1} = e^{-x}$, converges to the unique root of $x = e^{-x}$ in $[0.4, 0.7]$, for any x_0 in that interval. This confirms the apparent convergence of Example 8.5. □

As in Example 8.9, it can take some ingenuity to determine an interval $[a, b]$ on which g is a contraction mapping and therefore Theorem 8.3 is not of great *practical* importance. In particular, the closure condition is often awkward to verify. In practice, we will not often choose a first order method to solve an equation $f(x) = 0$. If f' can easily be evaluated we would probably prefer to use Newton's method because of its second order convergence. If it is difficult to evaluate f', we would opt for a 'bracketing' method such as Muller's method, which in general converges faster than the first order method based on contraction mapping. On occasions, we may even wish to use the bisection method because of its simplicity, despite the relatively large number of evaluations of f which it requires.

Problems

Section 8.2

8.1 Show that the equation $1 - x - \sin x = 0$ has a root in $[0, 1]$. How many iterations of the bisection method are required to estimate this root with an error not greater than $\frac{1}{2} \times 10^{-4}$?

8.2 Write a computer program to implement Algorithm 8.1 (the bisection method). Test it by estimating the positive root of $x^2 - x - 1 = 0$.

Section 8.3

8.3 Write computer programs for the *regula falsi* and secant methods. Apply them to estimate the root of $1 - x - \sin x = 0$, beginning with $x_0 = 0$ and $x_1 = 1$.

8.4 Suppose $f(x_0) > 0$ and $f(x_1) < 0$ and let f'' exist and be positive on $[x_0, x_1]$. (For example $f(x) = 1 - x - \sin x$ with $x_0 = 0$, $x_1 = 1$, as in the previous problem.) By drawing a graph, show that x_0 remains unchanged when the *regula falsi* method is used.

8.5 Estimate the smallest positive root of $8x^4 - 8x^2 + 1 = 0$ as follows. Take $x_0 = 0.3$, $x_1 = 0.4$ and use one iteration of the secant method followed by one iteration of Muller's method. (Note that the equation is $T_4(x) = 0$ and the smallest positive root is therefore $\cos(3\pi/8)$. See §4.10.)

Section 8.4

8.6 Compare the amount of calculation required to find the root of $e^x + 10x - 2 = 0$ to three decimal places by (i) the bisection method beginning with $x_0 = 0$, $x_1 = 1$ and (ii) the iterative scheme $x_{r+1} = (2 - e^{x_r})/10$, $r = 0, 1, \ldots,$ with $x_0 = 0$.

8.7 Find the root of the equation $2^x - 5x + 2 = 0$ by using the iterative scheme $x_{r+1} = (2 + 2^{x_r})/5$, with $x_0 = 0$. Prove that the sequence (x_r) is monotonic increasing.

8.8 A function f is such that $f'(x)$ exists and $M \geqslant f'(x) \geqslant m > 0$ for all x. Prove that the sequence (x_r) generated by $x_{r+1} = x_r - \lambda f(x_r)$, $r = 0, 1, \ldots,$ and arbitrary x_0, is convergent to the zero of f for any choice of λ in the range $0 < \lambda < 2/M$.

Section 8.5

8.9 Choose a constant λ so that the process

$$x_{r+1} = (\lambda x_r + 1 - \sin x_r)/(1 + \lambda)$$

will give a rapidly convergent method for finding the root near $x = 0.5$ of $1 - x - \sin x = 0$. Calculate the first few iterates, beginning with $x_0 = 0.5$.

8.10 Show that the number x^*_{r+1}, defined by (8.23), obtained by applying Aitken acceleration to the three successive iterates x_{r-1}, x_r and x_{r+1}, may also be expressed as

$$x^*_{r+1} = x_{r-1} - \frac{(\Delta x_{r-1})^2}{\Delta^2 x_{r-1}}.$$

8.11 If $x_r = a_1 + a_2 + \cdots + a_r$, show that the result of applying Aitken acceleration to x_{n-2}, x_{n-1} and x_n is

$$x^*_n = x_{n-2} + \frac{a^2_{n-1}}{a_{n-1} - a_n},$$

if $a_{n-1} \neq a_n$.

8.12 Working to five decimal places, apply the result of Problem 8.11, with $n = 12$, to estimate

$$\log 2 = 1 - \tfrac{1}{2} + \tfrac{1}{3} - \tfrac{1}{4} + \cdots.$$

8.13 Let (x^*_r) and (y^*_r) denote sequences obtained by applying Aitken acceleration to (x_r) and (y_r) respectively. If $x_r = c + y_r$ for all r (where c is constant), show that $x^*_r = c + y^*_r$. (For example, if $x_0 = 2.791$, $x_1 = 2.737$, $x_2 = 2.723$, we can apply Aitken acceleration to the last two digits of each number.)

Section 8.6

8.14 Obtain a quadratic equation to which the iterative method

$$x_{r+1} = (bx^2_r + 2cx_r)/(c - ax^2_r)$$

relates. Show that, if $c \neq 0$ and $ax^2 \neq c$ at a root, the method is exactly second order. What happens if $c = 0$?

8.15 Apply Newton's method to find the root of $2^x - 5x + 2 = 0$, taking $x_0 = 0$.

8.16 Given some number $x_0 > 0$, verify that the iterative scheme $x_{r+1} = c/x_r$, $r = 0, 1, \ldots$, produces the periodic sequence x_0, c/x_0, x_0, c/x_0, ... Show that, although this is obviously not a viable method for calculating $c^{1/2}$, Aitken acceleration on the three iterates x_0, c/x_0, x_0 yields the number $\frac{1}{2}(x_0 + c/x_0)$, which is equivalent to one iteration of Newton's method.

8.17 Derive an iterative process for finding the pth root of a number $c > 0$ by applying Newton's method to the equation $x^p - c = 0$.

8.18 Show that, if $c > 0$, all the conditions of Theorem 8.2 hold for the function $f(x) = x^2 - c$ on an interval $[a, b]$ with $0 < a < c^{1/2}$ and $b > \frac{1}{2}(a + c/a)$.

8.19 Deduce from (8.37) and another similar result that the sequence of iterates (x_r), produced by the Newton square root method, satisfy

$$\frac{x_r - c^{1/2}}{x_r + c^{1/2}} = \left(\frac{x_0 - c^{1/2}}{x_0 + c^{1/2}}\right)^{2^r}.$$

Hence show that, for any $x_0 > 0$, the sequence (x_r) converges to $c^{1/2}$.

8.20 What happens in Newton's square root process if we choose $x_0 < 0$?

8.21 If $\frac{1}{4} \leqslant c < 1$, let x_1 and x_2 denote the iterates produced by Newton's method for finding $c^{1/2}$, beginning with $x_0 = \frac{2}{3}c + \frac{17}{48}$. Noting that $x_0 > \frac{1}{2}$ as $c \geqslant \frac{1}{4}$, deduce from (8.37) that $|x_1 - c^{1/2}| < (x_0 - c^{1/2})^2$. Hence, using (8.38), show that $|x_2 - c^{1/2}| < 2 \times 10^{-7}$.

8.22 If $f(x) = (x - \alpha)^2 h(x)$ and $h(\alpha) \neq 0$, show that $f'(\alpha) = 0$ and that, in this case, Newton's method is *not* a second order process for finding α. Find a value of λ so that the following modified version of Newton's method is at least second order for this function:

$$x_{r+1} = x_r - \lambda f(x_r)/f'(x_r).$$

8.23 Apply Newton's method to the equation $1/x - c = 0$. (Notice that this provides an algorithm for computing the reciprocal of a number c without performing the arithmetical operation of division.)

8.24 One way of obtaining iterative methods of solving $f(x) = 0$ is to use an approximating function as in Muller's method but based on Taylor polynomials. If x_n is the nth iterate, we construct the Taylor polynomial at x_n,

$$p_k(x) = f(x_n) + \frac{(x - x_n)}{1!} f'(x_n) + \cdots + \frac{(x - x_n)^k}{k!} f^{(k)}(x_n),$$

assuming the derivatives exist. We now take x_{n+1} to be a root of $p_k(x) = 0$. Show that for $k = 1$ we obtain Newton's method and that for $k = 2$

$$x_{n+1} = x_n + \frac{-f'(x_n) \pm [f'(x_n)^2 - 2f(x_n)f''(x_n)]^{1/2}}{f''(x_n)},$$

the + or − sign being taken according to whether $f'(x_n)$ is positive or negative. Show that the last method is at least third order if $f'(\alpha) \neq 0$ and $f''(\alpha) \neq 0$, where α is the root. The method is difficult to use as we must calculate a square root in each iteration. A better third order method is given in the next problem.

8.25 Show that the iterative method

$$x_{n+1} = x_n - \frac{f(x_n)}{f'(x_n)} - \frac{f''(x_n)}{2f'(x_n)}\left(\frac{f(x_n)}{f'(x_n)}\right)^2,$$

used to solve $f(x)=0$, is of order 3 (at least) provided $f'(\alpha)\neq 0$ where α is the root. Obtain a third order square root process.

8.26 Suppose $y=f(x)$ has a unique inverse function $x=\phi(y)$. If ϕ is sufficiently differentiable, we have the Taylor series

$$\phi(y)=\phi(y_n)+(y-y_n)\phi'(y_n)+\frac{(y-y_n)^2}{2!}\phi''(y_n)+\cdots.$$

By expressing ϕ' and ϕ'' in terms of the derivatives of f, derive the iterative method quoted in Problem 8.25.

Section 8.7

8.27 If $x_{r+1}=g(x_r)$, $r=0,1,\ldots,$ and g and x_0 satisfy the conditions of Theorem 8.3, deduce that

$$|x_{r+1}-x_r|\leqslant L|x_r-x_{r-1}|,\qquad r=1,2,\ldots,$$

and hence that

$$|x_{r+1}-x_r|\leqslant L^r|x_1-x_0|,\qquad r=0,1,\ldots.$$

8.28 Applying the second inequality derived in Problem 8.27 to the inequality

$$|x_{m+n}-x_n|\leqslant|x_{m+n}-x_{m+n-1}|+\cdots+|x_{n+1}-x_n|,$$

deduce that, for $m>0$, $n\geqslant 0$,

$$|x_{m+n}-x_n|\leqslant\frac{L^n}{1-L}|x_1-x_0|.$$

Letting $m\to\infty$, obtain the estimate

$$|\alpha-x_n|\leqslant\frac{L^n}{1-L}|x_1-x_0|$$

for the error between the iterate x_n and the root α of the equation $x=g(x)$.

8.29 Deduce from Problem 8.28 that

$$|x_{n+1}-\alpha|\leqslant\frac{L}{1-L}|x_{n+1}-x_n|.$$

(This enables us to use $x_{n+1}-x_n$ to decide when to stop iterating.)

8.30 By summing the geometric series on the right, show that the positive root of the equation

$$x^k=1+x+x^2+\cdots+x^{k-1},\qquad k\geqslant 2,$$

also satisfies $x = 2 - x^{-k}$. Show that the conditions of Theorem 8.3 apply to this last equation on the interval $[2 - 2^{1-k}, 2]$. For the case $k = 6$, find the root to three decimal places. (It may be assumed that $(1 - 2^{-k})^k \geqslant \frac{1}{2}$ for $k = 1, 2, \ldots$.)

8.31 Suppose that on some interval I, g and g' are continuous and that $x = g(x)$ has a root $\alpha \in I$ with $|g'(\alpha)| < 1$. Show that g is a contraction mapping on some sub-interval of I containing α.

8.32 Suppose that for all x, g and g' exist with $|g'(x)| \leqslant L < 1$ and that $y = g(x)$ has a unique inverse function $x = \psi(y)$ such that $\psi'(y)$ exists for all y. Show that $\psi(y)$ is not a contraction mapping, regardless of the interval chosen.

8.33 Show that we cannot expect the iterative method

$$x_{r+1} = c/x_r^{p-1}$$

to be successful in finding the pth root (p a positive integer other than $+1$) of a positive number c.

Chapter **9**

LINEAR EQUATIONS

9.1 Introduction

In this chapter we shall be considering the solution of simultaneous linear algebraic equations. We shall consider separately (in Chapter 12) the more difficult problem of solving non-linear equations, as the methods for these are quite different.

If three unknowns x, y and z are related by equations of the form

$$\begin{aligned} a_1x + b_1y + c_1z &= d_1 \\ a_2x + b_2y + c_2z &= d_2 \\ a_3x + b_3y + c_3z &= d_3 \end{aligned} \tag{9.1}$$

where a_i, b_i, c_i and d_i $(i = 1, 2, 3)$ are given constants, we say that (9.1) is a system of three linear equations. Since we have three equations in three unknowns, we shall usually expect that there is a unique solution, that is, there are unique values of x, y and z satisfying (9.1). We shall see that this is not always true and the intention in the first part of this chapter is to revise basic material on the solution of linear equations including conditions for the existence of such solutions. We find it convenient to express linear equations in terms of *matrices* and therefore we introduce the latter first.

9.2 Matrices

An $m \times n$ matrix $\mathbf{A}$ is an ordered set of mn real (complex) numbers a_{ij} written in m rows and n columns:

$$\mathbf{A} = \begin{bmatrix} a_{11} & a_{12} & \cdots & a_{1n} \\ a_{21} & a_{22} & \cdots & a_{2n} \\ \vdots & & & \vdots \\ a_{m1} & a_{m2} & \cdots & a_{mn} \end{bmatrix}.$$

The *element* a_{ij} is in the ith row and jth column and **A** is said to be of *order* or *dimension* $m \times n$. We usually use bold capital letters to denote matrices and italic small letters to denote elements. If $m = n$ we say that **A** is a *square* matrix of order or dimension n. We refer to the elements a_{ii}, $i = 1, 2, \ldots, n$ as the *diagonal* of a square matrix.

A *column vector* is a matrix with just one column. For example,

$$\begin{bmatrix} x_1 \\ x_2 \\ \vdots \\ x_m \end{bmatrix}$$

is of order $m \times 1$ and is called a *column vector of dimension m*. We usually denote such a column vector by a small bold letter, for example, $\mathbf{x}$. We write $\mathbb{R}^m$ to denote the set of all m-dimensional column vectors with real elements. ($\mathbb{C}^m$ is used to denote the corresponding set of vectors with complex elements.) $\mathbb{R}^n$ is usually referred to as n – *dimensional real space*.

A *row vector* is a matrix with just one row. For example,

$$[x_1 \; x_2 \; \ldots \; x_n]$$

is of order $1 \times n$ and is called a *row vector of dimension n*. We usually denote such a vector by $\mathbf{x}^{\mathrm{T}}$. The superscript T indicates that the vector is a row vector obtained by writing the elements of an n-dimensional column vector $\mathbf{x}$ in a row. We call $\mathbf{x}^{\mathrm{T}}$ the *transpose* of $\mathbf{x} \in \mathbb{R}^n$.

Two matrices are said to be *equal* if they are of the same order and corresponding elements are equal, that is, $\mathbf{A} = \mathbf{B}$ if $a_{ij} = b_{ij}$ for all i and j. We add or subtract two matrices of the same dimensions by adding or subtracting corresponding elements. Thus $\mathbf{C} = \mathbf{A} \pm \mathbf{B}$ if $c_{ij} = a_{ij} \pm b_{ij}$ for all i and j. Note from the definition that addition is both *commutative*, $\mathbf{A} + \mathbf{B} = \mathbf{B} + \mathbf{A}$, and *associative*, $\mathbf{A} + (\mathbf{B} + \mathbf{C}) = (\mathbf{A} + \mathbf{B}) + \mathbf{C}$. (The *zero matrix* of order $m \times n$ is denoted by $\mathbf{0}$ and has zero for each element, whence $\mathbf{A} + \mathbf{0} = \mathbf{A}$.)

If λ is a real (or complex) number we define $\lambda\mathbf{A}$ to be the matrix with elements λa_{ij}, that is, each element of $\mathbf{A}$ is multiplied (scaled) by λ which is called a *scalar* in this context. We usually write $-\mathbf{A}$ in place of $(-1)\mathbf{A}$ and note that $\mathbf{A} + (-\mathbf{A}) = \mathbf{0}$, $0\mathbf{A} = \mathbf{0}$.

Multiplication of matrices is more complicated. We start with the product of a row vector and a column vector. If $\mathbf{x}, \mathbf{y} \in \mathbb{R}^n$ we define

$$\mathbf{y}^{\mathrm{T}}\mathbf{x} = y_1x_1 + y_2x_2 + \cdots + y_nx_n = \sum_{i=1}^{n} y_ix_i.$$

The result is a single number and is called the *scalar* or *inner product* of $\mathbf{y}$ and $\mathbf{x}$.

If $\mathbf{A}$ is an $m \times n$ matrix and $\mathbf{x} \in \mathbb{R}^n$, we define

$$\mathbf{Ax} = \begin{bmatrix} a_{11} & a_{12} & \cdots & a_{1n} \\ a_{21} & a_{22} & \cdots & a_{2n} \\ \vdots & \vdots & & \vdots \\ a_{m1} & a_{m2} & \cdots & a_{mn} \end{bmatrix} \begin{bmatrix} x_1 \\ x_2 \\ \vdots \\ x_n \end{bmatrix} = \begin{bmatrix} a_{11}x_1 + a_{12}x_2 + \cdots + a_{1n}x_n \\ a_{21}x_1 + a_{22}x_2 + \cdots + a_{2n}x_n \\ \vdots \\ a_{m1}x_1 + a_{m2}x_2 + \cdots + a_{mn}x_n \end{bmatrix}.$$

The result is a vector $\mathbf{z} \in \mathbb{R}^m$ with ith element

$$z_i = \sum_{j=1}^{n} a_{ij}x_j = [i\text{th row of } \mathbf{A}] \begin{bmatrix} x_1 \\ \vdots \\ x_n \end{bmatrix}.$$

We can now write the equations (9.1) as

$$\mathbf{Ax} = \mathbf{d},$$

where

$$\mathbf{A} = \begin{bmatrix} a_1 & b_1 & c_1 \\ a_2 & b_2 & c_2 \\ a_3 & b_3 & c_3 \end{bmatrix}, \qquad \mathbf{x} = \begin{bmatrix} x \\ y \\ z \end{bmatrix}, \qquad \mathbf{d} = \begin{bmatrix} d_1 \\ d_2 \\ d_3 \end{bmatrix}.$$

If $\mathbf{A}$ is an $m \times n$ matrix and $\mathbf{B}$ is an $n \times p$ matrix, we define their product $\mathbf{C} = \mathbf{AB}$ to be the $m \times p$ matrix with (i, j)th element

$$c_{ij} = \sum_{k=1}^{n} a_{ik}b_{kj} = [i\text{th row of } \mathbf{A}] \begin{bmatrix} j\text{th} \\ \text{column} \\ \text{of } \mathbf{B} \end{bmatrix}.$$

You will notice that the number of elements in a row of $\mathbf{A}$ must be the same as the number of elements in a column of $\mathbf{B}$ (number of columns in $\mathbf{A}$ = number of rows in $\mathbf{B}$). The multiplication rules for vectors are special cases of this more general rule.

Multiplication is not necessarily commutative even if both $\mathbf{AB}$ and $\mathbf{BA}$ are defined; that is, $\mathbf{AB} \neq \mathbf{BA}$ in general (see Problem 9.2). Thus the order in which two matrices are multiplied is important and we say that, for a product $\mathbf{AB}$, the matrix $\mathbf{A}$ is *postmultiplied* by $\mathbf{B}$ or that $\mathbf{B}$ is *premultiplied* by $\mathbf{A}$.

Multiplication is associative so that if $\mathbf{A}$, $\mathbf{B}$ and $\mathbf{C}$ are of suitable dimensions

$$\mathbf{A}(\mathbf{BC}) = (\mathbf{AB})\mathbf{C}. \tag{9.2}$$

Addition is distributive over multiplication so that provided the matrices are of suitable dimensions

$$\mathbf{A}(\mathbf{B}+\mathbf{C})=\mathbf{AB}+\mathbf{AC} \tag{9.3}$$

and

$$(\mathbf{A}+\mathbf{B})\mathbf{C}=\mathbf{AC}+\mathbf{BC}. \tag{9.4}$$

We define the $n \times n$ *unit matrix* $\mathbf{I}$ as

$$\mathbf{I}=\begin{bmatrix} 1 & 0 & \dots & 0 \\ 0 & 1 & \ddots & \vdots \\ \vdots & \ddots & \ddots & 0 \\ 0 & \dots & 0 & 1 \end{bmatrix}.$$

It may be verified from the multiplication rule that, if $\mathbf{A}$ is any $m \times n$ matrix and $\mathbf{I}$ is the $n \times n$ unit matrix, then

$$\mathbf{AI}=\mathbf{A}$$

and, if $\mathbf{I}$ is the $m \times m$ unit matrix,

$$\mathbf{IA}=\mathbf{A}.$$

Under certain circumstances a square matrix $\mathbf{A}$ has an inverse written as $\mathbf{A}^{-1}$ which satisfies

$$\mathbf{AA}^{-1}=\mathbf{I} \quad \text{and} \quad \mathbf{A}^{-1}\mathbf{A}=\mathbf{I}. \tag{9.5}$$

If such an inverse $\mathbf{A}^{-1}$ exists we say that $\mathbf{A}$ is *non-singular*. The inverse of the product of square matrices $\mathbf{A}$ and $\mathbf{B}$ satisfies

$$(\mathbf{AB})^{-1}=\mathbf{B}^{-1}\mathbf{A}^{-1}. \tag{9.6}$$

We sometimes find it convenient to write matrices in a block form. The $m \times n$ matrix $\mathbf{A}$ may be written as

$$\mathbf{A}=\begin{bmatrix} \mathbf{A}_{11} & \mathbf{A}_{12} \\ \mathbf{A}_{21} & \mathbf{A}_{22} \end{bmatrix}$$

where $\mathbf{A}_{11}$ is $r \times p$, $\mathbf{A}_{12}$ is $r \times (n-p)$, $\mathbf{A}_{21}$ is $(m-r) \times p$ and $\mathbf{A}_{22}$ is $(m-r) \times (n-p)$. The restrictions on dimensions are necessary so that, for example, $\mathbf{A}_{11}$ and $\mathbf{A}_{21}$ have the same number of columns. We call $\mathbf{A}_{11}$, $\mathbf{A}_{12}$, $\mathbf{A}_{21}$ and $\mathbf{A}_{22}$ *submatrices* of $\mathbf{A}$.

It is easily seen that if

$$\mathbf{B}=\begin{bmatrix} \mathbf{B}_{11} & \mathbf{B}_{12} \\ \mathbf{B}_{21} & \mathbf{B}_{22} \end{bmatrix}$$

is divided into blocks of the same dimensions as those of $\mathbf{A}$, then

$$\mathbf{A}+\mathbf{B}=\begin{bmatrix}\mathbf{A}_{11}+\mathbf{B}_{11} & \mathbf{A}_{12}+\mathbf{B}_{12}\\ \mathbf{A}_{21}+\mathbf{B}_{21} & \mathbf{A}_{22}+\mathbf{B}_{22}\end{bmatrix}.$$

It is not so obvious, however, that provided $\mathbf{B}$ and its blocks are of suitable dimensions

$$\mathbf{AB}=\begin{bmatrix}(\mathbf{A}_{11}\mathbf{B}_{11}+\mathbf{A}_{12}\mathbf{B}_{21}) & (\mathbf{A}_{11}\mathbf{B}_{12}+\mathbf{A}_{12}\mathbf{B}_{22})\\ (\mathbf{A}_{21}\mathbf{B}_{11}+\mathbf{A}_{22}\mathbf{B}_{21}) & (\mathbf{A}_{21}\mathbf{B}_{12}+\mathbf{A}_{22}\mathbf{B}_{22})\end{bmatrix}. \tag{9.7}$$

The proof of this last result is left for Problem 9.9.

9.3 Linear equations

We start by considering a simple example.

Example 9.1 Solve the set of equations

$$2x_1+2x_2+3x_3=\ \ 3 \tag{9.8a}$$

$$4x_1+7x_2+7x_3=\ \ 1 \tag{9.8b}$$

$$-2x_1+4x_2+5x_3=-7. \tag{9.8c}$$

We will solve these by an *elimination* process. First we use equation (9.8a) to eliminate x_1 from the other two equations. If we subtract twice (9.8a) from (9.8b) and add (9.8a) to (9.8c), we obtain the equations

$$\begin{aligned}2x_1+2x_2+3x_3&=\ \ 3\\ 3x_2+x_3&=-5\\ 6x_2+8x_3&=-4.\end{aligned}$$

We now eliminate x_2 from the third equation by subtracting twice the second equation. We obtain

$$2x_1+2x_2+3x_3=\ \ 3 \tag{9.9a}$$

$$3x_2+x_3=-5 \tag{9.9b}$$

$$6x_3=\ \ 6. \tag{9.9c}$$

Immediately from (9.9c) we deduce that $x_3=1$. On substituting this value of x_3 in (9.9b) we obtain $3x_2=-6$, so that $x_2=-2$. Finally, from (9.9a) we have

$$2x_1=3-2x_2-3x_3=4,$$

whence $x_1 = 2$. The equations (9.9) are called *triangular equations* and the process of working in reverse order to solve them is called *back substitution.* □

The method of Example 9.1 is known as *Gauss elimination* and may be extended to the m equations in n unknowns $x_1, x_2, \ldots, x_n$,

$$\begin{aligned} a_{11}x_1 + a_{12}x_2 + \cdots + a_{1n}x_n &= b_1 \\ a_{21}x_1 + a_{22}x_2 + \cdots + a_{2n}x_n &= b_2 \\ \vdots \qquad\qquad\qquad & \quad \vdots \\ a_{m1}x_1 + a_{m2}x_2 + \cdots + a_{mn}x_n &= b_m \end{aligned} \tag{9.10}$$

or

$$\mathbf{Ax} = \mathbf{b},$$

where the coefficients a_{ij} and b_i are given. We detach the coefficients and write them as the $m \times (n+1)$ matrix

$$[\mathbf{A}:\mathbf{b}] = \begin{bmatrix} a_{11} & a_{12} & \ldots & a_{1n}:b_1 \\ a_{21} & a_{22} & \cdots & a_{2n}:b_2 \\ \vdots & & & \vdots \\ a_{m1} & a_{m2} & \ldots & a_{mn}:b_m \end{bmatrix}. \tag{9.11}$$

The following operations may be made on the matrix (9.11) without affecting the solution of the corresponding equations.

(i) Rows of (9.11) may be interchanged. This merely changes the order in which the equations (9.10) are written.
(ii) Any row of (9.11) may be multiplied by a non-zero scalar. This is equivalent to multiplying an equation in (9.10) by the scalar.
(iii) Any row of (9.11) may be added to another. This is equivalent to adding one equation to another in (9.10).
(iv) The first n columns of (9.11) may be interchanged provided corresponding unknowns are interchanged. This is equivalent to changing the order in which the unknowns are written in (9.10). For example, if we interchange the first two columns of (9.11), the first two terms in each of the equations (9.10) are interchanged and the order of the unknowns becomes $x_2, x_1, x_3, x_4, \ldots, x_n$.

We use these operations to reduce the equations to a simpler form. Firstly we ensure that the leading element a_{11} is non-zero by making row and/or column interchanges. We now subtract multiples of the first row from the other rows so as to make the rest of the coefficients in the first column zero.

Thus, assuming that no interchanges were required at the start, we obtain the matrix

$$\begin{bmatrix} a_{11} & a_{12} & \dots & a_{1n} : b_1 \\ 0 & a'_{22} & \cdots & a'_{2n} : b'_2 \\ 0 & a'_{32} & \dots & a'_{3n} : b'_3 \\ \vdots & \vdots & & \vdots \quad \vdots \\ 0 & a'_{m2} & \dots & a'_{mn} : b'_m \end{bmatrix}$$

where, for example, the second row is obtained by subtracting $(a_{21}/a_{11}) \times$ the first row. Thus

$$a'_{22} = a_{22} - (a_{21}/a_{11})a_{12}$$

and

$$b'_m = b_m - (a_{m1}/a_{11})b_1.$$

We have eliminated x_1 from the 2nd to the mth equations.

We now ensure that the element a'_{22} is non-zero by making row and/or column interchanges, using rows 2 to m or columns 2 to n. Such interchanges do not affect the positions of zeros in the first column. We next subtract multiples of the second row from later rows to produce zeros in the rest of the second column, that is, we eliminate the second unknown from the 3rd to the mth equations.

We repeat this process and at the pth stage we have a matrix of the form:

$$\begin{bmatrix} \alpha_{11} & \alpha_{12} & \dots & \dots & \dots & \alpha_{1n} & :\beta_1 \\ 0 & \alpha_{22} & \dots & \dots & \dots & \alpha_{2n} & :\beta_2 \\ \vdots & \ddots & \ddots & & & \vdots & \vdots \\ \vdots & & 0 & \alpha_{pp} & \dots & \alpha_{pn} & :\beta_p \\ \vdots & & \vdots & \alpha_{p+1,p} & \dots & \alpha_{p+1,n} & :\beta_{p+1} \\ \vdots & & \vdots & \vdots & & \vdots & \vdots \\ 0 & \dots & 0 & \alpha_{mp} & \dots & \alpha_{mn} & :\beta_m \end{bmatrix} \tag{9.12}$$

The steps to be followed next are:

(i) interchange rows p to m or columns p to n so that α_{pp} is non-zero,
(ii) subtract multiples of row p from rows $p+1$ to m so that elements in the column below become zeros.

You will notice that neither of these steps will affect the positions of zeros in the earlier columns. At this stage α_{pp} is called the *pivot* and row p the *pivotal row*, corresponding to the *pivotal equation*.

We stop for one of three reasons:

- *either* it is not possible to make $\boldsymbol{\alpha}_{pp}$ non-zero by interchanges, as all the $\boldsymbol{\alpha}_{ij}$ in rows p to m are zero
- *or* $p = m$, when we have exhausted the equations
- *or* $p = n$, when we have exhausted the unknowns.

The final form of the matrix is

$$\begin{bmatrix} u_{11} & u_{12} & u_{13} & \dots & u_{1r} & u_{1,r+1} & \dots & u_{1n} : c_1 \\ 0 & u_{22} & u_{23} & \dots & u_{2r} & u_{2,r+1} & \dots & u_{2n} : c_2 \\ \vdots & & & & \vdots & \vdots & & \vdots \;\; \vdots \\ 0 & \dots & \dots & 0 & u_{rr} & u_{r,r+1} & \dots & u_{rn} : c_r \\ 0 & \dots & \dots & \dots & 0 & 0 & \dots & 0 \;\; : c_{r+1} \\ \vdots & & & & \vdots & \vdots & & \vdots \;\; \vdots \\ 0 & \dots & \dots & \dots & 0 & 0 & \dots & 0 \;\; : c_m \end{bmatrix} \tag{9.13}$$

which we may write in the block form

$$\begin{bmatrix} \mathbf{U} & \mathbf{B} : \mathbf{c} \\ \mathbf{0} & \mathbf{0} : \mathbf{d} \end{bmatrix}. \tag{9.14}$$

$\mathbf{U}$ is the $r \times r$ *upper triangular* matrix

$$\mathbf{U} = \begin{bmatrix} u_{11} & \dots & \dots & u_{1r} \\ 0 & u_{22} & & \vdots \\ \vdots & \ddots & \ddots & \vdots \\ 0 & \dots & 0 & u_{rr} \end{bmatrix},$$

$\mathbf{B}$ is $r \times (n - r)$ and the zero matrix below $\mathbf{B}$ is $(m - r) \times (n - r)$. The pivots u_{ii}, $i = 1, \dots, r$, are non-zero.

We suppose that, as a result of column interchanges, the order of the unknowns $x_1, x_2, \dots, x_n$ has been changed and that the unknowns corresponding to (9.13) are $y_1, y_2, \dots, y_n$. If

$$\mathbf{y} = \begin{bmatrix} y_1 \\ \vdots \\ y_r \end{bmatrix} \quad \text{and} \quad \mathbf{z} = \begin{bmatrix} y_{r+1} \\ \vdots \\ y_n \end{bmatrix}$$

the equations may, from (9.14), be written in the block form

$$\begin{bmatrix} \mathbf{U} & \mathbf{B} \\ \mathbf{0} & \mathbf{0} \end{bmatrix} \begin{bmatrix} \mathbf{y} \\ \mathbf{z} \end{bmatrix} = \begin{bmatrix} \mathbf{c} \\ \mathbf{d} \end{bmatrix}. \tag{9.15}$$

Thus

$$\mathbf{Uy} + \mathbf{Bz} = \mathbf{c} \tag{9.16a}$$

and

$$\mathbf{0y} + \mathbf{0z} = \mathbf{d}. \tag{9.16b}$$

The original equations (9.10) and the reduced equations (9.16) are equivalent, in that a solution of (9.16) must also be a solution of (9.10) (possibly after some re-ordering of the unknowns) and vice versa. We therefore investigate the existence of solutions of (9.16) in detail.

The integer r in (9.13) is called the *rank* of the original coefficient matrix $\mathbf{A}$. The existence of solutions will depend on this rank. If $r = m = n$, (9.16) is the triangular set of equations

$$\mathbf{Uy} = \mathbf{c}$$

where

$$\mathbf{U} = \begin{bmatrix} u_{11} & \cdots & u_{1n} \\ 0 & \ddots & \vdots \\ \vdots & \ddots & \vdots \\ 0 & \cdots & u_{nn} \end{bmatrix}, \qquad \mathbf{c} = \begin{bmatrix} c_1 \\ \vdots \\ c_n \end{bmatrix}, \qquad \mathbf{y} = \begin{bmatrix} y_1 \\ \vdots \\ y_n \end{bmatrix}$$

with $u_{ii} \neq 0$, $i = 1, 2, \ldots, n$. We can solve these equations by back substitution. From the last equation

$$u_{nn} y_n = c_n,$$

we find

$$y_n = c_n / u_{nn}.$$

From the penultimate equation

$$u_{n-1,n-1} y_{n-1} + u_{n-1,n} y_n = c_{n-1},$$

we find

$$y_{n-1} = (c_{n-1} - u_{n-1,n} y_n) / u_{n-1,n-1}$$

and so on. We work back through the equations, introducing new unknowns one at a time and hence find all the unknowns in reverse order. Thus, for $r = m = n$, the unknowns are all uniquely determined, as we would normally expect for n equations in n unknowns. We then say that the equations (9.10) are *consistent* with *unique* solution or just simply *non-singular*. We also find that the square coefficient matrix $\mathbf{A}$ is non-singular.

If $r < m$ in (9.13) and $\mathbf{d} \neq \mathbf{0}$† in (9.16), that is, at least one of the c_{r+1}, $c_{r+2}, \ldots, c_m$ is non-zero, we cannot satisfy (9.16b) for *any* choice of $\mathbf{y}$ and $\mathbf{z}$.

† By $\mathbf{d} \neq \mathbf{0}$ we mean that $\mathbf{d}$ is a vector other than the zero vector. It is possible for some components of $\mathbf{d}$ to be zero, but not all components may be zero. This is consistent with the definition of equality of matrices.

There is no solution of (9.16) and hence of (9.10) and we say that the equations are *inconsistent.*

If $r<m$ in (9.13) and $\mathbf{d}=\mathbf{0}$ then (9.16b) is satisfied for any $\mathbf{y}$ and $\mathbf{z}$ and (9.16a) may be written as

$$\mathbf{U}\mathbf{y}=\mathbf{c}-\mathbf{B}\mathbf{z}. \tag{9.17}$$

Similarly when $r=m<n$, (9.16b) does not apply and we obtain only (9.17). We may choose the elements of $\mathbf{z}$ arbitrarily. This fixes the right side of (9.17) and we may then solve these r triangular equations by back substitution to find $\mathbf{y}$. We say that the equations (9.10) are *consistent* with an *infinity of solutions*. A particular $n-r$ of the unknowns (those corresponding to $\mathbf{z}$) may be chosen arbitrarily; the remaining r unknowns are then determined uniquely.

We will restrict ourselves to the case $n=m$ so that hopefully the equations are consistent with a unique solution which can be determined numerically. Of course the equations may still turn out to have no solution as is illustrated by the second of the following examples. For convenience all coefficients are integers; this will not be the case in most problems.

Example 9.2 Solve

$$\begin{aligned}
x_1+3x_2-2x_3-\ 4x_4&=\ 3\\
2x_1+6x_2-7x_3-10x_4&=-2\\
-x_1-\ x_2+5x_3+\ 9x_4&=\ 14\\
-3x_1-5x_2\qquad\quad +15x_4&=-6.
\end{aligned}$$

We detach coefficients and subtract multiples of the first equation from the others to eliminate x_1:

$$\begin{bmatrix} 1 & 3 & -2 & -4: & 3\\ 2 & 6 & -7 & -10: & -2\\ -1 & -1 & 5 & 9: & 14\\ -3 & -5 & 0 & 15: & -6\end{bmatrix}\quad \begin{bmatrix} 1 & 3 & -2 & -4: & 3\\ 0 & 0 & -3 & -2: & -8\\ 0 & 2 & 3 & 5: & 17\\ 0 & 4 & -6 & 3: & 3\end{bmatrix}.$$

We interchange the second and third columns so that the second element in the second row is non-zero and use this row to eliminate the second variable:

$$\begin{bmatrix} 1 & -2 & 3 & -4: & 3\\ 0 & -3 & 0 & -2: & -8\\ 0 & 3 & 2 & 5: & 17\\ 0 & -6 & 4 & 3: & 3\end{bmatrix}\quad \begin{bmatrix} 1 & -2 & 3 & -4: & 3\\ 0 & -3 & 0 & -2: & -8\\ 0 & 0 & 2 & 3: & 9\\ 0 & 0 & 4 & 7: & 19\end{bmatrix}.$$

The last step is to eliminate the third variable from the fourth equation and we obtain the triangular set

$$\begin{bmatrix} 1 & -2 & 3 & -4: & 3 \\ 0 & -3 & 0 & -2: & -8 \\ 0 & 0 & 2 & 3: & 9 \\ 0 & 0 & 0 & 1: & 1 \end{bmatrix}.$$

Thus by back substitution

$$\begin{aligned} y_4 &= 1, & & \\ 2y_3 &= 9 - 3y_4, & y_3 &= 3, \\ -3y_2 &= -8 - 0.y_3 + 2y_4, & y_2 &= 2, \\ y_1 &= 3 + 2y_2 - 3y_3 + 4y_4, & y_1 &= 2. \end{aligned}$$

We have interchanged the second and third columns and, therefore,

$$x_1 = y_1, \qquad x_2 = y_3, \qquad x_3 = y_2, \qquad x_4 = y_4.$$

Hence

$$\mathbf{x} = \begin{bmatrix} 2 \\ 3 \\ 2 \\ 1 \end{bmatrix}.$$

The equations are non-singular with rank $r = n = 4$. □

Example 9.3 Consider

$$\begin{bmatrix} 1 & -1 & 1 \\ 1 & 2 & -1 \\ 1 & 5 & -3 \end{bmatrix} \begin{bmatrix} x_1 \\ x_2 \\ x_3 \end{bmatrix} = \begin{bmatrix} 2 \\ 0 \\ 3 \end{bmatrix}.$$

We detach coefficients and the steps of the reduction process are

$$\begin{bmatrix} 1 & -1 & 1: & 2 \\ 0 & 3 & -2: & -2 \\ 0 & 6 & -4: & 1 \end{bmatrix} \quad \begin{bmatrix} 1 & -1 & 1: & 2 \\ 0 & 3 & -2: & -2 \\ 0 & 0 & 0: & 5 \end{bmatrix}.$$

The final third equation is now clearly inconsistent and there is no solution. Note that the coefficient matrix is of rank 2.

If in (9.13) we have rank $r < m = n$, we say that the equations (9.10) are *singular* and that the square matrix $\mathbf{A}$ is singular. Notice that singular equations have either an infinity of solutions or no solutions. You will also notice that an $n \times n$ matrix $\mathbf{A}$ is non-singular if, and only if, it is possible to choose n non-zero pivots u_{11}, u_{22}, ..., u_{nn} in the elimination process. It is always possible to reduce an $n \times n$ matrix $\mathbf{A}$ to an upper triangular matrix $\mathbf{U}$ using the elimination process, although for $r < n$ some of the diagonal elements of $\mathbf{U}$ are zero. It can be shown that if a total of j interchanges of rows and columns are required in the process, the product $d = (-1)^j u_{11} u_{22} \dots u_{nn}$ is independent of the order in which the pivotal equations and unknowns are chosen. We call d the *determinant* of the $n \times n$ matrix $\mathbf{A}$. Thus $\mathbf{A}$ is non-singular if, and only if, its determinant is non-zero. (See, for example, Johnson *et al.* 1993.)

We now consider the particular system of equations

$$\mathbf{Ax} = \mathbf{0}$$

where $\mathbf{A}$ is $n \times n$. These will always be consistent as the right-hand side will remain zero throughout the reduction process, and, even for $r < n$, $\mathbf{d} = \mathbf{0}$ in (9.16b). Thus if $\mathbf{A}$ is singular there are an infinity of solutions of this problem and, in particular, there will be a solution $\mathbf{x} \neq \mathbf{0}$. If $\mathbf{A}$ is non-singular, $r = n$ and the solution $\mathbf{x} = \mathbf{0}$ is unique. We conclude that $\mathbf{A}$ is singular if, and only if, there exists a vector $\mathbf{x} \neq \mathbf{0}$, such that $\mathbf{Ax} = \mathbf{0}$. Thus if $\mathbf{A}$ is singular, its columns are linearly dependent (cf. Definition 5.4).

Finally note that if a square matrix $\mathbf{A}$ is non-singular then there is an inverse $\mathbf{A}^{-1}$ and we could solve linear equations by determining this inverse. On premultiplying both sides of

$$\mathbf{Ax} = \mathbf{b}$$

by $\mathbf{A}^{-1}$ we obtain

$$\mathbf{x} = \mathbf{A}^{-1}\mathbf{Ax} = \mathbf{A}^{-1}\mathbf{b}.$$

Such a method of calculating $\mathbf{x}$ is rarely to be recommended because elimination is a much more efficient algorithm.

9.4 Pivoting

Example 9.4 Consider the solution of

$$10^{-12}x + y + z = 2 \qquad (9.18a)$$

$$x + 2y - z = 2 \qquad (9.18b)$$

$$-x + y + z = 1 \qquad (9.18c)$$

if, during the calculation, numbers may be stored to an accuracy of only 10 decimal digits in floating point form. The solution to the accuracy required is seen to be $x = y = z = 1$. However, the elimination process produces

$$\begin{aligned} 10^{-12}x + \quad y + \quad z &= 2 \\ -10^{12}y - 10^{12}z &= -2 \times 10^{12} \\ 10^{12}y + 10^{12}z &= 2 \times 10^{12}. \end{aligned}$$

The last two equations are obtained by adding $-10^{12} \times$ (9.18a) and $10^{12} \times$ (9.18a) with rounding. The next step yields

$$\begin{aligned} 10^{-12}x + \quad y + \quad z &= 2 \\ -10^{12}y - 10^{12}z &= -2 \times 10^{12} \\ 0.z &= 0. \end{aligned}$$

These suggest that z can have any arbitrary value and, by back substitution, $y = 2 - z$ and $x = 0$.

If instead we had arranged the equations as

$$\begin{aligned} x + 2y - z &= 2 \\ 10^{-12}x + \quad y + z &= 2 \\ -x + \quad y + z &= 1 \end{aligned}$$

after one step they become

$$\begin{aligned} x + 2y - z &= 2 \\ y + z &= 2 \\ 3y \quad &= 3. \end{aligned}$$

Because of rounding, the second equation is not changed apart from the removal of x. We obtain the solution $y = 1$, $z = 1$, $x = 1$, on working back through the equations. □

The fault in the first attempt at solving Example 9.4 was that a small *pivot* (the coefficient of x in the first equation), meant that very large multiples of the first equation were added to the others, which were therefore 'swamped' by the first equation, after rounding.

The trouble can be avoided if interchanges are made before each step to ensure that the pivots are as large as possible. When we have reached the stage displayed in (9.12), we look through the block of elements

$$\begin{bmatrix} \alpha_{pp} & \cdots & \alpha_{pn} \\ \vdots & & \vdots \\ \alpha_{mp} & \cdots & \alpha_{mn} \end{bmatrix}$$

and pick out one with largest modulus. This is then moved to the pivot position (p, p) by appropriate row and column interchanges. This process is called *complete pivoting*.

Usually elimination with *partial pivoting* is employed, when only row interchanges are permitted. At the stage of (9.12), only the column

$$\begin{bmatrix} \alpha_{pp} \\ \vdots \\ \alpha_{mp} \end{bmatrix}$$

is inspected for an element with largest modulus. Normally we are seeking the unique solution of n non-singular equations in n unknowns and, therefore, column interchanges should not be necessary to reduce the equations to triangular form. Alternatively, we could use a method in which only column interchanges are allowed. At the stage of (9.12), the row

$$[\alpha_{pp} \dots \alpha_{pn}]$$

is inspected for an element with largest modulus.

Example 9.4 above may seem exceptional because of one very small coefficient. Of course this was deliberately chosen as an extreme case but the lessons learned from it do hold for more reasonable equations. As a second example, we consider a more innocent looking problem.

Example 9.5 Solve

$$\begin{aligned} 0.50x + \ 1.1y + \ 3.1z &= 6.0 \\ 2.0x + \ 4.5y + 0.36z &= 0.020 \\ 5.0x + 0.96y + \ 6.5z &= 0.96. \end{aligned} \tag{9.19}$$

The coefficients are given to two significant decimal digits. In an attempt to minimize rounding errors we will carry one 'guarding' digit and work out multipliers and intermediate coefficients to three significant digits. Without interchanges we obtain

$$\begin{bmatrix} 0.500 & 1.10 & 3.10: & 6.00 \\ 0 & 0.100 & -12.0\ : & -24.0 \\ 0 & -10.0 & -24.5\ : & -59.0 \end{bmatrix} \quad \begin{bmatrix} 0.500 & 1.10 & 3.10: & 6.00 \\ 0 & 0.100 & -12.0\ : & -24.0 \\ 0 & 0 & -1220\ : & -2460 \end{bmatrix}.$$

Back substitution yields

$$z \simeq 2.02, \qquad y \simeq 2.40, \qquad x \simeq -5.80.$$

With partial pivoting involving only row interchanges we obtain, on interchanging the first and third equations,

$$\begin{bmatrix} 5.00 & 0.960 & 6.50: & 0.960 \\ 0 & 4.12 & -2.24: & -0.364 \\ 0 & 1.00 & 2.45: & 5.90 \end{bmatrix} \quad \begin{bmatrix} 5.00 & 0.960 & 6.50: & 0.960 \\ 0 & 4.12 & -2.24: & -0.364 \\ 0 & 0 & 2.99: & 5.99 \end{bmatrix}.$$

Back substitution yields

$$z \simeq 2.00, \qquad y \simeq 1.00, \qquad x \simeq -2.60.$$

By substituting these values in equations (9.19), it may be confirmed that this is the exact solution. Later we shall consider in more detail the effects of rounding errors on the accuracy of the elimination process.

We are now in a position to consider a complete algorithm for solving linear equations. This is a reasonably accurate algorithm but as will be seen from further examples and analysis, improvements may be made. In particular, we will see in Chapter 10 that scaling of individual equations is also important if only partial pivoting is used.

Algorithm 9.1 This solves ($n \geqslant 2$) linear equations in n unknowns.

```
for k := 1 to n − 1
  max := |a_kk|, pivrow := k
  for i := k + 1 to n                          [seek max |a_kk|, ..., |a_nk|]
    if |a_ik| > max then pivrow := i and max := |a_ik|
  next i
  if max < ε stop                   [as the matrix is singular or nearly so]
  [At this point the next pivotal row is given by pivrow and we
  interchange rows k and pivrow.}
  for j := k to n
    swop a_kj and a_pivrow,j
  next j
  swop b_k and b_pivrow
  [Carry out reduction.]
  pivot := a_kk
  for i := k + 1 to n
    m := a_ik/pivot          [multiple of row k to be subtracted from row i]
    for j := k + 1 to n
      a_ij := a_ij − m.a_kj
    next j
    b_i := b_i − m.b_k
  next i
next k
if |a_nn| < ε stop                  [as the matrix is singular or nearly so]
[Now perform back substitution.]
x_n := b_n/a_nn
for i := n − 1 down to 1
  s := b_i
  for j := n down to i + 1
    s := s − a_ij x_j
  next j
  x_i := s/a_ii
next i
```

The equations are reported as nearly singular if a pivot would have magnitude less than a fixed $\varepsilon > 0$. (In practice a more sophisticated test should be used.) The matrix **A** is reduced to upper triangular form but the algorithm does not bother to insert zeros below the diagonal. Partial pivoting by rows is included. □

9.5 Analysis of elimination method

Again we shall assume that we have n equations in n unknowns

$$\mathbf{A}\mathbf{x} = \mathbf{b}$$

where **A** is $n \times n$. After one step of the elimination process, **A** becomes

$$\mathcal{A}_1 = \begin{bmatrix} a_{11} & a_{12} & \dots & a_{1n} \\ 0 & a'_{22} & \dots & a'_{2n} \\ \vdots & \vdots & & \vdots \\ 0 & a'_{n2} & \dots & a'_{nn} \end{bmatrix},$$

assuming there are no interchanges. We have *subtracted* $(a_{i1}/a_{11}) \times$ first row from the ith row. We can restore **A** by operating on $\mathcal{A}_1$, *adding* $(a_{i1}/a_{11}) \times$ first row to the ith row for $i = 2, 3, \dots, n$. This is equivalent to

$$\mathbf{A} = \mathcal{L}_1 \mathcal{A}_1$$

where

$$\mathcal{L}_1 = \begin{bmatrix} 1 & 0 & \dots & & 0 \\ l_{21} & 1 & \ddots & & \vdots \\ \vdots & 0 & & & \vdots \\ \vdots & \vdots & \ddots & & 0 \\ l_{n1} & 0 & \dots & 0 & 1 \end{bmatrix},$$

and the $l_{i1} = a_{i1}/a_{11}$, $i = 2, \dots, n$, are the multipliers. Similarly the new right side $\mathbf{b}_1$ is related to its old value by

$$\mathbf{b} = \mathcal{L}_1 \mathbf{b}_1.$$

The second step (assuming there are no interchanges) in the elimination process is to form

$$\mathcal{A}_2 = \begin{bmatrix} a_{11} & a_{12} & \dots & \dots & a_{1n} \\ 0 & a'_{22} & \dots & \dots & a'_{2n} \\ 0 & 0 & a''_{33} & \dots & a''_{3n} \\ \vdots & \vdots & \vdots & & \vdots \\ 0 & 0 & a''_{n3} & \dots & a''_{nn} \end{bmatrix}$$

by subtracting $(a'_{i2}/a'_{22}) \times$ second row from the ith row for $i = 3, 4, \dots, n$. Again we may restore $\mathscr{A}_1$ by an inverse process and

$$\mathscr{A}_1 = \mathscr{L}_2 \mathscr{A}_2$$

where

$$\mathscr{L}_2 = \begin{bmatrix} 1 & 0 & . & . & . & 0 \\ 0 & 1 & . & & & . \\ . & l_{32} & 1 & . & & . \\ . & . & 0 & . & . & . \\ . & . & . & . & . & 0 \\ 0 & l_{n2} & 0 & . & 0 & 1 \end{bmatrix}$$

and $l_{i2} = (a'_{i2}/a'_{22})$, $i = 3, \dots, n$, are the multipliers of the second row. Thus

$$\mathbf{A} = \mathscr{L}_1 \mathscr{A}_1 = \mathscr{L}_1 \mathscr{L}_2 \mathscr{A}_2.$$

Similarly

$$\mathbf{b} = \mathscr{L}_1 \mathbf{b}_1 = \mathscr{L}_1 \mathscr{L}_2 \mathbf{b}_2$$

where $\mathbf{b}_2$ is the right side after two steps.

Provided there are no interchanges, by repeating the above we find that

$$\mathbf{A} = \mathscr{L}_1 \mathscr{L}_2 \dots \mathscr{L}_{n-1} \mathbf{U} \tag{9.20}$$

and

$$\mathbf{b} = \mathscr{L}_1 \mathscr{L}_2 \dots \mathscr{L}_{n-1} \mathbf{c}. \tag{9.21}$$

$\mathbf{U}$ is the final upper triangular matrix, $\mathbf{c}$ is the final right side and

$$\mathscr{L}_j = \begin{bmatrix} 1 & & & & & & & \\ & . & & & & & & \\ 0 & & . & & & \mathbf{0} & & \\ . & . & & . & & & & \\ & & 0 & & 1 & & & \\ . & & & & & 1 & & \\ . & & . & l_{j+1,j} & & & & \\ . & & . & . & 0 & & . & \\ . & & . & . & & . & . & . \\ 0 & . & 0 & l_{nj} & 0 & . & . \; 0 & 1 \end{bmatrix} \tag{9.22}$$

where l_{ij} is the multiple of the jth row subtracted from the ith row during the reduction process.

Now

$$\mathcal{L}_1\mathcal{L}_2\ldots\mathcal{L}_{n-1} = \begin{bmatrix} 1 & & & & \\ l_{21} & 1 & & \mathbf{0} & \\ l_{31} & l_{32} & 1 & & \\ \vdots & \vdots & \ddots & \ddots & \\ l_{n1} & l_{n2} & \cdots & l_{n,n-1} & 1 \end{bmatrix} = \mathbf{L}, \tag{9.23}$$

say, which is a lower triangular matrix (see Problem 9.21). We therefore find that, provided the equations are non-singular and no interchanges are required,

$$\mathbf{A} = \mathbf{LU} \tag{9.24}$$

and

$$\mathbf{b} = \mathbf{Lc} \tag{9.25}$$

where $\mathbf{L}$ is a *lower triangular matrix* with units on the diagonal and $\mathbf{U}$ is an *upper triangular matrix*. We say that $\mathbf{A}$ has been factorized into triangular factors $\mathbf{L}$ and $\mathbf{U}$.

Example 9.6 For the equations of Example 9.1,

$$\mathcal{L}_1 = \begin{bmatrix} 1 & 0 & 0 \\ 2 & 1 & 0 \\ -1 & 0 & 1 \end{bmatrix} \quad \text{and} \quad \mathcal{L}_2 = \begin{bmatrix} 1 & 0 & 1 \\ 0 & 1 & 0 \\ 0 & 2 & 1 \end{bmatrix}.$$

Thus

$$\mathbf{L} = \mathcal{L}_1\mathcal{L}_2 = \begin{bmatrix} 1 & 0 & 0 \\ 2 & 1 & 0 \\ -1 & 2 & 1 \end{bmatrix}.$$

From the final set of equations we find

$$\mathbf{U} = \begin{bmatrix} 2 & 2 & 3 \\ 0 & 3 & 1 \\ 0 & 0 & 6 \end{bmatrix} \quad \text{and} \quad \mathbf{c} = \begin{bmatrix} 3 \\ -5 \\ 6 \end{bmatrix}.$$

It is easily verified that

$$\mathbf{LU} = \begin{bmatrix} 2 & 2 & 3 \\ 4 & 7 & 7 \\ -2 & 4 & 5 \end{bmatrix} = \mathbf{A}$$

and

$$\mathbf{Lc} = \begin{bmatrix} 3 \\ 1 \\ -7 \end{bmatrix} = \mathbf{b}.$$

□

If interchanges are used during the reduction process, then instead of determining triangular factors of $\mathbf{A}$, we determine triangular factors of $\mathbf{A}^*$ where $\mathbf{A}^*$ is obtained by making the necessary row and column interchanges on $\mathbf{A}$.

9.6 Matrix factorization

The elimination process for n equations in n unknowns

$$\mathbf{Ax} = \mathbf{b} \tag{9.26}$$

may also be described as a *factorization* method. We seek factors $\mathbf{L}$ (lower triangular with units on the diagonal) and $\mathbf{U}$ (upper triangular) such that

$$\mathbf{A} = \mathbf{LU},$$

when (9.26) becomes

$$\mathbf{LUx} = \mathbf{b}.$$

First we find $\mathbf{c}$ by solving the lower triangular equations

$$\mathbf{Lc} = \mathbf{b}$$

by *forward substitution*. We have

$$\begin{aligned}
&1.c_1 && = b_1\\
&l_{21}c_1 + 1.c_2 && = b_2\\
&l_{31}c_1 + l_{32}c_2 + 1.c_3 && = b_3\\
&\vdots && \vdots\\
&l_{n1}c_1 + l_{n2}c_2 + \cdots + 1.c_n && = b_n.
\end{aligned}$$

We find c_1 from the first equation, then c_2 from the second and so on. Secondly we find $\mathbf{x}$ by back substitution in the triangular equations

$$\mathbf{Ux} = \mathbf{c}.$$

The solution of (9.26) is this vector $\mathbf{x}$, since

$$\mathbf{Ax} = (\mathbf{LU})\mathbf{x} = \mathbf{L}(\mathbf{Ux}) = \mathbf{Lc} = \mathbf{b}.$$

We can see from the above that, given any method of calculating factors $\mathbf{L}$ and $\mathbf{U}$ such that $\mathbf{A} = \mathbf{LU}$, we can solve the equations. One way of finding the factors $\mathbf{L}$ and $\mathbf{U}$ is to build them up from submatrices. We will use this method to describe how an $n \times n$ matrix $\mathbf{A}$ may be factorized in the form

$$\mathbf{A} = \mathbf{LDV} \tag{9.27}$$

where

$$\mathbf{D} = \begin{bmatrix} d_1 & & & \\ & d_2 & & \mathbf{0} \\ & & \ddots & \\ & \mathbf{0} & & \\ & & & d_n \end{bmatrix}$$

is a *diagonal matrix* with $d_i \neq 0$, $i = 1, 2, \ldots, n$, $\mathbf{L}$ is a lower triangular matrix with units on the diagonal as before and $\mathbf{V}$ is an upper triangular matrix with units on the diagonal. We will show that, under certain circumstances, such a factorization exists and is unique. In this factorization we call the diagonal elements of $\mathbf{D}$ the *pivots*. We shall see later that these pivots are identical with those of the elimination method.

Let $\mathbf{A}_k$ denote the kth leading submatrix of $\mathbf{A}$. By this we mean the submatrix consisting of the first k rows and columns of $\mathbf{A}$. Note that $\mathbf{A}_1 = [a_{11}]$ and $\mathbf{A}_n = \mathbf{A}$. Suppose that for some value of k we already have a factorization

$$\mathbf{A}_k = \mathbf{L}_k \mathbf{D}_k \mathbf{V}_k$$

of the form (9.27), where $\mathbf{L}_k$, $\mathbf{D}_k$ and $\mathbf{V}_k$ are matrices of the appropriate kinds. We now seek $\mathbf{L}_{k+1}$, $\mathbf{D}_{k+1}$ and $\mathbf{V}_{k+1}$ such that

$$\mathbf{A}_{k+1} = \mathbf{L}_{k+1} \mathbf{D}_{k+1} \mathbf{V}_{k+1}.$$

We write

$$\mathbf{A}_{k+1} = \begin{bmatrix} \mathbf{A}_k & \mathbf{c}_{k+1} \\ \mathbf{r}_{k+1}^{\mathrm{T}} & a_{k+1,k+1} \end{bmatrix}, \qquad \text{where } \mathbf{c}_{k+1} = \begin{bmatrix} a_{1,k+1} \\ \vdots \\ a_{k,k+1} \end{bmatrix}$$

and

$$\mathbf{r}_{k+1}^{\mathrm{T}} = [a_{k+1,1} \quad a_{k+1,2} \quad \ldots \quad a_{k+1,k}].$$

We seek an element d_{k+1} and vectors

$$\mathbf{l}_{k+1}^{\mathrm{T}} = [l_{k+1,1} \quad l_{k+1,2} \quad \ldots \quad l_{k+1,k}], \qquad \mathbf{v}_{k+1} = \begin{bmatrix} v_{1,k+1} \\ \vdots \\ v_{k,k+1} \end{bmatrix}$$

such that

$$\mathbf{A}_{k+1} = \begin{bmatrix} \mathbf{A}_k & \mathbf{c}_{k+1} \\ \mathbf{r}_{k+1}^{\mathrm{T}} & a_{k+1,k+1} \end{bmatrix} = \begin{bmatrix} \mathbf{L}_k & \mathbf{0} \\ \mathbf{l}_{k+1}^{\mathrm{T}} & 1 \end{bmatrix} \begin{bmatrix} \mathbf{D}_k & \mathbf{0} \\ \mathbf{0}^{\mathrm{T}} & d_{k+1} \end{bmatrix} \begin{bmatrix} \mathbf{V}_k & \mathbf{v}_{k+1} \\ \mathbf{0}^{\mathrm{T}} & 1 \end{bmatrix}.$$

On multiplying out the right side and equating blocks, we find that we require

$$\mathbf{A}_k = \mathbf{L}_k \mathbf{D}_k \mathbf{V}_k \tag{9.28}$$

$$\mathbf{c}_{k+1} = \mathbf{L}_k \mathbf{D}_k \mathbf{v}_{k+1} \tag{9.29}$$

$$\mathbf{r}_{k+1}^{\mathrm{T}} = \mathbf{l}_{k+1}^{\mathrm{T}} \mathbf{D}_k \mathbf{V}_k \tag{9.30}$$

$$a_{k+1,k+1} = \mathbf{l}_{k+1}^{\mathrm{T}} \mathbf{D}_k \mathbf{v}_{k+1} + d_{k+1}. \tag{9.31}$$

Equation (9.28) is satisfied by hypothesis. We can solve (9.29) for $\mathbf{v}_{k+1}$ by forward substitution in the lower triangular equations

$$(\mathbf{L}_k \mathbf{D}_k)\mathbf{v}_{k+1} = \mathbf{c}_{k+1}. \tag{9.32}$$

We use forward substitution in the triangular equations

$$\mathbf{l}_{k+1}^{\mathrm{T}}(\mathbf{D}_k \mathbf{V}_k) = \mathbf{r}_{k+1}^{\mathrm{T}} \tag{9.33}$$

to find $\mathbf{l}_{k+1}^{\mathrm{T}}$. Having found $\mathbf{v}_{k+1}$ and $\mathbf{l}_{k+1}^{\mathrm{T}}$, we use (9.31) to determine d_{k+1}.

We have a practical method of factorizing $\mathbf{A}$ since clearly

$$\mathbf{A}_1 = [a_{11}] = [1][a_{11}][1] = \mathbf{L}_1 \mathbf{D}_1 \mathbf{V}_1$$

and the process only involves solving triangular equations. The $\mathbf{L}_k$, $\mathbf{D}_k$ and $\mathbf{V}_k$ are successively determined for $k = 1, 2, \ldots, n$ and are leading submatrices of $\mathbf{L}$, $\mathbf{D}$ and $\mathbf{V}$.

Example 9.7 Factorize

$$\mathbf{A} = \begin{bmatrix} 1 & 2 & 3 & -1 \\ 2 & -1 & 9 & -7 \\ -3 & 4 & -3 & 19 \\ 4 & -2 & 6 & -21 \end{bmatrix}.$$

We carry out the factorization of the leading submatrices $\mathbf{A}_k$, for $k = 1, 2, 3, 4$, in turn, as follows.

$$\mathbf{A}_1 = [1], \quad \mathbf{L}_1 = [1], \quad \mathbf{D}_1 = [1], \quad \mathbf{V}_1 = [1].$$

$$\mathbf{A}_2 = \begin{bmatrix} 1 & 2 \\ 2 & -1 \end{bmatrix}, \quad \mathbf{L}_2 = \begin{bmatrix} 1 & 0 \\ 2 & 1 \end{bmatrix}, \quad \mathbf{D}_2 = \begin{bmatrix} 1 & 0 \\ 0 & -5 \end{bmatrix}, \quad \mathbf{V}_2 = \begin{bmatrix} 1 & 2 \\ 0 & 1 \end{bmatrix}.$$

$$\mathbf{A}_3 = \begin{bmatrix} 1 & 2 & 3 \\ 2 & -1 & 9 \\ -3 & 4 & -3 \end{bmatrix}, \quad \mathbf{L}_3 = \begin{bmatrix} 1 & 0 & 0 \\ 2 & 1 & 0 \\ -3 & -2 & 1 \end{bmatrix},$$

$$\mathbf{D}_3 = \begin{bmatrix} 1 & 0 & 0 \\ 0 & -5 & 0 \\ 0 & 0 & 12 \end{bmatrix}, \quad \mathbf{V}_3 = \begin{bmatrix} 1 & 2 & 3 \\ 0 & 1 & -\frac{3}{5} \\ 0 & 0 & 1 \end{bmatrix}.$$

$\mathbf{A}_4 = \mathbf{A}$;

$$(\mathbf{L}_3\mathbf{D}_3)\mathbf{v}_4 = \begin{bmatrix} 1 & 0 & 0 \\ 2 & -5 & 0 \\ -3 & 10 & 12 \end{bmatrix}\mathbf{v}_4 = \mathbf{c}_4 = \begin{bmatrix} -1 \\ -7 \\ 19 \end{bmatrix},$$

and thus

$$\mathbf{v}_4 = \begin{bmatrix} -1 \\ 1 \\ \frac{1}{2} \end{bmatrix};$$

$$\mathbf{l}_4^{\mathrm{T}}(\mathbf{D}_3\mathbf{V}_3) = \mathbf{l}_4^{\mathrm{T}}\begin{bmatrix} 1 & 2 & 3 \\ 0 & -5 & 3 \\ 0 & 0 & 12 \end{bmatrix} = \mathbf{r}_4^{\mathrm{T}} = [4 \quad -2 \quad 6]$$

and thus

$$\mathbf{l}_4^{\mathrm{T}} = [4 \;\; 2 \;\; -1];$$

$$\begin{aligned} d_4 &= a_{4,4} - \mathbf{l}_4^{\mathrm{T}}\mathbf{D}_3\mathbf{v}_4 \\ &= -21 - [4 \quad 2 \quad -1]\begin{bmatrix} -1 \\ -5 \\ 6 \end{bmatrix} \\ &= -21 + 20 = -1. \end{aligned}$$

Thus

$$\mathbf{A} = \begin{bmatrix} 1 & & & \mathbf{0} \\ 2 & 1 & & \\ -3 & -2 & 1 & \\ 4 & 2 & -1 & 1 \end{bmatrix}\begin{bmatrix} 1 & & & \mathbf{0} \\ & -5 & & \\ & & 12 & \\ \mathbf{0} & & & -1 \end{bmatrix}\begin{bmatrix} 1 & 2 & 3 & -1 \\ & 1 & -\frac{3}{5} & 1 \\ & & 1 & \frac{1}{2} \\ \mathbf{0} & & & 1 \end{bmatrix}. \qquad \square$$

The process will fail if, and only if, any of the pivots d_k become zero as then we cannot solve (9.32) and (9.33). This can only happen if one of the leading submatrices is singular. To see this, suppose that $d_1, d_2, \ldots, d_k$ are non-zero but that when we solve (9.31) we find $d_{k+1} = 0$. We now have

$$\begin{aligned} \mathbf{A}_{k+1} &= \begin{bmatrix} \mathbf{L}_k & \mathbf{0} \\ \mathbf{l}_{k+1}^{\mathrm{T}} & 1 \end{bmatrix}\begin{bmatrix} \mathbf{D}_k & \mathbf{0} \\ \mathbf{0}^{\mathrm{T}} & 0 \end{bmatrix}\begin{bmatrix} \mathbf{V}_k & \mathbf{v}_{k+1} \\ \mathbf{0}^{\mathrm{T}} & 1 \end{bmatrix} \\ &= \begin{bmatrix} \mathbf{L}_k & \mathbf{0} \\ \mathbf{l}_{k+1}^{\mathrm{T}} & 1 \end{bmatrix}\begin{bmatrix} \mathbf{D}_k\mathbf{V}_k & \mathbf{D}_k\mathbf{v}_{k+1} \\ \mathbf{0}^{\mathrm{T}} & 0 \end{bmatrix}. \end{aligned}$$

Using a process equivalent to the elimination method of § 9.3, we have reduced A_{k+1} to the second of these factors. This is of the form (9.14) without the right-hand side vectors **c** and **d**, where the upper triangular matrix $\mathbf{D}_k\mathbf{V}_k$ is $k \times k$. Thus $\mathbf{A}_{k+1}$ is of rank k which is less than the dimension $k+1$ and, therefore, $\mathbf{A}_{k+1}$ is singular.

The above argument and method of finding factors forms a constructive proof† of the following theorem.

Theorem 9.1 If an $n \times n$ matrix **A** is such that all n of its leading submatrices are non-singular, then there exists a unique factorization of the form

$$\mathbf{A} = \mathbf{LDV}, \tag{9.34}$$

where **L** is lower triangular with units on the diagonal, **V** is upper triangular with units on the diagonal and **D** is diagonal with non-zero diagonal elements.

Proof The proof of both existence and uniqueness is based on induction. The factors exist and are unique for the 1×1 leading submatrix $\mathbf{A}_1$. Given a unique factorization of the kth leading submatrix $\mathbf{A}_k$ we deduce from (9.29), (9.30) and (9.31) that $\mathbf{A}_{k+1}$, has a unique factorization. Thus the factors of $\mathbf{A} = \mathbf{A}_n$, exist and are unique. □

9.7 Compact elimination methods

We do not usually bother to determine the diagonal matrix **D** of (9.34) explicitly and instead factorize **A** so that

$$\mathbf{A} = \mathbf{L}(\mathbf{DV}) \tag{9.35}$$

or

$$\mathbf{A} = (\mathbf{LD})\mathbf{V}. \tag{9.36}$$

In the former we calculate **L** with units on the diagonal and (**DV**). In the latter we calculate **V** with units on the diagonal and (**LD**). The elimination method of § 9.3 is equivalent to (9.35) and, on comparing with (9.24), we see that

$$\mathbf{U} = \mathbf{DV}.$$

From the diagonal,

$$u_{ii} = d_i, \qquad i = 1, 2, \ldots, n$$

†A proof in which an existence statement is proved by showing how to construct the objects involved. See p. 1.

and thus the pivots in the elimination method are the same as those for factorization.

In compact elimination methods, we seek a factorization of the form (9.35) or (9.36), but instead of building up factors from leading submatrices described in the last section, we find it more convenient to calculate rows and columns of **L** and **V** alternatively. For the factorization (9.35), we calculate the ith row of $(\mathbf{DV})$ followed by the ith column of **L** for $i = 1, 2, \ldots, n$. To obtain (9.36) we calculate the ith column of $(\mathbf{LD})$ followed by the ith row of **V**. The following algorithm describes the process.

Algorithm 9.2 Compact elimination without pivoting to factorize an $n \times n$ matrix **A** into a lower triangular matrix **L** with units on the diagonal and an upper triangular matrix **U** $(=\mathbf{DV})$. (As no pivoting is included, the algorithm does not check whether any of the pivots u_{ii} become zero or very small in magnitude and thus there is no check whether the matrix or any leading submatrix is singular or nearly so.)

[row 1 of **U**] := [row 1 of **A**]
[column 1 of **L**] := [column 1 of **A**]$/u_{11}$
for i:=2 **to** n
  for $k := i$ **to** n

$$u_{ik} := a_{ik} - [l_{i1} \ldots l_{i,i-1}] \begin{bmatrix} u_{1k} \\ \vdots \\ u_{i-1,k} \end{bmatrix}$$

  next k
  $l_{ii} := 1$
  if $i < n$ **then**
    for $k := i+1$ **to** n

$$l_{ki} := \left(a_{ki} - [l_{k1} \ldots l_{k,i-1}] \begin{bmatrix} u_{1i} \\ \vdots \\ u_{i-1,i} \end{bmatrix} \right) / u_{ii}$$

    next k
next i □

The process used in the last algorithm is exactly equivalent to elimination except that intermediate values are not recorded; hence the name *compact elimination* method. Because there are no intermediate coefficients the compact method can be programmed to give less rounding errors than simple elimination. However, it is necessary to include partial pivoting in the compact method to increase accuracy. This can be achieved by suitable modification of Algorithm 9.2.

It can be seen from (9.34), (9.35), (9.36) and Algorithms 9.1 and 9.2 that there are various ways in which we may factorize **A** and various ways in which we may order the calculations.

In all factorization methods it is necessary to carry out forward and back substitution steps to solve linear equations. The following algorithm describes the process for factorization of the form (9.35).

Algorithm 9.3 Forward and back substitution to solve $\mathbf{LUx} = \mathbf{b}$, where **L** is lower triangular with units on the diagonal and **U** $(= \mathbf{DV})$ is upper triangular.

$y_1 := b_1$
for $i := 2$ **to** n

$$y_i := b_i - [l_{i1} \dots l_{i,i-1}] \begin{bmatrix} y_1 \\ \vdots \\ y_{i-1} \end{bmatrix}$$

next i
[**y** is the solution of $\mathbf{Ly} = \mathbf{b}$]
$x_n := y_n / u_{nn}$
for $i := n-1$ **down to** 1

$$x_i := \left(y_i - [u_{i,i+1} \dots u_{in}] \begin{bmatrix} x_{i+1} \\ \vdots \\ x_n \end{bmatrix} \right) / u_{ii}$$

next i □

9.8 Symmetric matrices

Given any $m \times n$ matrix, we call the $n \times m$ matrix obtained by interchanging the rows and columns of **A** the *transpose* of **A** and write it as $\mathbf{A}^{\mathrm{T}}$. Thus

$$\mathbf{A}^{\mathrm{T}} = \begin{bmatrix} a_{11} & a_{21} & \dots & a_{m1} \\ a_{12} & a_{22} & \dots & a_{m2} \\ \vdots & & & \vdots \\ a_{1n} & a_{2n} & \dots & a_{mn} \end{bmatrix}$$

and a_{ij} is the element in the jth row and ith column of $\mathbf{A}^{\mathrm{T}}$. This extends the definition of the transpose of a vector, given in § 9.2, to a general $m \times n$ matrix.

For any two matrices **A** and **B** whose product is defined, we find (see Problem 9.29) that

$$(\mathbf{AB})^{\mathrm{T}} = \mathbf{B}^{\mathrm{T}}\mathbf{A}^{\mathrm{T}}. \tag{9.37}$$

An $n \times n$ matrix **A** is said to be *symmetric* if

$$\mathbf{A}^{\mathrm{T}} = \mathbf{A},$$

that is, $a_{ij} = a_{ji}$ for all i and j. When using a computer, we need store only approximately half the elements of a symmetric matrix as the elements below the diagonal are the same as those above. We therefore look for an elimination method which preserves symmetry when solving linear equations.

We have seen how to factorize **A** so that

$$\mathbf{A} = \mathbf{LDV} \tag{9.38}$$

where, provided the factorization exists, **L**, **D** and **V** are unique. Now

$$\mathbf{A}^{\mathrm{T}} = (\mathbf{LDV})^{\mathrm{T}} = \mathbf{V}^{\mathrm{T}}\mathbf{D}^{\mathrm{T}}\mathbf{L}^{\mathrm{T}} = \mathbf{V}^{\mathrm{T}}\mathbf{D}\mathbf{L}^{\mathrm{T}},$$

as any diagonal matrix **D** is symmetric. If **A** is symmetric

$$\mathbf{A} = \mathbf{A}^{\mathrm{T}} = \mathbf{V}^{\mathrm{T}}\mathbf{D}\mathbf{L}^{\mathrm{T}}, \tag{9.39}$$

$\mathbf{V}^{\mathrm{T}}$ is a lower triangular matrix with units on the diagonal and $\mathbf{L}^{\mathrm{T}}$ is an upper triangular matrix with units on the diagonal. On comparing (9.38) and (9.39) we deduce from the uniqueness of **L** and **V** that

$$\mathbf{L} = \mathbf{V}^{\mathrm{T}} \quad \text{and} \quad \mathbf{L}^{\mathrm{T}} = \mathbf{V},$$

These two are equivalent and (9.39) becomes

$$\mathbf{A} = \mathbf{LDL}^{\mathrm{T}}. \tag{9.40}$$

If

$$\mathbf{D} = \begin{bmatrix} d_1 & & \mathbf{0} \\ & d_2 & & \\ & & \ddots & \\ \mathbf{0} & & & d_n \end{bmatrix} \quad \text{and} \quad \mathbf{E} = \begin{bmatrix} \sqrt{d_1} & & \mathbf{0} \\ & \sqrt{d_2} & & \\ & & \ddots & \\ \mathbf{0} & & & \sqrt{d_n} \end{bmatrix}$$

then†

$$\mathbf{E}^2 = \mathbf{D}$$

so that

$$\begin{aligned} \mathbf{A} &= \mathbf{LEEL}^{\mathrm{T}} = \mathbf{LEE}^{\mathrm{T}}\mathbf{L}^{\mathrm{T}} \\ &= (\mathbf{LE})(\mathbf{LE})^{\mathrm{T}}. \end{aligned} \tag{9.41}$$

Thus

$$\mathbf{A} = \mathbf{MM}^{\mathrm{T}}$$

---

† We write $\mathbf{E}^2$ for **E.E**.

where

$$\mathbf{M} = \mathbf{LE}$$

is a lower triangular matrix and will have real elements only if all the diagonal elements of **D** are positive.

To calculate **M** we adapt the process described in § 9.6. Equation (9.28) becomes

$$\mathbf{A}_k = \mathbf{M}_k \mathbf{M}_k^{\mathrm{T}} \tag{9.42}$$

where $\mathbf{M}_k$ is the kth leading submatrix of **M**. Both (9.32) and (9.33) become

$$\mathbf{M}_k \mathbf{m}_{k+1} = \mathbf{c}_{k+1} \tag{9.43}$$

where

$$\mathbf{A}_{k+1} = \begin{bmatrix} \mathbf{A}_k & \mathbf{c}_{k+1} \\ \mathbf{c}_{k+1}^{\mathrm{T}} & a_{k+1,k+1} \end{bmatrix} \quad \text{and} \quad \mathbf{M}_{k+1} = \begin{bmatrix} \mathbf{M}_k & \mathbf{0} \\ \mathbf{m}_{k+1}^{\mathrm{T}} & m_{k+1,k+1} \end{bmatrix}.$$

We solve (9.43) for $\mathbf{m}_{k+1}$ by forward substitution. Finally (9.31) becomes

$$(m_{k+1,k+1})^2 = a_{k+1,k+1} - \mathbf{m}_{k+1}^{\mathrm{T}} \mathbf{m}_{k+1}. \tag{9.44}$$

Example 9.8 Solve

$$\begin{bmatrix} 4 & 2 & -2 \\ 2 & 2 & -3 \\ -2 & -3 & 14 \end{bmatrix} \begin{bmatrix} x_1 \\ x_2 \\ x_3 \end{bmatrix} = \begin{bmatrix} 10 \\ 5 \\ 4 \end{bmatrix}.$$

From $\mathbf{A}_1 = [4]$, $\quad \mathbf{M}_1 = [2]$.

From $\mathbf{A}_2 = \begin{bmatrix} 4 & 2 \\ 2 & 2 \end{bmatrix}$, $\quad \mathbf{M}_2 = \begin{bmatrix} 2 & 0 \\ 1 & 1 \end{bmatrix}$.

From $\mathbf{A}_3 = \begin{bmatrix} 4 & 2 & -2 \\ 2 & 2 & -3 \\ -2 & -3 & 14 \end{bmatrix}$, $\quad \begin{bmatrix} 2 & 0 \\ 1 & 1 \end{bmatrix} \begin{bmatrix} m_{31} \\ m_{32} \end{bmatrix} = \begin{bmatrix} -2 \\ -3 \end{bmatrix}$.

Therefore,

$$\begin{bmatrix} m_{31} \\ m_{32} \end{bmatrix} = \begin{bmatrix} -1 \\ -2 \end{bmatrix} \quad \text{and} \quad m_{33}^2 = 14 - [-1 \quad -2] \begin{bmatrix} -1 \\ -2 \end{bmatrix} = 9.$$

Hence

$$\mathbf{M} = \begin{bmatrix} 2 & 0 & 0 \\ 1 & 1 & 0 \\ -1 & -2 & 3 \end{bmatrix}$$

and $\mathbf{Ax} = \mathbf{b}$ becomes $\mathbf{MM}^{\mathrm{T}}\mathbf{x} = \mathbf{b}$. We seek $\mathbf{c}$ such that

$$\mathbf{Mc} = \begin{bmatrix} 10 \\ 5 \\ 4 \end{bmatrix}.$$

By forward substitution, we obtain

$$\mathbf{c} = \begin{bmatrix} 5 \\ 0 \\ 3 \end{bmatrix}.$$

Finally, by back substitution in

$$\mathbf{M}^{\mathrm{T}}\mathbf{x} = \mathbf{c},$$

we obtain

$$\mathbf{x} = \begin{bmatrix} 2 \\ 2 \\ 1 \end{bmatrix}.$$

□

The above method is difficult to implement if some of the elements of the diagonal matrix $\mathbf{D}$ are negative, as this means using complex numbers in finding $\mathbf{M}$. Since each of the elements of $\mathbf{M}$ will be either real or purely imaginary, there is, however, no need to use general complex arithmetic. In such cases it is probably best to ignore symmetry and proceed as for general matrices. Another difficulty, which may arise with the method preserving symmetry, is that pivots may become small or even zero and it is difficult to make row or column interchanges without destroying the symmetry. This is only possible if, whenever rows are interchanged, the corresponding columns are also interchanged and vice versa.

There is an important class of matrices, called *positive definite* matrices, for which $\mathbf{M}$ will always exist and be real. An $n \times n$ real symmetric matrix $\mathbf{A}$ is said to be positive definite if, given any real vector $\mathbf{x} \neq \mathbf{0}$, then

$$\mathbf{x}^{\mathrm{T}}\mathbf{Ax} > 0.$$

Note that $\mathbf{x}^{\mathrm{T}}\mathbf{Ax}$ is a 1×1 matrix. Positive definite matrices occur in a variety of problems, for example least squares approximation calculations (see Problem 9.39).

We note first that the unit matrix is positive definite, as it is symmetric and

$$\mathbf{x}^{\mathrm{T}}\mathbf{Ix} = \mathbf{x}^{\mathrm{T}}\mathbf{x} = x_1^2 + x_2^2 + \cdots + x_n^2.$$

The latter must be positive if the real vector $\mathbf{x}$ is non-zero.

Secondly we observe that, if a matrix $\mathbf{A}$ is positive definite, then it must be non-singular. For, suppose that $\mathbf{A}$ is singular so that, by the last part of §9.3, there is a vector $\mathbf{x} \neq \mathbf{0}$ for which

$$\mathbf{Ax} = \mathbf{0}.$$

Then clearly

$$\mathbf{x}^{\mathrm{T}}\mathbf{Ax} = 0$$

which would contradict the positive definite property.

Theorem 9.2 An $n \times n$ real symmetric matrix $\mathbf{A}$ may be factorized in the (Choleski) form

$$\mathbf{A} = \mathbf{MM}^{\mathrm{T}}, \tag{9.45}$$

where $\mathbf{M}$ is a lower triangular real matrix with non-zero elements on the diagonal if, and only if, $\mathbf{A}$ is positive definite.

Proof If $\mathbf{A}$ can be factorized in the form (9.45), then

$$\mathbf{x}^{\mathrm{T}}\mathbf{Ax} = \mathbf{x}^{\mathrm{T}}\mathbf{MM}^{\mathrm{T}}\mathbf{x} = \mathbf{y}^{\mathrm{T}}\mathbf{y}$$

where

$$\mathbf{y} = \mathbf{M}^{\mathrm{T}}\mathbf{x}.$$

If $\mathbf{x} \neq \mathbf{0}$ we must have $\mathbf{y} \neq \mathbf{0}$ as $\mathbf{M}$ is non-singular and thus

$$\mathbf{x}^{\mathrm{T}}\mathbf{Ax} = \mathbf{y}^{\mathrm{T}}\mathbf{y} > 0,$$

showing that $\mathbf{A}$ is positive definite.

Conversely, if $\mathbf{A}$ is positive definite all the leading submatrices are positive definite and hence non-singular (see Problem 9.35). By Theorem 9.1, $\mathbf{A}$ may be factorized in the form

$$\mathbf{A} = \mathbf{LDV}$$

and since this factorization is unique and $\mathbf{A}$ is symmetric

$$\mathbf{L}^{\mathrm{T}} = \mathbf{V}.$$

It remains to be shown that the diagonal elements of $\mathbf{D}$ are positive. Choose $\mathbf{x}$ so that

$$\mathbf{L}^{\mathrm{T}}\mathbf{x} = \mathbf{e}_j$$

where $\mathbf{e}_j = [0 \ldots 0 \;\; 1 \;\; 0 \ldots 0]^{\mathrm{T}}$, with the unit in the jth position, i.e. $\mathbf{e}_j$ is the jth column of $\mathbf{I}$. We can always find such an $\mathbf{x} \neq \mathbf{0}$ as $\mathbf{L}^{\mathrm{T}}$ is non-singular. Now since $\mathbf{A}$ is positive definite,

$$\begin{aligned} 0 < \mathbf{x}^{\mathrm{T}}\mathbf{Ax} = \mathbf{x}^{\mathrm{T}}\mathbf{LDL}^{\mathrm{T}}\mathbf{x} &= (\mathbf{L}^{\mathrm{T}}\mathbf{x})^{\mathrm{T}}\mathbf{D}(\mathbf{L}^{\mathrm{T}}\mathbf{x}) \\ &= \mathbf{e}_j^{\mathrm{T}}\mathbf{De}_j = d_j, \end{aligned}$$

where d_j is the jth diagonal element of $\mathbf{D}$. Since this holds for $j = 1, 2, \ldots, n$, it follows that all the diagonal elements of $\mathbf{D}$ are positive. □

What is not obvious from the above theorem is that in fact the Choleski factorization of a positive definite matrix is an accurate process and that pivoting is not required to maintain accuracy.

Algorithm 9.4 (Choleski factorization $\mathbf{A} = \mathbf{M}\mathbf{M}^{\mathrm{T}}$ of a positive definite symmetric matrix $\mathbf{A}$ where $\mathbf{M}$ is lower triangular) The algorithm does not check that $\mathbf{A}$ is positive definite but this could easily be included by checking for positivity all quantities whose square roots are required.

$m_{11} := \sqrt{a_{11}}$
for $i := 2$ **to** n
  for $j := 1$ **to** $i - 1$

$$m_{ij} := \left(a_{ij} - [m_{i1} \ldots m_{i,j-1}] \begin{bmatrix} m_{j1} \\ \vdots \\ m_{j,j-1} \end{bmatrix} \right) \Big/ m_{jj}$$

  [Omit inner product when $j = 1$.]
  next j

$$m_{ii} := \sqrt{\left(a_{ii} - [m_{i1} \ldots m_{i,i-1}] \begin{bmatrix} m_{i1} \\ \vdots \\ m_{i,i-1} \end{bmatrix} \right)}$$

next i □

9.9 Tridiagonal matrices

An $n \times n$ matrix $\mathbf{A}$ is said to be *tridiagonal* (or *triple band*) if it takes the form

$$\mathbf{A} = \begin{bmatrix} b_1 & c_1 & & & & \\ a_2 & b_2 & c_2 & & \mathbf{0} & \\ & a_3 & b_3 & c_3 & & \\ & & \cdot & \cdot & \cdot & \\ & \mathbf{0} & & \cdot & \cdot & c_{n-1} \\ & & & & a_n & b_n \end{bmatrix} \tag{9.46}$$

Thus the (i, j)th elements are zero for $j > i + 1$ and $j < i - 1$.

The elimination method can be considerably simplified if the coefficient matrix of a linear set of equations is tridiagonal. Consider

$$\mathbf{A}\mathbf{x} = \mathbf{d} \tag{9.47}$$

where $\mathbf{A}$ is of the form (9.46). Assuming $b_1 \neq 0$, the first step consists of eliminating x_1 from the second equation by subtracting a_2/b_1 times the first equation from the second equation. There is no need to change the 3rd to nth equations in the elimination of x_1. After $i-1$ steps, assuming no interchanges are required, the equations take the form

$$\begin{bmatrix} \beta_1 & c_1 & & & & & \\ 0 & \beta_2 & c_2 & & & \mathbf{0} & \\ & \ddots & \ddots & \ddots & & & \\ & & 0 & \beta_i & c_i & & \\ & & & a_{i+1} & b_{i+1} & c_{i+1} & \\ & \mathbf{0} & & & a_{i+2} & b_{i+2} & c_{i+2} \\ & & & & \ddots & \ddots & \ddots \end{bmatrix} \mathbf{x} = \begin{bmatrix} \delta_1 \\ \vdots \\ \delta_i \\ d_{i+1} \\ \vdots \\ d_n \end{bmatrix}.$$

We now eliminate x_i from the $(i+1)$th equation by subtracting a_{i+1}/β_i times the ith equation from the $(i+1)$th equation. The other equations are not changed. The final form of the equations is

$$\begin{bmatrix} \beta_1 & c_1 & & & \\ & \beta_2 & c_2 & & \mathbf{0} \\ & & \ddots & \ddots & \\ & \mathbf{0} & & \ddots & c_{n-1} \\ & & & & \beta_n \end{bmatrix} \mathbf{x} = \begin{bmatrix} \delta_1 \\ \\ \vdots \\ \\ \delta_n \end{bmatrix} \tag{9.48}$$

and these are easily solved by back substitution.

Algorithm 9.5 Solve tridiagonal equations of the form (9.46) and (9.47). There is no test for a zero pivot or singular matrix (see below).

$\beta_1 := b_1$
$\delta_1 := d_1$
for $i := 1$ **to** $n-1$
  $m_i := a_{i+1}/\beta_i$
  $\beta_{i+1} := b_{i+1} - m_i c_i$
  $\delta_{i+1} := d_{i+1} - m_i \delta_i$
next i
$x_n := \delta_n/\beta_n$

for $i := n - 1$ **down to** 1
 $x_i := (\delta_i - c_i x_{i+1})/\beta_i$
next i □

If **A** is positive definite we can be certain that the algorithm will not fail because of a zero pivot. In Problem 9.42, simple conditions on the elements a_i, b_i and c_i are given which ensure that **A** is positive definite. These conditions hold for the tridiagonal matrix **M** in (6.36) for cubic splines. We shall also see in Chapter 14 that tridiagonal equations occur in numerical methods of solving boundary value problems and that in many such applications **A** is positive definite (see Problem 14.5).

9.10 Rounding errors in solving linear equations

We have seen that, if we are not careful, rounding errors may seriously affect the accuracy of the calculation of a solution of a system of linear equations. Examples 9.4 and 9.5 illustrate the importance of pivoting. We should also consider the effect of rounding errors even in apparently well behaved cases. A large number of additions and multiplications are used in the elimination process and we need to investigate whether the error caused by repeated rounding will build up to serious proportions, particularly for large systems of equations.

We shall use *backward error analysis* in our investigation. This approach was pioneered by J. H. Wilkinson (1919–86) and is suitable for various different algebraic problems.

We will assume that, instead of calculating $\mathbf{x}$, the solution of

$$\mathbf{Ax} = \mathbf{b} \tag{9.49}$$

where **A** is $n \times n$ and non-singular, we actually calculate $\mathbf{x} + \boldsymbol{\delta}\mathbf{x}$, because of rounding errors. We will show that $\mathbf{x} + \boldsymbol{\delta}\mathbf{x}$ is the (exact) solution of the perturbed set of equations

$$(\mathbf{A} + \boldsymbol{\delta}\mathbf{A})(\mathbf{x} + \boldsymbol{\delta}\mathbf{x}) = \mathbf{b} = \boldsymbol{\delta}\mathbf{b}. \tag{9.50}$$

In this section we will derive *a posteriori* bounds on the elements of the $n \times n$ perturbation matrix $\boldsymbol{\delta}\mathbf{A}$ and the perturbation column vector $\boldsymbol{\delta}\mathbf{b}$. Backward error analysis of this type does not immediately provide bounds on the final error $\boldsymbol{\delta}\mathbf{x}$ but in the next chapter we will obtain a relation between $\boldsymbol{\delta}\mathbf{x}$ and the perturbations $\boldsymbol{\delta}\mathbf{A}$ and $\boldsymbol{\delta}\mathbf{b}$.

Much of the calculation in the factorization method consists of determining inner products of the form

$$u_1 v_1 + u_2 v_2 + \cdots + u_r v_r.$$

Most computers are able to work to extra precision for at least some of their calculations and because inner products form such an important part in solving linear equations we will assume that inner products are computed with double precision before rounding to a single precision result. The better computer packages for linear algebra use such extra precision for inner products. Thus with floating point arithmetic, we produce

$$(u_1v_1 + u_2v_2 + \cdots + u_rv_r)(1 + \varepsilon)$$

where bounds on ε are determined by the type of rounding (see Appendix).

In the compact elimination method (see Algorithm 9.2), we seek triangular factors $\mathbf{L}$ and $\mathbf{U}$ ($= \mathbf{DV}$) such that

$$\mathbf{A} = \mathbf{LU} \tag{9.51}$$

where $\mathbf{L}$ has units on the diagonal. Let us suppose that because of rounding errors we actually compute matrices $\mathbf{L}'$ and $\mathbf{U}'$. We now seek $\boldsymbol{\delta}\mathbf{A}$ such that

$$\mathbf{A} + \boldsymbol{\delta}\mathbf{A} = \mathbf{L}'\mathbf{U}'. \tag{9.52}$$

To compute l_{ij} for $i > j > 1$ we try to use

$$l_{i1}u_{1j} + l_{i2}u_{2j} + \cdots + l_{ij}u_{jj} = a_{ij}$$

in the form

$$l_{ij} = (a_{ij} - l_{i1}u_{1j} - \cdots - l_{i,j-1}u_{j-1,j})/u_{jj}.$$

There will be two rounding errors; one due to the formation of the numerator (an inner product) and the other due to division by u_{jj}. We obtain, therefore,

$$l'_{ij} = \frac{(a_{ij} - l'_{i1}u'_{1j} - \cdots - l'_{i,j-1}u'_{j-1,j})}{u'_{jj}}(1 + \varepsilon_1)(1 + \varepsilon_2)$$

where ε_1 and ε_2 are due to rounding. Thus

$$l'_{ij}\left(1 - \frac{\varepsilon_1 + \varepsilon_2 + \varepsilon_1\varepsilon_2}{(1 + \varepsilon_1)(1 + \varepsilon_2)}\right) = (a_{ij} - l'_{i1}u'_{1j} - \cdots - l'_{i,j-1}u'_{j-1,j})/u'_{jj}$$

which we can write as

$$l'_{i1}u'_{1j} + \cdots + l'_{ij}u'_{jj} = a_{ij} + l'_{ij}u'_{jj}\nu \tag{9.53}$$

where

$$\nu = \frac{\varepsilon_1 + \varepsilon_2 + \varepsilon_1\varepsilon_2}{(1 + \varepsilon_1)(1 + \varepsilon_2)}.$$

Hence the amount added to a_{ij} by the two rounding errors is

$$\delta a_{ij} = l'_{ij}u'_{jj}\nu.$$

If $|\varepsilon_1|$ and $|\varepsilon_2|$ are bounded by ε then

$$|\nu| < \frac{\varepsilon(2+\varepsilon)}{(1-\varepsilon)^2} = \varepsilon'', \tag{9.54}$$

say. Note that $\varepsilon'' \simeq 2\varepsilon$, since in practice ε is much smaller than 1, and that if partial pivoting is used all the multipliers l_{ij} satisfy $|l_{ij}| \leqslant 1$.

The calculation of u_{ij} for $1 \leqslant i \leqslant j$ is made using

$$l_{i1}u_{1j} + \cdots + l_{i,i-1}u_{i-1,j} + 1u_{ij} = a_{ij}. \tag{9.55}$$

Since only an inner product and no division is required to find u_{ij} there will be less errors than in the calculation of **L**. However, for uniformity we will assume the same error bound.

We can make a similar analysis of the solution of

$$\mathbf{Ly} = \mathbf{b} \tag{9.56}$$

and

$$\mathbf{Ux} = \mathbf{y}. \tag{9.57}$$

We find each element of **y** in (9.56) from a single inner product. We replace

$$l_{i1}y_1 + \cdots + l_{i,j-1}y_{j-1} + 1y_j = b_j$$

by

$$y_j' = (b_j - l_{i1}'y_1' - \cdots - l_{i,j-1}'y_{j-1}')(1+\varepsilon_1)$$

where ε_1 is due to one rounding error. Thus

$$l_{i1}'y_1' + \cdots + l_{i,j-1}'y_{j-1}' + 1y_j' = b_j + \left(\frac{\varepsilon_1}{1+\varepsilon_1}\right)y_j'$$

and the perturbation of b_j is bounded by

$$\left|\left(\frac{\varepsilon_1}{1+\varepsilon_1}\right)y_j'\right| < \varepsilon' y$$

where

$$\varepsilon' = \frac{\varepsilon}{(1-\varepsilon)} \simeq \varepsilon,$$

ε is a bound on $|\varepsilon_1|$ and y is an upper bound on $|y_1'|, \ldots, |y_n'|$. We have shown that $\mathbf{y}'$ satisfies

$$\mathbf{L}'\mathbf{y}' = \mathbf{b} + \delta\mathbf{b}_1 \tag{9.58}$$

in place of (9.56).

Similarly we replace (9.57) by

$$\mathbf{U}'\mathbf{x}' = \mathbf{y}' + \boldsymbol{\delta}\mathbf{y}. \tag{9.59}$$

The components of $\boldsymbol{\delta}\mathbf{y}$ satisfy

$$|\delta y_i| < xu\varepsilon''$$

where ε'' is defined by (9.54) and x and u are upper bounds on the magnitude of the elements of $\mathbf{x}'$ and $\mathbf{U}'$ respectively. Again there are two rounding errors: one is due to the formation of an inner product and the other to division by a pivot. To relate $\boldsymbol{\delta}\mathbf{y}$ to the original equations, we premultiply (9.59) by $\mathbf{L}'$ to give

$$\mathbf{L}'\mathbf{U}'\mathbf{x}' = \mathbf{L}'(\mathbf{y}' + \boldsymbol{\delta}\mathbf{y}) = \mathbf{b} + \boldsymbol{\delta}\mathbf{b}_1 + \boldsymbol{\delta}\mathbf{b}_2 \tag{9.60}$$

where

$$\boldsymbol{\delta}\mathbf{b}_2 = \mathbf{L}'\boldsymbol{\delta}\mathbf{y}.$$

The ith element of $\boldsymbol{\delta}\mathbf{b}_2$ has magnitude

$$\begin{aligned} &|l'_{i1}\delta y_1 + \cdots + l'_{i,i-1}\delta y_{i-1}\delta y_{i-1} + 1\delta y_i| \\ &\quad \leqslant |\delta y_1| + |\delta y_2| + \cdots + |\delta y_i| \\ &\quad \leqslant ixu\varepsilon'' \end{aligned}$$

assuming that partial pivoting is used.

We summarize these results as a theorem.

Theorem 9.3 Suppose that the compact elimination method is used and that

(i) arithmetic is floating point with relative errors bounded by ε,
(ii) inner products are computed with only one rounding error,
(iii) partial pivoting is used,
(iv) the elements of the final upper triangular matrix $\mathbf{U}'$, the reduced right side $\mathbf{y}'$ and the calculated approximate solution $\mathbf{x}'$ satisfy

$$|u'_{ij}| \leqslant u, \quad |y'_i| \leqslant y, \quad |x'_i| \leqslant x, \qquad 1 \leqslant i, j \leqslant n.$$

Then the computed approximation $\mathbf{x}' = \mathbf{x} + \boldsymbol{\delta}\mathbf{x}$ satisfies the perturbed equations

$$(\mathbf{A} + \boldsymbol{\delta}\mathbf{A})(\mathbf{x} + \boldsymbol{\delta}\mathbf{x}) = \mathbf{b} + \boldsymbol{\delta}\mathbf{b}_1 + \boldsymbol{\delta}\mathbf{b}_2 \tag{9.61}$$

where each element of $|\boldsymbol{\delta}\mathbf{A}|$ is bounded by $u\varepsilon''$, each element of $|\boldsymbol{\delta}\mathbf{b}_1|$ is bounded by $y\varepsilon'$ and

$$|\boldsymbol{\delta}\mathbf{b}_2| \leqslant xu\varepsilon'' \begin{bmatrix} 1 \\ 2 \\ \vdots \\ n \end{bmatrix}$$

with

$$\varepsilon' = \frac{\varepsilon}{1-\varepsilon} \quad \text{and} \quad \varepsilon'' = \frac{\varepsilon(2+\varepsilon)}{(1-\varepsilon)^2}. \tag{9.62}$$

(We use $|\mathbf{A}|$ to denote the matrix with elements $|a_{ij}|$, that is, the matrix of absolute values of elements of $\mathbf{A}$. By $\mathbf{A} \leqslant \mathbf{B}$, we mean that $a_{ij} \leqslant b_{ij}$ for all i and j.) Note that, for small ε,

$$\varepsilon' \simeq \varepsilon \quad \text{and} \quad \varepsilon'' \simeq 2\varepsilon.$$

□

The bounds in Theorem 9.3 as stated are not the best that can be derived but they do illustrate the effects of rounding errors. The largest perturbation is likely to be $\boldsymbol{\delta}\mathbf{b}_2$ which is due to errors in the back substitution process. Since these errors are in the final part of the calculation of the solution they are unlikely to have any really significant effect on the computed solution and are indeed all that one can expect from the final rounded form of the solution.

Strictly speaking, the bounds of Theorem 9.3 are *a posteriori* bounds, as they depend on the magnitude of elements determined in the calculation. However, the bounds may be used to decide, *a priori*, to what accuracy the calculations should be made, assuming the equations are 'well-behaved'. If the elements of $\mathbf{U}'$ and $\mathbf{A}$ are of similar magnitudes, so that

$$a = \max_{i,j} |a_{ij}| \simeq u,$$

the perturbation $\boldsymbol{\delta}\mathbf{A}$, in (9.61), has elements whose moduli cannot be much greater than that of a double rounding error determined by the floating point representation of a. Similarly, elements of $\boldsymbol{\delta}\mathbf{b}_1$ may not be much greater than a single rounding error. For a well-behaved set of equations there is therefore no point in computing with more than one extra 'guarding' decimal digit (or two binary digits) during the main part of the reduction process. Even without guarding digits, the error introduced in the main part of the process will be only a little more than that due to representing the coefficients by floating point numbers.

The calculation for the simple elimination process of § 9.3, in which many more intermediate coefficients are determined, is identical to that of the compact elimination method without double precision calculation of inner products. Extra rounding errors are introduced in computing intermediate coefficients. There are similar perturbations to those in Theorem 9.3 but the bounds on their elements are much larger. Rounding errors are introduced in forming each term in an inner product. If all initial, intermediate and final coefficients are bounded by a, we find (see Forsythe and Moler, 1967) that, provided pivoting is used,

$$|\delta \mathbf{A}| \leqslant a\varepsilon' \begin{bmatrix} 0 & 0 & 0 & 0 & \ldots & 0 & 0 \\ 1 & 2 & 2 & 2 & \ldots & 2 & 2 \\ 1 & 3 & 4 & 4 & \ldots & 4 & 4 \\ 1 & 3 & 5 & 6 & \ldots & 6 & 6 \\ \vdots & & & & & & \vdots \\ 1 & 3 & 5 & 7 & \ldots & (2n-4) & (2n-4) \\ 1 & 3 & 5 & 7 & \ldots & (2n-3) & (2n-2) \end{bmatrix}$$

where ε' is given by (9.62) and ε is the usual error for single precision floating point arithmetic. The perturbations $\delta \mathbf{b}_1$ and $\delta \mathbf{b}_2$ are similarly bounded.

A comparison of the perturbation bounds for the simple elimination and compact elimination methods provides a strong reason for preferring the latter when calculating on a computer capable of forming double precision inner products. It must be stressed that, in spite of the apparent extra length of simple elimination, the methods are identical in terms of computing effort, which consists primarily of approximately $n^3/3$ multiplications. One advantage of simple elimination is that complete pivoting is possible, which may be desirable for troublesome sets of equations.

So far we have not discussed the effects of perturbations caused by rounding errors, on the accuracy of the computed solution, that is, we have not discussed $\delta \mathbf{x}$ in (9.50). We have already seen in §9.4 that these effects may be very serious but that pivoting does help. In the next chapter we consider the sensitivity of equations to small changes in the coefficients by determining bounds on $\delta \mathbf{x}$ in terms of $\delta \mathbf{A}$ and $\delta \mathbf{b}$.

Problems

Section 9.2

9.1 If

$$\mathbf{A} = \begin{bmatrix} 2 & 3 & -1 \\ -1 & 2 & 4 \end{bmatrix}, \quad \mathbf{B} = \begin{bmatrix} 1 & 2 & 0 \\ 0 & 1 & 1 \\ 1 & -1 & 3 \end{bmatrix} \quad \text{and} \quad \mathbf{C} = \begin{bmatrix} 2 & 1 & -1 \\ 3 & 4 & -2 \\ -1 & 0 & 1 \end{bmatrix},$$

verify by calculation that

$$\mathbf{A}(\mathbf{B} + \mathbf{C}) = \mathbf{AB} + \mathbf{AC}.$$

9.2 Show that $\mathbf{BC} \neq \mathbf{CB}$ where $\mathbf{B}$ and $\mathbf{C}$ are as defined in Problem 9.1.

9.3 Produce an example which shows that it is possible for two matrices $\mathbf{A}$ and $\mathbf{B}$ to be such that

$$\mathbf{AB} = \mathbf{0},$$

although neither of the matrices is a zero matrix.

9.4 Produce an example which shows that it is possible for three matrices to be such that

$$\mathbf{AC} = \mathbf{BC}$$

even though **A** and **B** are not equal and $\mathbf{C} \neq \mathbf{0}$.

9.5 An $n \times n$ matrix **A** is said to be lower triangular if $a_{ij} = 0$ for all $j > i$.

(i) Show that the product of two lower triangular matrices is lower triangular.
(ii) Show that the product of two lower triangular matrices with units on the diagonal ($a_{ii} = 1$, $i = 1, 2, \ldots, n$) is a lower triangular matrix with units on the diagonal.
(iii) Show that the product of two lower triangular matrices with zeros on the diagonal ($a_{ii} = 0$, $i = 1, 2, \ldots, n$) is a lower triangular matrix with zeros on the diagonal.

9.6 If the product **AB** exists, show that

$$[i\text{th row of } \mathbf{AB}] = [i\text{th row of } \mathbf{A}]\mathbf{B}$$

and

$$\begin{bmatrix} j\text{th} \\ \text{column} \\ \text{of } \mathbf{AB} \end{bmatrix} = \mathbf{A} \begin{bmatrix} j\text{th} \\ \text{column} \\ \text{of } \mathbf{B} \end{bmatrix}.$$

9.7 If **A** is an $n \times n$ matrix and **D** an $n \times n$ diagonal matrix with diagonal elements $d_1, \ldots, d_n$, show that

$$[i\text{th row of } \mathbf{DA}] = d_i[i\text{th row of } \mathbf{A}].$$

What can be said about the elements of **AD**?

9.8 Suppose

$$\mathbf{A} = \begin{bmatrix} a & b \\ c & d \end{bmatrix} \quad \text{and} \quad \mathbf{B} = \begin{bmatrix} e & f \\ g & h \end{bmatrix}.$$

Find e, f, g, h in terms of a, b, c, d such that $\mathbf{AB} = \mathbf{I}$. Confirm that such values exist if, and only if, $ad \neq bc$ and that with such values we also have $\mathbf{BA} = \mathbf{I}$.

9.9 Verify the block multiplication rule (9.7) by considering (i, j)th elements.

Section 9.3

9.10 Show, using the elimination process, that the following equations have a unique solution and find this solution.

$$\mathbf{Ax} = \mathbf{b}$$

where

$$\mathbf{A} = \begin{bmatrix} 1 & 2 & -2 & 1 \\ 2 & 5 & -2 & 3 \\ -2 & -2 & 5 & 3 \\ 1 & 3 & 3 & 2 \end{bmatrix}, \qquad \mathbf{b} = \begin{bmatrix} 4 \\ 7 \\ -1 \\ 0 \end{bmatrix}.$$

Check your result by substituting it back in the original equations.

9.11 Use elimination with only row interchanges to show that the following equations have a unique solution and find this solution.

$$\begin{bmatrix} 1 & 1 & 3 & 2 \\ 1 & 1 & 4 & 3 \\ 2 & 1 & 1 & 2 \\ 2 & -1 & 4 & 5 \end{bmatrix} \mathbf{x} = \begin{bmatrix} 4 \\ 7 \\ 3 \\ 8 \end{bmatrix}.$$

Check your result as in Problem 9.10.

9.12 Repeat Problem 9.11 but use only column interchanges.

9.13 Show that the following equations are of rank 2 and inconsistent.

$$\begin{aligned} x + 2y + 3z &= 3 \\ x - y - 4z &= -5 \\ 3x + 3y + 2z &= 2. \end{aligned}$$

9.14 Show that the following equations are consistent if, and only if, $a = +1$ or $a = -1$.

$$\begin{aligned} x + y + z &= 1 + a^2 \\ x + 2y + 3z &= -2a \\ x + 3y + 4z &= -4a \\ x + 2y + 2z &= 2(1 - a). \end{aligned}$$

9.15 Write in an algorithmic form the elimination process without interchanges for a system of n non-singular linear equations in n unknowns (see Algorithm 9.1).

9.16 If the matrix of the linear equations of (5.13) is singular, show that there exist numbers c_i, not all zero, such that

$$\sum_{i=0}^{n} c_i \int_a^b \psi_i(x)\psi_j(x)\,\mathrm{d}x = 0, \qquad 0 \leqslant j \leqslant n.$$

Deduce that

$$\int_a^b \left[\sum_{i=0}^{n} c_i \psi_i(x) \right]^2 \mathrm{d}x = 0,$$

which is impossible if the ψ_i are linearly independent, and hence show that the normal equations (5.13) have a unique solution.

Section 9.4

9.17 Repeat Example 9.4 with the same ten digit accuracy but making only column interchanges to pick suitable pivots.

9.18 Repeat Example 9.5 with the same three digit accuracy but making only column interchanges to pick suitable pivots.

9.19 Use elimination to solve

$$\begin{aligned} 0.20x_1 + 1.2x_2 + 1.6x_3 &= 4.5 \\ 1.2x_1 + 7.1x_2 - 6.0x_3 &= -4.3 \\ 4.2x_1 - 2.8x_2 + 10x_3 &= 5.3 \end{aligned}$$

working to only three significant digits and

(i) without pivoting,
(ii) with partial pivoting.

9.20 Write a computer program to implement Algorithm 9.1 and test your program on Examples 9.1, 9.4, 9.5 and the equations of Problem 9.19.

Section 9.5

9.21 If

$$\mathbf{M}_j = \begin{bmatrix} 1 & & & & & & \\ l_{21} & \ddots & & & \mathbf{0} & & \\ \vdots & & 1 & & & & \\ \vdots & & l_{j+1,j} & 1 & & & \\ \vdots & & \vdots & 0 & \ddots & & \\ \vdots & & \vdots & \vdots & \ddots & \ddots & \\ l_{n1} & \cdots & l_{n,j} & 0 & \cdots & 0 & 1 \end{bmatrix}$$

show that

$$\mathbf{M}_{j+1} = \mathbf{M}_j \mathcal{L}_j$$

where $\mathcal{L}_j$ is defined by (9.22). Hence verify (9.23) by induction.

9.22 Find the matrices $\mathcal{L}_1$, $\mathcal{L}_2$, $\mathcal{L}_3$, $\mathbf{L}$, $\mathbf{U}$ and $\mathbf{c}$ of § 9.5 for the equations of Problem 9.10. Verify that (9.24) and (9.25) are valid for your results.

Section 9.6

9.23 Use the method of § 9.6 to find factors **L**, **D** and **V** of the coefficient matrix **A** of Problem 9.10.

9.24 Show that the coefficient matrix in Problem 9.11 cannot be factorized in the form **LDV**.

Section 9.7

9.25 Use Algorithm 9.2 (compact elimination) to factorize the matrix of Example 9.7.

9.26 Use Algorithms 9.2 and 9.3 to solve the equations of Problem 9.10 by hand.

9.27 Write a program to solve linear equations by implementing Algorithms 9.2 and 9.3. Test your program on the equations in Example 9.1, Problem 9.10 and Problem 9.19.

9.28 We can find the inverse of a non-singular matrix by solving linear equations. If $\mathbf{AX}=\mathbf{I}$ and $\mathbf{x}_j$ is the jth column of $\mathbf{X}$ then

$$\mathbf{Ax}_j=\mathbf{e}_j \tag{9.63}$$

where $\mathbf{e}_j$ is the jth column of **I**. We can solve the set of linear equations (9.63) by elimination. By taking $j=1,2,\ldots,n$ we obtain n different sets of equations and by writing the solutions as columns of a matrix we calculate $\mathbf{X}=\mathbf{A}^{-1}$. Use this method to find the inverse of the matrix

$$\mathbf{A}=\begin{bmatrix}-1 & 8 & -2\\ -6 & 49 & -10\\ -4 & 34 & -5\end{bmatrix}.$$

For simplicity do not use interchanges in the elimination process and note that the same factors **L** and **U** apply to each of the three sets of equations.

Section 9.8

9.29 By considering (i,j)th elements show that if matrices **A**, **B** are such that **AB** exists then

$$(\mathbf{AB})^{\mathrm{T}}=\mathbf{B}^{\mathrm{T}}\mathbf{A}^{\mathrm{T}}.$$

9.30 If the product **ABC** exists, by writing $\mathbf{ABC}=\mathbf{A}(\mathbf{BC})$, show that $(\mathbf{ABC})^{\mathrm{T}}=\mathbf{C}^{\mathrm{T}}\mathbf{B}^{\mathrm{T}}\mathbf{A}^{\mathrm{T}}$.

9.31 If **A** and **B** are symmetric $n \times n$ matrices, show that

(i) $(\mathbf{A}+\mathbf{B})$ is symmetric,
(ii) **AB** is not necessarily symmetric. (Produce a counter-example.)

9.32 An $n \times n$ matrix **A** is said to be *skew-symmetric* if

$$\mathbf{A}^{\mathrm{T}} = -\mathbf{A}.$$

Show that all the diagonal elements of such a matrix are zero. Hence, by considering the $(1, 1)$ element, prove that it is not possible to factorize such a matrix in the form **LDV** where **D** has non-zero diagonal elements.

9.33 Find a matrix **M** such that (9.45) is valid for

$$\mathbf{A} = \begin{bmatrix} 4 & -2 & -4 \\ -2 & 17 & 10 \\ -4 & 10 & 9 \end{bmatrix}$$

and hence solve the equations

$$\mathbf{Ax} = \begin{bmatrix} 10 \\ 3 \\ -7 \end{bmatrix}.$$

9.34 If **A** is any non-singular symmetric matrix, show that $\mathbf{A}^2$ is a positive definite matrix. (*Hint*: $\mathbf{x}^{\mathrm{T}}\mathbf{A}^2\mathbf{x} = \mathbf{x}^{\mathrm{T}}\mathbf{A}^{\mathrm{T}}\mathbf{Ax}$.)

9.35 Show that if **A** is a positive definite matrix, all of its leading sub-matrices are positive definite. (*Hint*: consider $\mathbf{x}^{\mathrm{T}}\mathbf{Ax}$, where **x** is a vector whose last $n-p$ elements are zero.)

9.36 Write a computer program which implements Algorithm 9.4 to perform Choleski factorization of a positive definite matrix. Test your program on the matrices of Example 9.8 and Problem 9.33.

9.37 Write in an algorithmic form the forward and back substitution process for solving $\mathbf{Mc} = \mathbf{b}$ and $\mathbf{M}^{\mathrm{T}}\mathbf{x} = \mathbf{c}$ where **M** is lower triangular and **b** is given.

Convert your algorithm to a computer program and combine it with your answer to Problem 9.36 to produce a program to solve $\mathbf{Ax} = \mathbf{b}$ where **A** is positive definite. Test your program on the equations in Example 9.8 and Problem 9.33.

9.38 Use the program developed in Problem 9.37 to solve $\mathbf{Ax} = \mathbf{b}$ where **A** is the positive definite matrix

$$\mathbf{A} = \begin{bmatrix} 420 & 210 & 140 & 105 \\ 210 & 140 & 105 & 84 \\ 140 & 105 & 84 & 70 \\ 105 & 84 & 70 & 60 \end{bmatrix} \quad \text{and} \quad \mathbf{b} = \begin{bmatrix} 875 \\ 539 \\ 399 \\ 319 \end{bmatrix}.$$

The exact solution is $\mathbf{x} = [1 \;\; 1 \;\; 1 \;\; 1]^T$ but because of rounding errors you may obtain a solution which is correct to only two or three significant decimal digits. (The precise accuracy depends on your computer.) These equations are difficult to solve accurately as $\mathbf{A} = 420\,\mathbf{H}_4$ where $\mathbf{H}_4$ is the 4×4 Hilbert matrix described in § 10.4. What happens to the solution if we take $b_3 = 400$ (instead of 399)?

9.39 Show that the least squares normal equations (5.10) may be written in matrix form as $\mathbf{\Psi\Psi}^T\mathbf{a} = \mathbf{\Psi f}$, where

$$\mathbf{a}^T = [a_0 \ldots a_n], \qquad \mathbf{f}^T = [f(x_0) \ldots f(x_N)]$$

and $\mathbf{\Psi}$ is the $(n+1) \times (N+1)$ matrix whose (i, j)th element is $\psi_{i-1}(x_{j-1})$. Show that if $\{\psi_0, \psi_1, \ldots, \psi_n\}$ is a Chebyshev set then $\mathbf{\Psi}^T\mathbf{b} \neq \mathbf{0}$ for all $(n+1)$-dimensional vectors $\mathbf{b} \neq \mathbf{0}$. Hence show (cf. Problem 9.34) that $\mathbf{\Psi\Psi}^T$ is positive definite and thus that the normal equations have a unique solution.

Section 9.9

9.40 Show that the process described in § 9.9 is equivalent to factorization of the tridiagonal matrix $\mathbf{A}$ into lower and upper triangular factors of the form

$$\mathbf{A} = \begin{bmatrix} 1 & & & & \\ m_1 & 1 & & \mathbf{0} & \\ & m_2 & 1 & & \\ & & \ddots & \ddots & \\ \mathbf{0} & & & m_{n-1} & 1 \end{bmatrix} \begin{bmatrix} \beta_1 & c_1 & & \mathbf{0} \\ & \ddots & \ddots & \\ & & \ddots & c_{n-1} \\ \mathbf{0} & & & \beta_n \end{bmatrix}.$$

9.41 Use Algorithm 9.5 to solve the equations

$$\begin{bmatrix} 2 & -1 & 0 & 0 & 0 \\ -1 & 2 & -1 & 0 & 0 \\ 0 & -1 & 2 & -1 & 0 \\ 0 & 0 & -1 & 2 & -1 \\ 0 & 0 & 0 & -1 & 2 \end{bmatrix} \mathbf{x} = \begin{bmatrix} 1 \\ 0 \\ 0 \\ 0 \\ 7 \end{bmatrix}.$$

Also find the triangular factors of the coefficient matrix as in Problem 9.40.

9.42 If $\mathbf{A}$ is the $n \times n$ symmetric matrix

$$\mathbf{A} = \begin{bmatrix} b_1 & c_1 & & & \\ c_1 & b_2 & c_2 & & \mathbf{0} \\ & c_2 & b_3 & c_3 & \\ & \ddots & \ddots & \ddots & \\ \mathbf{0} & & \ddots & \ddots & c_{n-1} \\ & & & c_{n-1} & b_n \end{bmatrix},$$

show that, for any n-dimensional vector $\mathbf{x}$,

$$\begin{aligned}\mathbf{x}^{\mathrm{T}}\mathbf{A}\mathbf{x} &= b_1x_1^2 + 2c_1x_1x_2 + b_2x_2^2 + 2c_2x_2x_3 + \cdots + b_{n-1}x_{n-1}^2 \\ &\quad + 2c_{n-1}x_{n-1}x_n + b_nx_n^2 \\ &= (b_1 - |c_1|)x_1^2 + |c_1|(x_1 \pm x_2)^2 + (b_2 - |c_1| - |c_2|)x_2^2 \\ &\quad + |c_2|(x_2 \pm x_3)^2 + \cdots + (b_{n-1} - |c_{n-2}| - |c_{n-1}|)x_{n-1}^2 \\ &\quad + |c_{n-1}|(x_{n-1} \pm x_n)^2 + (b_n - |c_{n-1}|)x_n^2,\end{aligned}$$

where the $\pm$ sign in $(x_i \pm x_{i+1})^2$ is taken to be that of c_i. Hence show that if $|c_i| > 0$, $i = 1, 2, \ldots, n-1$, and, defining $c_0 = c_n = 0$,

$$b_i \geqslant |c_i| + |c_{i-1}|, \qquad i = 1, 2, \ldots, n,$$

with strict inequality in the last relation for at least one i, then $\mathbf{x}^{\mathrm{T}}\mathbf{A}\mathbf{x} > 0$ for all $\mathbf{x} \neq \mathbf{0}$ and thus $\mathbf{A}$ is positive definite.

Show that the coefficient matrix in Problem 9.41 is positive definite.

Section 9.10

9.43 It is possible to mimic the behaviour of a computer of limited accuracy by multiplying the result of *every* arithmetic operation (addition, subtraction, multiplication and division) by an amount $1 + \varepsilon$ for a suitably small ε.

(i) Amend the program requested in Problem 9.20 so that the result of every arithmetic operation is multiplied by 1.005 (the maximum error for three significant decimal digit working). Try the amended program on the various equations tested earlier, including those of Problem 9.38, ignoring symmetry.

(ii) Carry out a similar amendment of the program of Problem 9.27 but in the case of inner products multiply only the final total by 1.005 and not individual terms. This simulates the use of extra precision for inner products. Try the amended program on the various equations encountered before.

Chapter **10**

MATRIX NORMS AND APPLICATIONS

10.1 Determinants, eigenvalues and eigenvectors

We will need to use eigenvalues to complete the error analysis of the solution of linear equations and to describe fully the use of iterative methods.

Definition 10.1 The *determinant* of an $n \times n$ matrix $\mathbf{A}$ is defined recursively as follows.

(i) If $\mathbf{A} = [a_{11}]$, that is a 1×1 matrix, then the determinant of $\mathbf{A}$, written as det $\mathbf{A}$, is a_{11}.

(ii) If $\mathbf{A}$ is $n \times n$, $n \geqslant 2$, we define the *minor* A_{ij} of $\mathbf{A}$ to be the determinant of the $(n-1) \times (n-1)$ matrix obtained by deleting the ith row and jth column of $\mathbf{A}$. We then define

$$\det \mathbf{A} = \sum_{i=1}^{n} (-1)^{i+j} a_{ij} A_{ij}, \qquad \text{for any } j, \quad 1 \leqslant j \leqslant n, \qquad (10.1)$$

or

$$\det \mathbf{A} = \sum_{j=1}^{n} (-1)^{i+j} a_{ij} A_{ij}, \qquad \text{for any } i, \quad 1 \leqslant i \leqslant n. \qquad (10.2)$$

Note that we obtain the same result (see, for example, Johnson *et al.*, 1993) regardless of whether we use (10.1) or (10.2) and regardless of the value of j in (10.1) or of i in (10.2). □

For a 2×2 matrix

$$\mathbf{A} = \begin{bmatrix} a_{11} & a_{12} \\ a_{21} & a_{22} \end{bmatrix}$$

we have

$$\det \mathbf{A} = a_{11}a_{22} - a_{12}a_{21}$$

and, for a 3×3 matrix,

$$\det \mathbf{A} = a_{11} \det\begin{bmatrix} a_{22} & a_{23} \\ a_{32} & a_{33} \end{bmatrix} - a_{12} \det\begin{bmatrix} a_{21} & a_{23} \\ a_{31} & a_{33} \end{bmatrix} + a_{13} \det\begin{bmatrix} a_{21} & a_{22} \\ a_{31} & a_{32} \end{bmatrix}.$$

We can show that adding one row (or column) of a matrix to another row (or column) does not change the determinant and that interchanging two rows (or columns) changes the sign of the determinant. Multiplying one row (or column) by a scalar λ multiplies the determinant by λ and thus

$$\det(\lambda \mathbf{A}) = \lambda^n \det \mathbf{A},$$

where n is the order of the matrix, since each row is scaled by λ. For the product of two square matrices we have

$$\det(\mathbf{AB}) = (\det \mathbf{A})(\det \mathbf{B}).$$

Proofs of these results may be found in Johnson *et al.* (1993). In Chapter 9 it was asserted that the product of pivots (arising from Gaussian elimination) is $\pm\det \mathbf{A}$ (the sign depends on how many row and/or column interchanges are made) and thus a matrix $\mathbf{A}$ is singular if, and only if, $\det \mathbf{A} = 0$.

Definition 10.2 An $n \times n$ matrix $\mathbf{A}$ is said to have *eigenvalue* $\lambda \in \mathbb{C}$ if there is a *non-zero* vector $\mathbf{x} \in \mathbb{C}^n$ such that

$$\mathbf{Ax} = \lambda \mathbf{x}. \tag{10.3}$$

The vector $\mathbf{x}$ is called an *eigenvector* of $\mathbf{A}$ corresponding to the eigenvalue λ.

□

Equations (10.3) are equivalent to

$$(\mathbf{A} - \lambda \mathbf{I})\mathbf{x} = \mathbf{0} \tag{10.4}$$

and these have a non-zero solution $\mathbf{x}$ if, and only if, $\mathbf{A} - \lambda \mathbf{I}$ is singular or, equivalently,

$$\det(\mathbf{A} - \lambda \mathbf{I}) = 0. \tag{10.5}$$

Equation (10.5) is called the *characteristic equation* of $\mathbf{A}$ and $\det(\mathbf{A} - \lambda \mathbf{I})$, which is a polynomial in λ of degree n, is called the *characteristic polynomial* of $\mathbf{A}$. There are n roots of (10.5), which we will denote by $\lambda_1, \lambda_2, \ldots, \lambda_n$. For a given eigenvalue λ_i, the corresponding eigenvector is not uniquely determined, for if $\mathbf{x}$ is an eigenvector then so is $\mu \mathbf{x}$ where μ is any non-zero scalar.

Example 10.1 If

$$\mathbf{A} = \begin{bmatrix} 4 & -3 & 3 \\ 2 & -1 & 1 \\ -4 & 4 & -4 \end{bmatrix}$$

then

$$\det(\mathbf{A} - \lambda\mathbf{I}) = \det\begin{bmatrix} 4-\lambda & -3 & 3 \\ 2 & -1-\lambda & 1 \\ -4 & 4 & -4-\lambda \end{bmatrix}$$

$$= (4-\lambda)[(-1-\lambda)(-4-\lambda) - 4]$$
$$+ 3[2(-4-\lambda) + 4] + 3[8 + 4(-1-\lambda)]$$
$$= \lambda(1-\lambda)(2+\lambda).$$

Thus the eigenvalues of **A** are $\lambda_1 = 0$, $\lambda_2 = 1$, $\lambda_3 = -2$. To find eigenvectors we substitute each eigenvalue in (10.4) in turn and solve equations for an $\mathbf{x} \neq \mathbf{0}$. For example, $\lambda = \lambda_2 = 1$ yields the linear equations

$$\begin{aligned} 3x_1 - 3x_2 + 3x_3 &= 0 \\ 2x_1 - 2x_2 + x_3 &= 0 \\ -4x_1 - 4x_2 + 5x_3 &= 0 \end{aligned}$$

and the general non-zero solution is $x_1 = q$, $x_2 = q$, $x_3 = 0$, $q \in \mathbb{R}$, $q \neq 0$. We can similarly solve the equations with $\lambda = 0$ and with $\lambda = -2$. We thus find eigenvectors corresponding to λ_1, λ_2 and λ_3:

$$\mathbf{x}_1 = \begin{bmatrix} 0 \\ p \\ p \end{bmatrix}, \qquad \mathbf{x}_2 = \begin{bmatrix} q \\ q \\ 0 \end{bmatrix}, \qquad \mathbf{x}_3 = \begin{bmatrix} -r \\ 0 \\ 2r \end{bmatrix},$$

where p, q, $r \in \mathbb{R}$ are non-zero. □

In general the eigenvalues and eigenvectors of an $n \times n$ real matrix are complex, but for real symmetric matrices the eigenvalues can be shown to be real and real eigenvectors can be chosen.

Definition 10.3 Two $n \times n$ matrices **A** and **B** are said to be *similar* if there exists a non-singular matrix **T** such that

$$\mathbf{A} = \mathbf{T}\,\mathbf{B}\,\mathbf{T}^{-1}$$

or equivalently

$$\mathbf{B} = \mathbf{T}^{-1}\,\mathbf{A}\,\mathbf{T}.$$

T is called a *similarity transformation* matrix. □

It is easily shown that similar matrices have common eigenvalues. If

$$\mathbf{A}\mathbf{x} = \lambda\mathbf{x}$$

where $\mathbf{x} \neq \mathbf{0}$ and **B** is similar to **A** then

$$\mathbf{T}\,\mathbf{B}\,\mathbf{T}^{-1}\mathbf{x} = \lambda\mathbf{x}$$

which implies

$$\mathbf{B}\,\mathbf{T}^{-1}\mathbf{x} = \mathbf{T}^{-1}\,\mathbf{T}\,\mathbf{B}\,\mathbf{T}^{-1}\mathbf{x} = \lambda\mathbf{T}^{-1}\mathbf{x}. \tag{10.6}$$

Now $\mathbf{T}^{-1}$ is non-singular so that $\mathbf{x} \neq \mathbf{0}$ implies $\mathbf{T}^{-1}\mathbf{x} \neq \mathbf{0}$ and hence λ is an eigenvalue of $\mathbf{B}$ with corresponding eigenvector $\mathbf{T}^{-1}\mathbf{x}$. Similarly if μ is an eigenvalue of $\mathbf{B}$ it can be shown to be an eigenvalue of $\mathbf{A}$. This proof of common eigenvalues is, however, not complete as we should also show that repeated eigenvalues of $\mathbf{A}$ are repeated eigenvalues of $\mathbf{B}$ and vice versa.

The most interesting matrix similar to a given matrix $\mathbf{A}$ is a diagonal matrix. It can be shown that many (but not all) real matrices are similar to a diagonal matrix and thus

$$\mathbf{A} = \mathbf{T}\Lambda\mathbf{T}^{-1} \tag{10.7}$$

where Λ is diagonal. Now $\mathbf{A}$ and Λ have the same eigenvalues and the eigenvalues of a diagonal matrix are simply the diagonal elements of that matrix. Thus the diagonal elements of Λ are precisely $\lambda_1, \lambda_2, \ldots, \lambda_n$ if such a similarity transformation $\mathbf{T}$ exists. From

$$\mathbf{A}\,\mathbf{T} = \mathbf{T}\,\Lambda$$

we have, on equating jth columns on each side,

$$\mathbf{A}\mathbf{x}_j = \lambda_j\mathbf{x}_j$$

where $\mathbf{x}_j$ is the jth column of $\mathbf{T}$. Thus the columns of $\mathbf{T}$ consist of eigenvectors of $\mathbf{A}$. We need to be able to choose n eigenvectors such that $\mathbf{T}$ is non-singular. Equivalently we must be able to choose n linearly independent eigenvectors for $\mathbf{T}$ to be non-singular (see Johnson *et al.*, 1993). It can be shown that this is always possible if $\mathbf{A}$ is a real symmetric matrix or if $\mathbf{A}$ has n distinct eigenvalues. In other cases $\mathbf{A}$ may not be similar to a diagonal matrix (see Problem 10.4).

If $\mathbf{A}$ is symmetric, a further simplification is that $\mathbf{T}$ may be chosen to be orthogonal ($\mathbf{T}^{-1} = \mathbf{T}^{\mathrm{T}}$). We then have

$$\mathbf{A} = \mathbf{T}\,\Lambda\,\mathbf{T}^{\mathrm{T}}.$$

Definition 10.4 The *spectral radius* $\rho(\mathbf{A})$ of a matrix $\mathbf{A}$ is

$$\rho(\mathbf{A}) = \max_{1 \leqslant i \leqslant n} |\lambda_i| \tag{10.8}$$

where $\lambda_1, \lambda_2, \ldots, \lambda_n$ are the eigenvalues of $\mathbf{A}$.

We will see that the spectral radius plays an important role in the theory of iterative methods for solving linear equations.

10.2 Vector norms

We shall require some measure of the 'magnitude' of a vector for satisfactory error analyses of methods of solving linear equations. You are probably familiar with the 'magnitude' (or 'length') of a vector in three-dimensional space. We generalize this concept to deal with n-dimensional vector spaces by defining a norm. (Note the similarity between the following definition and Definition 5.1 in Chapter 5, where we also use norms.)

Definition 10.5 For any vector $\mathbf{x} \in \mathbb{C}^n$, we define the *norm* of $\mathbf{x}$, written as $\|\mathbf{x}\|$, to be a *real* number satisfying the following conditions.

(i) $\|\mathbf{x}\| > 0$, unless $\mathbf{x} = \mathbf{0}$, and $\|\mathbf{0}\| = 0$,
(ii) For any scalar λ, and any vector $\mathbf{x}$, $\|\lambda\mathbf{x}\| = |\lambda| \, \|\mathbf{x}\|$.
(iii) For any vectors $\mathbf{x}$ and $\mathbf{y}$,

$$\|\mathbf{x}+\mathbf{y}\| \leqslant \|\mathbf{x}\| + \|\mathbf{y}\|. \quad (10.9) \quad \square$$

The last condition is known as the *triangle inequality* and from this it may also be shown (Problem 10.5) that

$$\|\mathbf{x}-\mathbf{y}\| \geqslant \big| \, \|\mathbf{x}\| - \|\mathbf{y}\| \, \big|. \quad (10.10)$$

We may also define the norm to be a mapping of the n-dimensional space into the real numbers such that (i)–(iii) hold.

There are many ways in which one may choose norms. One of the most common is the *Euclidean norm* (or *length*) $\|\mathbf{x}\|_2$ defined by

$$\begin{aligned} \|\mathbf{x}\|_2 &= (|x_1|^2 + |x_2|^2 + \cdots + |x_n|^2)^{1/2} \\ &= \left[\sum_{i=1}^{n} |x_i|^2\right]^{1/2} = [\bar{\mathbf{x}}^{\mathrm{T}}\mathbf{x}]^{1/2} \end{aligned} \quad (10.11)$$

where $\bar{\mathbf{x}}$ is the complex conjugate of $\mathbf{x}$. Conditions (i) and (ii) above are clearly satisfied. The verification that condition (iii) holds is more difficult and is left as Problem 10.6. In three-dimensional space, (iii) corresponds to stating that the length of one side of a triangle cannot exceed the sum of the lengths of the other two sides.

Other possible norms include

$$\begin{aligned} \|\mathbf{x}\|_1 &= |x_1| + |x_2| + \cdots + |x_n| \\ &= \sum_{i=1}^{n} |x_i| \end{aligned} \quad (10.12)$$

and

$$\|\mathbf{x}\|_\infty = \max_{1 \leqslant i \leqslant n} |x_i|. \quad (10.13)$$

These are particularly useful for numerical methods as the corresponding matrix norms (see § 10.3) are easily computed. All of these norms may be considered special cases of

$$\|\mathbf{x}\|_p = \left[\sum_{i=1}^{n} |x_i|^p\right]^{1/p}, \tag{10.14}$$

where $p \geqslant 1$. The norms $\|\cdot\|_1$ and $\|\cdot\|_2$ correspond to $p=1$ and $p=2$ respectively. The *maximum norm* $\|\cdot\|_\infty$ may be considered the result of allowing p to increase indefinitely. The reader is urged to verify that (10.12) and (10.13) do define norms, by showing that conditions (i)–(iii) are satisfied. (See Problem 10.7.) It is more difficult to verify that, for a general value of $p \geqslant 1$, (10.14) satisfies condition (iii). It is easily verified that conditions (i) and (ii) are satisfied. In most of this chapter we shall use $\|\cdot\|_1$ or $\|\cdot\|_\infty$ and, where it does not matter which norm is employed, we shall simply write $\|\cdot\|$. Of course we do not mix different types of norm as otherwise (10.9) may not hold.

Example 10.2 The norms of

$$\mathbf{x} = \begin{bmatrix} 1 \\ -2 \\ 3 \end{bmatrix} \quad \text{and} \quad \mathbf{y} = \begin{bmatrix} 0 \\ 2 \\ 3 \end{bmatrix}$$

are

$$\begin{aligned} &\|\mathbf{x}\|_1 = 6, && \|\mathbf{y}\|_1 = 5, \\ &\|\mathbf{x}\|_2 = \sqrt{14}, && \|\mathbf{y}\|_2 = \sqrt{13}, \\ &\|\mathbf{x}\|_\infty = \|\mathbf{y}\|_\infty = 3. && \end{aligned}$$

□

You will notice that two different vectors may have the same norm, so that

$$\|\mathbf{x}\| = \|\mathbf{y}\| \quad \not\Rightarrow \quad \mathbf{x} = \mathbf{y}, \tag{10.15}$$

However, from condition (i),

$$\|\mathbf{x} - \mathbf{y}\| = 0 \quad \Rightarrow \quad \mathbf{x} - \mathbf{y} = \mathbf{0} \quad \Rightarrow \quad \mathbf{x} = \mathbf{y}. \tag{10.16}$$

We say that a sequence of n-dimensional vectors $(\mathbf{x}_m)_{m=0}^{\infty}$ converges to a vector $\boldsymbol{\alpha}$ if, for $i = 1, 2, \ldots, n$, the sequence formed from the ith elements of the $\mathbf{x}_m$ converges to the ith element of $\boldsymbol{\alpha}$. We write

$$\lim_{m \to \infty} \mathbf{x}_m = \boldsymbol{\alpha}.$$

Lemma 10.1 For any type of vector norm

$$\lim_{m \to \infty} \mathbf{x}_m = \boldsymbol{\alpha}$$

if, and only if,

$$\lim_{m\to\infty} \|\mathbf{x}_m - \boldsymbol{\alpha}\| = 0.$$

Proof The lemma is clearly true for norms of the form (10.14) (including $p = \infty$) because of the continuous dependence of the norms on each element and condition (i). The proof for general norms is more difficult and may be found in Isaacson and Keller (1966). □

Note that

$$\lim_{m\to\infty} \|\mathbf{x}_m\| = \|\boldsymbol{\alpha}\| \tag{10.17}$$

is not sufficient to ensure that $(\mathbf{x}_m)$ converges to $\boldsymbol{\alpha}$. Indeed, such a sequence may not even converge. Consider, for example,

$$\begin{bmatrix} 1 \\ 1 \end{bmatrix}, \begin{bmatrix} -1 \\ 1 \end{bmatrix}, \begin{bmatrix} 1 \\ 1 \end{bmatrix}, \begin{bmatrix} -1 \\ 1 \end{bmatrix}, \begin{bmatrix} 1 \\ 1 \end{bmatrix}, \ldots .$$

This satisfies (10.17), with

$$\boldsymbol{\alpha} = \begin{bmatrix} 1 \\ 1 \end{bmatrix},$$

for any choice of the norms (10.14).

10.3 Matrix norms

Definition 10.6 For an $n \times n$ matrix $\mathbf{A}$, we define the *matrix norm* $\|\mathbf{A}\|$ *subordinate* to a given vector norm to be

$$\|\mathbf{A}\| = \sup_{\|\mathbf{x}\|=1} \|\mathbf{A}\mathbf{x}\|.$$ □

The supremum is taken over all n-dimensional vectors $\mathbf{x}$ with unit norm. The supremum is attained (the proof of this is beyond the scope of this book except for the norms (10.11), (10.12) and (10.13)) and, therefore, we write

$$\|\mathbf{A}\| = \max_{\|\mathbf{x}\|=1} \|\mathbf{A}\mathbf{x}\|. \tag{10.18}$$

Lemma 10.2 A subordinate matrix norm satisfies the following properties.

(i) $\|\mathbf{A}\| > 0$, unless $\mathbf{A} = \mathbf{0}$, and $\|\mathbf{0}\| = 0$.
(ii) For any scalar λ and any $\mathbf{A}$, $\|\lambda\mathbf{A}\| = |\lambda|\,\|\mathbf{A}\|$.
(iii) For any two matrices $\mathbf{A}$ and $\mathbf{B}$,

$$\|\mathbf{A}+\mathbf{B}\| \leqslant \|\mathbf{A}\| + \|\mathbf{B}\|.$$

(iv) For any n-dimensional vector $\mathbf{x}$ and any $\mathbf{A}$,

$$\|\mathbf{Ax}\| \leqslant \|\mathbf{A}\| \cdot \|\mathbf{x}\|.$$

(v) For any two matrices $\mathbf{A}$ and $\mathbf{B}$,

$$\|\mathbf{AB}\| \leqslant \|\mathbf{A}\| \cdot \|\mathbf{B}\|.$$

Proof

(i) If $\mathbf{A} \neq \mathbf{0}$, there is an $\mathbf{x} \neq \mathbf{0}$ such that $\mathbf{Ax} \neq \mathbf{0}$. We may scale $\mathbf{x}$ so that $\|\mathbf{x}\| = 1$ and we still have $\mathbf{Ax} \neq \mathbf{0}$. Thus, from (10.18), $\|\mathbf{A}\| > 0$. If $\mathbf{A} = \mathbf{0}$ then $\mathbf{Ax} = \mathbf{0}$ for all $\mathbf{x}$ and therefore $\|\mathbf{A}\| = 0$.

(ii) From condition (ii) for vector norms, it follows that

$$\|\lambda \mathbf{A}\| = \max_{\|\mathbf{x}\|=1} \|\lambda \mathbf{Ax}\| = \max_{\|\mathbf{x}\|=1} |\lambda| \cdot \|\mathbf{Ax}\| = |\lambda| \max_{\|\mathbf{x}\|=1} \|\mathbf{Ax}\|.$$

(iii) From condition (iii) for vector norms,

$$\begin{aligned}\|\mathbf{A}+\mathbf{B}\| &= \max_{\|\mathbf{x}\|=1} \|(\mathbf{A}+\mathbf{B})\mathbf{x}\| \leqslant \max_{\|x\|=1} [\|\mathbf{Ax}\| + \|\mathbf{Bx}\|] \\ &\leqslant \max_{\|\mathbf{x}\|=1} \|\mathbf{Ax}\| + \max_{\|\mathbf{x}\|=1} \|\mathbf{Bx}\| = \|\mathbf{A}\| + \|\mathbf{B}\|.\end{aligned}$$

(iv) For any $\mathbf{x} \neq \mathbf{0}$,

$$\|\mathbf{Ax}\| = \|\mathbf{x}\| \cdot \left\| \mathbf{A}\left(\frac{1}{\|\mathbf{x}\|}\mathbf{x}\right)\right\| \leqslant \|\mathbf{x}\| \cdot \|\mathbf{A}\|,$$

as

$$\left\| \frac{1}{\|\mathbf{x}\|}\mathbf{x} \right\| = 1$$

and $\|\mathbf{A}\|$ occurs for the maximum in (10.18). The result is trivial for $\mathbf{x} = \mathbf{0}$.

(v)

$$\begin{aligned}\|\mathbf{AB}\| &= \max_{\|\mathbf{x}\|=1} \|\mathbf{ABx}\| = \max_{\|\mathbf{x}\|=1} \|\mathbf{A}(\mathbf{Bx})\| \\ &\leqslant \max_{\|\mathbf{x}\|=1} \|\mathbf{A}\| \cdot \|\mathbf{Bx}\|\end{aligned}$$

by (iv) above. Thus

$$\|\mathbf{AB}\| \leqslant \|\mathbf{A}\| \max_{\|\mathbf{x}\|=1} \|\mathbf{Bx}\| = \|\mathbf{A}\| \cdot \|\mathbf{B}\|. \qquad \square$$

Matrix norms other than those subordinate to vector norms may also be defined. Such matrix norms are real numbers which satisfy conditions (i), (ii), (iii) and (v) of Lemma 10.2. Throughout this chapter we shall use only subordinate matrix norms and, therefore, condition (iv) of Lemma 10.2 is valid.

We now obtain more explicit expressions for the matrix norms subordinate to the vector norms of (10.12), (10.13) and (10.11). From (10.12)

$$\begin{aligned}\|\mathbf{Ax}\|_1 &= \sum_{i=1}^{n}\left|\sum_{j=1}^{n} a_{ij}x_j\right| \\ &\leqslant \sum_{i=1}^{n}\sum_{j=1}^{n}|a_{ij}||x_j| = \sum_{j=1}^{n}\left(\sum_{i=1}^{n}|a_{ij}|\right)|x_j| \\ &\leqslant \sum_{j=1}^{n}\left(\max_{1\leqslant k\leqslant n}\sum_{i=1}^{n}|a_{ik}|\right)|x_j| = \left(\max_{1\leqslant j\leqslant n}\sum_{i=1}^{n}|a_{ij}|\right)\left(\sum_{j=1}^{n}|x_j|\right).\end{aligned}$$

The second factor of the last expression is unity if $\|\mathbf{x}\|_1 = 1$ and thus

$$\|\mathbf{A}\|_1 = \max_{\|\mathbf{x}\|_1=1}\|\mathbf{Ax}\|_1 \leqslant \max_{1\leqslant j\leqslant n}\sum_{i=1}^{n}|a_{ij}|. \tag{10.19}$$

Suppose the maximum in the last expression is attained for $j=p$. We now choose $\mathbf{x}$ such that

$$x_p = 1 \quad \text{and} \quad x_j = 0, \qquad j \neq p,$$

when $\|\mathbf{x}\|_1 = 1$ and

$$\|\mathbf{Ax}\|_1 = \sum_{i=1}^{n}|a_{ip}|.$$

Equality is obtained in (10.19) for this choice of $\mathbf{x}$ and, therefore,

$$\|\mathbf{A}\|_1 = \max_{1\leqslant j\leqslant n}\sum_{i=1}^{n}|a_{ij}|.$$

Thus $\|\mathbf{A}\|_1$ is the maximum column sum of the absolute values of elements.

Similarly, from (10.13),

$$\begin{aligned}\|\mathbf{A}\|_\infty &= \max_{\|\mathbf{x}\|_\infty=1}\|\mathbf{Ax}\|_\infty \\ &\leqslant \max_{\|\mathbf{x}\|_\infty=1}\left[\max_i\left|\sum_{j=1}^{n}a_{ij}x_j\right|\right] \\ &\leqslant \max_i\sum_{j=1}^{n}|a_{ij}|,\end{aligned} \tag{10.20}$$

as $|x_j| \leqslant 1$ for $\|\mathbf{x}\|_\infty = 1$. Suppose the maximum in (10.20) occurs for $i=p$. Assuming $\mathbf{A}$ is real, we now choose $\mathbf{x}$ such that

$$x_j = \text{sign } a_{pj}.$$

We obtain equality in (10.20) and, therefore,

$$\|\mathbf{A}\|_\infty = \max_i \sum_{j=1}^{n} |a_{ij}|.$$

(If $\mathbf{A}$ is complex we make a different choice of $\mathbf{x}$. See Problem 10.9.) Thus $\|\mathbf{A}\|_\infty$ is the maximum row sum of the absolute values of elements.

To consider the Euclidean norm (10.11) we start with (10.18) but assume $\mathbf{x}$ is real (to simplify the analysis):

$$\|\mathbf{Ax}\|_2^2 = (\mathbf{Ax})^{\mathrm{T}}\mathbf{Ax} = \mathbf{x}^{\mathrm{T}}\mathbf{A}^{\mathrm{T}}\mathbf{Ax} = \mathbf{x}^{\mathrm{T}}\mathbf{Bx}$$

where $\mathbf{B} = \mathbf{A}^{\mathrm{T}}\mathbf{A}$ is a symmetric matrix. Thus there exists an orthogonal matrix $\mathbf{T}$ such that

$$\mathbf{x}^{\mathrm{T}}\mathbf{Bx} = \mathbf{x}^{\mathrm{T}}\mathbf{T}\boldsymbol{\Lambda}\mathbf{T}^{\mathrm{T}}\mathbf{x} = \mathbf{y}^{\mathrm{T}}\boldsymbol{\Lambda}\mathbf{y}$$

where $\boldsymbol{\Lambda}$ is the diagonal matrix formed from the eigenvalues of $\mathbf{B}$ and $\mathbf{y} = \mathbf{T}^{\mathrm{T}}\mathbf{x}$. Since $\mathbf{T}$ is orthogonal, $\mathbf{T}^{\mathrm{T}}$ is orthogonal and $\|\mathbf{x}\|_2 = \|\mathbf{y}\|_2$ (see Problem 10.11). Thus $\|\mathbf{y}\|_2 = 1$, so that

$$\sum_{i=1}^{n} y_i^2 = 1.$$

Now

$$\|\mathbf{Ax}\|_2^2 = \mathbf{y}^{\mathrm{T}}\boldsymbol{\Lambda}\mathbf{y} = \lambda_1 y_1^2 + \lambda_2 y_2^2 + \cdots + \lambda_n y_n^2. \tag{10.21}$$

As $\mathbf{B} = \mathbf{A}^{\mathrm{T}}\mathbf{A}$ is semi-positive definite all its eigenvalues are non-negative (see Problem 10.12). We assume that they are arranged in the order $\lambda_1 \geqslant \lambda_2 \geqslant \cdots \geqslant \lambda_n \geqslant 0$. Hence from (10.21)

$$\|\mathbf{Ax}\|_2^2 \leqslant \lambda_1(y_1^2 + y_2^2 + \cdots + y_n^2) = \lambda_1. \tag{10.22}$$

Now choose $\mathbf{x}$ such that $\mathbf{y} = [1\ \ 0\ \ 0\ \ \ldots\ \ 0]^{\mathrm{T}}$, when $\|\mathbf{x}\|_2 = \|\mathbf{y}\|_2 = 1$, and from (10.21) we have equality in (10.22).

We deduce that

$$\|\mathbf{A}\|_2 = \lambda_1^{1/2} = [\rho(\mathbf{A}^{\mathrm{T}}\mathbf{A})]^{1/2}. \tag{10.23}$$

Note that if $\mathbf{A}$ is symmetric $\mathbf{A}^{\mathrm{T}}\mathbf{A} = \mathbf{A}^2$, whence $\rho(\mathbf{A}^2) = (\rho(\mathbf{A}))^2$ and thus

$$\|\mathbf{A}\|_2 = \rho(\mathbf{A}).$$

The following lemma will be useful in subsequent error analyses.

Lemma 10.3 If $\|\mathbf{A}\| < 1$ then $\mathbf{I} + \mathbf{A}$ and $\mathbf{I} - \mathbf{A}$ are non-singular and

$$\frac{1}{1 + \|\mathbf{A}\|} \leqslant \|(\mathbf{I} \pm \mathbf{A})^{-1}\| \leqslant \frac{1}{1 - \|\mathbf{A}\|}. \tag{10.24}$$

Proof We restrict ourselves to $\mathbf{I}+\mathbf{A}$ since we can always replace $\mathbf{A}$ by $-\mathbf{A}$. We first note from (10.18) that

$$\|\mathbf{I}\| = \max_{\|\mathbf{x}\|=1} \|\mathbf{Ix}\| = \max_{\|\mathbf{x}\|=1} \|\mathbf{x}\| = 1.$$

If $\mathbf{I}+\mathbf{A}$ is singular, there is a vector $\mathbf{x} \neq \mathbf{0}$ such that

$$(\mathbf{I}+\mathbf{A})\mathbf{x} = \mathbf{0}.$$

We will assume that $\mathbf{x}$ is scaled so that $\|\mathbf{x}\| = 1$, when

$$\mathbf{Ax} = -\mathbf{Ix} = -\mathbf{x}$$

and

$$\|\mathbf{Ax}\| = |-1| \cdot \|\mathbf{x}\| = 1.$$

From (10.18), it follows that $\|\mathbf{A}\| \geqslant 1$. Thus $\mathbf{I}+\mathbf{A}$ is non-singular for $\|\mathbf{A}\| < 1$.

From

$$\mathbf{I} = (\mathbf{I}+\mathbf{A})^{-1}(\mathbf{I}+\mathbf{A}) \tag{10.25}$$

and Lemma 10.2, we have

$$1 = \|\mathbf{I}\| \leqslant \|(\mathbf{I}+\mathbf{A})^{-1}\| \; \|\mathbf{I}+\mathbf{A}\| \leqslant \|(\mathbf{I}+\mathbf{A})^{-1}\| \, (\|\mathbf{I}\| + \|\mathbf{A}\|).$$

Dividing by the last factor gives the left side of (10.24).

We rearrange (10.25) as

$$(\mathbf{I}+\mathbf{A})^{-1} = \mathbf{I} - (\mathbf{I}+\mathbf{A})^{-1}\mathbf{A}$$

and, on using Lemma 10.2,

$$\|(\mathbf{I}+\mathbf{A})^{-1}\| \leqslant \|\mathbf{I}\| + \|(\mathbf{I}+\mathbf{A})^{-1}\| \cdot \|\mathbf{A}\|.$$

Thus

$$(1 - \|\mathbf{A}\|)\,\|(\mathbf{I}+\mathbf{A})^{-1}\| \leqslant 1$$

and, as $\|\mathbf{A}\| < 1$, the right side of (10.24) follows. □

The convergence of a sequence of matrices $(\mathbf{A}_m)_{m=0}^{\infty}$ is defined in an identical manner to that for vectors and we write

$$\lim_{m\to\infty} \mathbf{A}_m = \mathbf{B}$$

if, for all i and j, the sequence of (i,j)th elements of the $\mathbf{A}_m$ converges to the (i,j)th element of $\mathbf{B}$. Corresponding to Lemma 10.1 we have the following.

Lemma 10.4 For any type of matrix norm

$$\lim_{m\to\infty} \mathbf{A}_m = \mathbf{B}$$

if and only if

$$\lim_{m \to \infty} \|\mathbf{A}_m - \mathbf{B}\| = 0.$$

Proof The lemma is clearly true for the norms $\|\cdot\|_1$ and $\|\cdot\|_\infty$ because of the continuous dependence of the norms on each element and (i) of Lemma 10.2. The proof for general norms is again more difficult and is given by Isaacson and Keller (1966). □

One important sequence is

$$\mathbf{I}, \mathbf{A}, \mathbf{A}^2, \mathbf{A}^3, \mathbf{A}^4, \ldots. \tag{10.26}$$

It follows from Lemma 10.2 (v) that

$$\|\mathbf{A}^m\| \leqslant \|\mathbf{A}\| \, \|\mathbf{A}^{m-1}\|$$

and, by induction,

$$\|\mathbf{A}^m\| \leqslant \|\mathbf{A}\|^m.$$

Hence, if for some norm

$$\|\mathbf{A}\| < 1,$$

then the sequence (10.26) converges to the zero matrix.

The *infinite series*

$$\mathbf{I} + \mathbf{A} + \mathbf{A}^2 + \mathbf{A}^3 + \cdots \tag{10.27}$$

is said to be convergent if the sequence of partial sums is convergent. The partial sums are

$$\mathbf{S}_m = \mathbf{I} + \mathbf{A} + \mathbf{A}^2 + \cdots + \mathbf{A}^m$$

and, if

$$\lim_{m \to \infty} \mathbf{S}_m = \mathbf{S},$$

we say that (10.27) is convergent with sum $\mathbf{S}$. Now,

$$\begin{aligned}(\mathbf{I} - \mathbf{A})\mathbf{S}_m &= (\mathbf{I} - \mathbf{A})(\mathbf{I} + \mathbf{A} + \mathbf{A}^2 + \cdots + \mathbf{A}^m) \\ &= (\mathbf{I} + \mathbf{A} + \cdots + \mathbf{A}^m) - (\mathbf{A} + \mathbf{A}^2 + \cdots + \mathbf{A}^{m+1}) \\ &= \mathbf{I} - \mathbf{A}^{m+1}\end{aligned}$$

and, if $\|\mathbf{A}\| < 1$, from Lemma 10.3 $\mathbf{I} - \mathbf{A}$ is non-singular, so that

$$\mathbf{S}_m = (\mathbf{I} - \mathbf{A})^{-1}(\mathbf{I} - \mathbf{A}^{m+1}).$$

Thus

$$\mathbf{S}_m - (\mathbf{I} - \mathbf{A})^{-1} = -(\mathbf{I} - \mathbf{A})^{-1}\mathbf{A}^{m+1}$$

and

$$\| \mathbf{S}_m - (\mathbf{I} - \mathbf{A})^{-1} \| \leq \| (\mathbf{I} - \mathbf{A})^{-1} \| \cdot \| \mathbf{A} \|^{m+1}.$$

For $\| \mathbf{A} \| < 1$, the right side tends to zero as m increases and by Lemma 10.4

$$\lim_{m \to \infty} \mathbf{S}_m = (\mathbf{I} - \mathbf{A})^{-1}.$$

Thus, for $\| \mathbf{A} \| < 1$, the series (10.27) converges with sum $(\mathbf{I} - \mathbf{A})^{-1}$.

Example 10.3 If

$$\mathbf{A} = \begin{bmatrix} 0.6 & 0.5 \\ 0.1 & 0.3 \end{bmatrix}$$

then

$$\| \mathbf{A} \|_1 = 0.8, \qquad \| \mathbf{A} \|_\infty = 1.1.$$

Using Lemma 10.3 with $\| \cdot \|_1$ we deduce that $\mathbf{I} + \mathbf{A}$ is non-singular and

$$\frac{1}{1.8} \leq \| (\mathbf{I} + \mathbf{A})^{-1} \|_1 \leq 5.$$

The sequence $(\mathbf{A}^m)_{m=0}^{\infty}$ converges to the zero matrix and the series $\Sigma\, \mathbf{A}^m$ is convergent with sum $(\mathbf{I} - \mathbf{A})^{-1}$. Notice that we cannot make these deductions using $\| \mathbf{A} \|_\infty$. □

10.4 Conditioning

Definition 10.7 We define the *condition number* of an $n \times n$ non-singular matrix $\mathbf{A}$ for the norm $\| \cdot \|_p$ to be

$$k_p(\mathbf{A}) = \| \mathbf{A} \|_p \| \mathbf{A}^{-1} \|_p.$$

If the particular choice of norm is immaterial, we omit the subscript p. □

The condition number of a matrix $\mathbf{A}$ gives a measure of how sensitive systems of equations, with coefficient matrix $\mathbf{A}$, are to small perturbations such as those caused by rounding. We shall see that, for large $k(\mathbf{A})$, perturbations may have a large effect on the solution.

Suppose that

$$\mathbf{A}\mathbf{x} = \mathbf{b}, \tag{10.28}$$

where $\mathbf{A}$ is non-singular, is perturbed so that

$$(\mathbf{A} + \boldsymbol{\delta}\mathbf{A})(\mathbf{x} + \boldsymbol{\delta}\mathbf{x}) = \mathbf{b} + \boldsymbol{\delta}\mathbf{b}.$$

Hence

$$\mathbf{A}\mathbf{x} + \mathbf{A}\, \boldsymbol{\delta}\mathbf{x} + \boldsymbol{\delta}\mathbf{A}(\mathbf{x} + \boldsymbol{\delta}\mathbf{x}) = \mathbf{b} + \boldsymbol{\delta}\mathbf{b}$$

and on using (10.28)

$$\boldsymbol{\delta}\mathbf{x} = -\mathbf{A}^{-1}\,\boldsymbol{\delta}\mathbf{A}(\mathbf{x}+\boldsymbol{\delta}\mathbf{x}) + \mathbf{A}^{-1}\,\boldsymbol{\delta}\mathbf{b}.$$

Thus, on using properties of norms,

$$\|\boldsymbol{\delta}\mathbf{x}\| \leqslant \|\mathbf{A}^{-1}\|\cdot\|\boldsymbol{\delta}\mathbf{A}\|\,(\,\|\mathbf{x}\| + \|\boldsymbol{\delta}\mathbf{x}\|\,) + \|\mathbf{A}^{-1}\|\cdot\|\boldsymbol{\delta}\mathbf{b}\|. \tag{10.29}$$

Now from (10.28)

$$\|\mathbf{b}\| \leqslant \|\mathbf{A}\|\cdot\|\mathbf{x}\|$$

and, therefore,

$$1 \leqslant \frac{\|\mathbf{A}\|\cdot\|\mathbf{x}\|}{\|\mathbf{b}\|}$$

so that

$$\begin{aligned}\|\mathbf{A}^{-1}\|\,\|\boldsymbol{\delta}\mathbf{b}\| &\leqslant \|\mathbf{A}^{-1}\|\,\|\mathbf{A}\|\,\|\mathbf{x}\|\,\frac{\|\boldsymbol{\delta}\mathbf{b}\|}{\|\mathbf{b}\|}\\ &= k(\mathbf{A})\,\|\mathbf{x}\|\,\frac{\|\boldsymbol{\delta}\mathbf{b}\|}{\|\mathbf{b}\|}.\end{aligned} \tag{10.30}$$

Also

$$\begin{aligned}\|\mathbf{A}^{-1}\|\,\|\boldsymbol{\delta}\mathbf{A}\| &= \|\mathbf{A}^{-1}\|\,\|\mathbf{A}\|\,\frac{\|\boldsymbol{\delta}\mathbf{A}\|}{\|\mathbf{A}\|}\\ &= ke\end{aligned}$$

where $k = k(\mathbf{A})$ and $e = \|\boldsymbol{\delta}\mathbf{A}\|/\|\mathbf{A}\|$. Hence in (10.29)

$$\|\boldsymbol{\delta}\mathbf{x}\|\,(1-ke) \leqslant ke\,\|\mathbf{x}\| + k\,\|\mathbf{x}\|\,\frac{\|\boldsymbol{\delta}\mathbf{b}\|}{\|\mathbf{b}\|}$$

and, provided $ke < 1$,

$$\frac{\|\boldsymbol{\delta}\mathbf{x}\|}{\|\mathbf{x}\|} \leqslant \frac{k}{1-ke}\left(e + \frac{\|\boldsymbol{\delta}\mathbf{b}\|}{\|\mathbf{b}\|}\right). \tag{10.31}$$

The quantity on the left of (10.31) may be considered a measure of the relative disturbance of **x**. The inequality provides a bound in terms of the relative disturbance $\|\boldsymbol{\delta}\mathbf{b}\|/\|\mathbf{b}\|$ of **b** and the relative disturbance $e = \|\boldsymbol{\delta}\mathbf{A}\|/\|\mathbf{A}\|$ of **A**. You will notice that the bound increases as $k(\mathbf{A})$ increases.

We have thus shown that, if the condition number of a matrix is large, the effects of rounding errors in the solution process may be serious. Even if a matrix or its inverse has large elements, the condition number is not necessarily large. It is easily seen that for any non-zero scalar λ,

$$k(\lambda\mathbf{A}) = k(\mathbf{A}).$$

Increasing (or decreasing) λ will increase the elements of $\lambda\mathbf{A}$ (or $(\lambda\mathbf{A})^{-1}$) but the condition number will not change. If $k(\mathbf{A}) \gg 1$ we say that $\mathbf{A}$ is *ill-conditioned.*

Example 10.4 The Hilbert matrices (see also § 5.4)

$$\mathbf{H}_n = \begin{bmatrix} 1 & \frac{1}{2} & \frac{1}{3} & \frac{1}{4} & \cdots & \frac{1}{n} \\ \frac{1}{2} & \frac{1}{3} & \frac{1}{4} & \frac{1}{5} & \cdots & \frac{1}{n+1} \\ \frac{1}{3} & \frac{1}{4} & \frac{1}{5} & \frac{1}{6} & \cdots & \frac{1}{n+2} \\ \vdots & & & & & \vdots \\ \frac{1}{n} & \frac{1}{n+1} & \frac{1}{n+2} & \frac{1}{n+3} & \cdots & \frac{1}{2n-1} \end{bmatrix},$$

$n = 1, 2, 3, \ldots,$ are notoriously ill-conditioned and $k(\mathbf{H}_n) \to \infty$ very rapidly as $n \to \infty$. For example,

$$\mathbf{H}_3 = \begin{bmatrix} 1 & \frac{1}{2} & \frac{1}{3} \\ \frac{1}{2} & \frac{1}{3} & \frac{1}{4} \\ \frac{1}{3} & \frac{1}{4} & \frac{1}{5} \end{bmatrix}, \qquad \mathbf{H}_3^{-1} = \begin{bmatrix} 9 & -36 & 30 \\ -36 & 192 & -180 \\ 30 & -180 & 180 \end{bmatrix},$$

$$\| \mathbf{H}_3 \|_1 = \| \mathbf{H}_3 \|_\infty = 11/6, \qquad \| \mathbf{H}_3^{-1} \|_1 = \| \mathbf{H}_3^{-1} \|_\infty = 408$$

and $k_1(\mathbf{H}_3) = k_\infty(\mathbf{H}_3) = 748$.

For n as large as 6, the ill-conditioning is extremely bad, with

$$k_1(\mathbf{H}_6) = k_\infty(\mathbf{H}_6) \simeq 29 \times 10^6.$$

Even for $n = 3$, the effects of rounding the coefficients are serious. For example the solution of

$$\mathbf{H}_3\mathbf{x} = \begin{bmatrix} \frac{11}{6} \\ \frac{13}{12} \\ \frac{47}{60} \end{bmatrix} \quad \text{is} \quad \mathbf{x} = \begin{bmatrix} 1 \\ 1 \\ 1 \end{bmatrix}.$$

If we round the coefficients in the equations to three correct significant decimal digits, we obtain

$$\begin{bmatrix} 1.00 & 0.500 & 0.333 \\ 0.500 & 0.333 & 0.250 \\ 0.333 & 0.250 & 0.200 \end{bmatrix} \mathbf{x} = \begin{bmatrix} 1.83 \\ 1.08 \\ 0.783 \end{bmatrix} \tag{10.32}$$

and these have as solution (correct to four significant figures)

$$\mathbf{x} = \begin{bmatrix} 1.090 \\ 0.4880 \\ 1.491 \end{bmatrix}. \tag{10.33}$$

The relative disturbance of the coefficients never exceeds 0.3% but the solution is changed by over 50%.

The main symptom of ill-conditioning is that the magnitudes of the pivots become very small even if pivoting is used. Consider, for example, the equations (10.32) in which the last two rows are interchanged if partial pivoting is employed. If we use the compact elimination method and work to three significant decimal digits with double precision calculation of inner products, we obtain the triangular matrices

$$\begin{bmatrix} 1 & 0 & 0 \\ 0.333 & 1 & 0 \\ 0.500 & 0.994 & 1 \end{bmatrix} \begin{bmatrix} 1.00 & 0.500 & 0.333 \\ 0 & 0.0835 & 0.0891 \\ 0 & 0 & -0.00507 \end{bmatrix}.$$

The last pivot, -0.00507, is very small in magnitude compared with other elements. □

Scaling equations (or unknowns) has an effect on the condition number of a coefficient matrix. It is often desirable to scale so as to reduce any disparity in the magnitude of coefficients. Such scaling does not always improve the accuracy of the elimination method but may be important, especially if only partial pivoting is employed, as the next example demonstrates.

Example 10.5 Consider the equations

$$\begin{bmatrix} 1 & 10^4 \\ 1 & 1 \end{bmatrix} \begin{bmatrix} x_1 \\ x_2 \end{bmatrix} = \begin{bmatrix} 10^4 \\ 2 \end{bmatrix}$$

for which

$$\mathbf{A}^{-1} = \left(\frac{1}{10^4 - 1} \right) \begin{bmatrix} -1 & 10^4 \\ 1 & -1 \end{bmatrix}$$

and

$$k_\infty(\mathbf{A}) = \frac{(10^4+1)^2}{10^4-1} \simeq 10^4.$$

The elimination method with partial pivoting does not involve interchanges, so that, working to three decimal digits, we obtain

$$\begin{aligned} x_1 + 10^4 x_2 &= 10^4 \\ -10^4 x_2 &= -10^4. \end{aligned}$$

On back substituting, we obtain the very poor result

$$x_2 = 1, \qquad x_1 = 0.$$

If the first equation is scaled by 10^{-4}, the coefficient matrix becomes

$$\mathbf{B} = \begin{bmatrix} 10^{-4} & 1 \\ 1 & 1 \end{bmatrix}, \qquad \text{with } \mathbf{B}^{-1} = \left(\frac{1}{1-10^{-4}}\right)\begin{bmatrix} -1 & 1 \\ 1 & -10^{-4} \end{bmatrix}$$

and

$$k_\infty(\mathbf{B}) = \frac{4}{1-10^{-4}} \simeq 4.$$

This time partial pivoting interchanges the rows, so that the equations reduce to

$$\begin{aligned} x_1 + x_2 &= 2 \\ x_2 &= 1. \end{aligned}$$

These yield $x_1 = x_2 = 1$, a good approximation to the solution. □

It must be stressed that the inequality (10.31) can rarely be used to provide a precise bound on $\|\boldsymbol{\delta}\mathbf{x}\|$ as only rarely is the condition number $k(\mathbf{A})$ known. When solving linear equations, it is usually impracticable to determine $k(\mathbf{A})$ as this requires a knowledge of $\mathbf{A}^{-1}$ or the eigenvalues of $\mathbf{A}$ (see Problem 10.23). The calculation of either would be longer than that for the original problem. However the inequality (10.31) when combined with the results of § 9.10 does provide qualitative information regarding $\boldsymbol{\delta}\mathbf{x}$, the error in the computed solution due to the effect of rounding error.

10.5 Iterative correction from residual vectors

Suppose that $\mathbf{x}_0$ is an approximation to the solution of non-singular equations

$$\mathbf{A}\mathbf{x} = \mathbf{b}.$$

Corresponding to $\mathbf{x}_0$ there is a *residual vector* $\mathbf{r}_0$ given by

$$\mathbf{r}_0 = \mathbf{A}\mathbf{x}_0 - \mathbf{b}. \tag{10.34}$$

We sometimes use $\mathbf{r}_0$ to assess the accuracy of $\mathbf{x}_0$ as an approximation to $\mathbf{x}$, but this may be misleading, especially if the matrix $\mathbf{A}$ is ill-conditioned, as the following example illustrates.

Example 10.6 The residual vector for the equations (10.32) when

$$\mathbf{x}_0 = \begin{bmatrix} 1 \\ 1 \\ 1 \end{bmatrix}, \quad \text{is} \quad \mathbf{r}_0 = \begin{bmatrix} 0.003 \\ 0.003 \\ 0 \end{bmatrix}.$$

Notice that for these equations

$$\frac{\|\mathbf{x} - \mathbf{x}_0\|_\infty}{\|\mathbf{x}\|_\infty} = \frac{0.512}{1.491} > 0.3, \tag{10.35}$$

whereas

$$\frac{\|\mathbf{r}_0\|_\infty}{\|\mathbf{b}\|_\infty} = \frac{0.003}{1.83} < 0.002. \tag{10.36}$$

The left of (10.35) is a measure of the relative accuracy of $\mathbf{x}_0$ and the left of (10.36) is a measure of the relative magnitude of $\mathbf{r}_0$. □

If a bound for $k(\mathbf{A})$ is known, then the residual vector may be used as a guide to the accuracy of the solution (see Problem 10.24). The residual vector is also useful in enabling us to make a correction to the approximation to the solution. If

$$\boldsymbol{\delta}\mathbf{x} = \mathbf{x} - \mathbf{x}_0, \tag{10.37}$$

then

$$\begin{aligned} \mathbf{r}_0 = \mathbf{A}\mathbf{x}_0 - \mathbf{b} &= \mathbf{A}(\mathbf{x} - \boldsymbol{\delta}\mathbf{x}) - \mathbf{b} \\ &= -\mathbf{A}\,\boldsymbol{\delta}\mathbf{x}, \end{aligned}$$

so that $\boldsymbol{\delta}\mathbf{x}$ is the solution of the system

$$\mathbf{A}\,\boldsymbol{\delta}\mathbf{x} = -\mathbf{r}_0. \tag{10.38}$$

We attempt to solve (10.38) and so determine $\mathbf{x} = \mathbf{x}_0 + \boldsymbol{\delta}\mathbf{x}$. In practice, rounding errors are introduced again and, therefore, only an approximation to $\boldsymbol{\delta}\mathbf{x}$ is computed. Hopefully, however, we obtain a better approximation to $\mathbf{x}$. We may iterate by repeating this process and, to investigate convergence, we need the following theorem. Note, however, that Theorem 10.2 gives a stronger result using eigenvalues.

Theorem 10.1 Suppose that for arbitrary $\mathbf{x}_0$ the sequence of n-dimensional vectors $(\mathbf{x}_m)_{m=0}^{\infty}$ satisfies

$$\mathbf{x}_{m+1} = \mathbf{M}\mathbf{x}_m + \mathbf{d}, \qquad m = 0, 1, 2, \ldots. \tag{10.39}$$

If, for some choice of norm,

$$\|\mathbf{M}\| < 1$$

then the sequence $(\mathbf{x}_m)$ converges to $\mathbf{x}$, the unique solution of

$$(\mathbf{I} - \mathbf{M})\mathbf{x} = \mathbf{d}. \tag{10.40}$$

Proof The theorem is a special case of the contraction mapping theorem 12.1, but because the proof is much simpler we give it here.

First we note from Lemma 10.3 that, for $\|\mathbf{M}\| < 1$, $\mathbf{I} - \mathbf{M}$ is non-singular and therefore (10.40) has a unique solution $\mathbf{x}$.

Now suppose that

$$\mathbf{x}_m = \mathbf{x} + \mathbf{e}_m$$

when, from (10.39),

$$\mathbf{x} + \mathbf{e}_{m+1} = \mathbf{M}(\mathbf{x} + \mathbf{e}_m) + \mathbf{d}$$

and, using (10.40),

$$\mathbf{e}_{m+1} = \mathbf{M}\mathbf{e}_m. \tag{10.41}$$

Hence

$$\|\mathbf{e}_{m+1}\| \leqslant \|\mathbf{M}\| \, \|\mathbf{e}_m\|. \tag{10.42}$$

Using induction, we have

$$\|\mathbf{e}_m\| \leqslant \|\mathbf{M}\|^m \|\mathbf{e}_0\|$$

and, as $\|\mathbf{M}\| < 1$,

$$\|\mathbf{x}_m - \mathbf{x}\| = \|\mathbf{e}_m\| \to 0 \qquad \text{as } m \to \infty,$$

that is,

$$\mathbf{x}_m \to \mathbf{x} \qquad \text{as } m \to \infty$$

□

We note from (10.42) that for $\mathbf{e}_0 \neq \mathbf{0}$ (and hence $\mathbf{e}_m \neq \mathbf{0}$ for all m),

$$\frac{\|\mathbf{e}_{m+1}\|}{\|\mathbf{e}_m\|} \leqslant \|\mathbf{M}\| < 1$$

and therefore we say that the convergence is of at least first order (cf. § 8.6).

We return to the problem of correcting an approximation $\mathbf{x}_0$ to a solution vector $\mathbf{x}$. We assume that $\mathbf{x}_0$ has been calculated by a factorization method with rounding errors, so that

$$(\mathbf{A}+\boldsymbol{\delta}\mathbf{A})=\mathbf{L}'\mathbf{U}' \tag{10.43}$$

and

$$(\mathbf{A}+\boldsymbol{\delta}\mathbf{A})\mathbf{x}_0=\mathbf{b}+\boldsymbol{\delta}\mathbf{b}.$$

We compute the residual vector $\mathbf{r}_0$ from

$$\mathbf{r}_0=\mathbf{A}\mathbf{x}_0-\mathbf{b} \tag{10.44}$$

using high accuracy arithmetic so that rounding errors in this step are negligible. This is not too difficult as each element of $\mathbf{r}_0$ consists of an inner product. We now attempt to solve (10.38) and so find $\boldsymbol{\delta}\mathbf{x}$. In practice, we use the known approximate factorization of $\mathbf{A}$, (10.43), but we do use more accurate arithmetic in the forward and back substitutions involving the right side $\mathbf{r}_0$, so that the perturbation of $\mathbf{r}_0$ due to rounding errors is negligible. Thus we calculate $\boldsymbol{\delta}\mathbf{x}_0$, say, the solution of

$$(\mathbf{A}+\boldsymbol{\delta}\mathbf{A})\boldsymbol{\delta}\mathbf{x}_0=-\mathbf{r}_0. \tag{10.45}$$

Finally we compute $\mathbf{x}_1$ from

$$\mathbf{x}_1=\mathbf{x}_0+\boldsymbol{\delta}\mathbf{x}_0, \tag{10.46}$$

which we again assume is achieved with negligible rounding error. In general $\mathbf{x}_1\neq\mathbf{x}$, due to the perturbation $\boldsymbol{\delta}\mathbf{A}$ in (10.45) and, therefore, we repeat the above process, starting with $\mathbf{x}_1$ instead of $\mathbf{x}_0$.

From (10.46), (10.45) and (10.44),

$$\begin{aligned}\mathbf{x}_1&=\mathbf{x}_0-(\mathbf{A}+\boldsymbol{\delta}\mathbf{A})^{-1}\mathbf{r}_0\\&=\mathbf{x}_0-(\mathbf{A}+\boldsymbol{\delta}\mathbf{A})^{-1}(\mathbf{A}\mathbf{x}_0-\mathbf{b})\\&=(\mathbf{A}+\boldsymbol{\delta}\mathbf{A})^{-1}(\mathbf{A}+\boldsymbol{\delta}\mathbf{A}-\mathbf{A})\mathbf{x}_0+(\mathbf{A}+\boldsymbol{\delta}\mathbf{A})^{-1}\mathbf{b}\\&=(\mathbf{A}+\boldsymbol{\delta}\mathbf{A})^{-1}\boldsymbol{\delta}\mathbf{A}.\mathbf{x}_0+(\mathbf{A}+\boldsymbol{\delta}\mathbf{A})^{-1}\mathbf{b}.\end{aligned}$$

If the process is repeated, we obtain a sequence of vectors $(\mathbf{x}_m)_{m=0}^{\infty}$ related by

$$\mathbf{x}_{m+1}=(\mathbf{A}+\boldsymbol{\delta}\mathbf{A})^{-1}\boldsymbol{\delta}\mathbf{A}.\mathbf{x}_m+(\mathbf{A}+\boldsymbol{\delta}\mathbf{A})^{-1}\mathbf{b},\qquad m=0,1,2,\ldots.$$

This is of the form (10.39) and, by Theorem 10.1, the sequence $(\mathbf{x}_m)$ will converge if

$$\|(\mathbf{A}+\boldsymbol{\delta}\mathbf{A})^{-1}\boldsymbol{\delta}\mathbf{A}\|<1. \tag{10.47}$$

This will be satisfied if the elements of $\boldsymbol{\delta}\mathbf{A}$ are sufficiently small, as

$$\|(\mathbf{A}+\boldsymbol{\delta}\mathbf{A})^{-1}\boldsymbol{\delta}\mathbf{A}\|\leqslant\|(\mathbf{A}+\boldsymbol{\delta}\mathbf{A})^{-1}\|\,\|\boldsymbol{\delta}\mathbf{A}\|.$$

The sequence $(\mathbf{x}_m)$ converges to the solution of

$$[\mathbf{I}-(\mathbf{A}+\boldsymbol{\delta}\mathbf{A})^{-1}\boldsymbol{\delta}\mathbf{A}]\mathbf{x}=(\mathbf{A}+\boldsymbol{\delta}\mathbf{A})^{-1}\mathbf{b},$$

that is, on premultiplying by $(\mathbf{A} + \boldsymbol{\delta}\mathbf{A})$,

$$\mathbf{Ax} = \mathbf{b}.$$

As was observed after Theorem 10.1, the convergence is of at least first order.

We normally apply this correction process only once or twice after an approximate solution has been calculated by one of the elimination methods. The intention is to reduce the effects of rounding errors.

Example 10.7 Consider the equations

$$\begin{bmatrix} 33 & 25 & 20 \\ 20 & 17 & 14 \\ 25 & 20 & 17 \end{bmatrix} \mathbf{x} = \begin{bmatrix} 78 \\ 51 \\ 62 \end{bmatrix},$$

which have solution

$$\mathbf{x} = [1 \;\; 1 \;\; 1]^{\mathrm{T}}.$$

If we use the simple elimination method with arithmetic accurate to only two significant decimal digits, except that we compute inner products in double precision, we obtain

$$\begin{bmatrix} 33 & 25 & 20 & : & 78 \\ 0 & 1.8 & 1.8 & : & 3.4 \\ 0 & 0 & 0.79 & : & 0.80 \end{bmatrix}.$$

The first three columns form $\mathbf{U}'$ and the last column is the reduced right side $(\mathbf{L}')^{-1}\mathbf{b}$, where the multipliers are elements of

$$\mathbf{L}' = \begin{bmatrix} 1 & 0 & 0 \\ 0.61 & 1 & 0 \\ 0.76 & 0.56 & 1 \end{bmatrix}.$$

The computed approximation to $\mathbf{x}$ is

$$\mathbf{x}_0 = [1.1 \;\; 0.89 \;\; 1.0]^{\mathrm{T}}.$$

We now find, working to four significant digits, that

$$\mathbf{r}_0 = [0.5500 \;\; 0.1300 \;\; 0.3000]^{\mathrm{T}}$$

which, when treated by the multipliers as above, becomes

$$(\mathbf{L}')^{-1}\mathbf{r}_0 = [0.5500 \;\; -0.2055 \;\; -0.002920]^{\mathrm{T}}.$$

On solving

$$\mathbf{U}'\boldsymbol{\delta}\mathbf{x}_0 = -(\mathbf{L}')^{-1}\mathbf{r}_0,$$

we obtain

$$\boldsymbol{\delta}\mathbf{x}_0 = [-0.1026 \;\; 0.1105 \;\; 0.003696]^{\mathrm{T}},$$

so that

$$\mathbf{x}_1 = \mathbf{x}_0 + \boldsymbol{\delta}\mathbf{x}_0 = [0.9974 \;\; 1.000 \;\; 1.004]^{\mathrm{T}}.$$

Notice that

$$\|\mathbf{x} - \mathbf{x}_0\|_\infty = 0.11,$$

whereas

$$\|\mathbf{x} - \mathbf{x}_1\|_\infty = 0.004,$$

showing that there has been a considerable improvement in the approximation.

A second application of the process yields

$$\boldsymbol{\delta}\mathbf{x}_1 = [0.002737 \;\; -0.0001538 \;\; -0.004034]^{\mathrm{T}}$$

so that if $\mathbf{x}_2 = \mathbf{x}_1 + \boldsymbol{\delta}\mathbf{x}_1$,

$$\|\mathbf{x} - \mathbf{x}_2\|_\infty < 0.0002.$$

□

We can also use an iterative process to improve the accuracy of a computed approximation to the inverse of a matrix. Details are given in Problem 10.26.

10.6 Iterative methods

We again consider the problem of solving n non-singular equations in n unknowns

$$\mathbf{A}\mathbf{x} = \mathbf{b}. \tag{10.48}$$

If $\mathbf{E}$ and $\mathbf{F}$ are $n \times n$ matrices such that

$$\mathbf{A} = \mathbf{E} - \mathbf{F}, \tag{10.49}$$

we call (10.49) a *splitting* of $\mathbf{A}$. For such a splitting, (10.48) may be written as

$$\mathbf{E}\mathbf{x} = \mathbf{F}\mathbf{x} + \mathbf{b}.$$

This form of the equations suggests an iterative procedure

$$\mathbf{E}\mathbf{x}_{m+1} = \mathbf{F}\mathbf{x}_m + \mathbf{b}, \qquad m = 0, 1, 2, \ldots, \tag{10.50}$$

for arbitrary $\mathbf{x}_0$. If the sequence is to be uniquely defined for a given $\mathbf{x}_0$, we require $\mathbf{E}$ to be non-singular, when

$$\mathbf{x}_{m+1} = \mathbf{E}^{-1}\mathbf{F}\mathbf{x}_m + \mathbf{E}^{-1}\mathbf{b}.$$

This is of the form (10.39) and Theorem 10.1 states that the sequence $(\mathbf{x}_m)_{m=0}^{\infty}$ converges if

$$\|\mathbf{E}^{-1}\mathbf{F}\| < 1. \tag{10.51}$$

It can also be seen that, in this case, $(\mathbf{x}_m)$ converges to the solution vector $\mathbf{x}$.

The next result, which is stronger than Theorem 10.1, gives more precise information on the behaviour of the error in iterative methods for linear equations.

Theorem 10.2 If $\mathbf{M}$ is similar to a diagonal matrix and $(\mathbf{x}_m)_{m=0}^{\infty}$ is a sequence of vectors satisfying (10.39) then this sequence converges to the solution of (10.40) for any choice of $\mathbf{x}_0 \in \mathbb{R}^n$ if, and only if, $\rho(\mathbf{M}) < 1$.

Proof From repeated use of (10.41)

$$\mathbf{e}_m = \mathbf{M}^m \mathbf{e}_0 \tag{10.52}$$

and, if $\mathbf{M} = \mathbf{T}\,\mathbf{\Lambda}\,\mathbf{T}^{-1}$, we have

$$\begin{aligned}\mathbf{M}^m &= \mathbf{T}\,\mathbf{\Lambda}\,\mathbf{T}^{-1}.\,\mathbf{T}\,\mathbf{\Lambda}\,\mathbf{T}^{-1}\ldots\mathbf{T}\,\mathbf{\Lambda}\,\mathbf{T}^{-1}\\ &= \mathbf{T}\,\mathbf{\Lambda}^m\,\mathbf{T}^{-1}.\end{aligned}$$

Hence

$$\mathbf{e}_m = \mathbf{T}\,\mathbf{\Lambda}^m\,\mathbf{T}^{-1}\,\mathbf{e}_0. \tag{10.53}$$

Now $\mathbf{\Lambda}$ is the diagonal matrix

$$\mathbf{\Lambda} = \begin{bmatrix} \lambda_1 & & & \mathbf{0} \\ & \lambda_2 & & \\ & & \ddots & \\ \mathbf{0} & & & \lambda_n \end{bmatrix}$$

where $\lambda_1, \lambda_2, \ldots, \lambda_n$ are the eigenvalues of $\mathbf{M}$ and thus

$$\mathbf{\Lambda}^m = \begin{bmatrix} \lambda_1^m & & & \mathbf{0} \\ & \lambda_2^m & & \\ & & \ddots & \\ \mathbf{0} & & & \lambda_n^m \end{bmatrix}. \tag{10.54}$$

From (10.53) and (10.54) we see that in general $\mathbf{e}_m \to \mathbf{0}$ if, and only if, $|\lambda_i| < 1$, $i = 1, 2, \ldots, n$. The process does not converge if $|\lambda_i| \geqslant 1$ for any i. □

We also see that in (10.54) the dominant term behaves like $(\rho(\mathbf{M}))^m$. Thus the rate of convergence depends on $\rho(\mathbf{M})$ and we should try to choose $\mathbf{E}$ and $\mathbf{F}$ so that $\rho(\mathbf{M})$ is as small as possible.

The important question is: what is a suitable choice of $\mathbf{E}$ and $\mathbf{F}$? If $\mathbf{E}=\mathbf{A}$ and $\mathbf{F}=\mathbf{0}$ we achieve nothing. We repeatedly solve (10.50) in order to determine $\mathbf{x}_{m+1}$ and we would need to use one of the elimination methods unless $\mathbf{E}$ is of a particularly simple form. If $\mathbf{E}$ is not of a simpler form than $\mathbf{A}$, the work involved for each iteration will be identical to that required in solving the original equations (10.48) by an elimination method from the start.

The simplest choice of $\mathbf{E}$ is a diagonal matrix, usually the diagonal of $\mathbf{A}$ provided all the diagonal elements are non-zero. We obtain the splitting

$$\mathbf{A}=\mathbf{D}-\mathbf{B},$$

where $\mathbf{D}$ is diagonal with non-zero diagonal elements and $\mathbf{B}$ has zeros on its diagonal. The relation (10.50) becomes

$$\mathbf{x}_{m+1}=\mathbf{D}^{-1}\mathbf{B}\mathbf{x}_m+\mathbf{D}^{-1}\mathbf{b}, \quad m=0,1,2,\ldots. \tag{10.55}$$

This is known as the *Jacobi iterative* method. $\mathbf{D}^{-1}$ is simply the diagonal matrix whose diagonal elements are the inverses of those of $\mathbf{D}$. The method is convergent if

$$\|\mathbf{D}^{-1}\mathbf{B}\|<1.$$

This condition is certainly satisfied if the matrix $\mathbf{A}$ is a *strictly diagonally dominant* matrix, that is, a matrix such that

$$|a_{ii}|>\sum_{\substack{j=1\\ j\neq i}}^{n}|a_{ij}| \qquad \text{for all } i=1,2,\ldots,n.$$

$\mathbf{D}^{-1}\mathbf{B}$ has off-diagonal elements $-a_{ij}/a_{ii}$ and zeros on the diagonal, so that for a strictly diagonally dominant matrix $\mathbf{A}$,

$$\|\mathbf{D}^{-1}\mathbf{B}\|_\infty=\max_{1\leqslant i\leqslant n}\sum_{\substack{j=1\\ j\neq i}}^{n}\left|\frac{a_{ij}}{a_{ii}}\right|=\max_{1\leqslant i\leqslant n}\frac{1}{|a_{ii}|}\sum_{\substack{j=1\\ j\neq i}}^{n}|a_{ij}|<1.$$

Alternatively we could obtain this result using Gerschgorin's theorems (see § 11.1) and Theorem 10.2.

Another suitable choice of $\mathbf{E}$ is a triangular matrix. We split $\mathbf{A}$ into

$$\mathbf{A}=(\mathbf{D}-\mathbf{L})-\mathbf{U},$$

where $\mathbf{D}$ is diagonal with non-zero diagonal elements, $\mathbf{L}$ is lower triangular with zeros on the diagonal and $\mathbf{U}$ is upper triangular with zeros on the diagonal.† We obtain in place of (10.50),

$$(\mathbf{D}-\mathbf{L})\mathbf{x}_{m+1}=\mathbf{U}\mathbf{x}_m+\mathbf{b}, \qquad m=0,1,2.\ldots. \tag{10.56}$$

† The $\mathbf{L}$ and $\mathbf{U}$ described here should not be confused with the triangular factors of $\mathbf{A}$.

This is known as the *Gauss–Seidel iterative* method. The coefficient matrix $(\mathbf{D}-\mathbf{L})$ is lower triangular and $\mathbf{x}_{m+1}$ is easily found by forward substitution. There is no need to compute $(\mathbf{D}-\mathbf{L})^{-1}\mathbf{U}$ explicitly.

The Gauss–Seidel method converges if

$$\|(\mathbf{D}-\mathbf{L})^{-1}\mathbf{U}\| < 1.$$

Again it can be shown that the Gauss–Seidel method is convergent if the original matrix is diagonally dominant. You will notice that both the Jacobi and Gauss–Seidel methods require the elements on the diagonal of $\mathbf{A}$ to be non-zero.

Algorithm 10.1 The Gauss–Seidel method for solving n linear equations. An initial approximation $\mathbf{x}_0$ to the solution must be given, but this need not be very accurate. We stop when $\|\mathbf{x}_{m+1}-\mathbf{x}_m\|_\infty$ is less than ε (see Problem 10.31).

set $\mathbf{x}=\mathbf{x}_0$
repeat
  $maxdiff := 0$
  for $i := 1$ **to** n

$$y := \left(b_i - \sum_{\substack{j=1 \\ j \neq i}}^{n} a_{ij}x_j\right) \Big/ a_{ii}$$

   if $|y - x_i| > maxdiff$ **then** $maxdiff := |y - x_i|$
   $x_i := y$
  next i
until $maxdiff < \varepsilon$
[$\mathbf{x}$ is the refined solution.] □

The algorithm for the Jacobi method is similar and the construction is left to the reader (Problem 10.29).

Example 10.8 Solve the equations

$$\begin{bmatrix} 5 & -1 & -1 & -1 \\ -1 & 10 & -1 & -1 \\ -1 & -1 & 5 & -1 \\ -1 & -1 & -1 & 10 \end{bmatrix} \mathbf{x} = \begin{bmatrix} -4 \\ 12 \\ 8 \\ 34 \end{bmatrix}$$

by an iterative process. (The solution is $\mathbf{x} = [1\ \ 2\ \ 3\ \ 4]^{\mathrm{T}}$.) The coefficient matrix is diagonally dominant and, therefore, the Jacobi method is convergent. The equations (10.55) are:

$$\mathbf{x}_{m+1} = \begin{bmatrix} 0 & 0.2 & 0.2 & 0.2 \\ 0.1 & 0 & 0.1 & 0.1 \\ 0.2 & 0.2 & 0 & 0.2 \\ 0.1 & 0.1 & 0.1 & 0 \end{bmatrix} \mathbf{x}_m + \begin{bmatrix} -0.8 \\ 1.2 \\ 1.6 \\ 3.4 \end{bmatrix}$$

and $\|\mathbf{D}^{-1}\mathbf{B}\|_\infty = 0.6$. Table 10.1 shows six successive iterates starting with $\mathbf{x}_0 = \mathbf{0}$ and working to three decimal places.

The Gauss–Seidel method (10.56) is

$$\begin{bmatrix} 5 & 0 & 0 & 0 \\ -1 & 10 & 0 & 0 \\ -1 & -1 & 5 & 0 \\ -1 & -1 & -1 & 10 \end{bmatrix} \mathbf{x}_{m+1} = \begin{bmatrix} 0 & 1 & 1 & 1 \\ 0 & 0 & 1 & 1 \\ 0 & 0 & 0 & 1 \\ 0 & 0 & 0 & 0 \end{bmatrix} \mathbf{x}_m + \begin{bmatrix} -4 \\ 12 \\ 8 \\ 34 \end{bmatrix}$$

and

$$(\mathbf{D} - \mathbf{L})^{-1}\mathbf{U} = \begin{bmatrix} 0 & 0.2 & 0.2 & 0.2 \\ 0 & 0.02 & 0.12 & 0.12 \\ 0 & 0.044 & 0.064 & 0.264 \\ 0 & 0.0264 & 0.0384 & 0.0584 \end{bmatrix}.$$

Thus $\|(\mathbf{D}-\mathbf{L})^{-1}\mathbf{U}\|_\infty = 0.6 < 1$ and this method is convergent also. Table 10.1 shows six successive iterates starting with $\mathbf{x}_0 = \mathbf{0}$ and working to three decimal places. □

You will notice that in Example 10.8 the iterates for the Gauss–Seidel method appear to converge faster than those for the Jacobi method. This is often the case and might be expected in that the Gauss–Seidel method makes use of the most recently calculated approximations to elements of $\mathbf{x}$ at all times. If $(x_i)_m$ denotes the ith component of $\mathbf{x}_m$, the Jacobi method is

$$(x_i)_{m+1} = \frac{1}{a_{ii}}[b_i - a_{i1}(x_1)_m - a_{i2}(x_2)_m - \cdots - a_{i,i-1}(x_{i-1})_m - a_{i,i+1}(x_{i+1})_m - \cdots - a_{i,n}(x_n)_m], \qquad i = 1, 2, \ldots, n,$$

Table 10.1 The iterative solution of the equations in Example 10.8

(a) Jacobi method

m	0	1	2	3	4	5
	0	−0.800	0.440	0.716	0.883	0.948
$\mathbf{x}_m$	0	1.200	1.620	1.840	1.929	1.969
	0	1.600	2.360	2.732	2.880	2.948
	0	3.400	3.600	3.842	3.929	3.969

(b) Gauss–Seidel method

m	0	1	2	3	4	5
	0	−0.800	0.476	0.889	0.977	0.995
$\mathbf{x}_m$	0	1.120	1.774	1.956	1.990	1.998
	0	1.664	2.770	2.949	2.989	2.998
	0	3.598	3.902	3.979	3.996	3.999

whereas the Gauss–Seidel method is

$$(x_i)_{m+1} = \frac{1}{a_{ii}} [b_i - a_{i1}(x_1)_{m+1} - a_{i2}(x_2)_{m+1} - \cdots - a_{i,i-1}(x_{i-1})_{m+1}$$

$$- a_{i,i+1}(x_{i+1})_m - \cdots - a_{i,n}(x_n)_m], \qquad i = 1, 2, \ldots, n.$$

The Jacobi method does not make use of the known components $(x_1)_{m+1}, \ldots, (x_{i-1})_{m+1}$ when calculating $(x_i)_{m+1}$. It must be stressed that the Gauss–Seidel method is not always better than the Jacobi method. Indeed the former may diverge whilst the latter converges.

We now consider the equation

$$\mathbf{D}\tilde{\mathbf{x}}_{m+1} = \mathbf{L}\mathbf{x}_{m+1} + \mathbf{U}\mathbf{x}_m + \mathbf{b}. \tag{10.57}$$

The Gauss–Seidel method (10.56) is obtained by putting $\mathbf{x}_{m+1} = \tilde{\mathbf{x}}_{m+1}$. Instead we let

$$\mathbf{x}_{m+1} = \omega\tilde{\mathbf{x}}_{m+1} + (1-\omega)\mathbf{x}_m, \tag{10.58}$$

where ω is a parameter to be fixed. On combining (10.57) and (10.58) we now have

$$\mathbf{D}\mathbf{x}_{m+1} = \omega\mathbf{L}\mathbf{x}_{m+1} + \omega\mathbf{U}\mathbf{x}_m + \omega\mathbf{b} + (1-\omega)\mathbf{D}\mathbf{x}_m$$

so that

$$(\mathbf{D} - \omega\mathbf{L})\mathbf{x}_{m+1} = ((1-\omega)\mathbf{D} + \omega\mathbf{U})\mathbf{x}_m + \omega\mathbf{b}. \tag{10.59}$$

We thus have an iterative method of the form (10.39) with iteration matrix **M** given by

$$\mathbf{M}_\omega = (\mathbf{D} - \omega\mathbf{L})^{-1}((1-\omega)\mathbf{D} + \omega\mathbf{U}). \tag{10.60}$$

Notice that $\mathbf{M}_\omega$ depends on ω. The iterative method (10.59) is called the *successive over-relaxation* (*SOR*) method. It can be shown that for convergence we must choose ω in the range $0 < \omega < 2$. We try to choose ω so that the sequence $(\mathbf{x}_m)$ converges as rapidly as possible. Thus we try to make $\rho(\mathbf{M}_\omega)$ as small as possible. For a wide range of problems arising from differential equations we find that a careful choice of ω can yield a substantial reduction in the magnitude of $\rho(\mathbf{M}_\omega)$. The optimum choice of ω is usually in the interval $(1, 2)$ but the precise value depends on the problem. A considerable amount of research has been devoted to this topic of which a simplified account may be found in Isaacson and Keller (1966). Of course for $\omega = 1$ we obtain the Gauss–Seidel method and, for $\omega \neq 0$, we can show the method (10.59) can be derived from a splitting of **A** (see Problem 10.34).

Rounding errors do not seriously affect the accuracy of a convergent iterative process unless $\|\mathbf{M}\| \simeq 1$ or $\rho(\mathbf{M}) \simeq 1$ (see Problem 10.35). One advantage of an iterative method is that the effects of an isolated error will rapidly decay as the iterations advance if $\|\mathbf{M}\|$ is small. As the number of completed iterations increases and the accuracy of the iterates improves, it is desirable to increase the accuracy of the calculation.

Problems

Section 10.1

10.1 Let **A** be the 3×3 matrix

$$\mathbf{A} = \begin{bmatrix} a_{11} & a_{12} & a_{13} \\ a_{21} & a_{22} & a_{23} \\ a_{31} & a_{32} & a_{33} \end{bmatrix}.$$

Show that

(i) adding one row of the matrix to another row does not change its determinant
(ii) interchanging two rows causes the sign of the determinant to change
(iii) multiplying one row of the matrix by a scalar λ causes the determinant to be multiplied by λ.

10.2 Show that the matrix

$$\mathbf{A} = \begin{bmatrix} 2 & 1 & 0 \\ 1 & 2 & 1 \\ 0 & 1 & 2 \end{bmatrix}$$

has eigenvalues $2 + \sqrt{2}, 2, 2 - \sqrt{2}$ with corresponding eigenvectors

$$\mathbf{x}_1 = \frac{1}{\sqrt{6}} \begin{bmatrix} 1 \\ \sqrt{2} \\ 1 \end{bmatrix}, \qquad \mathbf{x}_2 = \frac{1}{\sqrt{2}} \begin{bmatrix} 1 \\ 0 \\ -1 \end{bmatrix}, \qquad \mathbf{x}_3 = \frac{1}{\sqrt{6}} \begin{bmatrix} 1 \\ -\sqrt{2} \\ 1 \end{bmatrix}.$$

10.3 Show that the matrix **A** of Problem 10.2 is similar to a diagonal matrix by confirming that $\mathbf{A} = \mathbf{T}\mathbf{\Lambda}\mathbf{T}^{\mathrm{T}}$, where

$$\mathbf{\Lambda} = \begin{bmatrix} 2+\sqrt{2} & 0 & 0 \\ 0 & 2 & 0 \\ 0 & 0 & 2-\sqrt{2} \end{bmatrix} \quad \text{and} \quad \mathbf{T} = \frac{1}{\sqrt{6}} \begin{bmatrix} 1 & \sqrt{3} & 1 \\ \sqrt{2} & 0 & -\sqrt{2} \\ 1 & -\sqrt{3} & 1 \end{bmatrix}.$$

Note that **T** is orthogonal and is the matrix whose columns are the above eigenvectors $\mathbf{x}_1$, $\mathbf{x}_2$ and $\mathbf{x}_3$.

10.4 Show that

$$\mathbf{A} = \begin{bmatrix} 0 & 1 \\ 0 & 0 \end{bmatrix}$$

has eigenvalues $\lambda_1 = 0$ and $\lambda_2 = 0$ and that **A** cannot be similar to a diagonal matrix.

Section 10.2

10.5 Prove that (10.10) is valid for all vectors **x** and **y** and any norm.

10.6 Show that if **x** and **y** are n-dimensional vectors with real components then, for any real number λ,

$$\sum_{i=1}^{n} x_i^2 + 2\lambda \sum_{i=1}^{n} x_i y_i + \lambda^2 \sum_{i=1}^{n} y_i^2 = \sum_{i=1}^{n} (x_i + \lambda y_i)^2 \geqslant 0.$$

If $\mathbf{y} \neq \mathbf{0}$ put $\lambda = -\Sigma_{i=1}^{n} x_i y_i / \Sigma_{i=1}^{n} y_i^2$ and thus prove the Cauchy–Schwarz inequality

$$(x_1 y_1 + \cdots + x_n y_n)^2 \leqslant (x_1^2 + \cdots + x_n^2)(y_1^2 + \cdots + y_n^2).$$

Hence prove that

$$\|\mathbf{x} + \mathbf{y}\|_2^2 \leqslant \|\mathbf{x}\|_2^2 + 2\|\mathbf{x}\|_2 \|\mathbf{y}\|_2 + \|\mathbf{y}\|_2^2$$

and thus verify (10.9) for the Euclidean norm of real vectors.

10.7 Prove that (10.12) and (10.13) do define norms, that is they satisfy the conditions of Definition 10.5.

10.8 Show that for $p = \frac{1}{2}$ we do not obtain a norm in (10.14) by finding an example for which condition (iii) fails.

Section 10.3

10.9 (assumes knowledge of complex numbers) Show that, for a complex matrix, we may obtain equality in (10.20) by choosing

$$x_j = e^{-i\theta_j},$$

where θ_j is the argument of a_{pj}, that is, $a_{pj} = |a_{pj}| \mathrm{e}^{i\theta_j}$ and the maximum occurs in (10.20) for $i = p$.

10.10 Show that, for all non-singular matrices **A** and **B**,

$$\|\mathbf{A}^{-1}\| \geqslant \frac{1}{\|\mathbf{A}\|},$$

$$\|\mathbf{A}^{-1} - \mathbf{B}^{-1}\| \leqslant \|\mathbf{A}^{-1}\| \|\mathbf{B}^{-1}\| \|\mathbf{A} - \mathbf{B}\|.$$

10.11 Show that if $\mathbf{x}$ and $\mathbf{y}$ are real vectors with $\mathbf{x} = \mathbf{Ty}$, where $\mathbf{T}$ is an orthogonal matrix, then $\|\mathbf{x}\|_2 = \|\mathbf{y}\|_2$. (*Hint*: use $\|\mathbf{x}\|_2 = [\mathbf{x}^T\mathbf{x}]^{1/2}$.)

10.12 An $n \times n$ *semi-positive definite* matrix $\mathbf{B}$ is a symmetric matrix such that $\mathbf{x}^T\mathbf{Bx} \geqslant 0$ for all $\mathbf{x} \in \mathbb{R}^n$. Prove that if $\mathbf{A}$ is any $n \times n$ matrix then $\mathbf{B} = \mathbf{A}^T\mathbf{A}$ is semi-positive definite.

Show that all the eigenvalues of a semi-positive definite matrix $\mathbf{B}$ are non-negative by choosing $\mathbf{x} = \mathbf{Te}_i$, where $\mathbf{e}_i$ is the ith column of the unit matrix and $\mathbf{T}$ is the orthogonal matrix which reduces $\mathbf{B}$ to diagonal form, that is, $\mathbf{B} = \mathbf{T\Lambda T}^T$ where $\mathbf{\Lambda}$ is diagonal. Note that $\mathbf{x}^T\mathbf{Bx} = \mathbf{x}^T\mathbf{T\Lambda T}^T\mathbf{x} = \mathbf{e}_i^T\mathbf{\Lambda e}_i = \lambda_i$.

10.13 Show that

$$\mathbf{B} = \begin{bmatrix} 1 & 0.4 & 0.7 \\ 0.1 & 1 & 0.2 \\ 0.3 & 0.2 & 1 \end{bmatrix}$$

is non-singular and find an upper bound for $\|\mathbf{B}^{-1}\|_1$ without calculating the inverse.

10.14 Show that if $\mathbf{A}$ and $(\mathbf{A} - \mathbf{B})$ are non-singular and $\|\mathbf{A}^{-1}\|\,\|\mathbf{B}\| < 1$ then

$$\|(\mathbf{A} - \mathbf{B})^{-1}\| \leqslant \frac{1}{\|\mathbf{A}^{-1}\|^{-1} - \|\mathbf{B}\|}.$$

10.15 If $p(x)$ is the polynomial

$$p(x) = a_0 + a_1x + a_2x^2 + \cdots + a_rx^r,$$

with non-negative coefficients ($a_i \geqslant 0$, $i = 0, 1, \ldots, r$) and

$$p(\mathbf{A}) = a_0\mathbf{I} + a_1\mathbf{A} + a_2\mathbf{A}^2 + \cdots + a_r\mathbf{A}^r,$$

show that

$$\|p(\mathbf{A})\| \leqslant p(\|\mathbf{A}\|).$$

10.16 If

$$\mathbf{A} = \begin{bmatrix} 0 & 0 \\ 2 & 0 \end{bmatrix},$$

show that the series $\mathbf{I} + \mathbf{A} + \mathbf{A}^2 + \mathbf{A}^3 + \cdots$ is convergent although

$$\|\mathbf{A}\|_1 = \|\mathbf{A}\|_\infty > 1.$$

10.17 Show that the sequence

$$\mathbf{I}, \mathbf{A}, \frac{1}{2!}\mathbf{A}^2, \frac{1}{3!}\mathbf{A}^3, \frac{1}{4!}\mathbf{A}^4, \ldots$$

is convergent to $\mathbf{0}$ for any choice of $\mathbf{A}$.

Section 10.4

10.18 Show that if **A** and **B** are $n \times n$ matrices their condition numbers satisfy the following for any choice of norm and scalar $\lambda \neq 0$.

(i) $k(\mathbf{A}) \geqslant 1$ (ii) $k(\mathbf{AB}) \leqslant k(\mathbf{A})k(\mathbf{B})$ (iii) $k(\lambda\mathbf{A}) = k(\mathbf{A})$.

10.19 Find the condition numbers of

$$\mathbf{A} = \begin{bmatrix} 1 & 2 \\ 1.001 & 2.001 \end{bmatrix}$$

for the 1 and ∞ norms. This matrix is ill-conditioned because the second row is almost a multiple of the first row.

10.20 The linear equations

$$\begin{bmatrix} 2.01 & 1.01 \\ 1 & 0.5 \end{bmatrix}\mathbf{x} = \begin{bmatrix} -0.01 \\ 0 \end{bmatrix}$$

have solution $\mathbf{x} = [1 \ \ -2]^{\mathrm{T}}$. Show that, if the elimination method is used with *all* calculations rounded to three significant digits, the equations reduce to

$$\begin{bmatrix} 2.01 & 1.01 \\ 0 & -0.00300 \end{bmatrix}\mathbf{x} = \begin{bmatrix} -0.01 \\ 0.00498 \end{bmatrix}$$

and hence the computed solution is

$$[0.831 \ \ -1.6]^{\mathrm{T}}.$$

10.21 The first row of a matrix is scaled so that

$$\mathbf{A} = \begin{bmatrix} 2\lambda & \lambda \\ 1 & 1 \end{bmatrix}.$$

Show that $k_\infty(\mathbf{A})$ is a minimum for $\lambda = \pm\frac{2}{3}$.

10.22 Show that if $(\mathbf{A} + \boldsymbol{\delta}\mathbf{A})^{-1}$ is computed as an approximation to $\mathbf{A}^{-1}$ and

$$e = \|\boldsymbol{\delta}\mathbf{A}\| / \|\mathbf{A}\|,$$

then

$$\frac{\|(\mathbf{A} + \boldsymbol{\delta}\mathbf{A})^{-1} - \mathbf{A}^{-1}\|}{\|\mathbf{A}^{-1}\|} \leqslant \frac{k(\mathbf{A})e}{1 - k(\mathbf{A})e},$$

provided that $k(\mathbf{A})e < 1$. This is a bound on the relative error of $(\mathbf{A} + \boldsymbol{\delta}\mathbf{A})^{-1}$. (*Hint*: start by proving the identity

$$\begin{aligned}(\mathbf{A} + \boldsymbol{\delta}\mathbf{A})^{-1} - \mathbf{A}^{-1} &= -(\mathbf{A} + \boldsymbol{\delta}\mathbf{A})^{-1}.\boldsymbol{\delta}\mathbf{A}.\mathbf{A}^{-1} \\ &= -(\mathbf{I} + \mathbf{A}^{-1}\boldsymbol{\delta}\mathbf{A})^{-1}\mathbf{A}^{-1}.\boldsymbol{\delta}\mathbf{A}.\mathbf{A}^{-1}\end{aligned}$$

and take norms, using Lemma 10.3.)

10.23 Show that if $\mathbf{A}$ is a non-singular symmetric matrix then

$$k_2(\mathbf{A}) = |\lambda_1| / |\lambda_n|$$

where $\lambda_1, \ldots, \lambda_n$ are the eigenvalues (necessarily real) of $\mathbf{A}$ ordered so that $|\lambda_1| \geqslant |\lambda_2| \geqslant \cdots \geqslant |\lambda_n|$.

Section 10.5

10.24 Show that, with $\mathbf{r}_0$ and $\boldsymbol{\delta}\mathbf{x}$ as in (10.34) and (10.37),

$$\frac{\|\boldsymbol{\delta}\mathbf{x}\|}{\|\mathbf{x}\|} \leqslant k(\mathbf{A}) \frac{\|\mathbf{r}_0\|}{\|\mathbf{b}\|}.$$

10.25 Use the correction process described in §10.5 to improve the computed approximate solution in Problem 10.20. Work to 3 significant digits except in the calculation of the residual vector when extra accuracy should be used.

10.26 We can improve the accuracy of a computed approximation to the inverse of a matrix $\mathbf{A}$ as follows. We calculate the sequence $\mathbf{W}_0, \mathbf{W}_1, \mathbf{W}_2, \ldots$, using

$$\mathbf{W}_{m+1} = \mathbf{W}_m(2\mathbf{I} - \mathbf{A}\mathbf{W}_m).$$

Prove that, if $\mathbf{E}_m = \mathbf{W}_m - \mathbf{A}^{-1}$ is the error after m steps, we have

$$\mathbf{E}_{m+1} = -\mathbf{E}_m\mathbf{A}\mathbf{E}_m$$

so that

$$\|\mathbf{E}_{m+1}\| \leqslant \|\mathbf{A}\| \; \|\mathbf{E}_m\|^2.$$

Deduce that the sequence $(\|\mathbf{E}_m\|)$ will tend to zero if $\|\mathbf{A}\| \, \|\mathbf{E}_0\| < 1$. (The process is at least second order.)

10.27 Write an algorithm for the compact elimination method followed by one correction of the solution using the method of § 10.5.

Section 10.6

10.28 Carry out three iterations of

(i) the Jacobi method,
(ii) the Gauss–Seidel method

of solving

$$\begin{bmatrix} -8 & 1 & 1 \\ 1 & -5 & 1 \\ 1 & 1 & -4 \end{bmatrix} \mathbf{x} = \begin{bmatrix} 1 \\ 16 \\ 7 \end{bmatrix}.$$

(Solution $\mathbf{x} = [-1 \;\; -4 \;\; -3]^{\mathrm{T}}$.)

In each case check that the methods converge and start with $\mathbf{x}_0 = [0 \;\; 0 \;\; 0]^{\mathrm{T}}$.

10.29 Write an algorithm for the Jacobi method of solving n linear equations. Note that, unlike the Gauss–Seidel method, it is necessary to keep two vectors, say $\mathbf{x} = \mathbf{x}_m$ and $\mathbf{y} = \mathbf{x}_{m+1}$.

10.30 Write computer programs for the Jacobi and the Gauss–Seidel methods. Test your programs on the equations in Example 10.8 and those in Problem 10.28. In each case determine which method seems to give the faster convergence, for example, by considering how many iterations are required to obtain an accuracy of $\pm 10^{-6}$.

10.31 If $(\mathbf{x}_m)_{m=0}^{\infty}$ is a sequence satisfying (10.39), use (10.52) to deduce that

$$\mathbf{x}_{m+1} - \mathbf{x}_m = (\mathbf{M} - \mathbf{I})\mathbf{M}^m \mathbf{e}_0.$$

Hence use Lemma 10.3 to deduce that if $\|\mathbf{M}\| < 1$ then

$$\|\mathbf{e}_m\| = \|\mathbf{M}^m \mathbf{e}_0\| \leqslant \frac{1}{1 - \|\mathbf{M}\|} \|\mathbf{x}_{m+1} - \mathbf{x}_m\|.$$

10.32 If

$$\mathbf{A} = \begin{bmatrix} 1 & -\frac{1}{2} \\ -\frac{1}{2} & 1 \end{bmatrix}$$

calculate the Jacobi iteration matrix $\mathbf{M}_J = \mathbf{D}^{-1}(\mathbf{L} + \mathbf{U})$ and the Gauss–Seidel iteration matrix $\mathbf{M}_G = (\mathbf{D} - \mathbf{L})^{-1}\mathbf{U}$. Show that $\rho(\mathbf{M}_J) = \frac{1}{2}$ and $\rho(\mathbf{M}_G) = \frac{1}{4}$.

10.33 With $\mathbf{A}$ as defined in Problem 10.32, calculate the successive over-relaxation matrix $\mathbf{M}_\omega$ of (10.60). Show that if $\omega = 4(2 - \sqrt{3})$ we have $\rho(\mathbf{M}_\omega) = 7 - 4\sqrt{3} \simeq 0.072$. (This is the optimum choice of ω for the matrix $\mathbf{A}$, that is, it minimizes $\rho(\mathbf{M}_\omega)$.)

10.34 Show that for $\omega \neq 0$ the successive over relaxation method (10.59) may be obtained from the splitting (see (10.49))

$$\mathbf{A} = \left(\frac{1}{\omega}\mathbf{D} - \mathbf{L}\right) - \left(\mathbf{U} - \left(1 - \frac{1}{\omega}\right)\mathbf{D}\right).$$

10.35 Suppose that an iterative process

$$\mathbf{x}_{m+1} = \mathbf{M}\mathbf{x}_m + \mathbf{d}, \qquad m = 0, 1, 2, \ldots, \tag{10.61}$$

with $\|\mathbf{M}\| < 1$, is affected by rounding errors so that a sequence $(\tilde{\mathbf{x}}_m)_{m=0}^{\infty}$ is computed with

$$\tilde{\mathbf{x}}_{m+1} = \mathbf{M}\tilde{\mathbf{x}}_m + \mathbf{d} + \mathbf{r}_m, \qquad m = 0, 1, 2, \ldots, \tag{10.62}$$

where the error vectors $\mathbf{r}_m$ satisfy $\|\mathbf{r}_m\| \leqslant \varepsilon$ and $\tilde{\mathbf{x}}_0 = \mathbf{x}_0$.

Show that

$$\|\bar{\mathbf{x}}_m - \mathbf{x}_m\| \leqslant \frac{\varepsilon}{1 - \|\mathbf{M}\|}, \qquad m = 0, 1, 2, \ldots .$$

(*Hint*: if $\boldsymbol{\rho}_m = \tilde{\mathbf{x}}_m - \mathbf{x}_m$, show from (10.61) and (10.62) that

$$\|\boldsymbol{\rho}_{m+1}\| \leqslant \|\mathbf{M}\| \; \|\boldsymbol{\rho}_m\| + \varepsilon$$

and thus, by an induction argument, that

$$\|\boldsymbol{\rho}_m\| \leqslant \frac{\varepsilon}{1 - \|\mathbf{M}\|}.$$

See also Lemma 13.1.)

Chapter **11**

MATRIX EIGENVALUES AND EIGENVECTORS

11.1 Relations between matrix norms and eigenvalues; Gerschgorin theorems

Let the eigenvalues $\lambda_1, \lambda_2, \ldots, \lambda_n$ of a given $n \times n$ matrix $\mathbf{A}$ be ordered so that $|\lambda_1| \geqslant |\lambda_2| \geqslant \cdots \geqslant |\lambda_n|$. We call λ_1 a *dominant* eigenvalue of $\mathbf{A}$ and thus we have (see Definition 10.4) $\rho(\mathbf{A}) = |\lambda_1|$. Now suppose that $\mathbf{x}_1$ is an eigenvector of $\mathbf{A}$ corresponding to λ_1, with $\mathbf{x}_1$ scaled so that $\|\mathbf{x}_1\| = 1$ for some norm. Then

$$\mathbf{A}\mathbf{x}_1 = \lambda_1 \mathbf{x}_1$$

and

$$\|\mathbf{A}\mathbf{x}_1\| = \|\lambda_1 \mathbf{x}_1\| = |\lambda_1| \cdot \|\mathbf{x}_1\| = |\lambda_1| = \rho(\mathbf{A}).$$

By Definition 10.6

$$\|\mathbf{A}\| \geqslant \|\mathbf{A}\mathbf{x}_1\|$$

and thus

$$\rho(\mathbf{A}) \leqslant \|\mathbf{A}\| \tag{11.1}$$

for any choice of norm. As we saw in §10.3, when $\mathbf{A}$ is symmetric,

$$\|\mathbf{A}\|_2 = \rho(\mathbf{A}). \tag{11.2}$$

One consequence of (11.1), (10.19) and (10.20) is that

$$\rho(\mathbf{A}) \leqslant \max_j \sum_{i=1}^{n} |a_{ij}| \tag{11.3}$$

and

$$\rho(\mathbf{A}) \leqslant \max_i \sum_{j=1}^{n} |a_{ij}| \tag{11.4}$$

so that $|\lambda_1|$ is bounded by the maximum column and row sums of absolute values of elements of **A**. We can obtain tighter bounds than these by using Gerschgorin's theorems.

Theorem 11.1 The eigenvalues of an $n \times n$ matrix **A** lie within the union of n (Gerschgorin) disks D_i in the complex plane with centre a_{ii} and radius

$$\Lambda_i = \sum_{\substack{j=1 \\ j \neq i}}^{n} |a_{ij}|, \qquad i = 1, 2, \ldots, n. \tag{11.5}$$

Thus D_i is the set of points $z \in \mathbb{C}$ (the complex plane) given by

$$|z - a_{ii}| \leqslant \Lambda_i, \qquad i = 1, 2, \ldots, n. \tag{11.6}$$

Proof Suppose $\mathbf{Ax} = \lambda \mathbf{x}$ with $\mathbf{x}$ scaled so that $\|\mathbf{x}\|_\infty = 1$, and suppose that $|x_k| = 1$. Consider the kth equation of the system $\mathbf{Ax} = \lambda \mathbf{x}$,

$$a_{k1}x_1 + a_{k2}x_2 + \cdots + a_{kk}x_k + \cdots + a_{kn}x_n = \lambda x_k$$

which may be written as

$$(\lambda - a_{kk})x_k = \sum_{\substack{j=1 \\ j \neq k}}^{n} a_{kj}x_j.$$

Hence

$$|\lambda - a_{kk}| . |x_k| \leqslant \sum_{\substack{j=1 \\ j \neq k}}^{n} |a_{kj}| . |x_j|$$

and thus

$$|\lambda - a_{kk}| \leqslant \sum_{\substack{j=1 \\ j \neq k}}^{n} |a_{kj}|, \tag{11.7}$$

since $|x_j| \leqslant |x_k| = 1$, $j = 1, 2, \ldots, n$. Hence λ lies in D_k. For each eigenvalue, there is a disk D_k for which the required inequality is satisfied. □

The following theorem provides an even tighter result, using the Gerschgorin disks.

Theorem 11.2 If s of the Gerschgorin disks in the complex plane form a connected domain which is isolated from the remaining $n - s$ disks, then there are precisely s eigenvalues of **A** within this connected domain.

Proof We use a continuity argument. Let

$$\mathbf{A} = \mathbf{D} + \mathbf{B} \tag{11.8}$$

where $\mathbf{D}$ is diagonal and $\mathbf{B}$ has zeros on the diagonal, that is, we split $\mathbf{A}$ into its diagonal elements and its off-diagonal elements. Let

$$\mathbf{A}(\alpha) = \mathbf{D} + \alpha\mathbf{B} \tag{11.9}$$

where $\alpha \in [0, 1]$. Clearly $\mathbf{A}(0) = \mathbf{D}$ and $\mathbf{A}(1) = \mathbf{A}$. We then argue that the eigenvalues of $\mathbf{A}(\alpha)$ must be *continuous* functions of α, since they are the roots of a polynomial equation whose coefficients are themselves polynomials in α. The eigenvalues of $\mathbf{A}(0)$, the diagonal matrix, are simply the diagonal entries $a_{11}, a_{22}, \ldots, a_{nn}$, which are the centres of the Gerschgorin disks. Without loss of generality, we suppose that the first s disks, with centres at $a_{11}, a_{22}, \ldots, a_{ss}$ and radii $\Lambda_1, \Lambda_2, \ldots, \Lambda_s$, form a connected domain which is isolated from the disks centred at $a_{s+1,s+1}, \ldots, a_{nn}$ and with radii $\Lambda_{s+1}, \ldots, \Lambda_n$. It follows that disks with the same centres but with radii $\alpha\Lambda_1, \alpha\Lambda_2, \ldots, \alpha\Lambda_s$ are isolated from those with radii $\alpha\Lambda_{s+1}, \ldots, \alpha\Lambda_n$, for $\alpha \in [0, 1]$. For $\alpha = 0$, the first s disks reduce to points $a_{11}, a_{22}, \ldots, a_{ss}$ and their union contains s eigenvalues. As α is increased, these s eigenvalues must remain within the union of the first s disks, since the eigenvalues are continuously dependent on α. Although it is possible for an eigenvalue to move out of one disk into another as α increases and the two disks intersect, it is not possible for an eigenvalue to 'jump' out of an isolated set into another disk. The result follows by increasing α to 1. □

Example 11.1 Use Gerschgorin's theorems to deduce as much as possible about the eigenvalues of the matrix

$$\mathbf{A} = \begin{bmatrix} -4 & 0 & 0 & 1 \\ 1 & 2 & 1 & 0 \\ 0 & 2 & 3 & 0 \\ 1 & 0 & 1 & 4 \end{bmatrix}.$$

We note that

D_1 has centre at -4, radius 1
D_2 has centre at 2, radius 2
D_3 has centre at 3, radius 2
D_4 has centre at 4, radius 2.

We conclude that there is one eigenvalue in D_1 as this is isolated from the other disks. The other three eigenvalues lie in the union of D_2, D_3 and D_4. □

Sometimes we can obtain further results by using the columns of **A** instead of, or as well as, rows. The eigenvalues of $\mathbf{A}^T$ are the same as those of **A**. If **A** is symmetric the eigenvalues are real and the Gerschgorin disks reduce to intervals on the real line. In some cases we can also use simple similarity transformations to enhance bounds obtained from Gerschgorin's theorems, as the next example shows.

Example 11.2 Use similarity transformations and Gerschgorin's theorems to obtain bounds on the eigenvalue near $+1$ of the matrix

$$\mathbf{A} = \begin{bmatrix} 1 & 0.1 & 0.2 \\ 0.1 & 2 & 0.1 \\ 0.2 & 0.1 & 3 \end{bmatrix}.$$

A is real symmetric and the Gerschgorin disks reduce to intervals which in this case are disjoint. We have

$$\begin{aligned} D_1&: \quad |\lambda - 1| \leqslant 0.3 \Rightarrow \lambda_1 \in [0.7, 1.3] \\ D_2&: \quad |\lambda - 2| \leqslant 0.2 \Rightarrow \lambda_2 \in [1.8, 2.2] \\ D_3&: \quad |\lambda - 3| \leqslant 0.3 \Rightarrow \lambda_3 \in [2.7, 3.3]. \end{aligned}$$

Let

$$\mathbf{D}_\alpha = \begin{bmatrix} \alpha & 0 & 0 \\ 0 & 1 & 0 \\ 0 & 0 & 1 \end{bmatrix}$$

for any $\alpha > 0$, when

$$\mathbf{D}_\alpha \mathbf{A} \mathbf{D}_\alpha^{-1} = \begin{bmatrix} 1 & 0.1\alpha & 0.2\alpha \\ 0.1/\alpha & 2 & 0.1 \\ 0.2/\alpha & 0.1 & 3 \end{bmatrix}.$$

Since $\mathbf{D}_\alpha \mathbf{A} \mathbf{D}_\alpha^{-1}$ has the same eigenvalues as **A**, the Gerschgorin intervals are now

$$\begin{aligned} D_1 &= [1 - 0.3\alpha, 1 + 0.3\alpha] \\ D_2 &= [2 - 0.1 - 0.1/\alpha, 2 + 0.1 + 0.1/\alpha] \\ D_3 &= [3 - 0.1 - 0.2/\alpha, 3 + 0.1 + 0.2/\alpha]. \end{aligned}$$

We seek better bounds on λ_1. As we decrease α, D_1 decreases in length but D_2 and D_3 increase. As long as D_1 does not overlap D_2 or D_3 we know that $\lambda_1 \in D_1$. We therefore decrease α until D_1 is about to overlap D_2 or D_3. This occurs first when

$$1 + 0.3\alpha = 2 - 0.1 - 0.1/\alpha,$$

that is

$$3\alpha^2 - 9\alpha + 1 = 0.$$

The smallest root is $\alpha = (9 - \sqrt{69})/6 = 0.1155\ldots$. We round up to make sure there is no overlap of intervals and take $\alpha = 0.116$, giving the intervals $D_1 = [0.9652, 1.0348]$, $D_2 = [1.0379, 2.9621]$, $D_3 = [1.1758, 4.8242]$. Thus $\lambda_1 \in [0.9652, 1.0348]$, so that λ_1 is known to be within ± 0.035 of $+1$, a much tighter result than can be obtained by Gerschgorin's theorems alone.

□

11.2 Simple and inverse iterative method

We suppose initially that a matrix **A** has the following properties.

(i) The eigenvalues $\lambda_1, \lambda_2, \ldots, \lambda_n$ satisfy $|\lambda_1| > |\lambda_2| \geqslant \cdots \geqslant |\lambda_n|$, so that λ_1 is *the* dominant eigenvalue of **A**.

(ii) There are n linearly independent eigenvectors $\mathbf{x}_1, \mathbf{x}_2, \ldots, \mathbf{x}_n$ scaled so that

$$\|\mathbf{x}_1\|_\infty = \|\mathbf{x}_2\|_\infty = \cdots = \|\mathbf{x}_n\|_\infty = 1.$$

Algorithm 11.1 Given an initial non-zero vector $\mathbf{y}_0 \in \mathbb{R}^n$ we compute the sequence $\mathbf{y}_1, \mathbf{y}_2, \ldots$ given by

$$\left.\begin{aligned} \mathbf{y}^*_{r+1} &= \mathbf{A}\mathbf{y}_r \\ \mathbf{y}_{r+1} &= \frac{1}{\|\mathbf{y}^*_{r+1}\|_\infty} \cdot \mathbf{y}^*_{r+1} \text{ (normalization)} \end{aligned}\right\} \qquad r = 0, 1, 2, \ldots.$$

In general (although there are certain exceptions)

$$\mathbf{y}_r \to \pm\mathbf{x}_1 \quad \text{and} \quad \mathbf{y}^*_{r+1} \simeq \lambda_1 \mathbf{y}_r, \qquad \text{as } r \to \infty.$$

Note that because of the normalization, $\|\mathbf{y}_r\|_\infty = 1$, $r = 1, 2, \ldots$. To justify the convergence we note that, as $\mathbf{x}_1, \mathbf{x}_2, \ldots, \mathbf{x}_n$ are linearly independent, any vector (see Johnson *et al.*, 1993) may be expressed as a linear combination of these eigenvectors. Thus there exist numbers $c_1, c_2, \ldots, c_n$ such that

$$\mathbf{y}_0 = c_1\mathbf{x}_1 + c_2\mathbf{x}_2 + \cdots + c_n\mathbf{x}_n. \tag{11.10}$$

Hence

$$\mathbf{y}^*_1 = \mathbf{A}\mathbf{y}_0 = c_1\lambda_1\mathbf{x}_1 + c_2\lambda_2\mathbf{x}_2 + \cdots + c_n\lambda_n\mathbf{x}_n$$

and thus

$$\mathbf{y}_1 = k_1(c_1\lambda_1\mathbf{x}_1 + c_2\lambda_2\mathbf{x}_2 + \cdots + c_n\lambda_n\mathbf{x}_n)$$

where k_1 is the scaling factor to ensure that $\|\mathbf{y}_1\|_\infty = 1$.

Similarly

$$\mathbf{y}_2 = k_2(c_1\lambda_1^2\mathbf{x}_1 + c_2\lambda_2^2\mathbf{x}_2 + \cdots + c_n\lambda_n^2\mathbf{x}_n)$$

where k_2 is a scaling factor chosen so that $\|\mathbf{y}_2\|_\infty = 1$. In general

$$\mathbf{y}_r = k_r(c_1\lambda_1^r\mathbf{x}_1 + c_2\lambda_2^r\mathbf{x}_2 + \cdots + c_n\lambda_n^r\mathbf{x}_n) \tag{11.11}$$

where k_r is a scaling factor. Hence

$$\mathbf{y}_r = k_r\lambda_1^r\left\{c_1\mathbf{x}_1 + c_2\left(\frac{\lambda_2}{\lambda_1}\right)^r\mathbf{x}_2 + \cdots + c_n\left(\frac{\lambda_n}{\lambda_1}\right)^r\mathbf{x}_n\right\} \tag{11.12}$$

and thus, if $c_1 \neq 0$,

$$\mathbf{y}_r \sim k_r\lambda_1^r c_1\mathbf{x}_1, \qquad \text{as } r \to \infty,$$

since $|\lambda_i/\lambda_1| < 1$, $i = 2, 3, \ldots, n$. Thus we see that $\mathbf{y}_r$ becomes parallel to $\mathbf{x}_1$ and $\mathbf{y}^*_{r+1} = \mathbf{A}\mathbf{y}_r \simeq \lambda_1\mathbf{y}_r$ as $r \to \infty$.

We make the following observations about the algorithm.

(i) The normalization of each $\mathbf{y}_r$ is not essential, in theory, but it does prevent $\|\mathbf{y}_r\|_\infty \to 0$ for $|\lambda_1| < 1$ or $\|\mathbf{y}_r\|_\infty \to \infty$ for $|\lambda_1| > 1$. Either of these outcomes would be inconvenient in practice.

(ii) There are various ways in which λ_1 can be estimated from $\mathbf{y}_r$ and $\mathbf{y}^*_{r+1}$. For example,

$$\lambda_1 \simeq \left(\frac{\text{element of } \mathbf{y}^*_{r+1} \text{ of largest modulus}}{\text{corresponding element of } \mathbf{y}_r}\right) \tag{11.13}$$

or, if the matrix is symmetric, we shall see that the Rayleigh quotient is much better.

(iii) It is important that $c_1 \neq 0$ so that $\mathbf{y}_0$ has a component in the direction of $\mathbf{x}_1$. If $c_1 = 0$ and there are no rounding errors we will not obtain convergence to λ_1 and $\mathbf{x}_1$ (see Problem 11.6).

(iv) If λ_1 is not the sole dominant eigenvalue (for example, if $|\lambda_1| = |\lambda_2| > |\lambda_3| \geqslant \cdots \geqslant |\lambda_n|$), there are various possible outcomes but, in general, the algorithm does not converge to λ_1 and $\mathbf{x}_1$.

(v) We can apply the Aitken extrapolation process to speed up convergence.

Table 11.1 Results for Example 11.3

$\mathbf{y}_0$	$\mathbf{y}_1^*$	$\mathbf{y}_1$	$\mathbf{y}_2^*$	$\mathbf{y}_2$	$\mathbf{y}_3^*$	$\mathbf{y}_3$	...	$\mathbf{y}_6^*$	$\mathbf{y}_6$
1	−3	−3/4	5/2	−2/3	7/3	−7/11	...	65/29	−13/21
−1	4	1	−15/4	1	−11/3	1	...	−105/29	1
1	−4	−1	15/4	−1	11/3	−1	...	105/29	−1
−1	3	3/4	−5/2	2/3	−7/3	7/11	...	−65/29	13/21
Estimated λ		−4	−3.75		−3.6667			−3.6207	

Example 11.3 Use the simple iteration Algorithm 11.1 to estimate the largest eigenvalue of the matrix

$$\mathbf{A} = \begin{bmatrix} -2 & 1 & 0 & 0 \\ 1 & -2 & 1 & 0 \\ 0 & 1 & -2 & 1 \\ 0 & 0 & 1 & -2 \end{bmatrix}.$$

(The dominant eigenvalue is $\lambda_1 = -3.618034$ with corresponding eigenvector $\mathbf{x}_1 = [0.618034 \;\; -1 \;\; +1 \;\; -0.618034]^{\mathrm{T}}$, to six decimal places.)

We take $\mathbf{y}_0 = [1 \;\; -1 \;\; 1 \;\; -1]^{\mathrm{T}}$ and ignore the symmetry of $\mathbf{A}$. The first few vectors of the algorithm are given in Table 11.1.

The estimated λ_1, is given by (11.13) above and, by the time we have computed the sixth iterate $\mathbf{y}_6$, the estimate of λ_1 is accurate to almost three decimal places. The vector $\mathbf{y}_6$ is also a fairly accurate estimate of $\mathbf{x}_1$, as $\mathbf{y}_6 = [-0.619 \;\; 1 \;\; -1 \;\; 0.619]^{\mathrm{T}}$ to three decimal places. □

The matrix in Example 11.3 is symmetric and a more accurate estimate of λ_1 may be obtained using the *Rayleigh quotient*

$$\lambda_{\mathrm{R}}(\mathbf{y}) = \frac{\mathbf{y}^{\mathrm{T}}\mathbf{A}\mathbf{y}}{\mathbf{y}^{\mathrm{T}}\mathbf{y}} \tag{11.14}$$

where $\mathbf{y}$ is an estimate of the eigenvector. It can be shown that if $\mathbf{y} = \mathbf{x} + \mathbf{e}$, where $\mathbf{x}$ is an eigenvector of $\mathbf{A}$ and $\mathbf{e}$ is an error vector, then

$$\lambda_{\mathrm{R}}(\mathbf{y}) = \lambda + O(\|\mathbf{e}\|^2) \tag{11.15}$$

where λ is the eigenvalue to which $\mathbf{x}$ corresponds. If $\|\mathbf{e}\|$ is small then λ_{R} will be a very close (second order) approximation to λ. A thorough error analysis of the Rayleigh quotient is given by Wilkinson (1988).

Example 11.4 Apply Rayleigh quotients to the vectors obtained in Example 11.3.

First note that $\mathbf{y}^*_{r+1} = \mathbf{A}\mathbf{y}_r$ and thus

$$\lambda_R(\mathbf{y}_r) = \frac{\mathbf{y}_r^T \mathbf{y}^*_{r+1}}{\mathbf{y}_r^T \mathbf{y}_r}. \tag{11.16}$$

We find that

$$\lambda_R(\mathbf{y}_2) = \frac{-94/9}{26/9} \simeq -3.6154$$

which is a much better approximation to λ_1 than we obtained after three iterations using (11.13). Also,

$$\lambda_R(\mathbf{y}_6) = -3.618033$$

which is very accurate. □

To find a subdominant eigenvalue we can use *deflation*. Having found the dominant eigenvalue λ_1 and eigenvector $\mathbf{x}_1$, we construct a new matrix which has the same eigenvalues as $\mathbf{A}$ except that λ_1 is replaced by zero. Problem 11.10 gives such a method for symmetric matrices.

It can be seen from the above discussion that there are several pitfalls with simple iteration. The eigenvalue λ_1 may not be very dominant and then convergence, which depends on how rapidly $(\lambda_2/\lambda_1)^r$ tends to zero, will be slow. The choice of $\mathbf{y}_0$ is important and, if c_1 in (11.10) is small, the convergence will be slow but, since we are trying to find $\mathbf{x}_1$, we may not know how to ensure that c_1 is reasonably large.

To a certain extent we can overcome these problems by using *inverse iteration*. The matrix $(\mathbf{A}-p\mathbf{I})^{-1}$, where $p \in \mathbb{R}$, has eigenvalues $(\lambda_1 - p)^{-1}$, $(\lambda_2 - p)^{-1}, \ldots, (\lambda_n - p)^{-1}$ and the same eigenvectors as $\mathbf{A}$ (see Problem 11.11). We now choose p close to an eigenvalue. By choosing p sufficiently close to λ_j, where $\lambda_i \neq \lambda_j$ for $i \neq j$ we can ensure that

$$|(\lambda_j - p)^{-1}| >> |(\lambda_i - p)^{-1}|, \quad i = 1, 2, \ldots, n, \quad i \neq j, \tag{11.17}$$

so that $(\lambda_j - p)^{-1}$ is clearly the dominant eigenvalue of $(\mathbf{A}-p\mathbf{I})^{-1}$. We iterate with $(\mathbf{A}-p\mathbf{I})^{-1}$ and update p every iteration or every few iterations, as our estimate of λ_j improves, and thus obtain very rapid convergence.

In practice we do not compute $(\mathbf{A}-p\mathbf{I})^{-1}$ but solve the equations

$$(\mathbf{A}-p\mathbf{I})\mathbf{y}^*_{r+1} = \mathbf{y}_r \tag{11.18}$$

by factorization. If $\mathbf{A}-p\mathbf{I} = \mathbf{L}_p\mathbf{U}_p$ we can use the triangular factors $\mathbf{L}_p$ and $\mathbf{U}_p$ again in the next iteration if p is left unchanged. If $p = \lambda_j$ exactly, the matrix $\mathbf{A}-p\mathbf{I}$ is singular and the method fails. In practice we rarely obtain λ_j exactly, and so $p \neq \lambda_j$; however, for p very close to λ_j the equations (11.18) are nearly singular and hence ill-conditioned. This is not usually important since at this stage we have obtained the desired close approximation to λ_j. The method can be used to find *any* eigenvalue of **A**, even if it is not distinct from the other eigenvalues. For a discussion of this and other cases, see Wilkinson (1988) and Gourlay and Watson (1973).

If **A** is symmetric we can use Rayleigh quotients again to estimate the eigenvalue. If **A** is tridiagonal we can use the simplified form of factorization described in § 9.9.

Example 11.5 Use the method of inverse iteration to find the eigenvalue of the matrix of Example 11.3 nearest to −4. We start with $\mathbf{y}_0 = [1 \ -1 \ 1 \ -1]^{\mathrm{T}}$ as before, choose $p = -4$ and use Rayleigh quotients. With $r = 0$, we obtain $\mathbf{y}_1^*$ by solving the system of linear equations

$$(\mathbf{A} - p\mathbf{I})\mathbf{y}_{r+1}^* = \mathbf{y}_r, \tag{11.19}$$

compute $\mathbf{y}_1$ from

$$\mathbf{y}_{r+1} = \frac{1}{\|\mathbf{y}_{r+1}^*\|_\infty}\,\mathbf{y}_{r+1}^* \tag{11.20}$$

with $r = 0$, and then obtain $\mu_{\mathrm{R},p}$ and λ_{est} from

$$\mu_{\mathrm{R},p} = \frac{\mathbf{y}_r^{\mathrm{T}}\mathbf{y}_{r+1}^*}{\mathbf{y}_r^{\mathrm{T}}\mathbf{y}_r} = \frac{\mathbf{y}_r^{\mathrm{T}}(\mathbf{A} - p\mathbf{I})^{-1}\mathbf{y}_r}{\mathbf{y}_r^{\mathrm{T}}\mathbf{y}_r} \tag{11.21}$$

and

$$\lambda_{\mathrm{est}} = \frac{1}{\mu_{\mathrm{R},p}} + p. \tag{11.22}$$

We then increase r by 1, replace p by λ_{est} and repeat the sequence of calculations defined by equations (11.19)–(11.22). The values of $\mathbf{y}_r$ and λ_{est} for the first three iterations are given in Table 11.2. Note that equations (11.19) and (11.20) are the usual relations for inverse iteration, in which the elimination method for tridiagonal equations is used. Equation (11.21) gives

Table 11.2 Results for Example 11.5

	$\mathbf{y}_0$	$\mathbf{y}_1$	$\mathbf{y}_2$	$\mathbf{y}_3$
	1	0.666667	−0.617647	−0.618034
	−1	−1	1	1
	1	1	−1	−1
	−1	−0.666667	0.617647	0.618034
p	−4	−3.6	−3.618056	—
$\mu_{R,p}$	2.5	−55.3846	46366.6	—
λ_{est}	−3.6	−3.618056	−3.618034	—

the Rayleigh quotient for the matrix $(\mathbf{A}-p\mathbf{I})^{-1}$ and, since this matrix has eigenvalues $1/(\lambda_i - p)$, (11.22) gives our next estimate of the eigenvalue of $\mathbf{A}$.

It would have been more efficient to normalize by using $\|\mathbf{y}^*_{r+1}\|_2$ in (11.20), as then $\mathbf{y}_r^{\mathrm{T}}\mathbf{y}_r = 1$ in (11.21). However we have used $\|\cdot\|_\infty$ so that the vectors obtained can be compared directly with those in Examples 11.3 and 11.4.

11.3 Sturm sequence method

We shall see in §11.4 that it is possible to reduce any symmetric matrix to a tridiagonal matrix with the same eigenvalues, using similarity transformations. We therefore concentrate on methods of finding the eigenvalues of the symmetric tridiagonal matrix

$$\mathbf{A} = \begin{bmatrix} \beta_1 & \alpha_1 & & & & \\ \alpha_1 & \beta_2 & \alpha_2 & & \mathbf{0} & \\ & \alpha_2 & \beta_3 & \alpha_3 & & \\ & & \ddots & \ddots & \ddots & \\ & \mathbf{0} & & \alpha_{n-2} & \beta_{n-1} & \alpha_{n-1} \\ & & & & \alpha_{n-1} & \beta_n \end{bmatrix}. \tag{11.23}$$

We assume that $\alpha_i \neq 0$, $i = 1, 2, \ldots, n-1$; otherwise we may split $\mathbf{A}$ into two parts and find the eigenvalues of the two tridiagonal principal submatrices. We could use inverse iteration as described in § 11.2 but, as we will see, there are better methods. We start by describing the Sturm sequence method, which locates all the eigenvalues of (11.23).

Let

$$f_{r+1}(\lambda) = \det \begin{bmatrix} \beta_1 - \lambda & \alpha_1 & & & \\ \alpha_1 & \beta_2 - \lambda & \alpha_2 & & \\ & \ddots & \ddots & \ddots & \\ & & \alpha_{r-1} & \beta_r - \lambda & \alpha_r \\ & & & \alpha_r & \beta_{r+1} - \lambda \end{bmatrix}. \tag{11.24}$$

On expanding by the last row we obtain

$$f_{r+1}(\lambda) = (\beta_{r+1} - \lambda) f_r(\lambda) - \alpha_r \det \begin{bmatrix} \beta_1 - \lambda & \alpha_1 & & & \\ \alpha_1 & \beta_2 - \lambda & \alpha_2 & & \mathbf{0} \\ & \ddots & \ddots & & \ddots \\ & \mathbf{0} & & \alpha_{r-2} & \beta_{r-1} - \lambda & 0 \\ & & & & \alpha_{r-1} & \alpha_r \end{bmatrix}.$$

and, on expanding the latter determinant by its last column, we obtain

$$f_{r+1}(\lambda) = (\beta_{r+1} - \lambda)f_r(\lambda) - \alpha_r^2 f_{r-1}(\lambda). \tag{11.25}$$

Note also that

$$f_n(\lambda) = \det(\mathbf{A} - \lambda\mathbf{I}). \tag{11.26}$$

In seeking zeros of f_n, we may use (11.25) recursively, as we will describe below. Let

$$\begin{aligned} f_0(\lambda) &= 1, \\ f_1(\lambda) &= (\beta_1 - \lambda)f_0(\lambda), \\ f_{r+1}(\lambda) &= (\beta_{r+1} - \lambda)f_r(\lambda) - \alpha_r^2 f_{r-1}(\lambda), \\ & \qquad r = 1, 2, \ldots, n-1. \end{aligned} \tag{11.27}$$

The values $f_0(\lambda), f_1(\lambda), \ldots, f_n(\lambda)$ form a *Sturm sequence.*

Theorem 11.3 Suppose there are $s(\lambda)$ changes of sign in the Sturm sequence

$$f_0(\lambda), f_1(\lambda), \ldots, f_n(\lambda)$$

(where, if $f_i(\lambda)$ is zero, it is deemed to have opposite sign to $f_{i-1}(\lambda)$), then there are precisely $s(\lambda)$ eigenvalues of $\mathbf{A}$ which are less than or equal to λ.

Proof Left as an exercise (Problem 11.14). Note that there cannot be two successive zeros in the Sturm sequence as $\alpha_i \neq 0$, $i = 1, 2, \ldots, n-1$. □

This result may be used to locate eigenvalues by bisection. Suppose that for some a, b (with $a < b$) we have $s(a) < s(b)$. Then there are $s(b) - s(a)$ eigenvalues in the interval $(a, b]$. Next we write $c = (a+b)/2$ and evaluate $s(c)$. There are $s(b) - s(c)$ eigenvalues in $(c, b]$ and $s(c) - s(a)$ eigenvalues in $(a, c]$. We can repeat this process, although convergence is slow. However, it could be used to locate eigenvalues which may then be further refined using inverse iteration.

Example 11.6 Locate all the eigenvalues of the matrix $\mathbf{A}$ in Example 11.3 to within ± 0.5 and refine the eigenvalue nearest to -1.5 to within ± 0.125 using the Sturm sequence and bisection process.

In this case

$$f_0(\lambda) = 1, f_1(\lambda) = (-2 - \lambda)f_0(\lambda)$$

and

$$f_{r+1}(\lambda) = (-2 - \lambda)f_r(\lambda) - f_{r-1}(\lambda), \qquad r = 1, 2, 3.$$

Table 11.3 Sturm sequences for Example 11.6

λ	−4	−3	−2	−1	0	−1.5	−1.25
f_0	1	1	1	1	1	1	1
f_1	2	1	0	−1	−2	−0.5	−0.75
f_2	3	0	−1	0	3	−0.75	−0.4375
f_3	4	−1	0	1	−4	0.875	1.078125
f_4	5	−1	1	−1	5	0.3125	−0.371094
$s(\lambda)$	0	1	2	3	4	2	3

From the first five columns of Table 11.3, $\lambda_1 \in (-4, -3)$, $\lambda_2 \in (-3, -2)$, $\lambda_3 \in (-2, -1)$ and $\lambda_4 \in (-1, 0)$. (Note that we can exclude endpoints of intervals since $f_4(\lambda)$ is always non-zero at the tabulated values.) We now try $\lambda = -1.5$ and deduce that $\lambda_3 \in (-1.5, -1)$. Then $\lambda = -1.25$ yields $\lambda_3 \in (-1.5, -1.25)$. □

One way of refining eigenvalues quickly is to use Newton's method. Since we seek roots of the equation $f_n(\lambda) = 0$, we use the iteration

$$\mu_{r+1} = \mu_r - f_n(\mu_r)/f'_n(\mu_r). \tag{11.28}$$

We can use the Sturm sequence property to obtain a good starting value μ_0. The function $f_n(\mu)$ is evaluated using the recurrence relation (11.27), which we can also differentiate (with respect to λ) to find f'_n. Thus

$$\begin{aligned} f'_0(\lambda) &= 0, \\ f'_1(\lambda) &= (\beta_1 - \lambda)f'_0(\lambda) - f_0(\lambda), \\ f'_{r+1}(\lambda) &= (\beta_{r+1} - \lambda)f'_r(\lambda) - \alpha_r^2 f'_{r-1}(\lambda) - f_r(\lambda), \\ & \qquad r = 1, 2, \dots, n-1. \end{aligned} \tag{11.29}$$

Of course the process is not satisfactory if we have a repeated eigenvalue as then the convergence is not second order. However, this can only happen if $\alpha_r = 0$ for some r (see Problem 11.15).

Example 11.7 We again consider the matrix of Example 11.3 and find the eigenvalue closest to -1.5. From the relations in Example 11.6, we obtain

$$f'_0 = 0,\ f'_1 = (-2-\lambda)f'_0 - f_0, \qquad f'_2 = (-2-\lambda)f'_1 - f'_0 - f_1,$$
$$f'_3 = (-2-\lambda)f'_2 - f'_1 - f_2, \qquad f'_4 = (-2-\lambda)f'_3 - f'_2 - f_3.$$

Since we know that $\lambda_3 \in (-1.5, -1.25)$ we will choose, as initial estimate, $\mu_0 = -1.375$. With $\lambda = \mu_0$ we find

$$\begin{aligned} f_0 &= 1 & f'_0 &= 0 \\ f_1 &= -0.625 & f'_1 &= -1 \\ f_2 &= -0.609375 & f'_2 &= 1.25 \\ f_3 &= 1.005859 & f'_3 &= 0.828125 \\ f_4 &= -0.019287 & f'_4 &= -2.773437. \end{aligned}$$

Thus

$$\mu_1 = \mu_0 - f_4(\mu_0)/f'_4(\mu_0) = -1.381954.$$

In fact $\lambda_3 = -1.381966$ and so μ_1 is a good approximation to λ_3. □

Having found an eigenvalue, it may seem tempting to try to find the corresponding eigenvector by solving

$$(\mathbf{A} - \lambda\mathbf{I})\mathbf{x} = \mathbf{0},$$

perhaps leaving out one equation to solve these singular equations. However, this is a very ill-conditioned process and should not be attempted. The recommended procedure is to use inverse iteration or the QR algorithm, which is described next.

11.4 The QR algorithm

The most powerful method of finding the eigenvalues of a symmetric tridiagonal matrix is the *QR algorithm*. Given a symmetric tridiagonal matrix $\mathbf{A}$, we factorize it as $\mathbf{A} = \mathbf{QR}$, where the matrix $\mathbf{Q}$ is orthogonal ($\mathbf{Q}^{-1} = \mathbf{Q}^{\mathrm{T}}$) and $\mathbf{R}$ is an upper triangular matrix. We then reverse the order of the factors and obtain a new matrix $\mathbf{B} = \mathbf{RQ}$. If we repeat this process we obtain rapid convergence to a matrix which is diagonal, except that there may be 2×2 submatrices on the diagonal. The eigenvalues of $\mathbf{A}$ are then simply the eigenvalues of those 2×2 submatrices together with the values of the remaining elements on the diagonal. (Thus, in the simplest case, where there are no 2×2 submatrices, the eigenvalues of $\mathbf{A}$ are just the diagonal elements.)

First note that the matrices $\mathbf{A}$ and $\mathbf{B}$ are similar, since

$$\mathbf{B} = \mathbf{RQ} \quad \text{and} \quad \mathbf{R} = \mathbf{Q}^{\mathrm{T}}\mathbf{A}$$

and so

$$\mathbf{B} = \mathbf{Q}^{\mathrm{T}}\mathbf{A}\mathbf{Q} \tag{11.30}$$

which is a similarity transformation. Hence $\mathbf{B}$ has the same eigenvalues as $\mathbf{A}$. We generate a sequence of matrices $\mathbf{A}_1 (=\mathbf{A}), \mathbf{A}_2, \mathbf{A}_3, \ldots$ factored as described above, such that

$$\mathbf{A}_i = \mathbf{Q}_i\mathbf{R}_i \tag{11.31}$$

$$\mathbf{A}_{i+1} = \mathbf{R}_i\mathbf{Q}_i = \mathbf{Q}_i^{\mathrm{T}}\mathbf{A}_i\mathbf{Q}_i, \qquad i = 1, 2, \ldots \tag{11.32}$$

and we find that $\mathbf{A}_i$ converges to a diagonal matrix as $i \to \infty$.

We shall see that, since $\mathbf{A}$ is symmetric tridiagonal, so is each $\mathbf{A}_i$. To find $\mathbf{Q}_i$ and thus $\mathbf{R}_i$ we use a sequence of rotation matrices.

Definition 11.1 For some fixed $i \neq j$ let $\mathbf{T}$ denote a matrix whose elements are those of the identity matrix $\mathbf{I}$ except for the four elements defined by

$$t_{ii} = t_{jj} = \cos\,\theta, \qquad t_{ij} = -t_{ji} = \sin\,\theta,$$

for some θ. $\mathbf{T}$ is called a *rotation matrix.* □

Note that a rotation matrix is orthogonal, that is, $\mathbf{T}\mathbf{T}^{\mathrm{T}} = \mathbf{I}$. It is easily seen (Problem 11.19) that $\mathbf{TA}$ differs from $\mathbf{A}$ only in the ith and jth rows. Likewise, $\mathbf{AT}$ differs from $\mathbf{A}$ only in the ith and jth columns. The factorization of $\mathbf{A}$ into $\mathbf{QR}$ is achieved by finding a sequence of rotation matrices $\mathbf{T}_1, \mathbf{T}_2, \ldots, \mathbf{T}_{n-1}$ so that $\mathbf{R}$ may be obtained from

$$\mathbf{R} = \mathbf{T}_{n-1}\mathbf{T}_{n-2}\ldots\mathbf{T}_1\mathbf{A} = \mathbf{Q}^{\mathrm{T}}\mathbf{A}, \tag{11.33}$$

where

$$\mathbf{Q} = (\mathbf{T}_{n-1}\mathbf{T}_{n-2}\ldots\mathbf{T}_1)^{\mathrm{T}} = \mathbf{T}_1^{\mathrm{T}}\mathbf{T}_2^{\mathrm{T}}\ldots\mathbf{T}_{n-1}^{\mathrm{T}}. \tag{11.34}$$

(The product of orthogonal matrices is orthogonal. See Problem 11.20.)

Suppose that at some stage we have reduced the symmetric tridiagonal matrix $\mathbf{A}$ to the form

$$\mathbf{C}_{k-1} = \mathbf{T}_{k-1}\mathbf{T}_{k-2}\ldots\mathbf{T}_1\mathbf{A}, \tag{11.35}$$

where

$$\mathbf{C}_{k-1} = \begin{bmatrix} p_1 & q_1 & r_1 & 0 & & & & & \\ 0 & \ddots & \ddots & \ddots & \ddots & & & & \\ 0 & \ddots & & & & & \mathbf{0} & & \\ & \ddots & p_{k-1} & q_{k-1} & r_{k-1} & 0 & & & \\ & & 0 & u_k & v_k & 0 & \ddots & & \\ & & 0 & \alpha_k & \beta_{k+1} & \alpha_{k+1} & \ddots & & \\ & & & \ddots & \ddots & \ddots & \ddots & & 0 \\ & & & & & & & & 0 \\ & \mathbf{0} & & & & & & & \alpha_{n-1} \\ & & & & & & 0 & \alpha_{n-1} & \beta_n \end{bmatrix}. \tag{11.36}$$

We now choose $\mathbf{T}_k$ so that the element α_k (just below the diagonal) is reduced to zero. We take $\mathbf{T}_k$ to be the rotation matrix

$$\mathbf{T}_k = \begin{bmatrix} \mathbf{I}_{k-1} & & & \mathbf{0} \\ & c_k & s_k & \\ & -s_k & c_k & \\ \mathbf{0} & & & \mathbf{I}_{n-k-1} \end{bmatrix}, \tag{11.37}$$

where the 2×2 matrix in the middle is in rows k and $k+1$ and columns k and $k+1$, with $c_k = \cos\theta_k$ and $s_k = \sin\theta_k$, for some θ_k. When we premultiply $\mathbf{C}_{k-1}$ by $\mathbf{T}_k$, only six elements of $\mathbf{C}_{k-1}$ are changed, namely those which are in rows k or $k+1$ *and* columns k, $k+1$ or $k+2$. In particular α_k (the element in row $k+1$ and column k) is replaced by $-s_k u_k + c_k\alpha_k$ and we choose θ_k so that

$$-s_k u_k + c_k \alpha_k = 0. \tag{11.38}$$

Since $s_k^2 + c_k^2 = 1$ we may choose

$$s_k = \frac{\alpha_k}{\sqrt{u_k^2 + \alpha_k^2}}, \qquad c_k = \frac{u_k}{\sqrt{u_k^2 + \alpha_k^2}}. \tag{11.39}$$

Thus α_k is replaced by zero and we need to follow up the changes made, on premultiplying $\mathbf{C}_{k-1}$ by $\mathbf{T}_k$, to the other five elements defined above. We repeat this process until we obtain

$$\mathbf{R} = \mathbf{C}_{n-1} = \mathbf{T}_{n-1}\mathbf{T}_{n-2}\ldots\mathbf{T}_1\mathbf{A}.$$

Note that, as $\mathbf{A}$ is transformed into $\mathbf{R}$, a third diagonal of non-zeros $r_1, r_2, \ldots, r_{n-2}$ appears in the upper triangular form at the same time as the diagonal with entries $\alpha_1, \alpha_2, \ldots, \alpha_{n-1}$ (below the principal diagonal) disappears, as these elements are replaced one-by-one by zeros.

We now form

$$\mathbf{B} = \mathbf{R}\mathbf{T}_1^{\mathrm{T}}\mathbf{T}_2^{\mathrm{T}}\ldots\mathbf{T}_{k-1}^{\mathrm{T}}. \tag{11.40}$$

Because of the nature of $\mathbf{T}_1, \mathbf{T}_2, \ldots, \mathbf{T}_{k-1}$, the matrix $\mathbf{B}$ has only one diagonal of non-zero elements below the principal diagonal. However, by (11.30), $\mathbf{B}$ is symmetric as $\mathbf{A}$ is symmetric and so $\mathbf{B}$ is tridiagonal. Thus in the repeated process defined by (11.31) and (11.32) each matrix $\mathbf{A}_i$ is tridiagonal.

We find that if each $\alpha_i \neq 0$ in (11.23) so that all the eigenvalues of $\mathbf{A}$ are distinct (Problem 11.15) and if all the eigenvalues are of distinct magnitude, the sequence of matrices $\mathbf{A}_1, \mathbf{A}_2, \mathbf{A}_3, \ldots$ converges to a diagonal matrix. The eigenvalues are then simply the diagonal entries. If we have a pair of eigenvalues of the same magnitude, for example if $\lambda_p = -\lambda_{p+1}$, then we obtain a diagonal matrix except that the diagonal elements λ_p, λ_{p+1} are replaced by a 2×2 submatrix

$$\begin{bmatrix} 0 & \lambda_p \\ \lambda_p & 0 \end{bmatrix}, \tag{11.41}$$

so again we can easily determine the corresponding eigenvalues. Note that the eigenvalues of (11.41) are λ_p and $-\lambda_p$. Finally, if one $\boldsymbol{\alpha}_i$ in (11.23) is zero then the eigenvalues are no longer necessarily distinct. However we can write

$$\mathbf{A} = \begin{bmatrix} \mathbf{A}_i' & 0 \\ 0 & \mathbf{A}_{n-i}' \end{bmatrix},$$

where $\mathbf{A}_i'$ and $\mathbf{A}_{n-i}'$ are tridiagonal matrices of order i and $n-i$ respectively. Since

$$\det(\mathbf{A} - \lambda\mathbf{I}) = \det(\mathbf{A}_i' - \lambda\mathbf{I}_i)\det(\mathbf{A}_{n-i}' - \lambda\mathbf{I}_{n-i})$$

the problem of finding the eigenvalues of $\mathbf{A}$ reduces to that of finding the eigenvalues of $\mathbf{A}_i'$ and $\mathbf{A}_{n-i}'$.

It can be shown that if the eigenvalues of $\mathbf{A}$ are of distinct magnitude, the rate of convergence of the off-diagonal elements to zero is given by

$$O\left(\left|\frac{\lambda_{j+1}}{\lambda_j}\right|^i\right), \tag{11.42}$$

where the eigenvalues of $\mathbf{A}$ are ordered $|\lambda_1| > |\lambda_2| > \cdots > |\lambda_n|$, and we wish these off-diagonal elements to converge to zero as rapidly as possible. To speed up convergence we introduce a shift, much as we did in inverse iteration in §11.2. We have, for some $p \in \mathbb{R}$,

$$\mathbf{A}_i - p\mathbf{I} = \mathbf{Q}_i\mathbf{R}_i \tag{11.43}$$

and

$$\mathbf{A}_{i+1} = \mathbf{R}_i\mathbf{Q}_i + p\mathbf{I}. \tag{11.44}$$

The rate of convergence of off-diagonal elements now becomes

$$O\left(\left|\frac{\lambda_{j+1} - p}{\lambda_j - p}\right|^i\right), \tag{11.45}$$

which gives much more rapid convergence to zero than (11.42) if p is close to λ_{j+1} but not close to λ_j.

The choice of p at each stage is important and there are various options. A good strategy is to look at the 2×2 matrix in the bottom right-hand corner of $\mathbf{A}_i$,

$$\begin{bmatrix} \beta_{n-1}^{(i)} & \alpha_{n-1}^{(i)} \\ \alpha_{n-1}^{(i)} & \beta_n^{(i)} \end{bmatrix}, \tag{11.46}$$

and take the eigenvalue of this matrix which is closest to $\beta_n^{(i)}$. If we do this then either $\alpha_{n-1}^{(i)} \to 0$, and we can pick out $\beta_n^{(i)}$ as an eigenvalue, or $\alpha_{n-2}^{(i)} \to 0$ and we are left with the 2×2 matrix (11.46) in the bottom right-hand corner whose two eigenvalues may be determined.

We now know one or two eigenvalues of $\mathbf{A}_i$ and, by omitting either the last row and column of $\mathbf{A}_i$ or the last two rows and columns, we continue with a smaller matrix to find other eigenvalues. We repeat the process, gradually reducing the order of the matrix until all the eigenvalues are determined.

Although the algorithm may appear complicated, it is a very effective method of finding all the eigenvalues of a symmetric tridiagonal matrix, as very few iterations are required per eigenvalue.

The QR algorithm also provides eigenvectors. If $\alpha_{n-1}^{(i)} \to 0$ in (11.46), we know that the eigenvector of $\mathbf{A}_i$ corresponding to the final value of $\beta_n^{(i)}$ (an eigenvalue) is of the form $[0 \; \ldots \; 0 \; 1]^{\mathrm{T}}$. If $\alpha_{n-2}^{(i)} \to 0$ we have eigenvectors $[0 \; \ldots \; 0 \; u \; v]^{\mathrm{T}}$, where $[u \; v]^{\mathrm{T}}$ are eigenvectors of the 2×2 submatrix (11.46). In both cases we know at least one eigenvector of $\mathbf{A}_i$ from which we can derive an eigenvector of the original matrix $\mathbf{A}$ via the orthogonal transformations used to derive $\mathbf{A}_i$. We have $\mathbf{x} = \mathbf{S}\mathbf{v}$, where $\mathbf{x}$ and $\mathbf{v}$ are eigenvectors of $\mathbf{A}$ and $\mathbf{A}_i$ respectively. The orthogonal matrix $\mathbf{S}$ is the product of all the orthogonal transformations used but, since the matrices decrease in dimension as the algorithm proceeds, some of these transformation matrices will need to be extended to the full dimensions of $\mathbf{A}$ by the addition of the appropriate rows and columns of the $n \times n$ unit matrix. Similarly, vectors need to be extended by the addition of extra zeros.

11.5 Reduction to tridiagonal form: Householder's method

So far we have seen how to solve the eigenproblem for symmetric tridiagonal matrices. We now consider a method of reducing a general symmetric matrix to tridiagonal form, using similarity transformations. We will then have a means of determining all the eigenvalues and eigenvectors of any symmetric matrix.

Given any $\mathbf{x}, \mathbf{y} \in \mathbb{R}^n$ with $\|\mathbf{x}\|_2 = \|\mathbf{y}\|_2$, we show how to construct an $n \times n$ orthogonal matrix $\mathbf{T}$ such that $\mathbf{y} = \mathbf{T}\mathbf{x}$. (Note that orthogonal transformations do not change the Euclidean norm of a vector. See Problem 10.11.) We will assume that $\mathbf{y} \neq \mathbf{x}$, since for the case $\mathbf{y} = \mathbf{x}$ we simply take $\mathbf{T} = \mathbf{I}$. Without loss of generality, we can multiply $\mathbf{x}$ and $\mathbf{y}$ by a common scalar so that $\|\mathbf{y} - \mathbf{x}\|_2 = 1$. Then a convenient choice of $\mathbf{T}$ is the *Householder transformation*

$$\mathbf{T} = \mathbf{I} - 2\mathbf{w}\mathbf{w}^{\mathrm{T}}, \qquad \text{where } \mathbf{w} = \mathbf{y} - \mathbf{x}. \tag{11.47}$$

This is named after A. S. Householder (1904–1993).

Note that $\| \mathbf{w} \|_2 = 1$. The matrix $\mathbf{T}$ is symmetric and orthogonal, since

$$\mathbf{T}^{\mathrm{T}} = (\mathbf{I} - 2\mathbf{w}\mathbf{w}^{\mathrm{T}})^{\mathrm{T}} = \mathbf{I} - 2\mathbf{w}\mathbf{w}^{\mathrm{T}} = \mathbf{T}$$

and

$$\mathbf{T}^{\mathrm{T}}\mathbf{T} = \mathbf{T}^2 = \mathbf{I} - 4\mathbf{w}\mathbf{w}^{\mathrm{T}} + (2\mathbf{w}\mathbf{w}^{\mathrm{T}})(2\mathbf{w}\mathbf{w}^{\mathrm{T}}) = \mathbf{I}$$

since $\mathbf{w}^{\mathrm{T}}\mathbf{w} = \| \mathbf{w} \|_2^2 = 1$. To verify that $\mathbf{y} = \mathbf{T}\mathbf{x}$, we write

$$\begin{aligned}\mathbf{T}\mathbf{x} - \mathbf{y} &= \mathbf{x} - 2\mathbf{w}\mathbf{w}^{\mathrm{T}}\mathbf{x} - \mathbf{y} \\ &= (\mathbf{x} - \mathbf{y}) - 2(\mathbf{y} - \mathbf{x})(\mathbf{y}^{\mathrm{T}} - \mathbf{x}^{\mathrm{T}})\mathbf{x} \\ &= (\mathbf{x} - \mathbf{y})\{(\mathbf{y}^{\mathrm{T}} - \mathbf{x}^{\mathrm{T}})(\mathbf{y} - \mathbf{x}) + 2(\mathbf{y}^{\mathrm{T}} - \mathbf{x}^{\mathrm{T}})\mathbf{x}\}\end{aligned}$$

since, as noted above, $(\mathbf{y}^{\mathrm{T}} - \mathbf{x}^{\mathrm{T}})(\mathbf{y} - \mathbf{x}) = \mathbf{w}^{\mathrm{T}}\mathbf{w} = 1$. Hence

$$\begin{aligned}\mathbf{T}\mathbf{x} - \mathbf{y} &= (\mathbf{x} - \mathbf{y})(\mathbf{y}^{\mathrm{T}} - \mathbf{x}^{\mathrm{T}})(\mathbf{y} - \mathbf{x} + 2\mathbf{x}) \\ &= (\mathbf{x} - \mathbf{y})(\mathbf{y}^{\mathrm{T}}\mathbf{y} - \mathbf{x}^{\mathrm{T}}\mathbf{x} + \mathbf{y}^{\mathrm{T}}\mathbf{x} - \mathbf{x}^{\mathrm{T}}\mathbf{y}) \\ &= \mathbf{0}\end{aligned}$$

as $\mathbf{y}^{\mathrm{T}}\mathbf{y} = \| \mathbf{y} \|_2^2 = \| \mathbf{x} \|_2^2 = \mathbf{x}^{\mathrm{T}}\mathbf{x}$ and $\mathbf{y}^{\mathrm{T}}\mathbf{x} = \mathbf{x}^{\mathrm{T}}\mathbf{y} = \sum_{i=1}^{n} x_i y_i$. Thus $\mathbf{y} = \mathbf{T}\mathbf{x}$, as required.

In practice, instead of scaling $\mathbf{x}$ and $\mathbf{y}$ so that $\| \mathbf{y} - \mathbf{x} \|_2 = 1$, we leave $\mathbf{x} \neq \mathbf{y}$ unscaled and choose

$$\mathbf{w} = \frac{1}{\| \mathbf{y} - \mathbf{x} \|_2} (\mathbf{y} - \mathbf{x})$$

in the Householder transformation $\mathbf{T} = \mathbf{I} - 2\mathbf{w}\mathbf{w}^{\mathrm{T}}$. To apply the above result we will choose $\mathbf{y}$ such that

$$y_2 = y_3 = \cdots = y_n = 0 \tag{11.48}$$

and thus, as $\| \mathbf{y} \|_2 = \| \mathbf{x} \|_2$ we must have

$$y_1 = \pm \| \mathbf{x} \|_2. \tag{11.49}$$

Since we have a choice, we usually take the sign of y_1 to be opposite to that of x_1. This may reduce the effect of rounding error in calculating $\mathbf{w}$.

Example 11.8 With $\mathbf{x} = [2 \ \ -2 \ \ 4 \ \ -1]^{\mathrm{T}}$, find $\mathbf{y}$ of the form (11.48) and an orthogonal $\mathbf{T}$ such that $\mathbf{y} = \mathbf{T}\mathbf{x}$.

Now $\| \mathbf{x} \|_2 = \sqrt{(4 + 4 + 16 + 1)} = 5$ and so $y_1 = \pm 5$. Although the choice is not important in this case, we take $y_1 = -5$ (with opposite sign to x_1). Hence

$$\begin{aligned}\mathbf{y} &= [-5 \ \ 0 \ \ 0 \ \ 0]^{\mathrm{T}}, \\ \mathbf{w} &= \frac{1}{\| \mathbf{y} - \mathbf{x} \|_2} \cdot (\mathbf{y} - \mathbf{x}) = \frac{1}{\sqrt{70}} [-7 \ \ 2 \ \ -4 \ \ 1]^{\mathrm{T}},\end{aligned}$$

and so

$$\mathbf{T} = \mathbf{I} - 2\mathbf{w}\mathbf{w}^{\mathrm{T}} = \mathbf{I} - \frac{2}{70}\begin{bmatrix} -7 \\ 2 \\ -4 \\ 1 \end{bmatrix}[-7 \ 2 \ -4 \ 1]$$

$$= \frac{1}{35}\begin{bmatrix} -14 & 14 & -28 & 7 \\ 14 & 31 & 8 & -2 \\ -28 & 8 & 19 & 4 \\ 7 & -2 & 4 & 34 \end{bmatrix}.$$

The reader should check that $\mathbf{y} = \mathbf{Tx}$ and $\mathbf{T}^{\mathrm{T}}\mathbf{T} = \mathbf{T}^2 = \mathbf{I}$. □

We can use a set of Householder transformations to reduce a symmetric matrix to tridiagonal form. Suppose $\mathbf{A}$ is $n \times n$ symmetric and let

$$\mathbf{c} = [a_{12} \ a_{13} \ \dots \ a_{1n}]^{\mathrm{T}} \in \mathbb{R}^{n-1}.$$

We start by finding the $(n-1) \times (n-1)$ Householder transformation $\mathbf{T}$ such that

$$\mathbf{Tc} = [\times \ 0 \ \dots \ 0]^{\mathrm{T}}.$$

Let

$$\mathbf{S} = \begin{bmatrix} 1 & \mathbf{0}^{\mathrm{T}} \\ \mathbf{0} & \mathbf{T} \end{bmatrix}$$

which is $n \times n$ symmetric and orthogonal. Now

$$\mathbf{S}^{\mathrm{T}}\mathbf{A}\mathbf{S} = \begin{bmatrix} 1 & \mathbf{0}^{\mathrm{T}} \\ \mathbf{0} & \mathbf{T} \end{bmatrix}\begin{bmatrix} a_{11} & \mathbf{c}^{\mathrm{T}} \\ \mathbf{c} & \mathbf{B} \end{bmatrix}\begin{bmatrix} 1 & \mathbf{0}^{\mathrm{T}} \\ \mathbf{0} & \mathbf{T} \end{bmatrix}$$

$$= \begin{bmatrix} a_{11} & \mathbf{c}^{\mathrm{T}}\mathbf{T} \\ \mathbf{Tc} & \mathbf{TBT} \end{bmatrix}$$

$$= \begin{bmatrix} \times & \times & 0 & \dots & 0 \\ \times & \times & \times & \dots & \times \\ 0 & \times & \times & \dots & \times \\ \vdots & \vdots & \vdots & & \vdots \\ 0 & \times & \times & \dots & \times \end{bmatrix},$$

where $\mathbf{B}$ is the $(n-1) \times (n-1)$ submatrix of $\mathbf{A}$, consisting of its last $n-1$ rows and columns. The symbol $\times$ is used in the last matrix and in the vector $\mathbf{Tc}$ above to denote a general (non-zero or zero) element.

We repeat the process, using transformations of the form

$$\mathbf{S}^{(r)} = \begin{bmatrix} \mathbf{I}_r & \mathbf{0}^{\mathrm{T}} \\ \mathbf{0} & \mathbf{T}^{(r)} \end{bmatrix}$$

where $\mathbf{I}_r$ is the $r \times r$ unit matrix and $\mathbf{T}^{(r)}$ is an $(n-r) \times (n-r)$ Householder transformation. We have

$$\mathbf{A}^{(r+1)} = \mathbf{S}^{(r)}\mathbf{A}^{(r)}\mathbf{S}^{(r)}$$

where

$$\mathbf{A}^{(r)} = \left[\begin{array}{cccc|cccc} & & & & 0 & 0 & \dots & 0 \\ & \mathbf{C}^{(r)} & & & \vdots & \vdots & & \vdots \\ & & & & 0 & 0 & \dots & 0 \\ & & & & \times & 0 & \dots & 0 \\ 0 & \dots & 0 & \times & & & & \\ 0 & \dots & 0 & 0 & & \mathbf{B}^{(r)} & & \\ \vdots & & \vdots & \vdots & & & & \\ 0 & \dots & 0 & 0 & & & & \end{array}\right],$$

$\mathbf{C}^{(r)}$ is $r \times r$ tridiagonal and $\mathbf{B}^{(r)}$ is $(n-r)\times(n-r)$ symmetric. Note that $\mathbf{T}^{(r)}$ reduces the first column of $\mathbf{B}^{(r)}$ to the form $[\times \;\; \times \;\; 0 \;\; \dots \;\; 0]^{\mathrm{T}}$. We begin with $\mathbf{A}^{(0)} = \mathbf{A}$ and repeat the above process for $r = 0, 1, \dots, n-3$, so that $n-2$ transformations are needed.

Example 11.9 Find the first Householder transformation in the reduction of a matrix of the form

$$\mathbf{A} = \begin{bmatrix} 1 & 2 & -2 & 4 & -1 \\ 2 & \times & \times & \times & \times \\ -2 & \times & \times & \times & \times \\ 4 & \times & \times & \times & \times \\ -1 & \times & \times & \times & \times \end{bmatrix}.$$

In this case $\mathbf{c}^{\mathrm{T}} = [2 \;\; -2 \;\; 4 \;\; -1]$, which is the vector of Example 11.8. Thus the first transformation is

$$\mathbf{S}^{(1)} = \frac{1}{35}\begin{bmatrix} 35 & 0 & 0 & 0 & 0 \\ 0 & -14 & 14 & -28 & 7 \\ 0 & 14 & 31 & 8 & -2 \\ 0 & -28 & 8 & 19 & 4 \\ 0 & 7 & -2 & 4 & 34 \end{bmatrix}.$$

□

If the eigenvectors of a symmetric matrix **A** are required, we should keep a record of all the orthogonal transformations used. The eigenvectors are then calculated from the final matrix, as described at the end of §11.4.

The final recommended algorithm for finding the eigenvalues and eigenvectors of a symmetric matrix is first to reduce the matrix to symmetric tridiagonal form using Householder transformations and secondly use the QR algorithm. If only one or two eigenvalues are required, it is probably best to use Newton's method in the second part of the process. In both cases the Sturm sequence may be used to locate eigenvalues of the tridiagonal matrix.

For general matrices (non-symmetric) the calculation of eigenvalues and eigenvectors is much more complicated since, in general, these are complex even when the original matrix is real. One method is based on reduction to *upper Hessenberg* form, in which all (i, j)th elements with $j < i - 1$ are zero. We no longer have the symmetry of zeros as in a tridiagonal matrix. Detailed discussion of the method is beyond the scope of this text. (See, for example, Gourlay and Watson, 1973.)

Problems

Section 11.1

11.1 Use Gerschgorin's theorems to locate as far as possible the (real) eigenvalues of the symmetric matrices

$$\begin{bmatrix} 2 & 1 & 0 \\ 1 & 2 & 1 \\ 0 & 1 & 2 \end{bmatrix}, \quad \begin{bmatrix} 6 & 1 & 1 \\ 1 & 4 & 1 \\ 1 & 1 & -5 \end{bmatrix}.$$

11.2 Use Gerschgorin's theorems on the following matrices and their transposes to locate their eigenvalues in the complex plane as far as possible.

$$\begin{bmatrix} 6 & 1 & 0 \\ 2 & 1 & 4 \\ 0 & 1 & -7 \end{bmatrix}, \quad \begin{bmatrix} 4 & 1 & 1 \\ 1 & 5 & 0 \\ 2 & 1 & 10 \end{bmatrix}.$$

11.3 Show, by using Gerschgorin's theorems, that the symmetric matrix

$$\mathbf{A} = \begin{bmatrix} -6 & 1 & 1 \\ 1 & 2 & -1 \\ 1 & -1 & 7 \end{bmatrix}$$

has an eigenvalue $\lambda_1 \in [5, 9]$. By finding $\mathbf{D}_\alpha \mathbf{A} \mathbf{D}_\alpha^{-1}$, where

$$\mathbf{D}_\alpha = \begin{bmatrix} 1 & 0 & 0 \\ 0 & 1 & 0 \\ 0 & 0 & \alpha \end{bmatrix},$$

and taking a suitable value of α, show that $\lambda_1 \in [6.4, 7.6]$.

11.4 Given

$$\mathbf{A} = \begin{bmatrix} 1 & 0.01 & -0.01 \\ 0.01 & -2 & 0.02 \\ -0.01 & 0.02 & 3 \end{bmatrix},$$

show by means of Gerschgorin's theorems that there is an eigenvalue $\lambda_2 = -2$ to within ± 0.03. By finding $\mathbf{D}_\alpha \mathbf{A} \mathbf{D}_\alpha^{-1}$ for a suitable diagonal matrix $\mathbf{D}_\alpha$, show that $\lambda_2 = -2$ to within ± 0.00013.

Section 11.2

11.5 Use the simple iteration Algorithm 11.1 with the matrix

$$\mathbf{A} = \begin{bmatrix} 4 & 1 & 1 \\ 1 & 5 & 2 \\ 1 & 2 & 6 \end{bmatrix}.$$

Start with the vector $\mathbf{y}_0 = [1 \;\; 1 \;\; 1]^T$, carry out 10 iterations and show that the dominant eigenvalue is near 8.05, using (11.13).

11.6 Show that if in (11.10) $c_1 = 0$, $c_2 \neq 0$ and the eigenvalues are such that $|\lambda_1| > |\lambda_2| > |\lambda_3| \geqslant \cdots \geqslant |\lambda_n|$, then

$$\mathbf{y}_r \sim m_r \lambda_2^r c_2 \mathbf{x}_2$$

in place of (11.12), where m_r is a scaling factor. Thus the simple iteration Algorithm 11.1 would yield the subdominant eigenvalue λ_2 and the corresponding eigenvector $\mathbf{x}_2$.

11.7 Try the simple iteration Algorithm 11.1 with the matrix of Example 11.3, with $\mathbf{y}_0 = [1 \;\; 1 \;\; 1 \;\; 1]^T$. Carry out five iterations without normalizing, taking $\mathbf{y}_{r+1} = \overset{*}{\mathbf{y}}_{r+1} = \mathbf{A}\mathbf{y}_r$. (This avoids any rounding errors.) Why do the ratios given by (11.13) fail to converge to the dominant eigenvalue? (*Hint*: use the result of Problem 11.6.)

11.8 What goes wrong and why, if we use the simple iteration Algorithm 11.1 on the matrix

$$\mathbf{A} = \begin{bmatrix} 3 & -2 & -1 \\ 1 & -2 & 1 \\ 1 & -1 & 0 \end{bmatrix},$$

starting with $\mathbf{y}_0 = [1 \;\; 1 \;\; 1]^T$? Try the algorithm again with $\mathbf{y}_0 = [1 \;\; -1 \;\; 1]^T$.

11.9 Repeat Problem 11.5 but use Rayleigh quotients (11.16) to estimate the dominant eigenvalue.

11.10 A symmetric matrix $\mathbf{A}$ has dominant eigenvalue λ_1 and corresponding eigenvector $\mathbf{x}_1$. Show that the matrix

$$\mathbf{B} = \mathbf{A} - \lambda_1 \mathbf{x}_1 \mathbf{x}_1^{\mathrm{T}}$$

has the same eigenvalues as $\mathbf{A}$ except that λ_1 is replaced by zero. Show also that if $\mathbf{x}_1, \mathbf{x}_2, \ldots, \mathbf{x}_n$ are *orthonormal* eigenvectors of $\mathbf{B}$ (that is, they are orthogonal and $\mathbf{x}_i^{\mathrm{T}}\mathbf{x}_i = 1$, $i = 1, 2, \ldots, n$) they are also eigenvectors of $\mathbf{A}$.

11.11 Show that if λ is an eigenvalue and $\mathbf{x}$ is a corresponding eigenvector of a matrix $\mathbf{A}$ and $\mathbf{A} - p\mathbf{I}$ is non-singular for a given $p \in \mathbb{R}$, then $(\lambda - p)^{-1}$ is an eigenvalue of $(\mathbf{A} - p\mathbf{I})^{-1}$ with corresponding eigenvector $\mathbf{x}$.

11.12 Use inverse iteration (11.18) with $p = 8$ for each iteration for the matrix of Problem 11.5. Start with $\mathbf{y}_0 = [1 \;\; 1 \;\; 1]^{\mathrm{T}}$ and use Rayleigh quotients to estimate the eigenvalue. Compare your results with those in Problem 11.9.

11.13 Use the inverse iteration method (11.19)–(11.22) on the matrix

$$\mathbf{A} = \begin{bmatrix} 5 & 2 & 0 \\ 2 & 4 & 1 \\ 0 & 1 & 3 \end{bmatrix},$$

starting with $\mathbf{y}_0 = [4 \;\; 3 \;\; 1]^{\mathrm{T}}$ and taking $p = 7$.

Section 11.3

11.14 A Sturm sequence $f_0(\lambda), f_1(\lambda), \ldots$ is defined by (11.27), where each $\alpha_r \neq 0$.

(i) Show, using a diagram, that the zero of $f_1(\lambda)$ lies strictly between the two zeros of $f_2(\lambda)$.

(ii) Using (i) and induction, prove that the zeros of $f_{r-1}(\lambda)$ strictly separate those of $f_r(\lambda)$. (*Hint*: consider the recurrence relation (11.27) involving f_r, f_{r-1} and f_{r-2} and deduce that the sign of $f_r(\lambda)$ is opposite to that of $f_{r-2}(\lambda)$ at each zero of $f_{r-1}(\lambda)$.) Also show by induction that, for $r \geq 1$, $f_r(\lambda) \to \infty$ as $\lambda \to -\infty$ and $f_r(\lambda) \to (-1)^r\infty$ as $\lambda \to \infty$, so that $f_r(\lambda)$ has two zeros 'outside' the zeros of $f_{r-1}(\lambda)$.

(iii) By using a diagram of $f_1(\lambda), \ldots, f_4(\lambda)$, prove Theorem 11.3 for the case $n = 4$. Indicate how you would extend this proof to the general case.

11.15 Show that if each $\alpha_r \neq 0$ then the eigenvalues of (11.23) are distinct. (*Hint*: use the method of Problem 11.14.)

11.16 Use the Sturm sequence property and bisection to estimate the three eigenvalues of the matrix of Problem 11.13 to within ± 0.5. Use Newton's method (11.28) to refine the smallest eigenvalue further, to within $\pm 10^{-3}$.

11.17 From Table 11.3 we deduce that there is one eigenvalue of the matrix in Example 11.3 within the interval $[-4, -3]$. Use Newton's method (11.28) to determine this eigenvalue to six decimal places, starting with $\mu_0 = -3.5$.

11.18 Use the Sturm sequence property to deduce that the matrix

$$\begin{bmatrix} -3 & 1 & 0 & 0 \\ 1 & -3 & 2 & 0 \\ 0 & 2 & -4 & 2 \\ 0 & 0 & 2 & -5 \end{bmatrix}$$

has one eigenvalue in $(-8, -6)$. Use Newton's method (11.28) to find this eigenvalue to six decimal places, starting with $\mu_0 = -7$.

Section 11.4

11.19 $\mathbf{T}$ is an $n \times n$ rotation matrix, as given by Definition 11.1. Prove that if $\mathbf{A}$ is any $n \times n$ matrix then $\mathbf{TA}$ differs from $\mathbf{A}$ only in the ith and jth rows. The ith row of $\mathbf{TA}$ consists of the elements

$$a_{ik} \cos\theta + a_{jk} \sin\theta \qquad k = 1, 2, \ldots, n,$$

and the jth row has elements

$$-a_{ik} \sin\theta + a_{jk} \cos\theta \qquad k = 1, 2, \ldots, n.$$

Prove a similar result for the columns of $\mathbf{AT}$.

11.20 Prove that the product of two orthogonal matrices is orthogonal.

Section 11.5

11.21 Find the Householder transformation $\mathbf{T}$ which reduces $\mathbf{x} = [1 \;\; 2 \;\; -2 \;\; -4]^{\mathrm{T}}$ to the form $\mathbf{y} = [y_1 \;\; 0 \;\; 0 \;\; 0]^{\mathrm{T}}$. Check that $\mathbf{y} = \mathbf{Tx}$ and that $\mathbf{T}^2 = \mathbf{I}$.

11.22 Reduce the following symmetric matrix to tridiagonal form using Householder's method.

$$\mathbf{A} = \begin{bmatrix} 5 & 3 & 4 \\ 3 & 2 & 1 \\ 4 & 1 & 2 \end{bmatrix}.$$

Chapter 12

SYSTEMS OF NON-LINEAR EQUATIONS

12.1 Contraction mapping theorem

We now consider a system of algebraic equations of the form

$$\mathbf{f}(\mathbf{x}) = \mathbf{0}, \tag{12.1}$$

where f is a mapping from $\mathbb{R}^n$ to $\mathbb{R}^m$. Thus $\mathbf{x} \in \mathbb{R}^n$, while $\mathbf{f}(\mathbf{x})$ and $\mathbf{0} \in \mathbb{R}^m$. The equations (12.1) may be written in full as

$$\begin{aligned} f_1(x_1, \ldots, x_n) &= 0 \\ f_2(x_1, \ldots, x_n) &= 0 \\ &\vdots \\ f_m(x_1, \ldots, x_n) &= 0. \end{aligned} \tag{12.2}$$

A special case of (12.1) is the system of linear equations

$$\mathbf{Ax} - \mathbf{b} = \mathbf{0}, \tag{12.3}$$

where $\mathbf{A}$ is an $m \times n$ matrix (see Chapters 9 and 10). As an example of a system of *non-linear* equations, we have

$$x_1 + 2x_2 - 3 = 0 \tag{12.4a}$$

$$2x_1^2 + x_2^2 - 5 = 0 \tag{12.4b}$$

which is of the form (12.2) with $m = n = 2$. Geometrically, the solutions of (12.4) are the points in the x_1x_2-plane where the straight line with equation $x_1 + 2x_2 - 3 = 0$ cuts the ellipse with equation $2x_1^2 + x_2^2 - 5 = 0$. We see from a graph that the system (12.4) has two solutions, corresponding to the points A and B of Fig. 12.1. However, there is no general rule which tells us how many solutions to expect. A general system of equations (12.2) may have no solution, an infinite number of solutions or any finite number of

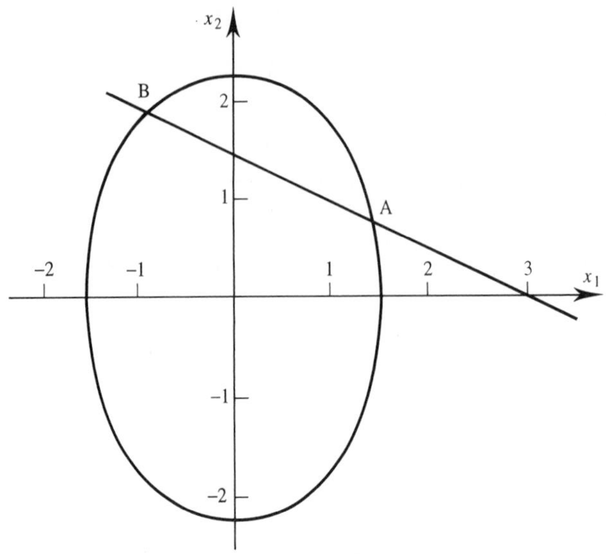

Fig. 12.1 The straight line and ellipse have equations (12.4a) and (12.4b) respectively.

solutions. For example, the system of two equations

$$x_1 - 1 = 0$$

$$(x_2 - 1)(x_2 - 2) \dots (x_2 - N) = 0$$

has exactly N solutions.

We restrict our attention to the case where $m = n$ in (12.2), so that the number of equations is the same as the number of unknowns, and write (12.1) in the form

$$\mathbf{x} = \mathbf{g}(\mathbf{x}). \tag{12.5}$$

This is an extension of the one variable form $x = g(x)$, which we used (see § 8.4) in discussing iterative methods for a single equation. Notice that (12.5) is also a generalization of the equation

$$\mathbf{x} = \mathbf{M}\mathbf{x} + \mathbf{d}$$

which we used (see §10.5 and §10.6) in solving *linear* equations by iterative methods.

Given the non-linear equations (12.5), we construct a sequence of vectors $(\mathbf{x}_m)_{m=0}^{\infty}$ from

$$\mathbf{x}_{r+1} = \mathbf{g}(\mathbf{x}_r),$$

$r = 0, 1, \dots,$ beginning with some initial vector $\mathbf{x}_0$. We are interested in whether the sequence $(\mathbf{x}_m)_{m=0}^{\infty}$ converges to a vector $\boldsymbol{\alpha}$ which is a solution of the equations (12.5). We recall from §10.2 that a sequence of vectors

$(\mathbf{x}_m)_{m=0}^{\infty}$ is said to converge to $\boldsymbol{\alpha}$ if

$$\lim_{m\to\infty} \mathbf{x}_m = \boldsymbol{\alpha}.$$

In Lemma 10.1 we saw that this is equivalent to the condition

$$\lim_{m\to\infty} \|\mathbf{x}_m - \boldsymbol{\alpha}\| = 0$$

for *any* vector norm, including the norms $\|\cdot\|_p$, $1 \leq p \leq \infty$.

Definition 12.1 A region R in n-dimensional space is said to be *closed* if every convergent sequence $(\mathbf{x}_m)$, with each $\mathbf{x}_m \in R$, is such that its limit $\boldsymbol{\alpha} \in R$. □

For example, in two-dimensional space with coordinates x_1, x_2, the region

$$a \leq x_1 \leq b, \qquad c \leq x_2 \leq d$$

is closed (see Problem 12.3). The region

$$a \leq x_1 \leq b, \qquad -\infty < x_2 < \infty$$

is also closed. But

$$a \leq x_1 < b, \qquad c \leq x_2 \leq d$$

is not closed as we can have sequences whose first coordinates converge to b.

Definition 12.2 A function $\mathbf{g}$, which maps real n-dimensional vectors into real n-dimensional vectors, is said to be a *contraction mapping* with respect to a norm $\|\cdot\|$ on a closed region R if

(i) $\mathbf{x} \in R \Rightarrow \mathbf{g}(\mathbf{x}) \in R$, (12.6)

(ii) $\|\mathbf{g}(\mathbf{x}) - \mathbf{g}(\mathbf{x}')\| \leq L\|\mathbf{x} - \mathbf{x}'\|$, with $0 \leq L < 1$, (12.7)

for all $\mathbf{x}$, $\mathbf{x}' \in R$. □

We have extended the closure and Lipschitz conditions to n variables and thus extended the notion of contraction mapping from Definition 8.1. We can now generalize Theorem 8.3.

Theorem 12.1 If there is a closed region $R \subset \mathbb{R}^n$ on which $\mathbf{g}$ is a contraction mapping with respect to some norm $\|\cdot\|$, then

(i) the equation $\mathbf{x} = \mathbf{g}(\mathbf{x})$ has a unique solution (say $\boldsymbol{\alpha}$) belonging to R,

(ii) for any $\mathbf{x}_0 \in R$, the sequence $(\mathbf{x}_r)$ defined by

$$\mathbf{x}_{r+1} = \mathbf{g}(\mathbf{x}_r), \qquad r = 0, 1, \ldots,$$

converges to $\boldsymbol{\alpha}$.

Proof. The only essential difference between this proof and that of Theorem 8.3 is that it is slightly more difficult, when $n > 1$, to show the *existence* of a solution. Otherwise, the proof is the same except that we write $\|\cdot\|$ in place of $|\cdot|$ throughout.

If $\mathbf{x}_0 \in R$, it follows from the closure condition that $\mathbf{x}_r \in R$, $r=0,1,\ldots$. Then, as in Problems 8.27 and 8.28 we obtain the inequality

$$\|\mathbf{x}_{m+k}-\mathbf{x}_m\| \leqslant \frac{L^m}{1-L}\|\mathbf{x}_1-\mathbf{x}_0\|, \tag{12.8}$$

for any fixed $m \geqslant 0$ and all $k=0,1,\ldots$. Thus, for any k, we have

$$\lim_{m\to\infty}\|\mathbf{x}_{m+k}-\mathbf{x}_m\|=0$$

and, from Lemma 10.1, this implies that the sequence of vectors $(\mathbf{x}_{m+k}-\mathbf{x}_m)_{m=0}^{\infty}$ converges to the zero vector. If $x_{m,i}$ denotes the ith element of the vector $\mathbf{x}_m$, it follows that, for each $i=1,2,\ldots,n$ the numbers $|x_{m+k,i}-x_{m,i}|$ may be made arbitrarily small *for all* $k\geqslant 0$ by choosing m sufficiently large. Thus, for any i, it follows from Cauchy's principle for convergence (see Theorem 2.8) that the sequence $x_{0,i}, x_{1,i}, x_{2,i}, \ldots$ converges to some number, say α_1. Therefore $(\mathbf{x}_m)$ converges to a vector $\boldsymbol{\alpha}$ and, since R is closed, $\boldsymbol{\alpha}\in R$. To show that $\boldsymbol{\alpha}$ is a solution of $\mathbf{x}=\mathbf{g}(\mathbf{x})$, we write

$$\boldsymbol{\alpha}-\mathbf{g}(\boldsymbol{\alpha})=\boldsymbol{\alpha}-\mathbf{x}_{m+1}+\mathbf{g}(\mathbf{x}_m)-\mathbf{g}(\boldsymbol{\alpha}),$$

for any $m\geqslant 0$. Thus

$$\begin{aligned}\|\boldsymbol{\alpha}-\mathbf{g}(\boldsymbol{\alpha})\| &\leqslant \|\boldsymbol{\alpha}-\mathbf{x}_{m+1}\| + \|\mathbf{g}(\mathbf{x}_m)-\mathbf{g}(\boldsymbol{\alpha})\| \\ &\leqslant \|\boldsymbol{\alpha}-\mathbf{x}_{m+1}\| + L\|\mathbf{x}_m-\boldsymbol{\alpha}\|.\end{aligned}$$

Since both terms on the right may be made arbitrarily small by choosing m sufficiently large, it follows that $\|\boldsymbol{\alpha}-\mathbf{g}(\boldsymbol{\alpha})\|=0$. Thus $\boldsymbol{\alpha}-\mathbf{g}(\boldsymbol{\alpha})=\mathbf{0}$ and $\boldsymbol{\alpha}$ is a solution of $\mathbf{x}=\mathbf{g}(\mathbf{x})$. The uniqueness is shown as in Theorem 8.3. □

We now suppose that on some region R, $\partial g_i/\partial x_j$, $1\leqslant i,j\leqslant n$, are continuous, and write $g_i(\mathbf{x})$ to denote $g_i(x_1,\ldots,x_n)$. Then by Taylor's theorem in several variables (Theorem 3.2) we have

$$g_i(\mathbf{x})-g_i(\mathbf{x}')=(x_1-x_1')\frac{\partial g_i(\boldsymbol{\xi})}{\partial x_1}+\cdots+(x_n-x_n')\frac{\partial g_i(\boldsymbol{\xi})}{\partial x_n}, \tag{12.9}$$

where $\mathbf{x}, \mathbf{x}' \in R$ and $\boldsymbol{\xi}$ is 'between' $\mathbf{x}$ and $\mathbf{x}'$, that is $\boldsymbol{\xi}=(1-\theta)\mathbf{x}+\theta\mathbf{x}'$ for some $\theta\in(0,1)$. We shall assume that $\boldsymbol{\xi}\in R$, which will be so if the line segment joining any two points of R is also in R. (A region satisfying this property is said to be *convex*.) If $\mathbf{G}$ denotes the $n\times n$ matrix whose (i,j)th element is

$$g_{ij}=\sup_{\mathbf{x}\in R}\left|\frac{\partial g_i(\mathbf{x})}{\partial x_j}\right|, \tag{12.10}$$

it follows that

$$|\mathbf{g}(\mathbf{x})-\mathbf{g}(\mathbf{x}')|\leqslant \mathbf{G}|\mathbf{x}-\mathbf{x}'|, \tag{12.11}$$

where, as in § 9.10, the inequality between the two vectors on each side of (12.11) holds element by element and $|\mathbf{u}|$ denotes a vector whose elements are the absolute values of the elements of the vector $\mathbf{u}$. Since for $1 \leqslant p \leqslant \infty$ the p-norms of $|\mathbf{u}|$ and $\mathbf{u}$ are equal, we deduce from (12.11) that

$$\|\mathbf{g}(\mathbf{x}) - \mathbf{g}(\mathbf{x}')\|_p \leqslant \|\mathbf{G}\|_p \|\mathbf{x} - \mathbf{x}'\|_p.$$

With the choice $p = 1$ or $p = \infty$, we can take the Lipschitz constant L as $\|\mathbf{G}\|_1$ or $\|\mathbf{G}\|_\infty$ respectively. As we saw in § 10.3 both of these matrix norms are easily evaluated. In practice, the only other norm we might wish to use is $\|\cdot\|_2$ and, as remarked in Chapter 10, the corresponding matrix norm is not easily evaluated. However, we can write (see Problem 12.4)

$$\|\mathbf{G}(\mathbf{x} - \mathbf{x}')\|_2 \leqslant \left[\sum_{i,j} g_{ij}^2\right]^{1/2} \|\mathbf{x} - \mathbf{x}'\|_2$$

so that in working with $\|\cdot\|_2$ we can take

$$L = \left[\sum_{i,j} g_{ij}^2\right]^{1/2}. \tag{12.12}$$

Example 12.1 We apply Theorem 12.1 to equations (12.4), although these are easily solved by eliminating x_2 between the two equations and solving the resulting quadratic equation for x_1. We rewrite equations (12.4) as

$$\begin{aligned} x_1 &= \left(\frac{5 - x_2^2}{2}\right)^{1/2} \\ x_2 &= \tfrac{1}{2}(3 - x_1) \end{aligned} \tag{12.13}$$

which are in the form $\mathbf{x} = \mathbf{g}(\mathbf{x})$. If R denotes the closed rectangular region $1 \leqslant x_1 \leqslant 2$, $0.5 \leqslant x_2 \leqslant 1.5$, we see from (12.13) that if $\mathbf{x} \in R$, $\mathbf{g}(\mathbf{x}) \in R$. We have

$$\frac{\partial g_1}{\partial x_2} = -x_2/(10 - 2x_2^2)^{1/2},$$

whose maximum modulus on R occurs for $x_2 = 1.5$, with value $3/\sqrt{22}$. Thus the matrix $\mathbf{G}$, defined by (12.10), is

$$\mathbf{G} = \begin{pmatrix} 0 & 3/\sqrt{22} \\ \frac{1}{2} & 0 \end{pmatrix}.$$

We have $\|\mathbf{G}\|_1 = \|\mathbf{G}\|_\infty = 3/\sqrt{22}$ and (12.12) gives $L = (29/44)^{1/2}$. Thus we have a contraction mapping on R with any of the three common norms and

Table 12.1 Results for Example 12.1

Iteration number	0	1	2	3	4	5	6
x_1	1.5	1.414	1.490	1.478	1.488	1.487	1.488
x_2	1.0	0.750	0.793	0.755	0.761	0.756	0.756

Theorem 12.1 is applicable. Taking $x_1 = 1.5$, $x_2 = 1$, the centre point of R, as the initial iterate, we obtain the next six iterates as shown in Table 12.1. Because of the simple form of the equations, it is possible to make the iterations computing values of only x_1 for even iterations and x_2 for odd iterations or vice versa. For the other solution of equations (12.4) (which is, of course, *outside* R) see Problem 12.7.

Example 12.2 Consider the equations

$$\begin{aligned} x_1 &= \tfrac{1}{12}(-1 + \sin x_2 + \sin x_3) \\ x_2 &= \tfrac{1}{3}(x_1 - \sin x_2 + \sin x_3) \\ x_3 &= \tfrac{1}{12}(1 - \sin x_1 + x_2). \end{aligned} \tag{12.14}$$

The closure condition is satisfied with R as the cube $-1 \leqslant x_1, x_2, x_3 \leqslant 1$. The matrix $\mathbf{G}$, defined by (12.10), is

$$\mathbf{G} = \frac{1}{12}\begin{bmatrix} 0 & 1 & 1 \\ 4 & 4 & 4 \\ 1 & 1 & 0 \end{bmatrix}$$

In this case, $\|\mathbf{G}\|_1 = \frac{1}{2}$, $\|\mathbf{G}\|_\infty = 1$ and, for $\|\cdot\|_2$, (12.12) gives $L = \sqrt{13}/6$ Thus we have a contraction mapping for norms $\|\cdot\|_1$ and $\|\cdot\|_2$. □

The concept of a contraction mapping, which has been applied in this section to a system of *algebraic* equations, applies also to other types of equations. For example, in Chapter 13, we apply it to a first order ordinary differential equation.

12.2 Newton's method

Suppose that $\boldsymbol{\alpha}$ is a solution of the system of equations $\mathbf{x} = \mathbf{g}(\mathbf{x})$ and that $\partial g_i(\boldsymbol{\alpha})/\partial x_j = 0$ for $1 \leqslant i, j \leqslant n$. We assume also that, on a closed region R containing $\boldsymbol{\alpha}$, the second derivatives $\partial^2 g_i/\partial x_j \partial x_k$ $(1 \leqslant i, j, k \leqslant n)$ are bounded and continuous. Then from Taylor's theorem we have

$$g_i(\mathbf{x}) - g_i(\boldsymbol{\alpha}) = \frac{1}{2}\sum_{j=1}^{n}\sum_{k=1}^{n}(x_j - \alpha_j)(x_k - \alpha_k)\,\frac{\partial^2 g_i(\boldsymbol{\xi})}{\partial x_j\,\partial x_k}. \tag{12.15}$$

As for (12.9), we assume that $\boldsymbol{\xi} \in R$ whenever $\mathbf{x} \in R$. If

$$\sup_{\mathbf{x} \in R} \left| \frac{\partial^2 g_i(\mathbf{x})}{\partial x_j \, \partial x_k} \right| \leqslant M, \qquad 1 \leqslant i, j, k \leqslant n,$$

then we deduce from (12.15) that

$$| g_i(\mathbf{x}) - g_i(\boldsymbol{\alpha}) | \leqslant \tfrac{1}{2} M n^2 \| \mathbf{x} - \boldsymbol{\alpha} \|^2_\infty,$$

since $| x_i - \alpha_i | \leqslant \| \mathbf{x} - \boldsymbol{\alpha} \|_\infty$, $1 \leqslant i \leqslant n$. It follows that

$$\| \mathbf{g}(\mathbf{x}) - \mathbf{g}(\boldsymbol{\alpha}) \|_\infty \leqslant \tfrac{1}{2} M n^2 \| \mathbf{x} - \boldsymbol{\alpha} \|^2_\infty. \tag{12.16}$$

Thus, if $\mathbf{x}_{r+1} = \mathbf{g}(\mathbf{x}_r)$,

$$\| \mathbf{x}_{r+1} - \boldsymbol{\alpha} \|_\infty \leqslant \tfrac{1}{2} M n^2 \| \mathbf{x}_r - \boldsymbol{\alpha} \|^2_\infty$$

and we have at least second order convergence of the iterative method.

We now construct an iterative method which has second order convergence. Consider the system of equations

$$\mathbf{f}(\mathbf{x}) = \mathbf{0} \tag{12.17}$$

and write

$$\mathbf{x}_m = \boldsymbol{\alpha} + \mathbf{e}_m, \tag{12.18}$$

where $\boldsymbol{\alpha}$ is a solution of (12.17). We have

$$f_i(\mathbf{x}_m) = e_{m,1} \frac{\partial f_i(\boldsymbol{\xi})}{\partial x_1} + \cdots + e_{m,n} \frac{\partial f_i(\boldsymbol{\xi})}{\partial x_n}, \qquad 1 \leqslant i \leqslant n, \tag{12.19}$$

where $e_{m,j}$ denotes the jth element of the vector $\mathbf{e}_m$. If we knew the values of $\boldsymbol{\xi}$ (which depend on i), we could solve the system of linear equations (12.19) to find $\mathbf{e}_m$ and thus find $\boldsymbol{\alpha}$ from (12.18). If instead we evaluate the partial derivatives $\partial f_i / \partial x_j$ at $\mathbf{x}_m$ and replace $\mathbf{e}_m$ in (12.19) by $\mathbf{x}_m - \mathbf{x}_{m+1}$, rather than $\mathbf{x}_m - \boldsymbol{\alpha}$, we obtain

$$f_i(\mathbf{x}_m) = (x_{m,1} - x_{m+1,1}) \frac{\partial f_i(\mathbf{x}_m)}{\partial x_1} + \cdots + (x_{m,n} - x_{m+1,n}) \frac{\partial f_i(\mathbf{x}_m)}{\partial x_n}, \qquad 1 \leqslant i \leqslant n.$$

We write this system of linear equations, with unknown vector $\mathbf{x}_{m+1}$, as

$$\mathbf{f}(\mathbf{x}_m) = \mathbf{J}(\mathbf{x}_m)(\mathbf{x}_m - \mathbf{x}_{m+1}), \tag{12.20}$$

where $\mathbf{J}(\mathbf{x})$ is the *Jacobian* matrix whose (i, j)th element is $\partial f_i(\mathbf{x}) / \partial x_j$. If $\mathbf{J}(\mathbf{x}_m)$ is non-singular, we have

$$\mathbf{x}_{m+1} = \mathbf{x}_m - [\mathbf{J}(\mathbf{x}_m)]^{-1} \mathbf{f}(\mathbf{x}_m), \tag{12.21}$$

which is called *Newton's* method. Note that we do not find the inverse of the Jacobian matrix $\mathbf{J}(\mathbf{x}_m)$ explicitly, as (12.21) might suggest: at each stage we solve the system of linear equations (12.20) to find $\mathbf{x}_m - \mathbf{x}_{m+1}$ and hence $\mathbf{x}_{m+1}$. When $n = 1$, (12.21) coincides with the familiar Newton method for a single equation. To verify that we have second order convergence, we write (12.21) as

$$\mathbf{x}_{m+1} = \mathbf{g}(\mathbf{x}_m),$$

where

$$\mathbf{g}(\mathbf{x}) = \mathbf{x} - [\mathbf{J}(\mathbf{x})]^{-1}\mathbf{f}(\mathbf{x}). \tag{12.22}$$

We denote by $\partial\mathbf{g}/\partial x_j$ the vector whose ith element is $\partial g_i/\partial x_j$. Then from (12.22)

$$\frac{\partial\mathbf{g}(\mathbf{x})}{\partial x_j} = \frac{\partial\mathbf{x}}{\partial x_j} - [\mathbf{J}(\mathbf{x})]^{-1}\frac{\partial\mathbf{f}(\mathbf{x})}{\partial x_j} - \left(\frac{\partial}{\partial x_j}[\mathbf{J}(\mathbf{x})]^{-1}\right)\mathbf{f}(\mathbf{x}). \tag{12.23}$$

The partial derivatives of $\mathbf{J}^{-1}$ (see Problem 12.9) satisfy the equation

$$\frac{\partial}{\partial x_j}(\mathbf{J}^{-1}) = -\mathbf{J}^{-1}\frac{\partial\mathbf{J}}{\partial x_j}\mathbf{J}^{-1}.$$

Thus these exist if $\mathbf{J}$ is non-singular and if the partial derivatives of $\mathbf{J}$ exist. The second condition is equivalent to asking for the existence of the second derivatives of $\mathbf{f}$. Premultiplying (12.23) by $\mathbf{J}(\mathbf{x})$, and noting that $\mathbf{f}(\boldsymbol{\alpha}) = \mathbf{0}$, we have

$$\left(\mathbf{J}(\mathbf{x})\frac{\partial\mathbf{g}(\mathbf{x})}{\partial x_j}\right)_{\mathbf{x}=\boldsymbol{\alpha}} = \left(\mathbf{J}(\mathbf{x})\frac{\partial\mathbf{x}}{\partial x_j} - \frac{\partial\mathbf{f}(\mathbf{x})}{\partial x_j}\right)_{\mathbf{x}=\boldsymbol{\alpha}}. \tag{12.24}$$

Since

$$\frac{\partial x_i}{\partial x_j} = \delta_{ij},$$

the Kronecker delta function, whose value is 1 when $i = j$ and zero otherwise, $\partial\mathbf{x}/\partial x_j$ is the jth column of the $n \times n$ unit matrix and we see that each element of the vector on the right of (12.24) is zero. Thus, from (12.24), if $\mathbf{J}(\boldsymbol{\alpha})$ is non-singular,

$$\left(\frac{\partial\mathbf{g}}{\partial x_j}\right)_{\mathbf{x}=\boldsymbol{\alpha}} = \mathbf{0}$$

and Newton's method is at least second order.

If for any $\mathbf{x}_m$, $\mathbf{J}(\mathbf{x}_m)$ is singular, then $\mathbf{x}_{m+1}$ is not defined. There is a theorem, proved by L. V. Kantorovich in 1937, which states conditions under which the sequence $(\mathbf{x}_m)$ is defined and converges to the solution $\boldsymbol{\alpha}$ (see Henrici, 1962).

Table 12.2 Results for Example 12.3

Iteration number	0	1	2	3
x_1	1.5	1.5	1.488095	1.488034
x_2	1.0	0.75	0.755952	0.755983

In §8.6 we saw how Newton's method for a single equation and unknown can be described in terms of tangents to the graph of the function involved. In the case of two equations in two unknowns, we can again visualize the process using graphs. This time we have two surfaces $z = f_1(x_1, x_2)$ and $z = f_2(x_1, x_2)$ and we are trying to find a point where both of these are zero, that is we seek a point at which both surfaces cut the plane $z = 0$. Given an approximate solution $x_{m,1}$, $x_{m,2}$, we construct the tangent planes to f_1 and f_2 at the point $(x_{m,1}, x_{m,2})$. We now calculate the point at which these two planes and the plane $z = 0$ intersect. This point is taken as the next iterate.

Example 12.3 For the equations (12.4), we have

$$\mathbf{J}(\mathbf{x}) = \begin{bmatrix} 1 & 2 \\ 4x_1 & 2x_2 \end{bmatrix}.$$

With $x_1 = 1.5$, $x_2 = 1.0$ initially, the next few iterates obtained by Newton's method (12.21) are as shown in Table 12.2. For the last iterate in the table, the residuals for equations (12.4) are less than 10^{-6}. □

Problems

Section 12.1

12.1 Show that the closure condition and contraction mapping property apply to the following equations on the region $0 \leqslant x_1, x_2, x_3 \leqslant 1$.

$$x_1 = \tfrac{1}{12}(x_2^2 + 2e^{-x_3})$$
$$x_2 = \tfrac{1}{6}(1 - x_1 + \sin x_3)$$
$$x_3 = \tfrac{1}{6}(x_1^2 + x_2^2 + x_3^2).$$

12.2 Show that the equations

$$x = \tfrac{1}{2}\cos y$$
$$y = \tfrac{1}{2}\sin x$$

have a unique solution. Find the solution to two decimal places.

12.3 If $\mathbf{a}$, $\mathbf{b}$ and $\mathbf{x}$ denote vectors with n real elements and R denotes the region $\mathbf{a} \leqslant \mathbf{x} \leqslant \mathbf{b}$, show that R is closed. (*Hint*: consider the sequences composed from corresponding components of vectors.)

12.4 If $\mathbf{A}$ is an $m \times n$ (real) matrix and $\mathbf{x}$ is an n-dimensional real vector show that

$$\begin{aligned}\|\mathbf{A}\mathbf{x}\|_2 &= \left(\sum_{i=1}^{m}\left(\sum_{j=1}^{n} a_{ij}x_j\right)^2\right)^{1/2} \\ &\leqslant \left(\sum_{i=1}^{m}\left(\sum_{j=1}^{n} a_{ij}^2 \sum_{j=1}^{n} x_j^2\right)\right)^{1/2} \\ &= \left(\sum_{i=1}^{m}\sum_{j=1}^{n} a_{ij}^2\right)^{1/2}\left(\sum_{j=1}^{n} x_j^2\right)^{1/2}\end{aligned}$$

using the Cauchy–Schwarz inequality (see Problem 10.6). Thus

$$\|\mathbf{A}\mathbf{x}\|_2 \leqslant N(\mathbf{A})\,\|\mathbf{x}\|_2$$

where

$$N(\mathbf{A}) = \left(\sum_{i,j} a_{ij}^2\right)^{1/2}.$$

(Because $N(\mathbf{A})$ satisfies all the properties mentioned in Lemma 10.2, it is sometimes called the *Schur* norm of the matrix $\mathbf{A}$. Note however that $N(\mathbf{A})$ is not a *subordinate* norm, as in Definition 10.6.)

12.5 Show that, working with the norm $\|\cdot\|_2$, the contraction mapping property holds for the following system of equations on the region $-1 \leqslant x_1, x_2, x_3 \leqslant 1$.

$$\begin{aligned}x_1 &= \tfrac{1}{3}(x_1 + x_2 - x_3) \\ x_2 &= \tfrac{1}{6}(2x_1 + \sin x_2 + x_3) \\ x_3 &= \tfrac{1}{6}(2x_1 - x_2 + \sin x_3).\end{aligned}$$

12.6 In Example 12.1, eliminate x_2 between the two equations and show that the sequence $(x_{2m,1})$ satisfies

$$x_{2m+2,1} = \tfrac{1}{4}[40 - 2(3 - x_{2m,1})^2]^{1/2}, \qquad m = 0, 1, \ldots.$$

If this relation is denoted by

$$y_{m+1} = g(y_m)$$

where $y_m = x_{2m,1}$, show that $|g'(y)| \leqslant \sqrt{2}/8$ for $1 \leqslant y \leqslant 2$. (Note that for the three common norms 1, 2 and ∞ we obtained in Example 10.1 Lipschitz constants of $3/\sqrt{22}$, $(29/44)^{1/2}$ and $3/\sqrt{22}$ respectively. The value $L=\sqrt{2}/8$ is in fairly close agreement with the apparent rate of convergence of alternate iterates in Table 12.1.)

Section 12.2

12.7 The equations (12.4) have a solution which is near to $x_1 = -1$, $x_2 = 2$. Use Newton's method to find this solution to two decimal places.

12.8 An $m \times n$ matrix $\mathbf{A}$ has elements $a_{ij}(x)$ which depend on x and an n-dimensional vector $\mathbf{y}$ has elements $y_j(x)$ dependent on x. If all elements of $\mathbf{A}$ and $\mathbf{y}$ are differentiable with respect to x, show that

$$\frac{\mathrm{d}}{\mathrm{d}x}(\mathbf{Ay}) = \frac{\mathrm{d}\mathbf{A}}{\mathrm{d}x}\mathbf{y} + \mathbf{A}\frac{\mathrm{d}\mathbf{y}}{\mathrm{d}x},$$

where $\mathrm{d}\mathbf{A}/\mathrm{d}x$ is an $m \times n$ matrix with elements $\mathrm{d}a_{ij}/\mathrm{d}x$ and $\mathrm{d}\mathbf{y}/\mathrm{d}x$ is an n-dimensional vector with elements $\mathrm{d}y_j/\mathrm{d}x$.

12.9 Assuming that $\mathbf{J}^{-1}(\mathbf{x})$ exists, by differentiating the product $\mathbf{JJ}^{-1} = \mathbf{I}$ partially with respect to x_j, show that

$$\frac{\partial}{\partial x_j}(\mathbf{J}^{-1}) = -\mathbf{J}^{-1}\frac{\partial \mathbf{J}}{\partial x_j}\mathbf{J}^{-1}.$$

12.10 Use Newton's method to obtain the solution of the system of equations in Problem 12.2, beginning with $x = y = 0$. Show that $\mathbf{J}^{-1}$ exists for all x and y.

12.11 What happens if Newton's method is applied to the system of equations

$$\mathbf{Ax} = \mathbf{b},$$

where $\mathbf{A}$ is an $n \times n$ non-singular matrix?

12.12 Given a system of two equations in two unknowns

$$F(x, y) = 0$$
$$G(x, y) = 0$$

show that Newton's method (see (12.21)) has the form

$$x_{m+1} = x_m - (FG_y - F_yG)/(F_xG_y - F_yG_x)$$
$$y_{m+1} = y_m - (F_xG - FG_x)/(F_xG_y - F_yG_x),$$

where the six functions F, G, F_x, G_x, F_y and G_y are all evaluated at the point (x_m, y_m).

12.13 Write a computer program which uses Newton's method to obtain a solution of the pair of simultaneous equations

$$x^2 + y^2 = 5$$
$$x^3 + y^3 = 2$$

taking $x = 2$ and $y = -1$ as the initial values in the iterative process. This system of equations has two solutions. Also find the other solution.

Chapter **13**

ORDINARY DIFFERENTIAL EQUATIONS

13.1 Introduction

Let $f(x, y)$ be a real valued function of two variables defined for $a \leqslant x \leqslant b$ and all real y. Suppose now that y is a real valued function defined on $[a, b]$. The equation

$$y' = f(x, y) \tag{13.1}$$

is called an ordinary differential equation of the first order. Any real valued function y, which is differentiable and satisfies (13.1) for $a \leqslant x \leqslant b$, is said to be a *solution* of this differential equation. In general such a solution, if it exists, will not be unique because an arbitrary constant is introduced on integrating (13.1). To make the solution unique we need to impose an extra condition on the solution. This condition usually takes the form of requiring $y(x)$ to have a specified value for a particular value of x, that is,

$$y(x_0) = s \tag{13.2}$$

where $x_0 \in [a, b]$ is the particular value of x and s is the specified value of $y(x_0)$. Equations (13.1) and (13.2) are collectively called an *initial value problem*. Equation (13.2) is called the initial condition and s the initial value of y.

Example 13.1 The general solution of

$$y' = ky, \tag{13.3}$$

where k is a given constant, is

$$y(x) = ce^{kx} \tag{13.4}$$

where c is an arbitrary constant. If, for example, we are given the extra condition

$$y(0) = s \tag{13.5}$$

then, on putting $x=0$ in (13.4), we see that we require $c=s$. The unique solution satisfying (13.3) and (13.5) is

$$y(x)=se^{kx}.$$

The differential equation (13.3) occurs in the description of many physical phenomena, for example, uninhibited bacteriological growth, a condenser discharging, a gas escaping under pressure from a container and radioactive decay. In all these examples, time is represented by x. The condition (13.5) gives the initial value of the physical quantity being measured, for example, the volume of bacteria at time $x=0$. Many more complex phenomena may be described by equations of the form (13.1). □

During most of this chapter we shall discuss numerical methods of solving (13.1) subject to (13.2). By this, we mean that we seek approximations to the numerical values of $y(x)$ for particular values of x or we seek a function, for example a polynomial, which approximates to y over some range of values of x. Only for rare special cases can we obtain an explicit form of the solution, as in Example 13.1. In most problems, $f(x, y)$ will be far too complicated for us to obtain an explicit solution and we must therefore resort to numerical methods. We shall also discuss *systems* of ordinary differential equations and higher order differential equations, in which derivatives of higher order than the first occur. The methods for both of these are natural extensions of those for a single first order equation.

The first question we ask of (13.1) and (13.2) is whether a solution y necessarily exists and, if not, what additional restrictions must be imposed on $f(x, y)$ to ensure this. The following example illustrates that a solution does not necessarily exist.

Example 13.2 For $0 \leqslant x \leqslant 2$, solve

$$y' = y^2 \quad \text{with} \quad y(0)=1.$$

The general solution of the differential equation is

$$y(x) = -\frac{1}{(x+c)},$$

where c is arbitrary. The initial condition implies $c=-1$, when

$$y(x) = \frac{1}{(1-x)}.$$

This solution is not defined for $x=1$ and therefore the given initial value problem has *no solution* on $[0,2]$. □

You will notice that in Example 13.2 the function $f(x, y)$ is the well-behaved function y^2. Obviously continuity of $f(x, y)$ is not sufficient to ensure the existence of a unique solution of (13.1) and (13.2). We find that a Lipschitz condition on $f(x, y)$ with respect to y is sufficient and we have the following theorem.

Theorem 13.1 The first order differential equation

$$y' = f(x, y)$$

with initial condition

$$y(x_0) = s,$$

where $x_0 \in [a, b]$, has a unique solution y on $[a, b]$ if $f(x, y)$ is continuous in x and if there exists L, independent of x, such that

$$|f(x, y) - f(x, z)| \leqslant L|y - z| \tag{13.6}$$

for all $x \in [a, b]$ and all real y and z.

Proof We shall consider an outline of the proof, which is based on the contraction mapping theorem. We may write the problem in the integral form

$$y(x) = s + \int_{x_0}^{x} f(x, y(x))\, dx. \tag{13.7}$$

We consider the sequence of functions $(y_m(x))_{m=0}^{\infty}$ defined recursively by

$$y_0(x) = s, \tag{13.8}$$

$$y_{m+1}(x) = s + \int_{x_0}^{x} f(x, y_m(x))\, dx, \qquad m = 0, 1, \ldots. \tag{13.9}$$

We find that, for $m \geqslant 1$,

$$\begin{aligned} |y_{m+1}(x) - y_m(x)| &\leqslant \int_{x_0}^{x} |f(x, y_m(x)) - f(x, y_{m-1}(x))|\, dx \\ &\leqslant \int_{x_0}^{x} L|y_m(x) - y_{m-1}(x)|\, dx. \end{aligned}$$

Now

$$|y_1(x) - y_0(x)| = |y_1(x) - s| = \left|\int_{x_0}^{x} f(x, s)\, dx\right| \leqslant M|x - x_0|,$$

where $M \geqslant |f(x, s)|$ for $x \in [a, b]$. M, an upper bound of $|f(x, s)|$, exists as f is continuous in x. Thus

$$|y_2(x) - y_1(x)| \leqslant \int_{x_0}^{x} L|y_1(x) - y_0(x)|\,dx$$

$$\leqslant \int_{x_0}^{x} LM|x - x_0|\,dx = \frac{LM}{2!}|x - x_0|^2$$

and

$$|y_3(x) - y_2(x)| \leqslant \int_{x_0}^{x} L|y_2(x) - y_1(x)|\,dx \leqslant \int_{x_0}^{x} \frac{L^2M}{2!}|x - x_0|^2\,dx$$

$$= \frac{L^2M}{3!}|x - x_0|^3.$$

We may prove by induction that

$$|y_{m+1}(x) - y_m(x)| \leqslant \frac{ML^m}{(m+1)!}|x - x_0|^{m+1}$$

for $m \geqslant 0$. The series

$$y_0(x) + (y_1(x) - y_0(x)) + (y_2(x) - y_1(x)) + \cdots$$

is thus uniformly convergent on $[a, b]$, as the absolute value of each term is less than the corresponding term of the uniformly convergent series

$$|s| + \frac{M}{L}\sum_{m=1}^{\infty}\frac{L^m|x - x_0|^m}{m!}.$$

The partial sums of the first series are $y_0(x),\ y_1(x),\ y_2(x), \ldots$ and this sequence has a limit, say $y(x)$. This limit $y(x)$ can be shown to be a differentiable function which is the solution of the initial value problem. By assuming the existence of two distinct solutions and obtaining a contradiction, the uniqueness of y can be shown. □

The sequence $(y_m(x))_{m=0}^{\infty}$ defined in the proof of the last theorem may be used to find approximations to the solution. The resulting process is called *Picard* iteration, but is only practicable for simple functions $f(x, y)$ for which the integration in (13.9) can be performed explicitly.

Example 13.3 Use Picard iteration to find an approximate solution of

$$y' = y, \qquad \text{with } y(0) = 1$$

for $0 \leqslant x \leqslant 2$.

In this case, $f(x, y) = y$ and thus

$$|f(x, y) - f(x, z)| = |y - z|,$$

so we choose $L = 1$. We start with $y_0(x) \equiv 1$ when, from (13.9),

$$y_1(x) = 1 + \int_0^x y_0(x)\,\mathrm{d}x = 1 + x,$$

$$y_2(x) = 1 + \int_0^x (1 + x)\,\mathrm{d}x = 1 + x + \frac{x^2}{2!},$$

$$y_3(x) = 1 + \int_0^x \left(1 + x + \frac{x^2}{2!}\right)\mathrm{d}x = 1 + x + \frac{x^2}{2!} + \frac{x^3}{3!},$$

and, in general,

$$y_r(x) = 1 + x + \frac{x^2}{2!} + \cdots + \frac{x^r}{r!}.$$

The exact solution in this case is

$$y(x) = \mathrm{e}^x = 1 + x + \frac{x^2}{2!} + \cdots$$

so that the approximations obtained by the Picard process are the partial sums of this infinite series. □

Example 13.4 We show that

$$y' = \frac{1}{1 + y^2}, \quad \text{with } y(x_0) = s$$

has a unique solution for $x \in [a, b]$. We have

$$f(x, y) = \frac{1}{1 + y^2}$$

and from the mean value theorem

$$f(x, y) - f(x, z) = (y - z)\,\frac{\partial f(x, \xi)}{\partial y},$$

where ξ lies between y and z. Thus to satisfy (13.6) we choose

$$L \geqslant \left|\frac{\partial f(x, y)}{\partial y}\right|$$

for all $x \in [a, b]$ and $y \in (-\infty, \infty)$, provided such an upper bound L exists. In this case,

$$\left|\frac{\partial f}{\partial y}\right| = \left|\frac{-2y}{(1+y^2)^2}\right|$$

and, as may be seen by sketching a graph, this is a maximum for $y = \pm 1/\sqrt{3}$. We therefore choose

$$L = \left|\frac{\partial f(x, 1/\sqrt{3})}{\partial y}\right| = \frac{3\sqrt{3}}{8}.$$

□

Most numerical methods of solving the initial value problem (13.1) and (13.2) use a *step-by-step* process. Approximations $y_0, y_1, y_2, \ldots$ are sought† for the values $y(x_0), y(x_1), y(x_2), \ldots$ of the solution y at distinct points $x = x_0, x_1, x_2, \ldots$. We calculate the y_n by means of a recurrence relation derived from (13.1), which expresses y_n in terms of $y_{n-1}, y_{n-2}, \ldots, y_0$. We take $y_0 = y(x_0) = s$, since this value is given. We usually take the points $x_0, x_1, \ldots$ to be equally spaced so that $x_r = x_0 + rh$, where $h > 0$ is called the *step size* or *mesh length*. We also take $a = x_0$ as we are only concerned with points $x_n \geqslant a$. (We could equally well take $h < 0$ but this may be avoided by the simple transformation $x' = -x$.)

The simplest suitable recurrence relation is derived by replacing the derivative in (13.1) by a forward difference. We obtain

$$\frac{y_{n+1} - y_n}{h} = f(x_n, y_n),$$

which on rearrangement yields *Euler's* method

$$y_0 = s, \tag{13.10}$$

$$y_{n+1} = y_n + hf(x_n, y_n), \qquad n = 0, 1, 2, \ldots. \tag{13.11}$$

The latter is a particular type of *difference* equation.

13.2 Difference equations and inequalities

In order to discuss the accuracy of approximations obtained by methods such as (13.11) in solving initial value problems, we must use some properties of difference equations. If $(F_n(w_0, w_1))$ is a sequence of

† We use the notation y_r to denote an approximation to $y(x_r)$, the exact solution at $x = x_r$, throughout the rest of this chapter.

functions defined for $n \in N$, some set of consecutive integers, and for all w_0 and w_1 in some interval I of the real numbers, then the system of equations

$$F_n(y_n, y_{n+1}) = 0, \tag{13.12}$$

for all $n \in N$, is called a *difference equation of order 1*. A sequence (y_n), with all $y_n \in I$ and satisfying (13.12) for all $n \in N$, is called a solution of this difference equation.

Example 13.5 If

$$F_n(w_0, w_1) = w_0 - w_1 + 2n$$

for all integers n, we obtain the difference equation

$$y_n - y_{n+1} + 2n = 0.$$

The general solution of this equation is

$$y_n = n(n-1) + c,$$

where c is an arbitrary constant. Such an arbitrary constant may be determined if the value of one element of the solution sequence is given. For example, the 'initial' value y_0 may be given, when clearly we must choose $c = y_0$. The arbitrary constant is analogous to that occurring in the integration of first order differential equations. □

The difference equation of order m is obtained from a sequence

$$(F_n(w_0, w_1, \ldots, w_m))$$

of functions of $m+1$ variables defined for all $n \in N$, a set of consecutive integers, and all $w_r \in I$, $r = 0, \ldots, m$, for some interval I. We seek a sequence (y_n) with all $y_n \in I$ and such that

$$F_n(y_n, y_{n+1}, \ldots, y_{n+m}) = 0,$$

for all $n \in N$. The simplest difference equation of order m is of the form

$$a_0 y_n + a_1 y_{n+1} + \cdots + a_m y_{n+m} = 0, \tag{13.13}$$

where a_0, a_1, …, a_m are independent of n. Equation (13.13) is called an mth *order homogeneous linear difference* equation with constant coefficients. It is homogeneous because if (y_n) is a solution, then so is (αy_n), where α is any scalar. Equation (13.13) is called linear because it consists of a linear combination of elements of (y_n). We illustrate the method of solution by first considering the second order equation,

$$a_0 y_n + a_1 y_{n+1} + a_2 y_{n+2} = 0. \tag{13.14}$$

We seek a solution of the form

$$y_n = z^n,$$

where z is independent of n. On substituting z^n in (13.14), we have

$$a_0 z^n + a_1 z^{n+1} + a_2 z^{n+2} = 0,$$

that is,

$$z^n(a_0 + a_1 z + a_2 z^2) = 0.$$

We conclude that for such a solution we require either

$$z^n = 0$$

or

$$a_0 + a_1 z + a_2 z^2 = 0. \tag{13.15}$$

The former only yields $z = 0$ and hence the trivial solution $y_n = 0$. The latter, a quadratic equation, is called the *characteristic* equation of the difference equation (13.14) and, in general, yields two possible values of z. We suppose for the moment that (13.15) has two distinct roots, z_1 and z_2. We then have two independent solutions

$$y_n = z_1^n \quad \text{and} \quad y_n = z_2^n.$$

It is easily verified that

$$y_n = c_1 z_1^n + c_2 + z_2^n, \tag{13.16}$$

where c_1 and c_2 are *arbitrary constants*, is also a solution of (13.14). This is the general solution of (13.14) for the case when the characteristic equation has distinct roots. You will notice that this general solution contains two arbitrary constants and those familiar with differential equations will know that two such constants occur in the solution of second order differential equations.

Now suppose that in (13.15)

$$a_1^2 = 4a_0 a_2 \quad \text{and} \quad a_2 \neq 0,$$

so that the characteristic equation has two equal roots. We only have one solution

$$y_n = z_1^n.$$

We try, as a possible second solution,

$$y_n = nz_1^n \tag{13.17}$$

in (13.14). The left side becomes

$$\begin{aligned} &a_0 n z_1^n + a_1(n+1)z_1^{n+1} + a_2(n+2)z_1^{n+2} \\ &\quad = z_1^n[n(a_0 + a_1 z_1 + a_2 z_1^2) + z_1(a_1 + 2a_2 z_1)] \\ &\quad = 0 \end{aligned}$$

as $z_1 = -\frac{1}{2}a_1/a_2$ is the repeated root of (13.15). Clearly (13.17) is a solution and the general solution of (13.14) is

$$y_n = c_1 z_1^n + c_2 n z_1^n. \tag{13.18}$$

The only remaining case is $a_2 = 0$, when the difference equation is of only first order and has general solution

$$y_n = c\left(-\frac{a_0}{a_1}\right)^n,$$

where c is arbitrary.

If $a_2 \neq 0$ we may write (13.14) as the recurrence relation

$$y_{n+2} = -(a_0/a_2)y_n - (a_1/a_2)y_{n+1}.$$

If we are to use this to calculate members of a sequence (y_n) we need two starting (or initial) values, for example, y_0 and y_1. Such starting values fix the arbitrary constants in the general solution. It can also be seen from this that (13.16) or (13.18) is the general solution in that any particular solution may be expressed in this form. Given any particular solution we can always use y_0 and y_1 to determine c_1 and c_2. The remaining elements of the solution are then uniquely determined by the recurrence relation above.

Example 13.6 The difference equation

$$y_n - 2\cos\theta . y_{n+1} + y_{n+2} = 0,$$

where θ is a constant, has characteristic equation

$$1 - 2\cos\theta . z + z^2 = 0$$

which has roots $z = \cos\theta \pm i\sin\theta = e^{\pm i\theta}$. The general solution is:

$$\begin{aligned} &y_n = c_1 e^{ni\theta} + c_2 e^{-ni\theta} = k_1 \cos n\theta + k_2 \sin n\theta, && \theta \neq m\pi, \\ &y_n = c_1 + c_2 n, && \theta = 2m\pi, \\ &y_n = (c_1 + c_2 n)(-1)^n, && \theta = (2m+1)\pi, \end{aligned}$$

where m is any integer and c_1, c_2 and, hence, k_1, k_2 are arbitrary. □

To determine the general solution of the mth order equation (13.13) we again try $y_n = z^n$. We require

$$a_0 z^n + a_1 z^{n+1} + \cdots + a_m z^{n+m} = 0$$

and thus either $z = 0$ or

$$p_m(z) \equiv a_0 + a_1 z + a_2 z^2 + \cdots + a_m z^m = 0. \tag{13.19}$$

Equation (13.19) is called the characteristic equation and p_m the characteristic polynomial of (13.13). In general, (13.19) has m roots $z_1, z_2, \ldots, z_m$ and, if these are distinct, the general solution of (13.13) is

$$y_n = c_1 z_1^n + c_2 z_2^n + \cdots + c_m z_m^n,$$

where the m constants $c_1, c_2, \ldots, c_m$ are arbitrary. Again we may expect m arbitrary constants in the general solution of (13.13), because we need m starting values when it is expressed as a recurrence relation. For example, $y_0, y_1, \ldots, y_{m-1}$ are needed to calculate $y_m, y_{m+1}, \ldots$.

If p_m has a repeated zero, for example if $z_2 = z_1$, we find that $y_n = nz_1^n$ is a solution. If the other zeros of p_m are distinct, we obtain the general solution

$$y_n = c_1 z_1^n + c_2 n z_1^n + c_3 z_3^n + \cdots + c_m z_m^n.$$

If a zero of p_m is repeated r times, for example if

$$z_1 = z_2 = \cdots = z_r,$$

the general solution is

$$y_n = c_1 z_1^n + c_2 n z_1^n + c_3 n^2 z_1^n + \cdots + c_r n^{r-1} z_1^n + c_{r+1} z_{r+1}^n + \cdots + c_m z_m^n,$$

again assuming that $z_{r+1}, \ldots, z_m$ are distinct.

Example 13.7 Find the solution of the third order equation

$$y_n - y_{n+1} - y_{n+2} + y_{n+3} = 0, \tag{13.20}$$

with $y_{-1} = 0$, $y_0 = 1$ and $y_1 = 2$.

The characteristic equation is

$$1 - z - z^2 + z^3 = 0$$

with roots $z = -1, 1, 1$. The general solution of (13.20) is

$$y_n = c_1(-1)^n + c_2 + nc_3$$

and thus we require

$$\begin{aligned} y_{-1} &= -c_1 + c_2 - c_3 = 0 \\ y_0 &= c_1 + c_2 \phantom{{}- c_3} = 1 \\ y_1 &= -c_1 + c_2 + c_3 = 2. \end{aligned}$$

We obtain

$$c_1 = 0, \qquad c_2 = c_3 = 1,$$

and the solution required is

$$y_n = 1 + n.$$ □

The non-homogeneous difference equation

$$a_0y_n + a_1y_{n+1} + \cdots + a_my_{n+m} = b_n \tag{13.21}$$

is solved by first finding the general solution of the homogeneous equation

$$a_0y_n + a_1y_{n+1} + \cdots + a_my_{n+m} = 0.$$

We denote this general solution by g_n. Secondly we find any particular solution $\tilde{y}_n$ of the complete equation (13.21) by some means (usually by inspection). The general solution of (13.21) is then

$$y_n = \tilde{y}_n + g_n.$$

There are m arbitrary constants in g_n and hence m in y_n. (This method is entirely analogous to the 'particular integral plus complementary function' solution of a linear differential equation.)

Example 13.8 Find the general solution of

$$2y_n + y_{n+1} - y_{n+2} = 1.$$

The homogeneous equation is

$$2y_n + y_{n+1} - y_{n+2} = 0,$$

which has general solution

$$y_n = c_1(-1)^n + c_2.2^n.$$

A particular solution is

$$\tilde{y}_n = \tfrac{1}{2}, \qquad \text{for all } n,$$

so that the general solution of the complete equation is

$$y_n = c_1(-1)^n + c_2.2^n + \tfrac{1}{2}.$$ □

In the error analysis of this chapter we shall meet *difference* (or *recurrent*) *inequalities*. We need the following lemma.

Lemma 13.1 Let the sequence $(u_n)_{n=0}^{\infty}$ of positive reals satisfy the difference inequality

$$u_{n+1} \leqslant u_n + \alpha_0u_n + \alpha_1u_{n-1} + \cdots + \alpha_mu_{n-m} + b \tag{13.22}$$

for $n = m, m+1, m+2, \ldots$, where $\alpha_r \geqslant 0$, $r = 0, 1, \ldots, m$, and $b \geqslant 0$. Then, if $A = \alpha_0 + \alpha_1 + \cdots + \alpha_m$,

$$u_n \leqslant \begin{cases} \left(\delta + \dfrac{b}{A}\right)\mathrm{e}^{nA} - \dfrac{b}{A}, & A > 0 \\ \delta + nb, & A = 0 \end{cases} \tag{13.23}$$

for $n = 0, 1, 2, \ldots$, where δ is any real number satisfying

$$\delta \geqslant u_r, \qquad r = 0, 1, \ldots, m.$$

Proof The proof in both cases is by induction. Suppose (13.23) is true when $A > 0$ for all $n \leqslant N$. On applying (13.22) with $n = N$ we see that

$$\begin{aligned} u_{N+1} &\leqslant \left(\delta + \frac{b}{A}\right) e^{NA} - \frac{b}{A} + \sum_{r=0}^{m} \alpha_r \left[\left(\delta + \frac{b}{A}\right) e^{(N-r)A} - \frac{b}{A}\right] + b \\ &\leqslant \left(\delta + \frac{b}{A}\right) e^{NA} + \sum_{r=0}^{m} \alpha_r \left(\delta + \frac{b}{A}\right) e^{NA} - \frac{b}{A} \\ &= \left(\delta + \frac{b}{A}\right) e^{NA}(1 + A) - \frac{b}{A} < \left(\delta + \frac{b}{A}\right) e^{(N+1)A} - \frac{b}{A} \end{aligned}$$

since $1 + A < e^A$. From the definition of δ we can see that (13.23) holds for $n \leqslant m$. The proof for $A = 0$ is simpler but similar. □

13.3 One-step methods

We return to the numerical solution of the initial value problem

$$y' = f(x, y)$$

with

$$y(x_0) = s,$$

where $x \in [x_0, b]$. We shall assume that f satisfies the conditions of Theorem 13.1, so that the existence of a unique solution is ensured. We shall also assume that f is 'sufficiently' differentiable, that is, all the derivatives used in the analysis exist. We look for a sequence y_0, y_1, $y_2, \ldots$, whose elements are approximations to $y(x_0)$, $y(x_1)$, $y(x_2)$, $\ldots$, where $x_n = x_0 + nh$. We seek a first order difference equation to determine (y_n). A first order difference equation yields a *one-step* process, in that we evaluate y_{n+1} from only y_n and do not use y_{n-1}, y_{n-2}, $\ldots$ explicitly.

One approach is to consider the Taylor series for y. We have

$$y(x_{n+1}) = y(x_n) + hy'(x_n) + \frac{h^2}{2!} y''(x_n) + \cdots + \frac{h^p}{p!} y^{(p)}(x_n) + h^{p+1} R_{p+1}(x_n), \tag{13.24}$$

where

$$R_{p+1}(x_n) = \frac{1}{(p+1)!} y^{(p+1)}(\xi) \quad \text{and} \quad x_n < \xi < x_{n+1}.$$

We find we can determine the derivatives of y from f and its partial derivatives. For example,

$$y'(x) = f(x, y(x))$$

and, on using the chain rule for partial differentiation,

$$\begin{aligned} y''(x) &= \frac{\partial f}{\partial x} + \frac{\partial f}{\partial y}\frac{dy}{dx} \\ &= f_x + f_y y' \\ &= f_x + f_y f. \end{aligned} \tag{13.25}$$

We have written f_x and f_y to denote the partial derivatives of f with respect to x and y.

If we define the functions $f^{(r)}(x, y)$ recursively by

$$f^{(r)}(x, y) = f_x^{(r-1)}(x, y) + f_y^{(r-1)}(x, y)f(x, y) \tag{13.26}$$

for $r \geqslant 1$, with $f^{(0)}(x, y) = f(x, y)$, then

$$y^{(r+1)}(x) = f^{(r)}(x, y(x)) \tag{13.27}$$

for $r \geqslant 0$. The proof of this is easily obtained by induction. The result is true for $r = 0$ and on differentiating (13.27) we have

$$\begin{aligned} y^{(r+2)}(x) &= f_x^{(r)}(x, y(x)) + f_y^{(r)}(x, y(x))\, y'(x) \\ &= f_x^{(r)}(x, y(x)) + f_y^{(r)}(x, y(x))f(x, y(x)) \\ &= f^{(r+1)}(x, y(x)), \end{aligned}$$

showing that we can extend the result to the $(r+2)$th derivative of y. Note that the $f^{(r)}(x, y)$ are *not* derivatives of f in the *usual* sense.

If we substitute (13.27) in (13.24), we obtain

$$y(x_{n+1}) = y(x_n) + hf(x_n, y(x_n)) + \cdots + \frac{h^p}{p!}f^{(p-1)}(x_n, y(x_n)) + h^{p+1}R_{p+1}(x_n). \tag{13.28}$$

The approximation obtained by neglecting the remainder term is

$$y_{n+1} = y_n + hf(x_n, y_n) + \cdots + \frac{h^p}{p!}f^{(p-1)}(x_n, y_n) \tag{13.29}$$

and the Taylor series method consists of using this recurrence relation for $n = 0, 1, 2, \ldots$ together with $y_0 = s$. The disadvantage of this method is that the $f^{(r)}(x, y)$ are difficult to determine. For example,

$$f^{(1)}(x, y) = f_x + f_y f, \tag{13.30}$$

$$\begin{aligned} f^{(2)}(x, y) &= f_x^{(1)} + f_y^{(1)} f \\ &= f_{xx} + 2f_{xy}f + f_{yy}f^2 + (f_x + f_y f)f_y. \end{aligned} \tag{13.31}$$

The method might, however, be useful for $p \leqslant 2$. For $p = 1$ we obtain Euler's method. For most problems the accuracy of the method increases with p since the remainder term decreases as p increases if the derivatives of $y(x)$ are well behaved.

Example 13.9 For the problem

$$y' = y \quad \text{with} \quad y(0) = 1,$$
$$f(x, y) = y, \qquad f^{(1)}(x, y) = f_x + f_y f = 0 + 1 \,.\, y = y,$$
$$f^{(2)}(x, y) = f_x^{(1)} + f_y^{(1)} f = y \quad \text{and} \quad f^{(r)}(x, y) = y.$$

From (13.29), the Taylor series method is

$$\begin{aligned} y_{n+1} &= y_n + hy_n + \frac{h^2}{2!} y_n + \cdots + \frac{h^p}{p!} y_n \\ &= \left(1 + h + \frac{h^2}{2!} + \cdots + \frac{h^p}{p!}\right) y_n \end{aligned}$$

with $y_0 = 1$. Thus we have

$$y_n = \left(1 + h + \cdots + \frac{h^p}{p!}\right)^n.$$

□

Example 13.10 For the problem

$$y' = \frac{1}{1 + y^2} \qquad \text{with } y(0) = 1,$$

we obtain

$$f(x, y) = \frac{1}{1 + y^2},$$

$$f^{(1)}(x, y) = f_x + f_y f = \frac{-2y}{(1 + y^2)^3},$$

$$f^{(2)}(x, y) = \frac{2(5y^2 - 1)}{(1 + y^2)^5}.$$

Thus, for $p = 3$, the Taylor series method (13.29) is

$$y_{n+1} = y_n + \frac{h}{1 + y_n^2} + \frac{h^2}{2!} \frac{(-2y_n)}{(1 + y_n^2)^3} + \frac{h^3}{3!} \frac{2(5y_n^2 - 1)}{(1 + y_n^2)^5},$$

with $y_0 = 1$.

□

A much more useful class of methods is derived by considering approximations to the first $p+1$ terms on the right of (13.28). We seek an expression of the form

$$y_{n+1} = y_n + h\sum_{i=1}^{r} \alpha_i k_i, \tag{13.32}$$

where $k_1 = f(x_n, y_n)$ and

$$k_i = f(x_n + h\lambda_i, y_n + h\mu_i k_{i-1}), \qquad i > 1.$$

We try to determine the coefficients α_i, λ_i and μ_i so that the first $p+1$ terms in the Taylor series expansion of (13.32) agree with the first $p+1$ terms in (13.28). Taylor's theorem for two variables gives

$$\begin{aligned} &f(x_n + h\lambda, y_n + h\mu k) \\ &\quad = f(x_n, y_n) + h\left(\lambda\frac{\partial}{\partial x} + \mu k\frac{\partial}{\partial y}\right)f(x_n, y_n) + \frac{h^2}{2!}\left(\lambda\frac{\partial}{\partial x} + \mu k\frac{\partial}{\partial y}\right)^2 f(x_n, y_n) \\ &\qquad + \cdots + \frac{h^{p-1}}{(p-1)!}\left(\lambda\frac{\partial}{\partial x} + \mu k\frac{\partial}{\partial y}\right)^{p-1} f(x_n, y_n) + h^p S_p(x_n, y_n, \lambda, \mu k), \end{aligned} \tag{13.33}$$

where $h^p S_p$ is the remainder term. It is clear that equating terms of (13.32), using the expansions (13.33), with the terms of the expansion of (13.28), using (13.26), is very tedious. We shall illustrate only the case $p = 2$, when (13.28) becomes

$$y(x_{n+1}) = y(x_n) + hf + \tfrac{1}{2}h^2(f_x + f_y f) + h^3 R_3(x_n), \tag{13.34}$$

where f, f_x and f_y have argument $(x_n, y(x_n))$. Clearly there are an infinity of ways of choosing r, α_i, λ_i and μ_i.

If we choose $r = 2$, (13.32) when expanded becomes

$$\begin{aligned} y_{n+1} = y_n &+ h(\alpha_1 + \alpha_2)f + h^2\alpha_2(\lambda_2 f_x + \mu_2 f f_y) \\ &+ h^3\alpha_2 S_2(x_n, y_n, \lambda_2, \mu_2 f), \end{aligned} \tag{13.35}$$

where f, f_x and f_y have arguments (x_n, y_n). On comparing (13.34) and (13.35) we see that we require

$$\alpha_1 + \alpha_2 = 1, \qquad \alpha_2\lambda_2 = \tfrac{1}{2}, \qquad \alpha_2\mu_2 = \tfrac{1}{2}. \tag{13.36}$$

One solution is

$$\alpha_1 = 0, \qquad \alpha_2 = 1, \qquad \lambda_2 = \mu_2 = \tfrac{1}{2},$$

when (13.32) becomes

$$y_{n+1} = y_n + hf(x_n + \tfrac{1}{2}h, y_n + \tfrac{1}{2}hf(x_n, y_n)). \tag{13.37}$$

This is known as the *modified Euler* method.

If we take $\alpha_1 = \alpha_2 = \frac{1}{2}$, then $\lambda_2 = \mu_2 = 1$ and we obtain

$$y_{n+1} = y_n + \tfrac{1}{2}h[f(x_n, y_n) + f(x_{n+1}, y_n + hf(x_n, y_n))], \tag{13.38}$$

which is known as the *simple Runge–Kutta* method.

For $p = 4$ and $r = 4$, we obtain the *classical Runge–Kutta* method

$$y_{n+1} = y_n + \tfrac{1}{6}h[k_1 + 2k_2 + 2k_3 + k_4], \tag{13.39}$$

where

$$\begin{aligned} k_1 &= f(x_n, y_n), \\ k_2 &= f(x_n + \tfrac{1}{2}h, y_n + \tfrac{1}{2}hk_1), \\ k_3 &= f(x_n + \tfrac{1}{2}h, y_n + \tfrac{1}{2}hk_2), \\ k_4 &= f(x_{n+1}, y_n + hk_3). \end{aligned}$$

This is not the only choice of method for $p = r = 4$. Ralston and Rabinowitz (1978) give a complete account of the derivation of this and other similar methods.

13.4 Truncation errors of one-step methods

The general one-step method of solving (13.1) may be expressed as

$$y_{n+1} = y_n + h\phi(x_n, y_n; h) \tag{13.40}$$

where $\phi(x, y; h)$ is called the *increment function*. For $h \neq 0$, we rewrite (13.40) as

$$\frac{y_{n+1} - y_n}{h} = \phi(x_n, y_n; h). \tag{13.41}$$

We are interested in the accuracy of (13.41) when considered as an approximation to (13.1). If $y(x)$ is the solution of (13.1) and (13.2), we define the *local truncation error*, $t(x; h)$, of (13.40) to be

$$t(x; h) = \frac{y(x+h) - y(x)}{h} - \phi(x, y(x); h), \tag{13.42}$$

where $x \in [x_0, b]$ and $h > 0$. We say that (13.41) is a *consistent approximation* to (13.1) if

$$t(x; h) \to 0, \qquad \text{as } h \to 0,$$

uniformly for $x \in [x_0, b]$. From (13.42) we see that, for a consistent method,

$$y'(x) = \phi(x, y(x); 0) = f(x, y(x)). \tag{13.43}$$

More precisely, we say that the method (13.40) is *consistent of order* p, if there exists $N \geqslant 0$, $h_0 > 0$ and a positive integer p such that

$$\sup_{x_0 \leqslant x \leqslant b} |t(x; h)| \leqslant Nh^p, \qquad \text{for all } h \in (0, h_0]. \tag{13.44}$$

The Taylor series method (13.29) is consistent of order p as, on comparing (13.28) and (13.29), we see from (13.42) that

$$t(x; h) = h^p R_{p+1}(x) = \frac{h^p}{(p+1)!} y^{(p+1)}(\xi), \tag{13.45}$$

where $x < \xi < x + h$. We choose

$$N = \frac{1}{(p+1)!} \max_{x_0 \leqslant x \leqslant b} |y^{(p+1)}(x)|. \tag{13.46}$$

The Runge–Kutta method (13.32) with $p+1$ terms of its expansion agreeing with the Taylor series will also be consistent of order p. It is difficult to obtain an explicit form of N. We just observe from (13.33) and (13.28) that the truncation error takes the form

$$t(x; h) = h^p (R_{p+1}(x) + Q_p(x, h)), \tag{13.47}$$

where Q_p depends on p, r and the coefficients α_i, λ_i, μ_i. Provided sufficient derivatives of y and f are bounded, we can choose a suitable N. Full details are given by Henrici (1962).

There are methods of the Runge–Kutta type for which it is possible to estimate the local truncation error t from the values calculated. The most commonly used method is due to R. H. Merson, and details of this and other methods may be found in Lambert (1991). Such estimates of the truncation error may be used to produce methods which automatically adjust the step size h according to the accuracy desired in the results.

13.5 Convergence of one-step methods

So far we have considered local truncation errors. These give a measure of the accuracy of the difference equation used to approximate to a differential equation. We have not considered how accurately the solution sequence (y_n), of the difference equation, approximates to the solution $y(x)$ of the differential equation. We shall also be interested in knowing whether the solution of the difference equation converges to the solution of the differential equation as the step size h is decreased. We call $y(x_n) - y_n$ the *global truncation error.*

We must define convergence carefully. If we look at the behaviour of y_n as $h \to 0$ and keep n fixed, we will not obtain a useful concept. Clearly

$$x_n = x_0 + nh \to x_0, \qquad \text{as } h \to 0, \quad \text{with } n \text{ fixed},$$

yet we are interested in obtaining solutions for values of x other than $x = x_0$. We must therefore consider the behaviour of y_n as $h \to 0$ with $x_n = x_0 + nh$ kept fixed. In order to obtain a solution at a fixed value of $x \neq x_0$, we must increase the number of steps required to reach x from x_0 if the step size h is decreased. If

$$\lim_{\substack{h \to 0 \\ x_n = x \text{ fixed}}} y_n = y(x) \tag{13.48}$$

for all $x \in [x_0, b]$, we say that the method is convergent. This definition is due to G. Dahlquist.

Example 13.11 The Taylor series method (13.29) applied to $y' = y$ with $y(0) = 1$, $x \in [0, b]$, is convergent. In Example 13.9 we obtained

$$y_n = \left(1 + h + \cdots + \frac{h^p}{p!}\right)^n.$$

Now by Taylor's theorem

$$e^h = 1 + h + \cdots + \frac{h^p}{p!} + \frac{h^{p+1}}{(p+1)!} e^{h'}, \quad \text{where } 0 < h' < h,$$

and thus

$$\begin{aligned} y_n &= \left(e^h - \frac{h^{p+1}}{(p+1)!} e^{h'}\right)^n \\ &= e^{nh}\left(1 - \frac{h^{p+1}}{(p+1)!} e^{h'-h}\right)^n. \end{aligned}$$

As $x_n = nh$ and $y(x_n) = e^{nh}$, we have for the global truncation error

$$\begin{aligned} |y(x_n) - y_n| &= e^{nh}\left|\left(1 - \frac{h^{p+1}}{(p+1)!} e^{h'-h}\right)^n - 1\right| \\ &\leqslant e^{x_n} \frac{nh^{p+1}}{(p+1)!} e^{h'-h} \end{aligned}$$

for h sufficiently small (see Example 2.13). Thus, as $e^{h'-h}<1$,

$$|y(x_n)-y_n| \leqslant e^{x_n} \frac{x_n h^p}{(p+1)!}. \tag{13.49}$$

Hence $|y(x_n)-y_n| \to 0$ as $h \to 0$ with x_n fixed.

We also remark that there exists M such that

$$|y(x_n)-y_n| \leqslant Mh^p. \tag{13.50}$$

We choose any

$$M \geqslant \max_{0 \leqslant x \leqslant b} \frac{xe^x}{(p+1)!}. \qquad \square$$

The following theorem provides an error bound for a consistent one-step method, provided the increment function ϕ satisfies a Lipschitz condition with respect to y. From this error bound it follows that such a method is convergent.

Theorem 13.2 The initial value problem

$$y' = f(x,y), \qquad y(x_0)=s, \qquad x \in [x_0, b],$$

is replaced by a one-step method:

$$\left.\begin{aligned} y_0 &= s \\ x_n &= x_0 + nh \\ y_{n+1} &= y_n + h\phi(x_n, y_n; h) \end{aligned}\right\} \quad n = 0, 1, 2, \ldots$$

where ϕ satisfies the Lipschitz condition

$$|\phi(x,y;h) - \phi(x,z;h)| \leqslant L_\phi |y-z| \tag{13.51}$$

for all $x \in [x_0, b]$, $-\infty < y, z < \infty$, $h \in (0, h_0]$, for some L_ϕ and $h_0 > 0$.

If the method is consistent of order p, so that the local truncation error defined by (13.42) satisfies (13.44), then the global truncation error $y(x_n) - y_n$ satisfies

$$|y(x_n)-y_n| \leqslant \begin{cases} \left(\dfrac{e^{(x_n-x_0)L_\phi}-1}{L_\phi}\right) Nh^p, & \text{for } L_\phi \neq 0, \\ (x_n - x_0)Nh^p, & \text{for } L_\phi = 0, \end{cases} \tag{13.52}$$

for all $x_n \in [x_0, b]$ and all $h \in (0, h_0]$, and thus the method is convergent, as (13.48) is satisfied.

Proof We let

$$e_n = |y(x_n) - y_n|.$$

From (13.42),

$$y(x_{n+1}) = y(x_n) + h\phi(x_n, y(x_n); h) + ht(x_n; h) \tag{13.53}$$

and

$$y(x_{n+1}) - y_{n+1} = y(x_n) - y_n + h\phi(x_n, y(x_n); h) - h\phi(x_n, y_n; h) + ht(x_n; h).$$

Thus

$$\begin{aligned} |y(x_{n+1}) - y_{n+1}| &\leqslant |y(x_n) - y_n| + h|\phi(x_n, y(x_n); h) - \phi(x_n, y_n; h)| \\ &\quad + h|t(x_n; h)| \\ &\leqslant |y(x_n) - y_n| + hL_\phi|y(x_n) - y_n| + Nh^{p+1}, \end{aligned} \tag{13.54}$$

where we have used (13.44) and the Lipschitz condition (13.51). Hence

$$e_{n+1} \leqslant (1 + hL_\phi)e_n + Nh^{p+1}.$$

This is a difference inequality of the type considered in Lemma 13.1. In this case $m = 0$ and

$$\delta = e_0 = |y(x_0) - y_0| = 0$$

and from (13.23), for $L_\phi \neq 0$,

$$\begin{aligned} e_n &\leqslant \frac{e^{nhL_\phi} - 1}{hL_\phi} Nh^{p+1} \\ &= \left(\frac{e^{(x_n - x_0)L_\phi} - 1}{L_\phi}\right) Nh^p. \end{aligned}$$

For $L_\phi = 0$, Lemma 13.1 yields

$$e_n \leqslant n \,.\, Nh^{p+1} = (x_n - x_0)Nh^p. \qquad \square$$

You will notice from (13.52) that, for all $h \in (0, h_0]$ and all $x_n \in [x_0, b]$, there is a constant $M \geqslant 0$ such that

$$|y(x_n) - y_n| \leqslant Mh^p. \tag{13.55}$$

A method for which the global error satisfies (13.55) is said to be *convergent of order p*. We shall normally expect that a method which is consistent of order p is also convergent of order p and we say simply that such a method is *of order p*.

It follows from (13.43) that, for a consistent method, (13.51) reduces to

$$|f(x, y) - f(x, z)| \leqslant L_\phi|y - z|$$

when $h=0$. Thus (13.51) implies the existence of a Lipschitz condition on $f(x, y)$, of the type required to ensure the existence of a unique solution of the differential equation.

In order to obtain expressions for L_ϕ for the Taylor series and Runge–Kutta methods, we assume that bounds

$$L_k \geqslant \sup_{\substack{x_0 \leqslant x \leqslant b \\ -\infty < y < \infty}} \left| \frac{\partial f^{(k)}(x, y)}{\partial y} \right|$$

exist for the values of k required in the following analysis.

As in Example 13.4 we observe, using the mean value theorem, that we may choose $L = L_0$. For the Taylor series method (13.29),

$$\phi(x, y; h) = f(x, y) + \frac{h}{2!} f^{(1)}(x, y) + \cdots + \frac{h^{p-1}}{p!} f^{(p-1)}(x, y)$$

and, on using the mean value theorem for each of the $f^{(k)}$, we obtain

$$|\phi(x, y; h) - \phi(x, z; h)| \leqslant \left(L_0 + \frac{h}{2!} L_1 + \frac{h^2}{3!} L_2 + \cdots + \frac{h^{p-1}}{p!} L_{p-1}\right)|y - z|.$$

We therefore choose

$$L_\phi = L_0 + \frac{h_0}{2!} L_1 + \cdots + \frac{h_0^{p-1}}{p!} L_{p-1}, \qquad \text{for } h \in (0, h_0]. \qquad (13.56)$$

For the simple Runge–Kutta method (13.38),

$$\phi(x, y; h) = \tfrac{1}{2}[f(x, y) + f(x + h, y + hf(x, y))].$$

Now with $L = L_0$ we have the Lipschitz condition on f,

$$|f(x, y) - f(x, z)| \leqslant L_0 |y - z|$$

and, for $h \in (0, h_0]$,

$$\begin{aligned} &|f(x + h, y + hf(x, y)) - f(x + h, z + hf(x, z))| \\ &\quad \leqslant L_0 |y + hf(x, y) - z - hf(x, z)| \\ &\quad \leqslant L_0[|y - z| + h|f(x, y) - f(x, z)|] \\ &\quad \leqslant L_0[|y - z| + hL_0|y - z|] \\ &\quad = L_0(1 + h_0 L_0)|y - z|. \end{aligned}$$

Thus

$$\begin{aligned} |\phi(x, y; h) - \phi(x, z; h)| &\leqslant \tfrac{1}{2}[L_0 + L_0(1 + h_0 L_0)]|y - z| \\ &= L_0(1 + \tfrac{1}{2} h_0 L_0)|y - z|, \end{aligned}$$

so we may choose

$$L_\phi = L_0(1 + \tfrac{1}{2} h_0 L_0). \tag{13.57}$$

For the classical Runge–Kutta method (13.39), we may choose

$$L_\phi = L_0\left(1 + \frac{h_0 L_0}{2} + \frac{h_0^2 L_0^2}{6} + \frac{h_0^3 L_0^3}{24}\right). \tag{13.58}$$

Full details are given by Henrici (1962).

Example 13.12 We reconsider Example 13.11, only this time we will obtain a global error bound, using Theorem 13.2.

We know that $y(x) = \mathrm{e}^x$ and thus from (13.46)

$$N = \frac{1}{(p+1)!} \max_{0 \leqslant x \leqslant b} |y^{(p+1)}(x)| = \frac{\mathrm{e}^b}{(p+1)!}.$$

From Example 13.9, $f^{(k)} = y$ for all k and thus $L_k = 1$ for all k. We therefore choose, from (13.56),

$$L_\phi = 1 + \frac{h_0}{2!} + \cdots + \frac{h_0^{p-1}}{p!}$$

and (13.52) becomes

$$|y(x_n) - y_n| \leqslant \left(\frac{\mathrm{e}^{x_n L_\phi} - 1}{L_\phi}\right) \frac{\mathrm{e}^b}{(p+1)!} h^p.$$

For $x \in [0, b]$ this bound is larger and hence not as good as that of (13.49). □

Example 13.13 Find a bound of the global error for Euler's method applied to the problem

$$y' = \frac{1}{1+y^2}, \qquad \text{with } y(0) = 1 \quad \text{and} \quad 0 \leqslant x \leqslant 1.$$

From (13.45) with $p = 1$ the truncation error is

$$t(x; h) = \frac{h}{2!} y''(\xi) = \frac{h}{2} f^{(1)}(\xi, y(\xi)).$$

By sketching a graph, we find that

$$f^{(1)} = f_x + f f_y = \frac{-2y}{(1+y^2)^3}$$

has maximum modulus for $y = 1/\sqrt{5}$ and thus

$$|f^{(1)}(x, y(x))| \leqslant \frac{25\sqrt{5}}{108}.$$

We therefore obtain

$$|t(x; h)| \leqslant h\,\frac{25\sqrt{5}}{216}.$$

Since $\phi(x, y; h) = f(x, y)$ for Euler's method, we choose $L_\phi = L = 3\sqrt{3}/8$ (see Example 13.4). The bound (13.52) becomes

$$|y_n - y(x_n)| \leqslant (\mathrm{e}^{x_n L_\phi} - 1)\,\frac{25\sqrt{5}}{81\sqrt{3}}\,h.$$

For $x_n = 1$,

$$|y_n - y(1.0)| < 0.365\,h.$$

This is a very pessimistic bound for the error. With $h = 0.1$ the absolute value of the actual error is 0.00687. □

From the last example it can be seen that Theorem 13.2 may not give a useful bound for the error. In practice this is often true and in any case it is usually impossible to determine (13.52) because of the difficulty of obtaining suitable values of N and L_ϕ. The importance of Theorem 13.2 is that it describes the general behaviour of the global errors as $h \to 0$.

13.6 Effect of rounding errors on one-step methods

One difficulty when computing with a recurrence relation such as

$$y_{n+1} = y_n + h\phi(x_n, y_n; h)$$

is that an error will be introduced at each step due to rounding. If we reduce h so as to reduce the truncation error, then for a given x_n we increase the number of steps and the number of rounding errors. For a given arithmetical accuracy, there is clearly a *minimum* step size h below which rounding errors will produce inaccuracies larger than those due to truncation errors. Suppose that instead of (y_n) we actually compute a sequence (z_n) with

$$z_{n+1} = z_n + h\phi(x_n, z_n; h) + \varepsilon_n, \tag{13.59}$$

where ε_n is due to rounding. From (13.53) and (13.59) we obtain, analogous to (13.54),

$$\begin{aligned}|y(x_{n+1}) - z_{n+1}| \leqslant{}& |y(x_n) - z_n| + h|\phi(x_n, y(x_n); h) - \phi(x_n, z_n; h)| \\ &+ h|t(x_n; h)| + |\varepsilon_n|.\end{aligned}$$

Thus, if $|\varepsilon_n| \leq \varepsilon$ for all n, a quantity ε/h must be added to Nh^p in (13.52) to bound $|y(x_n) - z_n|$. The choice of method and arithmetical accuracy to which we work should therefore depend on whether ε/h is appreciable in magnitude when compared with Nh^p.

13.7 Methods based on numerical integration; explicit methods

On integrating the differential equation

$$y' = f(x, y)$$

between limits x_n and x_{n+1}, we obtain

$$y(x_{n+1}) = y(x_n) + \int_{x_n}^{x_{n+1}} f(x, y(x))\, dx. \tag{13.60}$$

The basis of many numerical methods of solving the differential equation is to replace the integral in (13.60) by a suitable approximation. For example, we may replace $f(x, y(x))$ by its Taylor polynomial constructed at $x = x_n$ and integrate this polynomial. This leads to the Taylor series method described earlier (see Problem 13.19). A more successful approach is to replace f by an interpolating polynomial. Suppose that we have already calculated approximations $y_0, y_1, \ldots, y_n$ to $y(x)$ at the equally spaced points $x_r = x_0 + rh$, $r = 0, 1, \ldots, n$. We write

$$f_r = f(x_r, y_r), \qquad r = 0, 1, \ldots, n;$$

these are approximations to $f(x_r, y(x_r))$.

We construct the interpolating polynomial p_m through the $m+1$ points (x_n, f_n), (x_{n-1}, f_{n-1}), $\ldots$, (x_{n-m}, f_{n-m}), where $m \leq n$. We use $p_m(x)$ as an approximation to $f(x, y(x))$ between x_n and x_{n+1} and replace (13.60) by

$$y_{n+1} = y_n + \int_{x_n}^{x_{n+1}} p_m(x)\, dx. \tag{13.61}$$

It should be remembered that $p_m(x)$, being based on the values $f_n, f_{n-1}, \ldots, f_{n-m}$, is not the exact interpolating polynomial for $f(x, y(x))$ constructed at the points $x_n, x_{n-1}, \ldots, x_{n-m}$.

By the backward difference formula (4.36),

$$p_m(x) = \sum_{j=0}^{m} (-1)^j \binom{-s}{j} \nabla^j f_n, \qquad \text{where } s = (x - x_n)/h.$$

Thus, as in §7.1,

$$\int_{x_n}^{x_{n+1}} p_m(x)\, dx = h \sum_{j=0}^{m} b_j\, \nabla^j f_n, \tag{13.62}$$

where

$$b_j = (-1)^j \int_0^1 \binom{-s}{j} \mathrm{d}s. \tag{13.63}$$

Note that the b_j are independent of m and n.

On substituting (13.62) in (13.61), we obtain the *Adams–Bashforth* algorithm for an initial value problem. To compute approximations to the solution $y(x)$ at equally spaced points $x_r = x_0 + rh$, $r = 0, 1, \ldots,$ given the starting approximations $y_0, y_1, \ldots, y_m$, we use the recurrence relation

$$y_{n+1} = y_n + h[b_0 f_n + b_1 \nabla f_n + \cdots + b_m \nabla^m f_n] \tag{13.64}$$

for $n = m, m+1, \ldots,$ where $f_r = f(x_r, y_r)$.

The starting values $y_0, y_1, \ldots, y_m$ are usually found by a one-step method. Table 13.1 exhibits some of the b_j. We shall see that (13.64) is of order $m+1$, that is, both the local and global truncation errors are of order $m+1$.

Table 13.1 The coefficients b_j

j	0	1	2	3	4
b_j	1	$\frac{1}{2}$	$\frac{5}{12}$	$\frac{3}{8}$	$\frac{251}{720}$

In practice we usually expand the differences in (13.64) so that it takes the form

$$y_{n+1} = y_n + h[\beta_0 f_n + \beta_1 f_{n-1} + \cdots + \beta_m f_{n-m}]. \tag{13.65}$$

The β_j now depend on m. For example, when $m = 1$ we obtain the second order method

$$y_{n+1} = y_n + \tfrac{1}{2} h(3f_n - f_{n-1}) \tag{13.66}$$

and, for $m = 3$, the fourth order method

$$y_{n+1} = y_n + \tfrac{1}{24} h(55f_n - 59f_{n-1} + 37f_{n-2} - 9f_{n-3}). \tag{13.67}$$

The following algorithm is for the fourth order method (13.67) on the interval $[a, b]$. Four starting values y_0, y_1, y_2, y_3 are required and these are found by a one-step method such as the fourth order Runge–Kutta method (13.39).

Algorithm 13.1 (Adams–Bashforth of order 4). Let F_0, F_1, F_2, F_3 represent the values of $f_n, f_{n-1}, f_{n-2}, f_{n-3}$ respectively. Choose a step h such that $(b-a)/h$ is an integer.

Start with y_0, y_1, y_2, y_3
$X := a + 3h; \qquad Y := y_3$
$F_0 := f(X, y_3)$
$F_1 := f(X - h, y_2)$
$F_2 := f(X - 2h, y_1)$
$F_3 := f(X - 3h, y_0)$
repeat
  $Y := Y + \frac{1}{24}h\,(55F_0 - 59F_1 + 37F_2 - 9F_3)$
  $F_3 := F_2; \qquad F_2 := F_1; \qquad F_1 := F_0$
  $X := X + h$
  $F_0 := f(X, Y)$
until $X = b$ □

You will notice that, in the above algorithm, f is evaluated only once in each step.

To determine the truncation error of the Adams–Bashforth formula, we rewrite (13.64) as

$$\frac{y_{n+1} - y_n}{h} = \sum_{j=0}^{m} b_j \nabla^j f_n \tag{13.68}$$

and consider this as an approximation to (13.1). Now from (7.22)

$$\int_{x_n}^{x_{n+1}} f(x, y(x))\,\mathrm{d}x = \int_{x_n}^{x_{n+1}} y'(x)\,\mathrm{d}x$$
$$= h\sum_{j=0}^{m} b_j \nabla^j y'(x_n) + h^{m+2} b_{m+1} y^{(m+2)}(\xi_n),$$

where $x_{n-m} < \xi_n < x_{n+1}$. Thus from (13.60)

$$\frac{y(x_{n+1}) - y(x_n)}{h} = \sum_{j=0}^{m} b_j \nabla^j f(x_n, y(x_n)) + h^{m+1} b_{m+1} y^{(m+2)}(\xi_n), \tag{13.69}$$

where $\nabla^j f(x_n, y(x_n)) = \nabla^{j-1} f(x_n, y(x_n)) - \nabla^{j-1} f(x_{n-1}, y(x_{n-1}))$.

The last term in (13.69) is the *local truncation error*. You will notice that this is of order $m + 1$. We expand the differences in (13.69) to obtain

$$\begin{aligned} y(x_{n+1}) = y(x_n) + h[\beta_0 f(x_n, y(x_n)) + \beta_1 f(x_{n-1}, y(x_{n-1})) + \cdots \\ + \beta_m f(x_{n-m}, y(x_{n-m}))] + h^{m+2} b_{m+1} y^{(m+2)}(\xi_n). \end{aligned} \tag{13.70}$$

We assume that $y^{(m+2)}(x)$ is bounded for $x \in [x_0, b]$, so there exists M_{m+2} such that

$$\max_{x \in [x_0, b]} |y^{(m+2)}(x)| \leq M_{m+2}. \tag{13.71}$$

We again let

$$e_n = |y(x_n) - y_n|$$

and, by the Lipschitz condition on f,

$$|f(x_n, y(x_n)) - f(x_n, y_n)| \leq L|y(x_n) - y_n| = Le_n.$$

On subtracting (13.65) from (13.70), we obtain

$$e_{n+1} \leq e_n + hL[|\beta_0|e_n + |\beta_1|e_{n-1} + \cdots + |\beta_m|e_{n-m}] + h^{m+2}|b_{m+1}|M_{m+2}. \tag{13.72}$$

We suppose that the starting values $y_0, \ldots, y_m$ are such that

$$e_r \leq \delta, \qquad \text{for } r = 0, 1, \ldots, m.$$

We now apply Lemma 13.1 with

$$\alpha_r = hL|\beta_r| \quad \text{and} \quad b = h^{m+2}|b_{m+1}|M_{m+2},$$

so that

$$A = \alpha_0 + \alpha_1 + \cdots + \alpha_m = hLB_m,$$

where

$$B_m = |\beta_0| + |\beta_1| + \cdots + |\beta_m|. \tag{13.73}$$

From (13.23) we obtain, remembering that $nh = x_n - x_0$,

$$|y(x_n) - y_n| = e_n \leq \left(\delta + \frac{h^{m+1}|b_{m+1}|M_{m+2}}{LB_m}\right) \exp((x_n - x_0)LB_m) - \frac{h^{m+1}|b_{m+1}|M_{m+2}}{LB_m}. \tag{13.74}$$

The inequality (13.74) provides a global truncation error bound for the Adams–Bashforth algorithm. This bound is again more of qualitative than quantitative interest. It is usually impossible to obtain useful bounds on the derivatives of $y(x)$; however, (13.74) does provide much information about the behaviour of the error as h is decreased. From the first term on the right-hand side, it can be seen that if the method is to be of order $m+1$, we require δ, a bound for the error in the starting values, to decrease like h^{m+1} as h tends to zero. Thus we should choose a one-step method of order $m+1$ to determine starting values.

Theorem 13.3 The Adams–Bashforth algorithm (13.64) is convergent of order $m+1$ if $y^{(m+2)}(x)$ is bounded for $x \in [x_0, b]$ and the starting values

$y_0, \ldots, y_m$ are such that

$$|y(x_r) - y_r| \leqslant Dh^{m+1}, \qquad r = 0, 1, \ldots, m,$$

for some $D \geqslant 0$ and all $h \in (0, h_0]$ where $h_0 > 0$. Furthermore, the global truncation errors satisfy (13.74), where M_{m+2} and B_m are defined by (13.71) and (13.73) and

$$\delta = \max_{0 \leqslant r \leqslant m} |y(x_r) - y_r|. \qquad \square$$

Example 13.14 As we have seen before, for simple equations it is possible to determine bounds on the derivatives of y and hence compute global error bounds. We will seek a global error bound for the second order Adams–Bashforth algorithm applied to the problem of Example 13.13. From Example 13.4, we have $L = 3\sqrt{3}/8$ and in Example 13.10 we found

$$f^{(2)}(x, y) = \frac{10y^2 - 2}{(1 + y^2)^5}.$$

Thus

$$|y^{(3)}(x)| = |f^{(2)}(x, y(x))| \leqslant \max_{-\infty < y < \infty} \left| \frac{10y^2 - 2}{(1 + y^2)^5} \right| = 2$$

and we choose $M_3 = 2$.

For the Adams–Bashforth algorithm of order 2,

$$m = 1, \qquad b_2 = \tfrac{5}{12}, \qquad B_1 = |\beta_0| + |\beta_1| = 2$$

and thus (13.74) yields

$$\begin{aligned} |y(x_n) - y_n| &\leqslant \left(\delta + h^2 \frac{5}{12} \cdot \frac{8}{3\sqrt{3}} \cdot \frac{2}{2}\right) \exp(3\sqrt{3}(x_n - 0)/4) \\ &\quad - h^2 \frac{5}{12} \cdot \frac{8}{3\sqrt{3}} \cdot \frac{2}{2} \\ &= \left(\delta + \frac{10h^2}{9\sqrt{3}}\right) e^{3\sqrt{3}x_n/4} - \frac{10h^2}{9\sqrt{3}}. \end{aligned} \tag{13.75}$$

This error bound is considerably larger than the actual error. For example, with $h = 0.1$, $n = 10$ and y_1 obtained by the simple Runge–Kutta method (which yields $\delta = 10^{-5}$), (13.75) becomes

$$|y(1.0) - y_{10}| \leqslant 0.0172.$$

In fact, to five decimal places,

$$y(1.0) = 1.40629 \quad \text{and} \quad y_{10} = 1.40584$$

so that

$$|y(1.0) - y_{10}| = 0.00045. \qquad \square$$

There is no reason why we should not consider the integral form of the differential equation over intervals other than x_n to x_{n+1}. For example, instead of (13.60), we may consider

$$y(x_{n+1}) = y(x_{n-1}) + \int_{x_{n-1}}^{x_{n+1}} f(x, y(x))\,\mathrm{d}x. \tag{13.76}$$

On replacing $f(x, y(x))$ by a polynomial as before, we derive the formula

$$y_{n+1} = y_{n-1} + h[b_0^* f_n + b_1^* \nabla f_n + \cdots + b_m^* \nabla^m f_n], \tag{13.77}$$

where

$$b_j^* = (-1)^j \int_{-1}^{1} \binom{-s}{j} \mathrm{d}s. \tag{13.78}$$

This method is attributed to E. J. Nyström.

The Adams–Bashforth and Nyström algorithms are known as *explicit* or *open* methods. The recurrence relations (13.64) and (13.77) do not involve $f(x_{n+1}, y_{n+1})$; they express y_{n+1} explicitly in terms of $y_n, y_{n-1}, \ldots, y_{n-m}$. The explicit nature of these algorithms arises from the fact that we use extrapolation when obtaining an approximation to $f(x, y(x))$ between x_n and x_{n+1}. As was seen in Chapter 4, extrapolation, that is interpolation outside the range of the interpolating points, is usually less accurate than interpolation inside this range. The class of formulas considered in the next section avoid extrapolation by introducing the extra point (x_{n+1}, f_{n+1}) in the interpolatory process.

13.8 Methods based on numerical integration; implicit methods

We again approximate to the integral in (13.60) and thus obtain a recurrence relation. Suppose that $q_{m+1}(x)$ is the interpolating polynomial through the points $(x_{n+1}, f_{n+1}), (x_n, f_n), \ldots, (x_{n-m}, f_{n-m})$. We regard $q_{m+1}(x)$ as an approximation to $f(x, y(x))$ between x_n and x_{n+1} and replace (13.60) by

$$y_{n+1} = y_n + \int_{x_n}^{x_{n+1}} q_{m+1}(x)\,\mathrm{d}x.$$

Notice that this approximation does not involve extrapolation.

By the backward difference formula (4.36),

$$q_{m+1}(x) = \sum_{j=0}^{m+1} (-1)^j \binom{-s}{j} \nabla^j f_{n+1},$$

where $s = (x - x_{n+1})/h$, and

$$\int_{x_n}^{x_{n+1}} q_{m+1}(x)\,dx = h \sum_{j=0}^{m+1} c_j \nabla^j f_{n+1},$$

where

$$c_j = (-1)^j \int_{-1}^{0} \binom{-s}{j} ds. \tag{13.79}$$

The latter are independent of m and n and Table 13.2 gives some of the values. We thus replace (13.60) by the approximation

$$y_{n+1} = y_n + h[c_0 f_{n+1} + c_1 \nabla f_{n+1} + \cdots + c_{m+1} \nabla^{m+1} f_{n+1}]. \tag{13.80}$$

Given $y_n, y_{n-1}, \ldots, y_{n-m}$, we seek y_{n+1} satisfying (13.80). In general, (13.80) does not yield an explicit expression for y_{n+1} as each of the terms $\nabla^j f_{n+1}$ involves $f(x_{n+1}, y_{n+1})$. Usually (13.80) must be solved as an algebraic equation in y_{n+1} by an iterative method. Let us suppose that $y_{n+1}^{(i)}$ is the ith iterate in this process. Then (13.80) is already in a convenient form for iteration of the kind

$$w_{i+1} = F(w_i)$$

where F is some given function and $w_i = y_{n+1}^{(i)}$. The $y_n, y_{n-1}, \ldots, y_{n-m}$ remain unchanged throughout this iterative process to calculate y_{n+1}. We will examine the convergence of this process in the next section. To calculate a good first approximation $y_{n+1}^{(0)}$, we employ one of the explicit formulas of the last section, for example, the Adams–Bashforth formula.

This leads to the *Adams–Moulton* algorithm for an initial value problem. To compute approximations to the solution $y(x)$ at equally spaced points $x_k = x_0 + kh$, $k = 0, 1, \ldots,$ given the starting approximations $y_0, y_1, \ldots, y_m$, we

Table 13.2 The coefficients c_j

j	0	1	2	3	4
c_j	1	$-\frac{1}{2}$	$-\frac{1}{12}$	$-\frac{1}{24}$	$-\frac{19}{720}$

use the following for $n = m, m+1, \ldots$.

$$\text{Predictor: } y_{n+1}^{(0)} = y_n + h[b_0 f_n + b_1 \nabla f_n + \cdots + b_m \nabla^m f_n] \tag{13.81a}$$

$$\text{Corrector: } y_{n+1}^{(i+1)} = y_n + h[c_0 f_{n+1}^{(i)} + c_1 \nabla f_{n+1}^{(i)} + \cdots + c_{m+1} \nabla^{m+1} f_{n+1}^{(i)}],$$

$$i = 0, 1, \ldots, (I-1), \tag{13.81b}$$

$$y_{n+1} = y_{n+1}^{(I)},$$

where

$$f_{n+1}^{(i)} = f(x_{n+1}, y_{n+1}^{(i)}), \qquad f_n = f(x_n, y_n)$$

and

$$\nabla^j f_{n+1}^{(i)} = \nabla^{j-1} f_{n+1}^{(i)} - \nabla^{j-1} f_n.$$

The coefficients b_j and c_j are defined by (13.63) and (13.79) respectively. Equation (13.81a) is called the *predictor* and (13.81b) the *corrector*. For each step, I inner iterations of the corrector (or I corrections) are performed and $y_{n+1}^{(I)}$ is considered a sufficiently accurate approximation to the y_{n+1} of (13.80). We shall prove later that (13.81b) yields a method of order $m+2$. The following algorithm describes the process which integrates a differential equation over the interval $[a, b]$.

Algorithm 13.2 (Adams–Moulton). Choose a step h such that $N = (b-a)/h$ is an integer.

Obtain starting values $y_0, \ldots, y_m$ by a one step method.
$n := m$
repeat
  calculate $y_{n+1}^{(0)}$ from the predictor
  $i := 0$
  repeat
    calculate $y_{n+1}^{(i+1)}$ from the corrector
    $i := i + 1$
  until $i = I$
  $y_{n+1} := y_{n+1}^{(I)}$
  $n := n + 1$
until $n = N$ □

We normally expand the differences so that the formulas become

$$\text{Predictor: } y_{n+1}^{(0)} = y_n + h[\beta_0 f_n + \beta_1 f_{n-1} + \cdots + \beta_m f_{n-m}] \tag{13.82a}$$

$$\text{Corrector: } y_{n+1}^{(i+1)} = y_n + h[\gamma_{-1} f_{n+1}^{(i)} + \gamma_0 f_n + \cdots + \gamma_m f_{n-m}]. \tag{13.82b}$$

The β_r and γ_r are dependent on m. For $m = 0$, we obtain the second order method

$$y_{n+1}^{(0)} = y_n + h f_n, \tag{13.83a}$$

$$y_{n+1}^{(i+1)} = y_n + \tfrac{1}{2} h (f_{n+1}^{(i)} + f_n) \tag{13.83b}$$

and for $m = 2$, the fourth order method

$$y_{n+1}^{(0)} = y_n + \tfrac{1}{12} h(23f_n - 16f_{n-1} + 5f_{n-2}), \tag{13.84a}$$

$$y_{n+1}^{(i+1)} = y_n + \tfrac{1}{24} h(9f_{n+1}^{(i)} + 19f_n - 5f_{n-1} + f_{n-2}). \tag{13.84b}$$

We may simplify the iteration procedure for the corrector by noting from (13.82b) that

$$y_{n+1}^{(i+1)} = y_{n+1}^{(i)} + h\gamma_{-1}(f_{n+1}^{(i)} - f_{n+1}^{(i-1)}), \qquad i = 1, 2, \ldots, (I-1). \tag{13.85}$$

Therefore we only need to employ the full formula (13.82b) for the first inner iteration. For subsequent iterations we may use (13.85). Although we prove in § 13.9 that the inner iteration process is convergent, we shall also show that we normally expect to make only one or two iterations.

In the rest of this section, we will discuss the truncation and global errors of the corrector and will assume that members of the sequence y_{m+1}, $y_{m+2}, \ldots$ satisfy (13.80) exactly. Of course in practice this will rarely be true, as only a finite number of inner iterations may be performed for each step. We leave the discussion of the additional errors due to this and rounding until the next section.

To derive the local truncation error, we rewrite (13.80) as

$$\frac{y_{n+1} - y_n}{h} = c_0 f_{n+1} + \cdots + c_{m+1} \nabla^{m+1} f_{n+1}$$

and consider this as an approximation to (13.1). From (7.24)

$$\begin{aligned}\int_{x_n}^{x_{n+1}} f(x, y(x))\, \mathrm{d}x &= \int_{x_n}^{x_{n+1}} y'(x)\, \mathrm{d}x \\ &= h \sum_{j=0}^{m+1} c_j \nabla^j y'(x_{n+1}) + h^{m+3} c_{m+2} y^{(m+3)}(\xi_n),\end{aligned}$$

where $x_{n-m} < \xi_n < x_{n+1}$. Thus from (13.60)

$$\frac{y(x_{n+1}) - y(x_n)}{h} = \sum_{j=0}^{m+1} c_j \nabla^j f(x_{n+1}, y(x_{n+1})) + h^{m+2} c_{m+2} y^{(m+3)}(\xi_n). \tag{13.86}$$

The last term is the local truncation error and is of order $m + 2$.

We expand the differences in (13.86) to obtain

$$\begin{aligned}y(x_{n+1}) = y(x_n) &+ h[\gamma_{-1} f(x_{n+1}, y(x_{n+1})) + \cdots \\ &+ \gamma_m f(x_{n-m}, y(x_{n-m}))] + h^{m+3} c_{m+2} y^{(m+3)}(\xi_n).\end{aligned} \tag{13.87}$$

As in the analysis of the Adams–Bashforth algorithm, we let

$$e_n = |\, y(x_n) - y_n \,|$$

and, on expanding and subtracting (13.80) from (13.87), we obtain

$$e_{n+1} \leqslant e_n + hL[\,|\gamma_{-1}|\,e_{n+1} + |\gamma_0|\,e_n + \cdots + |\gamma_m|\,e_{n-m}] \\ + h^{m+3}\,|c_{m+2}|\,M_{m+3}, \tag{13.88}$$

where $|y^{(m+3)}(x)| \leqslant M_{m+3}$ for $x \in [x_0, b]$. If h is sufficiently small, so that

$$1 - hL\,|\gamma_{-1}| > 0, \tag{13.89}$$

then

$$\begin{aligned} e_{n+1} &\leqslant \frac{1}{1 - hL|\gamma_{-1}|}\,\{e_n + hL[|\gamma_0|e_n + \cdots + |\gamma_m|e_{n-m}] + h^{m+3}|c_{m+2}|M_{m+3}\} \\ &= e_n + \frac{hL}{1 - hL|\gamma_{-1}|}\,\{(|\gamma_0| + |\gamma_{-1}|)e_n + |\gamma_1|e_{n-1} + \cdots + |\gamma_m|e_{n-m}\} \\ &\quad + \frac{h^{m+3}|c_{m+2}|M_{m+3}}{1 - hL|\gamma_{-1}|}. \end{aligned}$$

On applying Lemma 13.1, we obtain

$$|y(x_n) - y_n| = e_n \leqslant \left(\delta + \frac{h^{m+2}|c_{m+2}|M_{m+3}}{LC_m}\right)\exp\left(\frac{(x_n - x_0)LC_m}{1 - hL|\gamma_{-1}|}\right) \\ - \frac{h^{m+2}|c_{m+2}|M_{m+3}}{LC_m}, \tag{13.90}$$

where

$$C_m = |\gamma_{-1}| + |\gamma_0| + \cdots + |\gamma_m| \tag{13.91}$$

and δ is the maximum starting error, so that

$$|y(x_r) - y_r| \leqslant \delta, \qquad r = 0, 1, \ldots, m.$$

The inequality (13.90) provides a global truncation error bound for the Adams–Moulton algorithm, assuming that the corrector is always satisfied exactly. Again, this result is qualitative rather than quantitative but it may be seen that the method is of order $m+2$ if the maximum starting error δ decreases like h^{m+2} as h is decreased. We should therefore choose a one-step method of order $m+2$ to determine starting values.

Theorem 13.4 The Adams–Moulton algorithm (13.81), with the corrector satisfied exactly for each step, is convergent of order $m+2$ if $y^{(m+3)}(x)$ is bounded for $x \in [x_0, b]$ and the starting values $y_0, \ldots, y_m$ are such that

$$|y(x_r) - y_r| \leqslant Dh^{m+2}, \qquad r = 0, 1, \ldots, m,$$

for some $D \geq 0$ and all $h \in (0, h_0]$ where $h_0 > 0$. Furthermore, the global truncation errors satisfy (13.90) provided $h < (L|\gamma_{-1}|)^{-1}$. □

You will observe that the order of the corrector in (13.81) is one higher than the order of the predictor. This is because we have chosen a pair which require an equal number of starting values. There is no reason why we should not use other explicit formulas as predictors, as these clearly do not affect the final accuracy of the corrector, provided I is suitably adjusted.

It should be noted that, for small h and a given value of δ, the error bound for the Adams–Moulton method is smaller than that of the Adams–Bashforth method of the same order, since for $m > 0$, we can show that $|c_m| < |b_m|$ (see Problem 13.32) and $C_m < B_{m+1}$. For the second order Adams–Moulton method the relevant constants are $c_2 = -\frac{1}{12}$ and $C_0 = 1$, whereas for the second order Adams–Bashforth method we have $b_2 = \frac{5}{12}$ and $B_1 = 2$. For nearly all problems the Adams–Moulton algorithm is far superior to the Adams–Bashforth algorithm of the same order, in that global truncation errors are much smaller. This might be expected since the Adams–Moulton method does not use extrapolation.

Again, we may obtain other corrector formulas by integrating the differential equation over two or more intervals. By integrating over two intervals, we obtain a class of formulas which includes Simpson's rule (see Problem 13.33).

Example 13.15 Obtain a global truncation error bound for the second order Adams–Moulton algorithm applied to the problem of Example 13.13, assuming that the corrector is satisfied exactly after each step. From Example 13.4, $L = 3\sqrt{3}/8$ and, from Example 13.14, $M_3 = 2$. Since $c_2 = -\frac{1}{12}$, $C_0 = 1$ and $\gamma_{-1} = \frac{1}{2}$, (13.90) becomes

$$|y(x_n) - y_n| \leq \left(\delta + h^2 \frac{1}{12} \cdot \frac{8}{3\sqrt{3}} \cdot \frac{2}{1}\right) \exp\left(\frac{x_n 3\sqrt{3}/8}{1 - h3\sqrt{3}/16}\right) - h^2 \frac{1}{12} \cdot \frac{8}{3\sqrt{3}} \cdot \frac{2}{1}$$

$$= \left(\delta + \frac{4h^2}{9\sqrt{3}}\right) \exp\left(\frac{6\sqrt{3}x_n}{16 - 3\sqrt{3}h}\right) - \frac{4h^2}{9\sqrt{3}}$$

provided $h < 16/(3\sqrt{3})$. For example, with $h = 0.1$, $\delta = 10^{-5}$ and $x_n = 1$, this bound is

$$|y(1.0) - y_{10}| \leq 0.00248.$$

Notice that this is much smaller than that in Example 13.14 for the second order Adams–Bashforth algorithm. □

13.9 Iterating with the corrector

We first verify that the inner iterative process (13.81b) for solving the corrector is convergent for small h. We rearrange the expanded form (13.82b) as

$$y_{n+1}^{(i+1)} = y_n + h[\gamma_0 f_n + \cdots + \gamma_m f_{n-m}] + h\gamma_{-1} f(x_{n+1}, y_{n+1}^{(i)}), \qquad (13.92)$$

which, on writing $w_i = y_{n+1}^{(i)}$, may be expressed as

$$w_{i+1} = F(w_i).$$

In Chapter 8, Theorem 8.3, it was shown that the sequence (w_i) will converge to a solution if F satisfies closure and a suitable Lipschitz condition on some interval. In this case, we take the interval to be all the real numbers, when clearly closure is satisfied, and we seek $L_F < 1$, such that

$$|F(w) - F(z)| \leqslant L_F |w - z| \qquad (13.93)$$

for all reals w and z. From (13.92), we have

$$\begin{aligned} |F(w) - F(z)| &= |h\gamma_{-1} f(x_{n+1}, w) - h\gamma_{-1} f(x_{n+1}, z)| \\ &\leqslant hL|\gamma_{-1}|\ |w - z| \end{aligned} \qquad (13.94)$$

where L is the usual Lipschitz constant for $f(x, y)$ with respect to y. We obtain a condition of the form (13.93) with $L_F < 1$ for

$$h < \frac{1}{|\gamma_{-1}|L}.$$

This does not impose a very severe restriction on h for most problems. You will notice that this is the same condition as (13.89) which was necessary for the validity of the error bound (13.90).

We also observe, for later use, that

$$\begin{aligned} |y_{n+1}^{(I+1)} - y_{n+1}^{(I)}| &= |F(y_{n+1}^{(I)}) - F(y_{n+1}^{(I-1)})| \\ &\leqslant hL|\gamma_{-1}|\ |y_{n+1}^{(I)} - y_{n+1}^{(I-1)}|. \end{aligned} \qquad (13.95)$$

We now investigate the effect of performing only a finite number of iterations with the corrector and we will assume that $y_{n+1}^{(I)}$ is the last iterate which is actually computed. From (13.82b)

$$y_{n+1}^{(I)} = y_n + h[\gamma_{-1} f_{n+1}^{(I)} + \gamma_0 f_n + \cdots + \gamma_m f_{n-m}] + y_{n+1}^{(I)} - y_{n+1}^{(I+1)}.$$

We let

$$\rho_n = y_{n+1}^{(I+1)} - y_{n+1}^{(I)}$$

and assume that

$$|\rho_n| \leqslant \rho, \qquad \text{for all } n.$$

We will also allow for rounding errors in the process and thus, instead of computing a sequence (y_n) which satisfies the corrector exactly for each n, we actually compute a sequence (z_n) with

$$z_{n+1} = z_n + h[\gamma_{-1} f(x_{n+1}, z_{n+1}) + \cdots + \gamma_m f(x_{n-m}, z_{n-m})] + \rho_n + \varepsilon_n$$

where ε_n is due to rounding. On subtracting (13.87) and putting

$$e_n = |y(x_n) - z_n|$$

we obtain

$$\begin{aligned} e_{n+1} \leqslant e_n + hL[|\gamma_{-1}| e_{n+1} + |\gamma_0| e_n + \cdots + |\gamma_m| e_{n-m}] \\ + h^{m+3} |c_{m+2}| M_{m+3} + \rho + \varepsilon, \end{aligned} \tag{13.96}$$

where $|\varepsilon_n| \leqslant \varepsilon$ for all n. By comparing (13.96) with (13.88), it can be seen that an amount $(\rho + \varepsilon)/h$ must be added to $h^{m+2}|c_{m+2}| M_{m+3}$ in the error bound (13.90). If we have an estimate for M_{m+3}, we can then decide how small ρ and ε should be. Since $h^{m+2}|c_{m+2}| M_{m+3}$ is a bound for the local truncation error of the corrector (see (13.87)), we are in fact comparing $(\rho + \varepsilon)/h$ with the truncation error and in the next section we describe how this may be estimated.

Having chosen ρ, we may use (13.95) to decide when to stop iterating with the corrector. We require

$$|y_{n+1}^{(l)} - y_{n+1}^{(l-1)}| \leqslant \frac{\rho}{h|\gamma_{-1}|L}. \tag{13.97}$$

If L is not known, as is usual, we can use the mean value theorem in (13.94) and replace L by $|\partial f/\partial y|$ at some point near $x = x_{n+1}$, $y = y_{n+1}$. We then use the estimate

$$L \simeq \left|\frac{\partial f}{\partial y}\right| \simeq \left|\frac{f_{n+1}^{(1)} - f_{n+1}^{(0)}}{y_{n+1}^{(1)} - y_{n+1}^{(0)}}\right|, \tag{13.98}$$

in place of L in (13.97).

13.10 Milne's method of estimating truncation errors

W. E. Milne first proposed a technique for estimating local truncation errors. For the technique to work the predictor must be of the same order as the corrector. However, it is only necessary to make one or two applications of the corrector for each step, provided h is small.

We consider the following formulas.

$$\text{Predictor: } y_{n+1}^{(0)} = y_n + h[b_0 f_n + b_1 \nabla f_n + \cdots + b_{m+1} \nabla^{m+1} f_n] \quad (13.99\text{a})$$

$$\text{Corrector: } y_{n+1}^{(i+1)} = y_n + h[c_0 f_{n+1}^{(i)} + c_1 \nabla f_{n+1}^{(i)} + \cdots + c_{m+1} \nabla^{m+1} f_{n+1}^{(i)}],$$
$$i = 0, 1, \ldots, (I-1), \quad (13.99\text{b})$$

$$y_{n+1} = y_{n+1}^{(I)}.$$

Apart from the necessity of an extra starting value, these formulas are implemented in an identical manner to (13.81).

From (13.69) and (13.86),

$$y(x_{n+1}) = y(x_n) + h \sum_{j=0}^{m+1} b_j \nabla^j f(x_n, y(x_n)) + h^{m+3} b_{m+2} y^{(m+3)}(\eta_n), \quad (13.100\text{a})$$

$$y(x_{n+1}) = y(x_n) + h \sum_{j=0}^{m+1} c_j \nabla^j f(x_{n+1}, y(x_{n+1})) + h^{m+3} c_{m+2} y^{(m+3)}(\xi_n),$$

$$(13.100\text{b})$$

where η_n and ξ_n are intermediate points. We assume that

$$y^{(m+3)}(\eta_n) \simeq y^{(m+3)}(\xi_n)$$

and write

$$T_n = h^{m+2} c_{m+2} y^{(m+3)}(\xi_n) \quad (13.101)$$

for the local truncation error of (13.100b). (An additional factor h is accounted for by writing (13.100b) in the form (13.86).) On subtracting (13.100a) from (13.100b), we obtain

$$\left(\frac{b_{m+2}}{c_{m+2}} - 1\right) T_n \simeq \sum_{j=0}^{m+1} c_j \nabla^j f(x_{n+1}, y(x_{n+1})) - \sum_{j=0}^{m+1} b_j \nabla^j f(x_n, y(x_n)). \quad (13.102)$$

We now use approximations

$$\nabla^j f(x_n, y(x_n)) \simeq \nabla^j f_n, \qquad \nabla^j f(x_{n+1}, y(x_{n+1})) \simeq \nabla^j f_{n+1}^{(0)},$$

to obtain

$$T_n \simeq \left(\frac{c_{m+2}}{b_{m+2} - c_{m+2}}\right)\left(\sum_{j=0}^{m+1} c_j \nabla^j f_{n+1}^{(0)} - \sum_{j=0}^{m+1} b_j \nabla^j f_n\right)$$
$$= \frac{1}{h}\left(\frac{c_{m+2}}{b_{m+2} - c_{m+2}}\right)(y_{n+1}^{(1)} - y_{n+1}^{(0)}), \quad (13.103)$$

on using (13.99a) and (13.99b). We may use this last result to estimate the local truncation error at each step.

If only one application of the corrector is made, $I = 1$ and we seek an estimate of

$$|\rho_n| = |y_{n+1}^{(2)} - y_{n+1}^{(1)}| \leqslant hL|\gamma_{-1}||y_{n+1}^{(1)} - y_{n+1}^{(0)}|.$$

Now, from (13.103) and (13.101),

$$h|\gamma_{-1}|L|y_{n+1}^{(1)} - y_{n+1}^{(0)}| \simeq h^2 L|\gamma_{-1}|\left|\frac{b_{m+2} - c_{m+2}}{c_{m+2}}\right||T_n|$$

$$\leqslant h^{m+4} L|\gamma_{-1}||b_{m+2} - c_{m+2}|M_{m+3},$$

where $M_{m+3} \geqslant |y^{(m+3)}(x)|$ for $x \in [x_0, b]$. Thus an *approximate* upper bound for the ρ_n is

$$\tilde{\rho} = h^{m+4} L|\gamma_{-1}|\ |b_{m+2} - c_{m+2}|M_{m+3}$$

and $\tilde{\rho}/h$ is small compared with

$$h^{m+2}|c_{m+2}|M_{m+3} > |T_n|,$$

if

$$h \ll \frac{1}{L|\gamma_{-1}|}\ \frac{|c_{m+2}|}{|b_{m+2} - c_{m+2}|}. \tag{13.104}$$

If this condition is satisfied, there should be no advantage in applying the corrector more than once. The effects of correcting only once are investigated more precisely by Henrici (1962). Lambert (1991) also describes the consequences of applying the corrector only a small number of times. Ralston and Rabinowitz (1978) show how the estimate (13.103) of the local truncation error may be used to improve the predicted value $y_{n+1}^{(0)}$ at each step.

Example 13.16 We consider the following fourth order pair:

$$\text{Predictor: } y_{n+1}^{(0)} = y_n + \tfrac{1}{24}h(55f_n - 59f_{n-1} + 37f_{n-2} - 9f_{n-3}), \tag{13.105a}$$

$$\text{Corrector: } y_{n+1}^{(i+1)} = y_n + \tfrac{1}{24}h(9f_{n+1}^{(i)} + 19f_n - 5f_{n-1} + f_{n-2}). \tag{13.105b}$$

Here $m = 2$, $b_{m+2} = b_4 = \frac{251}{720}$ and $c_{m+2} = c_4 = -\frac{19}{720}$. The local truncation error of the predictor is

$$h^{m+2} b_{m+2} y^{(m+3)}(\eta_n) = \tfrac{251}{720} h^4 y^{(5)}(\eta_n)$$

and that of the corrector is

$$T_n = h^{m+2} c_{m+2} y^{(m+3)}(\xi_n) = -\tfrac{19}{720} h^4 y^{(5)}(\xi_n).$$

Thus, from (13.103),

$$T_n \simeq \frac{1}{h}\left(\frac{c_4}{b_4 - c_4}\right)(y_{n+1}^{(1)} - y_{n+1}^{(0)}) = \frac{1}{h}\left(-\frac{19}{270}\right)(y_{n+1}^{(1)} - y_{n+1}^{(0)}). \tag{13.106}$$

Table 13.3 Results for Example 13.17

x_n	$y_n^{(0)}$	$f_n^{(0)}$	$y_n^{(1)}$	$f_n^{(1)}$	$T_{n-1} \times 10^8$	$y(x_n)$	Error $\times 10^8$
0			1	0.5		1.0	0
0.1			1.04879039	0.47619926		1.04879039	0
0.2			1.09531337	0.45460510		1.09531337	0
0.3			1.13977617	0.43495475		1.13977617	0
0.4	1.18236027	0.41701839	1.18236150	0.41701788	−87	1.18236140	−10
0.5	1.22322920	0.40059453	1.22323034	0.40059409	−80	1.22323016	−18
0.6	1.26252409	0.38550990	1.26252510	0.38550953	−71	1.26252485	−25
0.7	1.30037111	0.37161390	1.30037199	0.37161358	−62	1.30037168	−31
0.8	1.33688243	0.35877598	1.33688319	0.35877571	−54	1.33688284	−35
0.9	1.37215806	0.34688284	1.37215872	0.34688262	−46	1.37215832	−40
1.0	1.40628744	0.33583594	1.40628800	0.33583576	−39	1.40628758	−42

If only one correction is made for each step, then from (13.104) with $|\gamma_{-1}| = \frac{9}{24}$ we require

$$h \ll \frac{8}{3L} \cdot \frac{19}{270} \simeq \frac{0.2}{L}. \qquad (13.107) \quad \square$$

Example 13.17 We apply the fourth order method of Example 13.16 to the problem in Example 13.13. From (13.107) with $L = 3\sqrt{3}/8$, we see that we require $h \ll 0.289$ if only one correction is made for each step. Table 13.3 gives results with $h = 0.1$; the T_n are obtained from (13.106). The starting values are calculated using the fourth order Runge–Kutta method. The last column gives the errors $y(x_n) - y_n^{(1)}$.

At first sight it might appear that for $x_n \geqslant 0.9$ the correction process makes the approximations worse, as $|y_n^{(1)} - y(x_n)| > |y_n^{(0)} - y(x_n)|$ for $n \geqslant 9$. However, omission of the corrector affects all approximations so that the fourth order Adams–Bashforth process does not give the column $y_n^{(0)}$. □

13.11 Numerical stability

We have seen that the inclusion of rounding errors does not radically change the global truncation error bounds for the methods introduced earlier in this chapter. However, the error bounds do not always provide a good description of the effects of rounding errors and starting errors.

We consider the calculation of members of a sequence $(y_n)_{n=0}^{\infty}$ by means of a recurrence relation (or difference equation) of the form

$$y_{n+1} = F(y_n, y_{n-1}, \ldots, y_{n-m}), \qquad n = m, m+1, \ldots, \qquad (13.108)$$

given starting values $y_0, y_1, \ldots, y_m$. We suppose that a single error is introduced so that, for some arbitrary r,

$$y_r^* = y_r + \varepsilon$$

and we compute the sequence

$$y_0, y_1, \ldots, y_{r-1}, y_r^*, y_{r+1}^*, \ldots.$$

We assume that this sequence satisfies (13.108) for all $n \geqslant m$ *except* $n = r - 1$. If $r \leqslant m$ the error will occur in the starting values and if $r > m$ the error will occur when (13.108) is computed for $n = r - 1$. We say that the recurrence relation (13.108) is *numerically unstable* if the errors $|y_n - y_n^*|$ are unbounded as $n \to \infty$.

Clearly an unstable recurrence relation will not provide a satisfactory numerical process, as we are certain to introduce rounding errors. Of course in practice we will introduce more than one rounding error, but in the analysis it is more instructive to consider the effect of an isolated error.

Example 13.18 Suppose we wish to calculate (y_n) from

$$y_{n+1} = 100.01y_n - y_{n-1} \tag{13.109}$$

with $y_0 = 1$, $y_1 = 0.01$. The general solution of (13.109) is

$$y_n = c_1 100^n + c_2\left(\frac{1}{100}\right)^n$$

where c_1 and c_2 are arbitrary. The solution which satisfies the initial conditions is

$$y_n = \left(\frac{1}{100}\right)^n,$$

for which $c_1 = 0$ and $c_2 = 1$. The introduction of errors may be considered equivalent to changing the particular solution being computed. This will have the effect of making c_1 non-zero, so that a term $c_1.100^n$ must be added to y_n. Clearly $c_1 . 100^n$ will quickly 'swamp' $c_2/100^n$. For example, if

$$y_0^* = y_0 + 10^{-6} = 1.000001,$$
$$y_1^* = y_1 = 0.01$$

then

$$y_2^* = 0.000099 \quad \text{and} \quad y_5^* \simeq -1.0001.$$

□

For an approximation to an initial value differential problem, the function F in the recurrence relation (13.108) will depend on h and $f(x, y)$. In defining stability we do *not* decrease h as $n \to \infty$ as was necessary when considering the convergence of y_n to $y(x_n)$. We are now interested in only the stability of recurrence relations resulting from differential equations. Because the precise form of the recurrence relation depends on $f(x, y)$, it is difficult to derive general results regarding the stability of such approximations. We find it instructive, however, to consider in detail the linear differential equation

$$y' = -Ay, \qquad \text{with } y(0) = 1, \tag{13.110}$$

where A is a positive constant, which has solution $y(x) = e^{-Ax}$. Note that this solution converges to zero as $x \to \infty$. Later in this section, we show that to a large extent the lessons to be learned from this problem will apply to other initial value problems. We shall certainly expect that any approximate solution of the test equation (13.110) should satisfy $y_n \to 0$ as $n \to \infty$ so that the behaviour of the exact solution is mimicked. If such a property is satisfied the method is said to be *absolutely stable.*

We start by considering the *midpoint* rule

$$y_{n+1} = y_{n-1} + 2hf_n \tag{13.111}$$

which is obtained by using

$$\frac{y_{n+1} - y_{n-1}}{2h} = f_n$$

as an approximation to (13.1) or, alternatively, from Nyström's formula (13.77) with $m = 0$ (see Problem 13.25). For (13.110), $f(x, y) = -Ay$ and the method becomes

$$y_{n+1} = -2hAy_n + y_{n-1}. \tag{13.112}$$

This is a second order difference equation with characteristic equation

$$z^2 + 2hAz - 1 = 0,$$

whose roots are

$$\left.\begin{matrix} z_1 \\ z_2 \end{matrix}\right\} = -Ah \pm (1 + A^2h^2)^{1/2}.$$

The general solution of (13.112) is

$$y_n = c_1 z_1^n + c_2 z_2^n. \tag{13.113}$$

We find that z_1^n behaves like $y(x)$ as h is decreased with x_n fixed, but that z_2^n has a completely different behaviour. On expanding the square root, we obtain

$$z_1 = -Ah + 1 + \tfrac{1}{2}(A^2h^2) + O(h^4).$$

By $O(h^4)$ we mean terms involving h^4 and higher powers. On comparison with

$$e^{-Ah} = 1 - Ah + \tfrac{1}{2}(A^2h^2) + O(h^3),$$

we see that

$$\begin{aligned} z_1 &= e^{-Ah} + O(h^3) \\ &= e^{-Ah}(1 + O(h^3)), \end{aligned}$$

since

$$e^{-Ah} = 1 + O(h).$$

Thus

$$\begin{aligned} z_1^n &= e^{-Anh}(1 + O(h^3))^n = e^{-Anh}(1 + n\,.\,O(h^3)) \\ &= e^{-Ax_n}(1 + O(h^2)), \end{aligned}$$

as $x_n = nh$. Now as $h \to 0$ with $x_n = nh$ fixed,

$$z_1^n \to e^{-Ax_n} = y(x_n)$$

and, for small values of h, z_1^n behaves like an approximation to $y(x)$. It can also be shown that $z_1^n \to 0$ as $n \to \infty$ for any fixed $h > 0$ and $A > 0$.

On the other hand

$$z_2 = -Ah - (1 + A^2h^2)^{1/2} < -1$$

since $Ah > 0$ and thus $|z_2| > 1$. In fact

$$\begin{aligned} z_2 &= -Ah - (1 + A^2h^2)^{1/2} \\ &= -Ah - 1 - \tfrac{1}{2}A^2h^2 + O(h^4) \\ &= -e^{Ah} + O(h^3) \\ &= -e^{Ah}(1 + O(h^3)) \end{aligned}$$

and thus

$$\begin{aligned} z_2^n &= (-1)^n e^{Anh}(1 + O(h^3))^n \\ &= (-1)^n e^{Ax_n}(1 + O(h^2)). \end{aligned}$$

As h is decreased with x_n fixed, z_2^n does *not* approximate to $y(x_n)$.

The general solution of (13.112) is, therefore,

$$y_n = c_1 e^{-Ax_n}(1 + O(h^2)) + c_2(-1)^n e^{Ax_n}(1 + O(h^2)) \qquad (13.114)$$

and for this to be an approximation to $y(x_n)$, we require

$$c_1 = 1 \quad \text{and} \quad c_2 = 0.$$

We would therefore like to compute the particular solution

$$y_n = z_1^n$$

of (13.112). In practice we cannot avoid errors and we obtain, say,

$$y_n^* = (1 + \delta_1)z_1^n + \delta_2 z_2^n.$$

In this case the error

$$|y_n - y_n^*| = |\delta_1 z_1^n + \delta_2 z_2^n| \to \infty, \qquad \text{as } n \to \infty$$

since $|z_2| > 1$. Therefore the method (13.111) is unstable for this differential equation.

The solution z_2^n is called the *parasitic* solution of the difference equation (13.112). It appears because we have replaced a *first* order differential equation by a *second* order difference equation. In this case the parasitic solution 'swamps' the solution we would like to determine.

Even if we could use exact arithmetic with the midpoint rule, we would not avoid numerical instability. We need starting values y_0 and y_1, and y_1

must be obtained by some other method. Let us suppose that this yields the exact solution $y(x_1)$. Clearly, with

$$y_1 = y(x_1) = e^{-Ah} \neq z_1,$$

we do not obtain $c_1 = 1$ and $c_2 = 0$ in (13.113).

Example 13.19 The initial value problem

$$y'(x) = -3y(x) \qquad \text{with } y(0) = 1$$

has solution $y(x) = e^{-3x}$. We use the midpoint rule (13.111) with $h = 0.1$ and starting values

$$y_0 = y(0) = 1 \quad \text{and} \quad y_1 = 0.7408.$$

The last value is $y(h) = e^{-3h}$ correct to four decimal places. Results are tabulated in Table 13.4. The oscillatory behaviour for $x_n > 1$ is typical of unstable processes and is due to the $(-1)^n$ factor in the second term of (13.114). □

The midpoint rule is thus at a major disadvantage when used for problems with exponentially decreasing solutions and will not produce accurate results although it has a smaller local truncation error (see Problem 13.38) than the second order Adams–Bashforth algorithm, which we consider next.

From

$$y_{n+1} = y_n + \frac{h}{2}(3f_n - f_{n-1})$$

with $f(x, y) = -Ay$ ($A > 0$, as before), we obtain the recurrence relation

$$y_{n+1} = (1 - \tfrac{3}{2}Ah)y_n + \tfrac{1}{2}Ahy_{n-1}, \tag{13.115}$$

which has characteristic polynomial

$$z^2 - (1 - \tfrac{3}{2}Ah)z - \tfrac{1}{2}Ah.$$

The roots are

$$\left.\begin{matrix} z_1 \\ z_2 \end{matrix}\right\} = \tfrac{1}{2}[(1 - \tfrac{3}{2}Ah) \pm (1 - Ah + \tfrac{9}{4}A^2h^2)^{1/2}].$$

On expanding the square root, we obtain

$$\begin{aligned} z_1 &= \tfrac{1}{2}[(1 - \tfrac{3}{2}Ah) + 1 + \tfrac{1}{2}(-Ah + \tfrac{9}{4}A^2h^2) - \tfrac{1}{8}(-Ah + \tfrac{9}{4}A^2h^2)^2 + O(h^3)] \\ &= 1 - Ah + \tfrac{1}{2}A^2h^2 + O(h^3) \\ &= e^{-Ah}(1 + O(h^3)). \end{aligned}$$

Thus

$$z_1^n = e^{-Anh}(1+O(h^3))^n$$
$$= e^{-Ax_n}(1+O(h^2)),$$

so that, for small h, z_1^n is an approximation to $y(x_n)$. We also find that for all values of $h>0$ and $A>0$ we have $|z_1|<1$ so that $z_1^n \rightarrow 0$ as $n\rightarrow\infty$.

We seek the solution of (13.115) of the form

$$y_n = z_1^n$$

but, in practice, obtain

$$y_n^* = (1+\delta_1)z_1^n + \delta_2 z_2^n.$$

Now

$$z_2 = \tfrac{1}{2}[(1-\tfrac{3}{2}Ah) - (1 - Ah + \tfrac{9}{4}A^2h^2)^{1/2}]$$

and detailed analysis shows that $z_2<-1$ if $Ah>1$, and $-1<z_2<0$ if $0<Ah<1$. Hence the method is absolutely stable only if $h<1/A$. We will certainly also need to satisfy this inequality if the truncation error is not to be excessive and, therefore, this is not a severe restriction for the differential equation $y' = -Ay$. However, this restriction may be severe for other problems.

Similar results may also be obtained for the equation

$$y' = -Ay + B, \tag{13.116}$$

where $A>0$ and B are constants. The constant B will not affect the homogeneous part of the difference approximations to (13.116) and it is

Table 13.4 An unstable process

x_n	y_n
0	1
0.1	0.7408
0.2	0.5555
0.3	0.4075
0.4	0.3110
0.5	0.2209
0.6	0.1785
0.7	0.1138
0.8	0.1102
0.9	0.0477
1.0	+0.0816
1.1	−0.0013
1.2	+0.0824
1.3	−0.0507
1.4	+0.1128

only necessary to add the particular solution $y_n = B/A$, to the general solution of the homogeneous equations (13.112) and (13.115). As $h \to 0$, with $nh = x_n$ fixed, this particular solution converges to $y(x) = B/A$, a particular solution of (13.116). For both of the methods investigated in this section, we would like the solution

$$y_n = z_1^n + B/A,$$

but actually obtain

$$y_n^* = (1 + \delta_1)z_1^n + \delta_2 z_2^n + B/A.$$

Thus the absolute stability conditions are identical to those for the problem (13.110).

If a first order difference equation is used to approximate a differential equation then there are no parasitic solutions to give stability difficulties. However, the only solution of the difference equation may not satisfy our absolute stability condition: $y_n \to 0$ as $n \to \infty$. If Euler's method is applied to (13.110), we obtain

$$\begin{aligned} y_{n+1} &= y_n - Ahy_n \\ &= (1 - Ah)y_n, \end{aligned}$$

so that

$$y_n = (1 - Ah)^n y_0.$$

We can see that $y_n \to 0$ only if $|1 - Ah| < 1$, that is, $h < 2/A$. In practice, for (13.110), this restriction on h should normally be satisfied for accuracy reasons even when A is large. For a large value of A, the exact solution decreases very rapidly and a small step size would be essential. However, consider the equation

$$y' = -A(y - 1)$$

which has solution $y(x) = ce^{-Ax} + 1$ where c is arbitrary. Euler's method is

$$y_{n+1} = y_n - Ah(y_n - 1)$$

so that

$$y_n = c_1(1 - Ah)^n + 1$$

where c_1 is arbitrary. Now $y(x) \to 1$ as $x \to \infty$, but $y_n \to 1$ as $n \to \infty$ with h fixed, only if $|1 - Ah| < 1$, that is, $h < 2/A$. This condition may be severe if A is very large. If the initial value $y(0)$ is near $+1$, the solution will not vary rapidly for $x > 0$ and there should be no need to use a small step size.

There are similar restrictions for all the other methods introduced in this book except the Adams–Moulton method of order 2. When this method is applied to (13.110) we obtain

$$y_{n+1} = y_n + \tfrac{1}{2}h(-Ay_n - Ay_{n+1})$$

so that

$$(1 + \tfrac{1}{2}Ah)y_{n+1} = (1 - \tfrac{1}{2}Ah)y_n.$$

Thus

$$y_n = \left(\frac{1 - \tfrac{1}{2}Ah}{1 + \tfrac{1}{2}Ah}\right)^n y_0$$

and $y_n \to 0$, as $n \to \infty$, for all $A > 0$ and fixed $h > 0$. Results for the Runge–Kutta methods are included in Problem 13.40 and a full discussion of stability is given by Lambert (1991).

The stability behaviour of approximations to the general equation

$$y' = f(x, y) \tag{13.117}$$

will to a certain extent be similar to those for the special problem (13.110). On expanding $f(x, y)$ using Taylor series at the point $x = X$, $y = Y$, we obtain

$$\begin{aligned} f(x, y) &= f(X, Y) + (y - Y)f_y(X, Y) + O(x - X) + O(y - Y)^2 \\ &\simeq yf_y(X, Y) + [f(X, Y) - Yf_y(X, Y)] \end{aligned}$$

near the point (X, Y). Thus locally (that is, over a small range of values of x) the problem (13.117) will behave like one of the form (13.116). Alternatively, we may argue that one would need to be rather optimistic to expect success in computing a decreasing solution of (13.117) by a method which is unstable for the particular case (13.110).

13.12 Systems and higher order equations

Suppose that

$$f_j = f_j(x, z_1, z_2, \ldots, z_k), \qquad j = 1, \ldots, k,$$

is a set of k real valued functions of $k + 1$ variables which are defined for $a \leqslant x \leqslant b$ and all real $z_1, \ldots, z_k$. The simultaneous equations

$$\left.\begin{aligned} y_1'(x) &= f_1(x, y_1(x), \ldots, y_k(x)) \\ y_2'(x) &= f_2(x, y_1(x), \ldots, y_k(x)) \\ &\vdots \\ y_k'(x) &= f_k(x, y_1(x), \ldots, y_k(x)) \end{aligned}\right\} \tag{13.118}$$

where $x \in [a, b]$, are called a *system* of ordinary differential equations. Any set of k differentiable functions†

$$y_1(x), \ldots, y_k(x)$$

satisfying (13.118) is called a solution.

In general, the solution of (13.118), if it exists, will not be unique unless we are given k extra conditions. These usually take the form

$$y_j(x_0) = s_j, \qquad j = 1, \ldots, k, \tag{13.119}$$

where the s_j are known and $x_0 \in [a, b]$. We are thus given the values of the $y_j(x)$ at the point $x = x_0$. The problem of solving (13.118) subject to (13.119) is again known as an initial value problem.

We write (13.118) and (13.119) in the form

$$\begin{aligned} \mathbf{y}' &= \mathbf{f}(x, \mathbf{y}), \\ \mathbf{y}(x_0) &= \mathbf{s}, \end{aligned} \tag{13.120}$$

where $\mathbf{y}(x)$, $\mathbf{y}'(x)$, $\mathbf{f}(x, \mathbf{y})$ and $\mathbf{s}$ are k-dimensional vectors, whose ith components are $y_i(x)$, $y_i'(x)$, $f_i(x, y_1(x), \ldots, y_k(x))$ and s_i respectively. The problem has a unique solution if $\mathbf{f}$ satisfies a Lipschitz condition of the form: there exists $L \geqslant 0$ such that, for all real vectors $\mathbf{y}$ and $\mathbf{z}$ of dimension k and all $x \in [a, b]$,

$$\| \mathbf{f}(x, \mathbf{y}) - \mathbf{f}(x, \mathbf{z}) \|_\infty \leqslant L \| \mathbf{y} - \mathbf{z} \|_\infty. \tag{13.121}$$

The norm is defined in § 10.2.

Numerical methods analogous to those for one equation may be derived in an identical manner. For the Taylor series method, we determine $\mathbf{f}^{(r)}(x, \mathbf{y})$ such that

$$\begin{aligned} \mathbf{y}(x_{n+1}) = \mathbf{y}(x_n) + h\mathbf{f}(x_n, \mathbf{y}(x_n)) + \frac{h^2}{2!}\mathbf{f}^{(1)}(x_n, \mathbf{y}(x_n)) + \cdots \\ + \frac{h^p}{p!}\mathbf{f}^{(p-1)}(x_n, \mathbf{y}(x_n)) + h^{p+1}\mathbf{R}_{p+1}(x_n) \end{aligned}$$

(cf. (13.28)), where $\mathbf{R}_{p+1}$ is a remainder term. For example,

$$\mathbf{f}^{(1)}(x, \mathbf{y}(x)) = \frac{\mathrm{d}}{\mathrm{d}x}\mathbf{f}(x, \mathbf{y}(x)) = \frac{\partial \mathbf{f}}{\partial x} + \sum_{j=1}^{k} \frac{\mathrm{d}y_j(x)}{\mathrm{d}x}\frac{\partial \mathbf{f}}{\partial y_j} = \frac{\partial \mathbf{f}}{\partial x} + \sum_{j=1}^{k} f_j \frac{\partial \mathbf{f}}{\partial y_j}, \tag{13.122}$$

† Note that the subscripts denote different functions and not values of approximations to values of a single function, as in earlier sections.

where the ith component of $\partial \mathbf{f}/\partial y_j$ is $\partial f_i/\partial y_j$. The formulas for $\mathbf{f}^{(r)}$ are very lengthy for $r \geqslant 2$. The Taylor series method of order 2 is thus

$$\begin{aligned} \mathbf{y}_0 &= \mathbf{s}, \\ \mathbf{y}_{n+1} &= \mathbf{y}_n + h\mathbf{f}(x_n, \mathbf{y}_n) + \frac{h^2}{2!}\mathbf{f}^{(1)}(x_n, \mathbf{y}_n), \end{aligned} \tag{13.123}$$

where $\mathbf{y}_n$ is a vector whose ith element $y_{i,n}$ is to be considered as an approximation to $y_i(x_n)$ and $x_n = x_0 + nh$.

We may derive Runge–Kutta methods by using Taylor series in several variables. Analogous to (13.39), we have the classical method

$$\mathbf{y}_{n+1} = \mathbf{y}_n + \tfrac{1}{6}h[\mathbf{k}_1 + 2\mathbf{k}_2 + 2\mathbf{k}_3 + \mathbf{k}_4],$$

where

$$\mathbf{k}_1 = \mathbf{f}(x_n, \mathbf{y}_n), \qquad \mathbf{k}_2 = \mathbf{f}(x_n + \tfrac{1}{2}h,\ \mathbf{y}_n + \tfrac{1}{2}h\mathbf{k}_1),$$
$$\mathbf{k}_3 = \mathbf{f}(x_n + \tfrac{1}{2}h,\ \mathbf{y}_n + \tfrac{1}{2}h\mathbf{k}_2), \qquad \mathbf{k}_4 = \mathbf{f}(x_{n+1}, \mathbf{y}_n + h\mathbf{k}_3).$$

Predictor–corrector methods are also identical to those for single equations. We merely replace the y_n and f_n in (13.82) by vectors $\mathbf{y}_n$ and $\mathbf{f}_n$.

Global truncation error bounds for all methods are very similar to those for a single equation. We replace the quantity $|y_n - y(x_n)|$ by $\|\mathbf{y}_n - \mathbf{y}(x_n)\|_\infty$.

Example 13.20 Consider the initial value problem

$$y_1'(x) = xy_1(x) - y_2(x)$$
$$y_2'(x) = -y_1(x) + y_2(x)$$

with

$$y_1(0) = s_1, \qquad y_2(0) = s_2.$$

We have

$$\mathbf{f}(x, \mathbf{y}) = \begin{bmatrix} f_1(x, y_1, y_2) \\ f_2(x, y_1, y_2) \end{bmatrix} = \begin{bmatrix} (xy_1 - y_2) \\ (-y_1 + y_2) \end{bmatrix}$$

and

$$\begin{aligned} \mathbf{f}^{(1)}(x, \mathbf{y}) &= \frac{\partial \mathbf{f}}{\partial x} + f_1 \frac{\partial \mathbf{f}}{\partial y_1} + f_2 \frac{\partial \mathbf{f}}{\partial y_2} \\ &= \begin{bmatrix} y_1 \\ 0 \end{bmatrix} + (xy_1 - y_2)\begin{bmatrix} x \\ -1 \end{bmatrix} + (-y_1 + y_2)\begin{bmatrix} -1 \\ 1 \end{bmatrix} \\ &= \begin{bmatrix} ((x^2 + 2)y_1 - (1 + x)y_2) \\ (-(1 + x)y_1 + 2y_2) \end{bmatrix}. \end{aligned}$$

The Taylor series method of order 2 is

$$y_{1,0} = s_1, \qquad y_{2,0} = s_2,$$

$$y_{1,n+1} = y_{1,n} + h(x_n y_{1,n} - y_{2,n}) + \frac{h^2}{2}\{(x_n^2+2)y_{1,n} - (1+x_n)y_{2,n}\},$$

$$y_{2,n+1} = y_{2,n} + h(-y_{1,n} + y_{2,n}) + \frac{h^2}{2}\{-(1+x_n)y_{1,n} + 2y_{2,n}\},$$

for $n = 0, 1, 2, \dots$.

The Adams–Moulton predictor–corrector method of order 2 (cf. (13.83)) is as follows.

Initial values: $y_{1,0} = s_1, \qquad y_{2,0} = s_2;$

$$\text{Predictor: }\begin{cases} y_{1,n+1}^{(0)} = y_{1,n} + hf_{1,n}, \\ y_{2,n+1}^{(0)} = y_{2,n} + hf_{2,n}; \end{cases}$$

$$\text{Corrector: }\begin{cases} y_{1,n+1}^{(i+1)} = y_{1,n} + \tfrac{1}{2}h(f_{1,n+1}^{(i)} + f_{1,n}), \\ y_{2,n+1}^{(i+1)} = y_{2,n} + \tfrac{1}{2}h(f_{2,n+1}^{(i)} + f_{2,n}), \end{cases} \qquad i = 0, 1, \dots, (I-1);$$

$$y_{1,n+1} = y_{1,n+1}^{(I)}, \qquad y_{2,n+1} = y_{2,n+1}^{(I)};$$

where $f_{j,n+1}^{(i)} = f_j(x_{n+1}, y_{1,n+1}^{(i)}, y_{2,n+1}^{(i)})$.

(The superscript i in the above is to distinguish different iterates and does not denote differentiation.) □

If the real valued function

$$f = f(x, z_1, z_2, \dots, z_m)$$

of m variables is defined for $x \in [a, b]$ and all real $z_1, \dots, z_m$, we say that

$$y^{(m)}(x) = f(x, y(x), y'(x), \dots, y^{(m-1)}(x)) \tag{13.124}$$

is a differential equation of order m. If the solution $y(x)$ is to be unique we need m extra conditions. If these take the form

$$y^{(r)}(x_0) = t_r, \qquad r = 0, 1, \dots, m-1, \tag{13.125}$$

where $x_0 \in [a, b]$ and the t_r are given, we again say that we have an initial value problem. You will notice that we are given the values of y and its first $m-1$ derivatives at one point $x = x_0$. In Chapter 14 we will discuss problems with conditions involving more than one value of x.

We make the substitutions

$$z_j(x) = y^{(j-1)}(x), \qquad j = 1, \dots, m,$$

so that

$$z_j'(x) = y^{(j)}(x) = z_{j+1}(x), \qquad j = 1, \ldots, m-1,$$

and thus rewrite the single equation (13.124) as the *system* of first order equations

$$\begin{aligned} z_1'(x) &= z_2(x) \\ z_2'(x) &= z_3(x) \\ &\vdots \\ z_{m-1}'(x) &= z_m(x) \\ z_m'(x) &= f(x, z_1(x), z_2(x), \ldots, z_m(x)). \end{aligned}$$

The conditions (13.125) become

$$z_1(x_0) = t_0, \qquad z_2(x_0) = t_1, \qquad \ldots, \qquad z_m(x_0) = t_{m-1},$$

and we have an initial value problem of the type considered at the beginning of this section.

Example 13.21 Consider

$$y'' = g(x, y, y')$$

with

$$y(x_0) = t_0, \qquad y'(x_0) = t_1.$$

Let

$$z_1(x) = y(x), \qquad z_2(x) = y'(x),$$

when

$$\begin{aligned} z_1'(x) &= z_2(x), \\ z_2'(x) &= g(x, z_1(x), z_2(x)). \end{aligned}$$

Euler's method is

$$\begin{aligned} \mathbf{z}_0 &= \mathbf{s}, \\ \mathbf{z}_{n+1} &= \mathbf{z}_n + h\mathbf{f}(x_n, \mathbf{z}_n). \end{aligned}$$

For this example, we obtain

$$\begin{aligned} z_{1,0} &= t_0, \qquad z_{2,0} = t_1, \\ z_{1,n+1} &= z_{1,n} + hz_{2,n}, \\ z_{2,n+1} &= z_{2,n} + hg(x_n, z_{1,n}, z_{2,n}). \end{aligned}$$

□

There are also some special methods for higher order differential equations, particularly second order equations. See Henrici (1962) and Lambert (1991).

We have already seen in §13.11 that for Euler's method applied to $y' = -Ay$ with $A > 0$, we obtain $y_n \to 0$ as $n \to \infty$ with h fixed, only if $h < 2/A$. If we have a *linear system* of the form

$$\mathbf{y}' = -\mathbf{A}\mathbf{y} \tag{13.126}$$

where $\mathbf{A}$ is an $m \times m$ matrix, then one component of the general solution vector has the form

$$y(x) = c_1 e^{-\lambda_1 x} + c_2 e^{-\lambda_2 x} + \cdots + c_m e^{-\lambda_m x} \tag{13.127}$$

where $\lambda_1, \lambda_2, \ldots, \lambda_m$ are the eigenvalues of $\mathbf{A}$. If $\lambda_r > 0$, $r = 1, 2, \ldots, m$, the solution shown in (13.127) will satisfy $y(x) \to 0$ as $x \to \infty$. It can be shown that with Euler's method we need to choose

$$h < 2/\lambda_r, \qquad \text{for } r = 1, 2, \ldots, m \tag{13.128}$$

if the solution of the difference equation is to decrease towards zero as the number of steps increases. If $0 < \lambda_1 \leqslant \lambda_2 \leqslant \cdots \leqslant \lambda_m$ in (13.127) then, for $c_1 \neq 0$, the first term will be the dominant term for large x. The inequalities (13.128) may impose a very severe restriction on h if λ_m is very much larger than λ_1. We illustrate this by the following example.

Example 13.22 Given

$$\begin{aligned} y'(x) &= -8y(x) + 7z(x) \\ z'(x) &= 42y(x) - 43z(x) \end{aligned} \tag{13.129}$$

with $y(0) = 1$ and $z(0) = 8$, the solution is

$$\begin{aligned} y(x) &= 2e^{-x} - e^{-50x} \\ z(x) &= 2e^{-x} + 6e^{-50x}. \end{aligned}$$

The equations (13.129) are of the form (13.126) where $\mathbf{A}$ has eigenvalues $\lambda_1 = 1$ and $\lambda_2 = 50$. For large $x > 0$ the dominant term in this solution is $2e^{-x}$ but we must choose $h < 0.04$ if Euler's method is to produce a decreasing solution. □

Equations such as those of Example 13.22 are called *stiff* equations. There are similar difficulties with all the methods we have discussed except the Adams–Moulton method of order 2 which, as we saw in §13.11, is always absolutely stable. The restrictions on h given in Problem 13.40 for the Runge–Kutta method and in §13.11 for the second order Adams–Bashforth method must be satisfied for each λ_r in (13.127). Lambert (1991) gives a full discussion of the solution of stiff equations.

13.13 Comparison of step-by-step methods

The types considered are

(T) Taylor series methods.
(RK) Runge–Kutta methods.

(E) Explicit methods, for example, the Adams–Bashforth method (13.64) and Algorithm 13.1.

(PC) Predictor–corrector methods, for example, the Adams–Moulton algorithm 13.2 and the method (13.99).

Ease of use

The main consideration is how simple they are to program for a digital computer.

(T) Difficult to establish suitable formulas, except for low order methods. No special starting procedure.

(RK) Very simple. No special starting procedure.

(E) Quite straightforward, but special starting procedure required.

(PC) Slightly difficult to implement, due to inner iterations, unless only one correction is made per step. Special starting procedure required.

Amount of computation

We usually measure this by the number of evaluations of f that are needed for each step. If f is at all complicated, most of the calculation time will be spent on evaluations of f.

(T) Many evaluations of f and derivatives required per step, for example, six evaluations for the third order method.

(RK) Comparatively many evaluations of f, for example, four for the fourth order method. For the commonly used methods, the number of evaluations equals the order of the method.

(E) One evaluation of f, regardless of the order.

(PC) Two or three evaluations of f, regardless of the order.

Truncation error

The truncation error gives a measure of the accuracy of a method for a range of problems but does not necessarily provide a correct comparison of the accuracy of methods for one particular problem. We consider the magnitude of the truncation error for a *given* order.

(T) Not very good.

(RK) May be quite good.

(E) Much poorer than (PC).

(PC) Good.

Error estimation

(T) Difficult.
(RK) Difficult, although there are some methods which give some indication of truncation errors.
(E) Difficult.
(PC) Fairly easy if Milne's method (§13.10) is used.

Stability and stiff systems

For all the methods introduced in this chapter, except the Adams–Moulton algorithm of order 2, the step size must be restricted for absolute stability and hence there are difficulties for stiff equations.

Step size adjustment

It is sometimes desirable to adjust the step size as the solution proceeds. This is particularly necessary if there are regions in which the solution has large derivatives, in which a small step size is required, and other regions in which derivatives are small, so that a large step size may be employed.

(T) Very easy to change the step size, as these are one-step methods.
(RK) As (T).
(E) Quite difficult, as special formulas are required.
(PC) As (E).

Choice of method

(T) Not recommended unless the problem has some special features, for example if f and its derivatives are known from other information.
(RK) Recommended for a quick calculation for which there is little concern about accuracy. Also recommended to obtain starting values for (E) and (PC).
(E) Only recommended for problems for which f is particularly difficult to evaluate.
(PC) Recommended (particularly (13.99)) for most problems as truncation error estimation is possible. The fourth order predictor–corrector (13.105) is probably one of the best methods, except for stiff equations. It involves comparatively few function evaluations (for one or two inner iterations) and provides truncation error estimates.

Problems

Section 13.1

13.1 Show that for $0 \leqslant x \leqslant 1$

$$y' = xy, \qquad \text{with } y(0) = 1,$$

has a unique solution and find an approximate solution by Picard iteration.

13.2 Show that

$$y' = \sin y, \qquad \text{with } y(x_0) = s,$$

has a unique solution on any interval containing x_0.

13.3 Given that $g(x)$ and $h(x)$ are continuous functions on some interval $[a, b]$ containing x_0, show that the linear differential equation

$$y' = g(x) \, . \, y + h(x), \qquad \text{with } y(x_0) = s,$$

has a unique solution on $[a, b]$.
(*Hint*: use the fact that $g(x)$ is bounded on $[a, b]$.)

Section 13.2

13.4 Find the general solution of the homogeneous difference equations:

(i) $y_{n+1} - ay_n = 0$, where a is independent of n;
(ii) $y_{n+2} - 4y_{n+1} + 4y_n = 0$;
(iii) $y_{n+3} - 2y_{n+2} - y_{n+1} + 2y_n = 0$;
(iv) $y_{n+3} - 5y_{n+2} + 8y_{n+1} - 4y_n = 0$;
(v) $y_{n+3} + 3y_{n+2} + 3y_{n+1} + y_n = 0$.

13.5 Find the general solutions of the difference equations:

(i) $y_{n+2} - 4y_{n+1} + 4y_n = 1$;
(ii) $y_{n+2} + y_{n+1} - 2y_n = 1$;
(iii) $y_{n+2} - 2y_{n+1} + y_n = n + 1$.

(*Hint*: find particular solutions by trying solutions which are polynomials in n, for example, $y_n = k_0$, $y_n = k_0 + k_1 n$, $y_n = k_0 + k_1 n + k_2 n^2$, where the k_i are independent of n.)

13.6 The sequence $(u_n)_{n=0}^{\infty}$ of positive reals satisfies the recurrent inequality

$$u_{n+1} \leqslant au_n + b$$

for $n = 0, 1, \ldots$, where $a, b \geqslant 0$ and $a \neq 1$. Prove by induction that

$$u_n \leqslant a^n u_0 + \left(\frac{a^n - 1}{a - 1}\right) b.$$

This is a stronger result than (13.23) when (13.22) holds with $m = 0$ and $a = 1 + \alpha_0$.

Section 13.3

13.7 Derive an explicit formula for the Taylor series method (13.29) with $p = 3$ for the differential equation of Problem 13.2.

13.8 The initial value problem,

$$y' = ax + b, \qquad \text{with } y(0) = 0,$$

has solution

$$y(x) = \tfrac{1}{2}ax^2 + bx.$$

If Euler's method is applied to this problem, show that the resulting difference equation has solution

$$y_n = \tfrac{1}{2}ax_n^2 + bx_n - \tfrac{1}{2}ahx_n,$$

where $x_n = nh$, and thus that

$$y(x_n) - y_n = \tfrac{1}{2}ahx_n.$$

13.9 Show that, for any choice of ν, the 'Runge–Kutta' method

$$\begin{gathered} y_{n+1} = y_n + \tfrac{1}{2}h[k_2 + k_3], \qquad (13.130) \\ k_1 = f(x_n, y_n), \qquad k_2 = f(x_n + \nu h,\ y_n + \nu hk_1), \\ k_3 = f(x_n + [1-\nu]h,\ y_n + [1-\nu]hk_1), \end{gathered}$$

is consistent of order 2. That is, show that the first three terms in the Taylor series expansion of (13.130) agree with the first three terms ($p = 2$) in (13.28).

13.10 Show that the 'Runge–Kutta' method

$$\begin{gathered} y_{n+1} = y_n + h[\tfrac{2}{9}k_1 + \tfrac{1}{3}k_2 + \tfrac{4}{9}k_3], \\ k_1 = f(x_n, y_n), \qquad k_2 = f(x_n + \tfrac{1}{2}h,\ y_n + \tfrac{1}{2}kh), \\ k_3 = f(x_n + \tfrac{3}{4}h,\ y_n + \tfrac{3}{4}k_2h), \end{gathered}$$

is consistent of order 3. (Consider four terms in the Taylor series expansion.)

13.11 Write a formal algorithm for the classical Runge–Kutta method (13.39) when applied to the initial value problem (13.1) and (13.2).

13.12 Solve the equation

$$y' = x^2 + x - y, \qquad \text{with } y(0) = 0$$

by the simple and classical Runge–Kutta methods to find an approximation to $y(0.6)$ using a step size $h=0.2$. Compare your solution with the exact solution

$$y(x) = -e^{-x} + x^2 - x + 1.$$

13.13 Use the classical Runge–Kutta method (13.39) to solve the initial value problem

$$y' = \sin y, \qquad \text{with } y(0) = 1.$$

Take $h = 0.2$ and obtain estimates of the solution for $0 \leqslant x \leqslant 1$.

Section 13.5

13.14 Show that a suitable value of the Lipschitz constant L_ϕ of (13.51) for the method of Problem 13.9 is

$$L_\phi = \tfrac{1}{2}L(2 + [\,|\nu| + |1-\nu|\,]Lh_0),$$

where L is the usual Lipschitz constant for f. Deduce that the method is convergent of order 2.

13.15 Show that a suitable value of the Lipschitz constant L_ϕ of (13.51) for the method of Problem 13.10 is

$$L_\phi = L + h_0 \frac{L^2}{2!} + h_0^2 \frac{L^3}{3!}.$$

13.16 Use Theorem 13.2 to find a global truncation error bound for Euler's method applied to the equation given in Problem 13.12 with $0 \leqslant x \leqslant 1$.

13.17 Find a global truncation error bound for the Taylor series method of order 2 applied to the initial value problem of Examples 13.10 and 13.13 with $0 \leqslant x \leqslant 1$. (*Hint*: show first that $|f^{(2)}(x, y)| \leqslant 2$.)

13.18 Find a global truncation error bound for Euler's method applied to the differential equation of Problem 13.2 with $x_0=0$, $h = 0.1$ and $0 \leqslant x \leqslant 1$.

Section 13.7

13.19 The Taylor polynomial of degree $p-1$ for $f(x, y(x))$ constructed at $x = x_n$ is

$$q(x) = f(x_n, y(x_n)) + \frac{(x-x_n)}{1!} f^{(1)}(x_n, y(x_n)) + \cdots + \frac{(x-x_n)^{p-1}}{(p-1)!} f^{(p-1)}(x_n, y(x_n)),$$

where the $f^{(r)}$ are as defined in § 13.3. Thus

$$\int_{x_n}^{x_{n+1}} q(x)\,dx = hf(x_n, y(x_n)) + \cdots + \frac{h^p}{p!} f^{(p-1)}(x_n, y(x_n)).$$

Show that, on approximating to the integral in (13.60) by the above integral of q and replacing $y(x_n)$ by y_n, we obtain the Taylor series algorithm for an initial value problem.

13.20 Determine the third order Adams–Bashforth algorithm in the form (13.65), that is, expand the differences.

13.21 Determine an approximation to $y(1.0)$, where $y(x)$ is the solution of

$$y' = 1 - y, \qquad \text{with } y(0) = 0,$$

using the Adams–Bashforth algorithm of order 2 with $h = 0.2$ and starting values $y_0 = 0$, $y_1 = 0.181$ $(= y(0.2)$ correct to three decimal places). Compare your calculated solution with

$$y(x) = 1 - e^{-x}.$$

13.22 Use the fourth order Adams–Bashforth algorithm 13.1 to solve the equation in Problem 13.13. Take $h = 0.2$ and consider $0 \leqslant x \leqslant 1$.

13.23 Evaluate the error bound (13.74) for the initial value problem and method of Problem 13.21. Compare the global error bound with the actual error for $x_n = 1.0$. (Initial error bound, $\delta = 0.0005$.)

13.24 Obtain a global truncation error bound for the Adams–Bashforth algorithm of order 2 applied to the equation in Problem 13.2. (*Hint*: show that $f^{(2)}(x, y) = \sin y \cos 2y$ has maximum modulus $+1$.)

13.25 Derive explicit forms of the Nyström method (13.77) for $m = 0, 1, 2$. (Expand the differences.)

Section 13.8

13.26 Determine the third order Adams–Moulton algorithm ($m = 1$) in the form (13.82).

13.27 Show that if a predictor–corrector method is used to solve the linear differential equation of Problem 13.3, then y_{n+1} may be found *explicitly* from the corrector formula.

13.28 Use the Adams–Moulton algorithm of order 2 with $h = 0.2$ to determine an approximation to $y(1.0)$, where $y(x)$ is the solution of the differential equation of Problem 13.21. (Note the result of Problem 13.27.)

13.29 Use the fourth order Adams–Moulton predictor–corrector method (13.84) to solve the equation in Problem 13.13. Take $h = 0.2$, consider $0 \leqslant x \leqslant 1$ and perform just one or two corrections per step ($I = 1$ or 2). Compare your results with those of Problems 13.13 and 13.22.

13.30 Evaluate the global truncation error bound (13.90) at $x_n = 1.0$ for the calculation of Problem 13.28. Compare the actual error with this error bound.

13.31 Obtain a global truncation error bound for the Adams–Moulton method of order 2 applied to the equation in Problem 13.2. (Compare with the result of Problem 13.24 which also contains a hint.)

13.32 Show that

$$\left|\int_0^1 \binom{-s}{j} \mathrm{d}s\right| > \left|\int_0^1 \binom{s}{j} \mathrm{d}s\right|, \qquad \text{for } j \geqslant 2,$$

and thus that $|b_j| > |c_j|$ for $j \geqslant 2$, where b_j and c_j are defined by (13.63) and (13.79) respectively.

13.33 Corrector formulas may be obtained by approximating to (13.76), using q_{m+1} as in §13.8. Show that, with both $m = 1$ and $m = 2$, there results the formula

$$y_{n+1} = y_{n-1} + \tfrac{1}{3}h[f_{n-1} + 4f_n + f_{n+1}].$$

This is called the Milne–Simpson formula and it reduces to Simpson's rule if f is independent of y.

Section 13.10

13.34 The predictor–corrector formulas

$$y_{n+1}^{(0)} = y_n + \tfrac{1}{2}h(3f_n - f_{n-1}),$$
$$y_{n+1}^{(i+1)} = y_n + \tfrac{1}{2}h(f_n + f_{n+1}^{(i)}),$$

obtained from (13.99) with $m = 0$, are both of order 2. Show that the local truncation error of the corrector as given by (13.101) is

$$T_n = -\tfrac{1}{12}h^2 y^{(3)}(\xi_n)$$

and that

$$T_n \simeq -\frac{1}{6h}(y_{n+1}^{(1)} - y_{n+1}^{(0)}).$$

Discuss the choice of h if only one correction is to be made for each step.

13.35 Write a formal algorithm for the method of Example 13.16. You may assume only one correction is made per step.

Section 13.11

13.36 Investigate the numerical stability of the following methods for the differential equation $y' = -Ay$ with $A > 0$.

(i) $y_{n+1} = y_n + \frac{1}{12}h[5f_{n+1} + 8f_n - f_{n-1}]$
(Adams–Moulton corrector of order 3.)

(ii) $y_{n+1} = y_{n-1} + \frac{1}{3}h[f_{n+1} + 4f_n + f_{n-1}]$
(Milne–Simpson formula. See Problem 13.33.)

(iii) $y_{n+1} = y_{n-1} + \frac{1}{2}h[f_{n+1} + 2f_n + f_{n-1}]$
(Double trapezoidal rule.)

13.37 Investigate the numerical stability of the following method for the test equation $y' = -Ay$ with $A > 0$.

$$y_{n+1} = y_{n-1} + \tfrac{1}{2}h(f_n + 3f_{n-1}).$$

Show that absolute errors $|y_n - y_n^*|$ (notation as in § 13.11) will decrease for sufficiently small h but that relative errors $|(y_n - y_n^*)/y_n|$ are unbounded as $n \to \infty$.

13.38 Show that the local truncation error of the midpoint rule (13.111) is $\frac{1}{3}h^2 y^{(3)}(\xi_n)$.

13.39 The equation $y' = -Ay + B$ has general solution $y(x) = ce^{-Ax} + B/A$ where c is arbitrary and thus $y(x) \to B/A$ as $x \to \infty$. If Euler's method is applied to this equation, show that $y_n \to B/A$ as $n \to \infty$ with the step size h fixed, only if $h < 2/A$.

13.40 The equation $y' = -Ay$ with $A > 0$ is to be solved. Show that $y_n \to 0$ as $n \to \infty$ with the step size h fixed (and thus there is absolute stability), if

(i) $h < 2/A$, for the simple Runge–Kutta method;

(ii) $\left| 1 - Ah + \dfrac{1}{2!}A^2h^2 - \dfrac{1}{3!}A^3h^3 + \dfrac{1}{4!}A^4h^4 \right| < 1,$

for the classical Runge–Kutta method.

Section 13.12

13.41 Show that, for the function $\mathbf{f}$ of the equations in Example 13.20 with $x \in [0, b]$ where $b > 1$, a suitable choice of L, the Lipschitz constant in (13.121), is $L = 1 + b$.

13.42 Determine, as in Example 13.20, the difference equations for the Taylor series method of order 2 applied to the two simultaneous equations

$$y_1'(x) = x^2 y_1(x) - y_2(x)$$
$$y_2'(x) = -y_1(x) + xy_2(x).$$

with $y_1(0) = s_1$ and $y_2(0) = s_2$.

13.43 Solve the equation

$$y'' + 4xy'y + 2y^2 = 0.$$

with $y(0) = 1$, $y'(0) = 0$, using Euler's method with $h = 0.1$, to obtain approximations to $y(0.5)$ and $y'(0.5)$.

13.44 Determine the difference equations for the classical Runge–Kutta method applied to the differential equation of Problem 13.43.

13.45 Derive the difference equations for the Taylor series method of order 2 applied to

$$y^{(3)} = 2y'' + x^2 y + 1 + x,$$

assuming that $y(x)$, $y'(x)$ and $y''(x)$ are known for $x = x_0$.

13.46 Determine the difference equations for the Adams–Moulton predictor–corrector method of order 2 applied to the differential equation in Problem 13.45.

Chapter **14**

BOUNDARY VALUE AND OTHER METHODS FOR ORDINARY DIFFERENTIAL EQUATIONS

14.1 Shooting method for boundary value problems

Two associated conditions are required for the second order equation

$$y'' = f(x, y, y') \tag{14.1}$$

if the solution y is to be unique. If these two conditions are imposed at two distinct values of x, we call the differential equation and conditions a *boundary value problem.* The conditions are called *boundary conditions*. The simplest type are

$$y(a) = \alpha \quad \text{and } y(b) = \beta, \tag{14.2}$$

where $a < b$ and α and β are known. Thus $y(x)$ is given at the distinct points $x = a$ and $x = b$. For such problems, we seek $y(x)$ for $x \in [a, b]$.

We may adapt the step-by-step methods employed for initial value problems to boundary value problems. Suppose that we have an estimate w of $y'(a)$, where $y(x)$ is the solution of (14.1) subject to (14.2). We now solve the initial value problem (14.1) with

$$y(a) = \alpha$$
$$y'(a) = w$$

using a step-by-step method and suppose that β_w is the resulting approximation to $y(b)$. It is most likely that $\beta_w \neq \beta$. We adjust w and repeat the calculation to try to produce $\beta_w = \beta$. Figure 14.1 describes a typical problem. In the upper curve, w is too large and in the lower too small. The broken line is the required solution. We call the process of starting from one end, $x = a$, to try to produce the correct value at the other end, $x = b$, a 'shooting' method.

Of course we should adjust w in a systematic fashion. It is clear that

$$\beta_w = F(w)$$

for some function $F(w)$ and we are therefore trying to find w such that

$$F(w) = \beta.$$

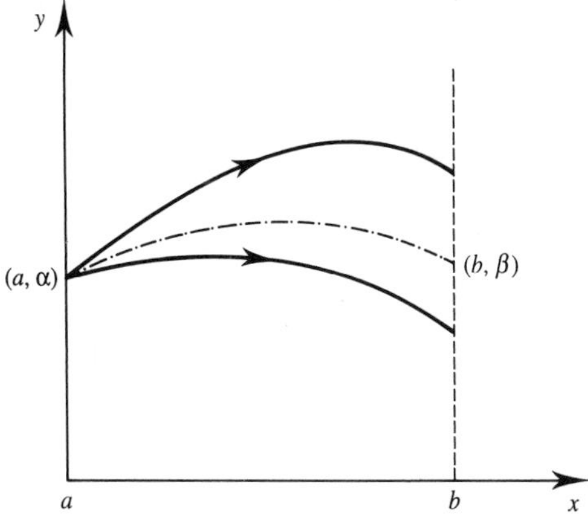

Fig. 14.1 The shooting method.

This is an algebraic equation and the bisection, *regula falsi* and Muller methods are suitable for its solution (see Chapter 8).

More general boundary conditions than (14.2) are

$$p_1 y(a) + q_1 y'(a) = \alpha_1, \tag{14.3a}$$

$$p_2 y(b) + q_2 y'(b) = \alpha_2, \tag{14.3b}$$

where p_i, q_i and α_i, $i = 1, 2$, are constants. The shooting method may be adapted to these conditions by choosing approximations to $y(a)$ and $y'(a)$ which satisfy (14.3a) exactly and calculating approximations to $y(b)$ and $y'(b)$ by a step-by-step method. These approximations will probably not satisfy (14.3b), so we adjust the approximations to $y(a)$ and $y'(a)$ and repeat the calculation.

14.2 Boundary value method

An alternative to step-by-step methods, when solving a boundary value problem, is to make difference approximations to derivatives and impose boundary conditions on the solution of the resulting difference equation. We illustrate the method first by considering the linear second order equation on $[a, b]$,

$$u(x)y'' + v(x)y' + w(x)y = f(x), \tag{14.4}$$

where u, v, w and f are continuous on $[a, b]$. We are also given that $u(x) > 0$ and $w(x) \leqslant 0$ on $[a, b]$. (Not all of these restrictions on u, v, w and f are necessary but they do simplify our analysis.) The given boundary conditions are again taken to be

$$y(a) = \alpha, \qquad y(b) = \beta \tag{14.5}$$

and we will assume that (14.4) and (14.5) have a unique solution $y(x)$ on $[a, b]$.

We divide the interval $[a, b]$ into $N+1$ intervals each of length h and seek approximations y_n to the solution $y(x_n)$ at the points $x_n = a + nh$, $n = 1, 2, \ldots, N$. We choose the mesh length $h = (b-a)/(N+1)$ so that $x_{N+1} = b$. We approximate to (14.4) at the point x_n by replacing derivatives by differences as in (7.78) and (7.81) and so obtain

$$u_n\left(\frac{y_{n+1} - 2y_n + y_{n-1}}{h^2}\right) + v_n\left(\frac{y_{n+1} - y_{n-1}}{2h}\right) + w_n y_n = f_n, \quad n = 1, 2, \ldots, N, \tag{14.6}$$

where $u_n = u(x_n)$, $v_n = v(x_n)$, $w_n = w(x_n)$ and $f_n = f(x_n)$. Corresponding to (14.5) we also have

$$y_0 = \alpha, \qquad y_{N+1} = \beta. \tag{14.7}$$

Equations (14.6) are N linear equations in the N unknowns $y_1, y_2, \ldots, y_N$ and may be written in the form

$$\mathbf{A}\mathbf{y} = \mathbf{f}, \tag{14.8}$$

where $\mathbf{A}$ is the $N \times N$ tridiagonal matrix

$$\mathbf{A} = \begin{bmatrix} \left(\dfrac{-2u_1}{h^2} + w_1\right) & \left(\dfrac{u_1}{h^2} + \dfrac{v_1}{2h}\right) & & & \mathbf{0} \\ \left(\dfrac{u_2}{h^2} - \dfrac{v_2}{2h}\right) & \left(\dfrac{-2u_2}{h^2} + w_2\right) & \left(\dfrac{u_2}{h^2} + \dfrac{v_2}{2h}\right) & & \\ & \ddots & \ddots & \ddots & \\ & & \ddots & \ddots & \left(\dfrac{u_{N-1}}{h^2} + \dfrac{v_{N-1}}{2h}\right) \\ \mathbf{0} & & & \left(\dfrac{u_N}{h^2} - \dfrac{v_N}{2h}\right) & \left(\dfrac{-2u_N}{h^2} + w_N\right) \end{bmatrix},$$

$$\mathbf{y} = \begin{bmatrix} y_1 \\ y_2 \\ \vdots \\ y_N \end{bmatrix} \quad \text{and} \quad \mathbf{f} = \begin{bmatrix} f_1 - \left(\dfrac{u_1}{h^2} - \dfrac{v_1}{2h}\right)\alpha \\ f_2 \\ f_3 \\ \vdots \\ f_{N-1} \\ f_N - \left(\dfrac{u_N}{h^2} + \dfrac{v_N}{2h}\right)\beta \end{bmatrix}.$$

Because of the tridiagonal nature of **A**, the equations (14.8) may easily be solved by an elimination method, as was seen in §9.9. It can be shown that if h is sufficiently small the equations are non-singular. Problem 14.5 gives a proof of this for the case $v(x) \equiv 0$.

If the boundary conditions are of the form (14.3) with $q_1 \neq 0$ and $q_2 \neq 0$ so that derivatives are involved, we must make an approximation to (14.3) using (7.78). At the end $x = a$ we introduce the point $x_{-1} = a - h$ and a 'fictitious' value y_{-1} at this point.† We replace (14.3a) by

$$p_1 y_0 + q_1 \left(\frac{y_1 - y_{-1}}{2h} \right) = \alpha_1, \qquad (14.9)$$

which on rearrangement yields

$$y_{-1} = \frac{2h(p_1 y_0 - \alpha_1)}{q_1} + y_1. \qquad (14.10)$$

Since y_0 is unknown we must also use (14.6) with $n = 0$, the difference approximation to (14.4) at $x = a$. We use (14.10) to eliminate the y_{-1} introduced and obtain

$$\left(\frac{-2u_0}{h^2} + w_0 + \frac{2hp_1}{q_1} \left(\frac{u_0}{h^2} - \frac{v_0}{2h} \right) \right) y_0 + \frac{2}{h^2} u_0 y_1 = f_0 + \frac{2h\alpha_1}{q_1} \left(\frac{u_0}{h^2} - \frac{v_0}{2h} \right), \qquad (14.11)$$

where again $u_0 = u(x_0)$, etc. We can obtain a similar equation at x_{N+1} relating y_N and y_{N+1} by approximating to (14.3b) and using (14.6) with $n = N + 1$. We now have linear equations of the form

$$\mathbf{Bz} = \mathbf{g},$$

where

$$\mathbf{z} = \begin{bmatrix} y_0 \\ y_1 \\ \vdots \\ y_{N+1} \end{bmatrix}.$$

It can be seen from (14.6) and (14.11) that the $(N + 2) \times (N + 2)$ coefficient matrix **B** is again tridiagonal and therefore the equations are easily solved.

Example 14.1 We use the above method for the linear boundary value problem on $[0, 1]$,

$$(1 + x^2)y'' - xy' - 3y = 6x - 3 \qquad (14.12)$$

---

† We cannot say that y_{-1} is an approximation to $y(a-h)$ since the latter is not defined.

with

$$y(0) - y'(0) = 1 \tag{14.13a}$$

$$y(1) = 2. \tag{14.13b}$$

We take $h = 0.2$ and seek y_n, $n = 0, 1, 2, 3, 4$, approximations to $y(x_n)$ where $x_n = 0.2n$.

We replace the boundary conditions (14.13) by

$$y_0 - \left(\frac{y_1 - y_{-1}}{0.4}\right) = 1 \tag{14.14}$$

and

$$y_5 = 2.$$

At $x = 0$ the approximation to (14.12) is

$$\frac{y_1 - 2y_0 + y_{-1}}{(0.2)^2} - 3y_0 = -3$$

and, on multiplying by $(0.2)^2$ and substituting for y_{-1} using (14.14) we obtain

$$-2.52y_0 + 2y_1 = -0.52.$$

The difference approximation to (14.12) is

$$(1 + 0.04n^2)\left(\frac{y_{n+1} - 2y_n + y_{n-1}}{(0.2)^2}\right) - 0.2n\left(\frac{y_{n+1} - y_{n-1}}{0.4}\right) - 3y_n = 1.2n - 3,$$

$$n = 1, 2, 3, 4.$$

We multiply these equations by $(0.2)^2$ and rearrange them to obtain the tridiagonal set

$$\begin{bmatrix} -2.52 & 2 & 0 & 0 & 0 \\ 1.06 & -2.20 & 1.02 & 0 & 0 \\ 0 & 1.20 & -2.44 & 1.12 & 0 \\ 0 & 0 & 1.42 & -2.84 & 1.30 \\ 0 & 0 & 0 & 1.72 & -3.40 \end{bmatrix} \begin{bmatrix} y_0 \\ y_1 \\ y_2 \\ y_3 \\ y_4 \end{bmatrix} = \begin{bmatrix} -0.52 \\ -0.072 \\ -0.024 \\ +0.024 \\ -3.048 \end{bmatrix}.$$

The solution, correct to five significant decimal digits, is $y_0 = 1.0132$, $y_1 = 1.0167$, $y_2 = 1.0693$, $y_3 = 1.2188$, $y_4 = 1.5130$. The required solution of the differential equation is $y(x) = 1 + x^3$, so that

$$y(x_0) = 1, \quad y(x_1) = 1.008, \quad y(x_2) = 1.064, \quad y(x_3) = 1.216, \quad y(x_4) = 1.512. \quad \square$$

We now return to the non-linear problem

$$y'' = f(x, y, y') \tag{14.15}$$

with $y(a) = \alpha$, $y(b) = \beta$. We again seek approximations y_n to $y(x_n)$ at the N equally spaced points $x_n = a + nh$, $n = 1, 2, \ldots, N$, with $h = (b - a)/(N + 1)$. The usual approximation to (14.15) is

$$\frac{y_{n+1} - 2y_n + y_{n-1}}{h^2} = f\left(x_n, y_n, \frac{y_{n+1} - y_{n-1}}{2h}\right), \qquad n = 1, 2, \ldots, N, \tag{14.16}$$

with $y_0 = \alpha$ and $y_{N+1} = \beta$. If the boundary conditions are of the form (14.3), we introduce extra equations as in the linear case. Equations (14.16) form N *non-linear* equations in the unknowns $y_1, y_2, \ldots, y_N$. The solution must usually be found by an iterative method. For example, we could try keeping the right side in (14.16) fixed and solving the resulting linear equations. We use this solution to compute new values for the right side and then solve the linear equations again. However, the resulting iterative process will not always converge. A much more satisfactory approach is to use Newton's method, especially as each of the equations (14.16) involves at most three unknowns so that the Jacobian matrix is tridiagonal (see Problem 14.7). Henrici (1962) discusses in detail the convergence of Newton's process when f is independent of y'.

So far we have not considered the global errors $|y_n - y(x_n)|$ and the possible convergence of these to zero as the mesh length h is decreased. In this context, we also will assume that f is independent of y' and will investigate only problems of the form

$$y'' = f(x, y) \tag{14.17}$$

with

$$y(a) = \alpha \quad \text{and} \quad y(b) = \beta. \tag{14.18}$$

The simplest difference approximation is

$$\frac{y_{n+1} - 2y_n + y_{n-1}}{h^2} = f(x_n, y_n), \qquad n = 1, 2, \ldots, N, \tag{14.19}$$

with

$$y_0 = \alpha, \qquad y_{N+1} = \beta. \tag{14.20}$$

If $y^{(4)}(x)$ is continuous on $[a, b]$ we note from (7.89) that

$$\frac{y(x_{n+1}) - 2y(x_n) + y(x_{n-1})}{h^2} = f(x_n, y(x_n)) + \frac{h^2}{12}\, y^{(4)}(\xi_n) \tag{14.21}$$

where $\xi_n \in (x_{n-1}, x_{n+1})$. The last term in (14.21) is called the local truncation error of the method (14.19). To obtain a bound on global errors, we need the following lemma, which is based on the so-called *maximum principle*.

Lemma 14.1 If w_n, $n=0,1,\ldots,N+1$, satisfy the difference equation

$$w_{n+1}-(2+c_n)w_n+w_{n-1}=d_n, \qquad n=1,2,\ldots,N,$$

where $c_n \geqslant 0$ and $d_n \geqslant 0$, $n=1,2,\ldots,N$, and $w_0 \leqslant 0$ and $w_{N+1} \leqslant 0$ are given, then

$$w_n \leqslant 0, \qquad n=1,2,\ldots,N.$$

Proof For $n=1,2,\ldots,N$,

$$\begin{aligned} w_n &= \frac{w_{n-1}+w_{n+1}}{2+c_n} - \frac{d_n}{2+c_n} \\ &\leqslant \frac{w_{n-1}+w_{n+1}}{2}. \end{aligned}$$

Hence w_n cannot exceed both of its neighbours, w_{n-1} and w_{n+1}, and, therefore, the maximum w_n must occur at one or both of the boundaries $n=0$ and $n=N+1$. Thus

$$w_n \leqslant w_0 \leqslant 0 \quad \text{or} \quad w_n \leqslant w_{N+1} \leqslant 0, \qquad 0 \leqslant n \leqslant N+1. \qquad \square$$

We will make the additional assumption that in (14.17) $f_y(x,y)$ exists and is continuous with $f_y(x,y) \geqslant 0$ for $x \in [a,b]$ and $-\infty < y < \infty$. Henrici (1962) shows that this condition is sufficient to ensure the existence of a solution of the boundary value problem.

Theorem 14.1 Given the boundary value problem (14.17) and (14.18), let y_n, $n=0,1,\ldots,N+1$, satisfy (14.19) and (14.20). If

$$M = \max_{x \in [a,b]} |y^{(4)}(x)|$$

exists and $\partial f/\partial y \geqslant 0$ for $x \in [a,b]$, $-\infty < y < \infty$, the global error satisfies

$$|y(x_n)-y_n| \leqslant \frac{Mh^2}{24}(x_n-a)(b-x_n). \tag{14.22}$$

Proof It is easily verified by direct substitution that the solution of

$$z_{n+1}-2z_n+z_{n-1}=\frac{h^4M}{12}, \qquad n=1,2,\ldots,N, \tag{14.23}$$

with

$$z_0 = z_{N+1} = 0,$$

is

$$z_n = \frac{-h^2 M}{24}(x_n - a)(b - x_n). \tag{14.24}$$

From (14.21) we have

$$y(x_{n+1}) - 2y(x_n) + y(x_{n+1}) = h^2 f(x_n, y(x_n)) + \tfrac{1}{12}h^4 y^{(4)}(\xi_n).$$

On subtracting $h^2 \times$ (14.19), putting $e_n = y(x_n) - y_n$ and using the mean value theorem, we obtain

$$e_{n+1} - 2e_n + e_{n-1} = h^2 g_n e_n + \tfrac{1}{12}h^4 y^{(4)}(\xi_n), \tag{14.25}$$

where $g_n = f_y(x_n, \eta_n) \geqslant 0$, with η_n lying between y_n and $y(x_n)$. On adding (14.23) and (14.25), we see that

$$(z_{n+1} + e_{n+1}) - (2 + h^2 g_n)(z_n + e_n) + (z_{n-1} + e_{n-1})$$
$$= \tfrac{1}{12}h^4(M + y^{(4)}(\xi_n)) - h^2 g_n z_n. \tag{14.26}$$

The right side of (14.26) is non-negative, as can be seen from $z_n \leqslant 0$, $g_n \geqslant 0$ and the definition of M. Also $z_0 + e_0 = 0$ and $z_{N+1} + e_{N+1} = 0$ and, on applying Lemma 14.1 to (14.26) with $w_n = z_n + e_n$, we deduce that

$$z_n + e_n \leqslant 0,$$

that is,

$$e_n \leqslant -z_n. \tag{14.27}$$

Similarly, by subtracting (14.25) from (14.23), we deduce that

$$z_n - e_n \leqslant 0$$

so that

$$e_n \geqslant z_n. \tag{14.28}$$

Combining (14.27), (14.28) and (14.24) provides the result (14.22). □

For many problems the bound in Theorem 14.1 is very much larger than the actual error. The bound also requires an upper bound of $|y^{(4)}(x)|$, which may not be readily available. The theorem does show, however, that the global error is $O(h^2)$, as h is decreased, and therefore we say that the method is of order 2.

As in the discussion of errors for initial value problems, we can include the effects of rounding errors. Suppose that, instead of finding $y_1, \ldots, y_N$, we actually calculate $w_1, \ldots, w_N$ satisfying

$$w_{n+1} - 2w_n + w_{n-1} = h^2 f(x_n, w_n) + \varepsilon_n,$$

where ε_n is the error introduced in solving the simultaneous algebraic equations. If for some ε, $|\varepsilon_n| \leqslant h^4\varepsilon$, $n=1,\ldots,N$, we must add ε to $M/12$ in the bound (14.22) to obtain a bound of $|w_n - y(x_n)|$.

There is a more accurate three-point difference approximation than (14.19) to (14.17), the special case of (14.1) when f is independent of y'. The formula

$$\frac{y_{n+1}-2y_n+y_{n-1}}{h^2} = \tfrac{1}{12}(f(x_{n+1},y_{n+1}) + 10f(x_n,y_n) + f(x_{n-1},y_{n-1})) \qquad (14.29)$$

has a truncation error $-h^4y^{(6)}(\xi_n)/240$, $\xi_n \in (x_{n-1}, x_{n+1})$. See Problem 14.2 for a proof that the error is $O(h^4)$. We can obtain a global error bound which is of $O(h^4)$ but is otherwise similar to that in Theorem 14.1. Since (14.29) again involves at most three unknowns in each equation, it is no more difficult to implement than (14.19) and is, therefore, to be preferred for problems of the form (14.17).

Example 14.2 We consider the simple linear equation on [0, 1]

$$y'' = y$$

with

$$y(0)=0, \qquad y(1)=1.$$

The solution is $y(x) = \sinh x/\sinh 1$.†

With $h=0.2$, the difference approximation (14.19) becomes

$$y_{n+1} - 2.04y_n + y_{n-1} = 0, \qquad n=1,2,3,4,$$

with $y_0=0$, $y_5=1$. Table 14.1 shows results including a comparison of the solutions of the difference equation and the differential equation. The global error bound is calculated from (14.22) with

$$M = \max_{0\leqslant x\leqslant 1} |y^{(4)}(x)| = \max_{0\leqslant x\leqslant 1}\left|\frac{\sinh x}{\sinh 1}\right| = 1.$$

Table 14.1 Solution of Example 14.2

n	y_n	$y(x_n)$	$\lvert y(x_n)-y_n\rvert \times 10^5$	Error bound (14.22) $\times 10^5$
1	0.17140	0.17132	8	27
2	0.34967	0.34952	15	40
3	0.54192	0.54174	18	40
4	0.75584	0.75571	13	27

† $\sinh x \equiv \frac{1}{2}(e^x - e^{-x})$.

The difference approximation (14.29) is in this case

$$y_{n+1} - 2y_n + y_{n-1} = \frac{0.04}{12}(y_{n+1} + 10y_n + y_{n-1}), \qquad n = 1, 2, 3, 4,$$

that is,

$$0.996667y_{n+1} - 2.033333y_n + 0.996667y_{n-1} = 0, \qquad n = 1, 2, 3, 4,$$

with $y_0 = 0$, $y_5 = 1$. The solution to five significant decimal digits is identical to $y(x_n)$, $n = 1, 2, 3, 4$. □

14.3 Extrapolation to the limit

In solving a differential equation we can often apply an extrapolation to the limit process to results using different step sizes as in Chapter 7 for numerical differentiation and integration. We consider, for example, the solution of the first order initial value problem (13.1) using Euler's method (13.11). We write $Y(x_n; h)$ to denote the approximation y_n to $y(x_n)$ when a step size h is used. We will assume that the global error may be expressed as a power series in h of the form

$$y(x) - Y(x; h) = A_1(x)h + A_2(x)h^2 + A_3(x)h^3 + \cdots, \qquad (14.30)$$

where $A_1(x)$, $A_2(x)$, ... are unknown functions of x but are independent of h. The existence of an expansion of the form (14.30) is dependent on there being no singularities in the differential equation and the existence of 'sufficient' partial derivatives of $f(x, y)$. Henrici (1962) proves the existence of $A_1(x)$ under these conditions. As in §7.2, we usually halve the step size and eliminate the first term on the right of (14.30). By repeating this process we can eliminate terms in h^2, h^3, ... Analogous to (7.32), we compute†

$$Y^{(m)}(x; h) = \frac{2^m Y^{(m-1)}(x; h/2) - Y^{(m-1)}(x; h)}{2^m - 1}, \qquad (14.31)$$

where $Y^{(0)}(x; h) = Y(x; h)$.

Example 14.3 We use Euler's method and repeated extrapolation to the limit to find approximations to $y(0.8)$, where $y(x)$ is the solution of

$$y' = x^2 - y$$

with

$$y(0) = 1.$$

† The superscripts m do not denote derivatives.

Table 14.2 Extrapolation to the limit (Example 14.3)

h	$Y^{(0)}$	$Y^{(1)}$	$Y^{(2)}$	$Y^{(3)}$
0.8	0.20000			
		0.64800		
0.4	0.42400		0.58485	
		0.60064		0.59101
0.2	0.51232		0.59024	
		0.59284		0.59069
0.1	0.55258		0.59063	
		0.59118		
0.05	0.57188			

The results are tabulated in Table 14.2. The layout is similar to that of Table 7.2. The solution of the differential equation is

$$y(x) = 2 - 2x + x^2 - e^{-x}$$

and, therefore, to five decimal places $y(0.8) = 0.59067$. □

In general, if a method is of order p, the power series for the global error will take the form

$$y(x) - Y(x; h) = A_p(x)h^p + A_{p+1}(x)h^{p+1} + A_{p+2}(x)h^{p+2} + \cdots, \tag{14.32}$$

if such a power series exists. We need to modify (14.31) by replacing 2^m by 2^{m+p-1} in both the denominator and numerator. Some methods will yield a power series in only even powers of h, so that

$$y(x) - Y(x; h) = E_1(x)h^2 + E_2(x)h^4 + E_3(x)h^6 + \cdots$$

for some functions $E_1, E_2, \ldots$. We must then use the relation (7.32) (with T replaced by Y). The Adams–Moulton corrector (13.83b) and the difference approximation (14.16) are both examples of such methods.

Finally, we must add a warning that extrapolation will produce misleading results if a series like (14.32) does not exist. This can be caused by singularities in the differential equation, that is, one of the coefficients in the equation or one or more of the solutions of the equation is singular at some point. It is also important that derivatives of the solutions should exist. There may be trouble even if the solution which misbehaves is not the particular solution we are seeking. We have already seen in numerical integration, §7.2, that singular derivatives at one point may have an adverse effect on extrapolation.

14.4 Deferred correction

It is often possible to estimate the truncation error of difference approximations to differential equations by using differences of the computed solution. We can then use these estimates to improve (or 'correct') the computed solution. This technique can be used for both initial value and boundary value problems and we will illustrate it by considering the solution of the boundary value problem (14.17) and (14.18) using the difference approximation (14.29). From Problem 14.2 it follows that, if $y(x)$ is the solution of the differential equation and $y(x)$ has a continuous eighth derivative,

$$\frac{y(x_{n+1}) - 2y(x_n) + y(x_{n-1})}{h^2} = \tfrac{1}{12}\,(f(x_{n+1}, y(x_{n+1})) + 10f(x_n, y(x_n)) + f(x_{n-1}, y(x_{n-1}))) - \frac{h^4}{240}\, y^{(6)}(x_n) + O(h^6). \tag{14.33}$$

Now from (7.84), using Taylor series, we see that

$$y^{(6)}(x_n) = \frac{1}{h^6}\,\Delta^6 y(x_{n-3}) + O(h^2). \tag{14.34}$$

Having computed the y_n, the solution of (14.29), we approximate to the truncation error in (14.33), using $-\Delta^6 y_{n-3}/(240h^2)$, and solve

$$\frac{z_{n+1} - 2z_n + z_{n-1}}{h^2} = \tfrac{1}{12}\,(f(x_{n+1}, z_{n+1}) + 10f(x_n, z_n) + f(x_{n-1}, z_{n-1})) - \frac{1}{240h^2}\,\Delta^6 y_{n-3}. \tag{14.35}$$

We expect z_n to be a better approximation to $y(x_n)$. In general, (14.35) gives a non-linear set of equations in the z_n, which are therefore difficult to compute (although no more difficult than the y_n from (14.29)). We can replace (14.35) by the linearization

$$\frac{e_{n+1} - 2e_n + e_{n-1}}{h^2} = \tfrac{1}{12}\,(f_y(x_{n+1}, y_{n+1}) \,.\, e_{n+1} + 10f_y(x_n, y_n) \,.\, e_n + f_y(x_{n-1}, y_{n-1}) \,.\, e_{n-1}) - \frac{1}{240h^2}\,\Delta^6 y_{n-3}, \tag{14.36}$$

where $f_y = \partial f/\partial y$ and e_n is to be the correction to y_n, that is we take $y_n + e_n$ as an improved approximation to $y(x_n)$. We derive (14.36) by subtracting (14.29) from (14.35) and using

$$f(x_n, z_n) - f(x_n, y_n) \simeq (z_n - y_n) f_y(x_n, y_n)$$

from the mean value theorem. We also replace $z_n - y_n$ by e_n.

It can be shown that, for both (14.35) and (14.36), the global error of the resulting process is $O(h^6)$, that is $|y(x_n) - z_n| = O(h^6)$ and $|y(x_n) - (y_n + e_n)| = O(h^6)$, compared with $O(h^4)$ for (14.29). It is possible to use a more accurate approximation to the truncation error in (14.33) and so increase the order of the method even more. To achieve this we also need to iterate. We compute a succession of approximations, each of which is used to estimate the truncation error from which the next approximation can be calculated. Fox (1957) gives many details of deferred correction.

With boundary conditions (14.18), we must apply (14.29) and (14.35) for $n = 1, 2, \ldots, N$ where $N = (b-a)/h - 1$. To use (14.35) we need values y_{-2}, y_{-1}, y_{N+2}, y_{N+3} and these are found by using (14.29) in a step-by-step procedure. For example, we use (14.29) with $n = 0$ to compute y_{-1} from y_0 and y_1, and then with $n = -1$, we compute y_{-2} from y_{-1} and y_0.

Example 14.4 Compute an approximation to the solution of

$$y'' = y + (2 - x^2)$$

with

$$y(0) = 0, \qquad y(4) = 17.$$

Use (14.29) with $h = 1$ and the deferred correction process (14.36). The linearization (14.36) is identical to (14.35) with $e_n = z_n - y_n$ because f is linear in y. The equations (14.29), $n = 1, 2, 3$, become

$$\begin{aligned}
(-2 - \tfrac{10}{12})y_1 + (1 - \tfrac{1}{12})y_2 \qquad\qquad &= \tfrac{10}{12} \\
(1 - \tfrac{1}{12})y_1 + (-2 - \tfrac{10}{12})y_2 + (1 - \tfrac{1}{12})y_3 &= -\tfrac{26}{12} \\
(1 - \tfrac{1}{12})y_2 + (-2 - \tfrac{10}{12})y_3 &= -\tfrac{86}{12} - (1 - \tfrac{1}{12}).17.
\end{aligned}$$

Table 14.3 Solution of Example 14.4 before correction

n	y	Δy	$\Delta^2 y$	$\Delta^3 y$	$\Delta^4 y$	$\Delta^5 y$	$\Delta^6 y$
−2	3.86762						
		−2.91045					
−1	0.95717		1.95328				
		−0.95717		0.04672			
0	0		2.00000		0.00001		
		1.04283		0.04673		0.05094	
1	1.04283		2.04673		0.05095		0.05569
		3.08956		0.09768		0.10663	
2	4.13239		2.14441		0.15758		0.17177
		5.23397		0.25526		0.27840	
3	9.36636		2.39967		0.43598		0.47571
		7.63364		0.69124		0.75411	
4	17.00000		3.09091		1.19009		
		10.72455		1.88133			
5	27.72455		4.97224				
		15.69679					
6	43.42134						

Table 14.4 Solution of Example 14.4 with correction

n	y_n	$e_n \times 10^5$	z_n	$y(x_n)$	$(y(x_n)-y_n)\times 10^5$	$(y(x_n)-z_n)\times 10^5$
1	1.04283	29	1.04312	1.04306	23	−6
2	4.13239	64	4.13303	4.13290	51	−13
3	9.36636	91	9.36727	9.36709	73	−18

Solving these gives the values y_1, y_2, y_3 shown correct to five decimal places in Table 14.3. We now find y_{-1}, y_{-2}, y_5 and y_6, using (14.29) with $n=0$, $-1, 4$ and 5 respectively. Thus, for example, with $n=4$,

$$\tfrac{11}{12}y_5 = (2+\tfrac{10}{12})y_4 - \tfrac{11}{12}y_3 - \tfrac{170}{12}.$$

From (14.36), the corrections satisfy

$$(1-\tfrac{1}{12})e_{n-1} + (-2-\tfrac{10}{12})e_n + (1-\tfrac{1}{12})e_{n+1} = -\tfrac{1}{240}\,\Delta^6\, y_{n-3},$$

$n=1,2,3$, with $e_0=e_4=0$. The solution of these equations and the corrected solution $z_n = y_n + e_n$ are shown in Table 14.4, which also gives a comparison with the exact solution $y(x) = x^2 + \sinh x/\sinh 4$. □

14.5 Chebyshev series method

We will show how to obtain a Chebyshev series approximation to the solution of a differential equation with polynomial coefficients. We shall obtain approximations which are valid over an interval and not just approximations to the solution at particular points. We need the following identities (see Problem 14.13) concerning products and indefinite integrals of Chebyshev polynomials:

$$x^s T_r(x) = 2^{-s}\sum_{j=0}^{s}\binom{s}{j} T_{r-s+2j}(x) \tag{14.37}$$

$$T_r(x)T_s(x) = \tfrac{1}{2}[T_{r+s}(x) + T_{r-s}(x)] \tag{14.38}$$

$$\int T_r(x)\,dx = \begin{cases} \dfrac{1}{2}\left[\dfrac{T_{r+1}(x)}{r+1} - \dfrac{T_{r-1}(x)}{r-1}\right], & r \neq 1 \\ \tfrac{1}{4}T_2(x), & r = 1 \end{cases} \tag{14.39}$$

$$T_r'(x) = \begin{cases} 0, & r = 0 \\ 2r(\tfrac{1}{2}T_0 + T_2 + \cdots + T_{r-1}), & r \text{ odd} \\ 2r(T_1 + T_3 + \cdots + T_{r-1}), & r \text{ even} \end{cases} \tag{14.40}$$

for $r, s = 0, 1, 2, \ldots$, where we define $T_{-r}(x) = T_r(x)$.

We consider first the initial value problem of $[-1, 1]$,

$$q_1(x)y'(x) + q_2(x)y(x) = q_3(x) \tag{14.41}$$

with

$$y(-1) = s, \tag{14.42}$$

where q_1, q_2 and q_3 are polynomials. On integrating (14.41) we obtain

$$q_1(x) \,.\, y(x) + \int (q_2(x) - q_1'(x))y(x)\,\mathrm{d}x = \int q_3(x)\,\mathrm{d}x + C, \tag{14.43}$$

where C is an arbitrary constant. We now express q_1, q_2 and q_3 in terms of Chebyshev polynomials and seek a series†

$$y(x) = \sum_{r=0}^{\infty}{}' a_r T_r(x)$$

which satisfies (14.43). On using (14.37)–(14.40) we can express (14.43) as

$$\sum_{r=0}^{\infty}{}' b_r T_r(x) = \sum_{r=0}^{m}{}' c_r T_r(x), \tag{14.44}$$

if $q_3 \in P_{m-1}$. Each coefficient b_r is some linear combination of the unknown a_r; the coefficients c_r are known, being determined by q_3, with the exception of c_0 which also depends on the arbitrary constant C. On equating coefficients b_r and c_r, we obtain an infinite set of linear equations in the a_r. We usually solve the first few of these equations, except for $r = 0$ as this involves the arbitrary constant, and so find an approximation to $y(x)$. We also impose the initial condition (14.42) on our approximation and so obtain a further linear equation in the a_r. The following example illustrates the process.

Example 14.5 Consider

$$y' + xy = x, \qquad -1 \leqslant x \leqslant 1, \tag{14.45}$$

with

$$y(-1) = 0. \tag{14.46}$$

On integrating, (14.45) becomes

$$y + \int xy\,\mathrm{d}y = \tfrac{1}{2}x^2 + C. \tag{14.47}$$

If

$$y = \sum_{r=0}^{\infty}{}' a_r T_r$$

† Σ' denotes summation with the first term halved.

then, from (14.37) and (14.39),

$$\int xy\,dx = \tfrac{1}{4}(a_1 - a_3)T_1 + \tfrac{1}{4}\sum_{r=2}^{\infty}\left(\frac{a_{r-2} - a_{r+2}}{r}\right)T_r.$$

Thus

$$y + \int xy\,dx = \tfrac{1}{2}a_0 + \tfrac{1}{4}(5a_1 - a_3)T_1 + \tfrac{1}{4}\sum_{r=2}^{\infty}\left(\frac{a_{r-2}}{r} + 4a_r - \frac{a_{r+2}}{r}\right)T_r. \quad (14.48)$$

Also,

$$\tfrac{1}{2}x^2 + C = (\tfrac{1}{4} + C) + \tfrac{1}{4}T_2. \quad (14.49)$$

On equating coefficients (except constant terms) in (14.48) and (14.49), we obtain

$$\left.\begin{aligned}
\tfrac{5}{4}a_1 - \tfrac{1}{4}a_3 &= 0\\
\tfrac{1}{8}a_0 + a_2 - \tfrac{1}{8}a_4 &= \tfrac{1}{4}\\
\tfrac{1}{12}a_1 + a_3 - \tfrac{1}{12}a_5 &= 0\\
\tfrac{1}{16}a_2 + a_4 - \tfrac{1}{16}a_6 &= 0\\
\text{and}\qquad\qquad &\\
\frac{1}{4r}a_{r-2} + a_r - \frac{1}{4r}a_{r+2} &= 0, \qquad r = 5, 6, 7, \ldots
\end{aligned}\right\} \quad (14.50)$$

We look for an approximation $p_4 \in P_4$ to y of the form

$$p_4(x) = \sum_{r=0}^{4}{}' \tilde{a}_r T_r(x),$$

where the $\tilde{a}_r$ satisfy the first four of the equations (14.50) with $\tilde{a}_5 = \tilde{a}_6 = \cdots = 0$. If we solve these equations, keeping $\tilde{a}_0$ as a free parameter, we find that

$$\begin{gathered}
\tilde{a}_1 = \tilde{a}_3 = 0\\
\tilde{a}_4 = -\tfrac{1}{16}\tilde{a}_2 = \tfrac{1}{129}(\tilde{a}_0 - 2).
\end{gathered}$$

Finally, p_4 satisfies the initial condition (14.46) if

$$\tfrac{1}{2}\tilde{a}_0 - \tilde{a}_1 + \tilde{a}_2 - \tilde{a}_3 + \tilde{a}_4 = 0,$$

that is,

$$\tfrac{1}{2}\tilde{a}_0 - \tfrac{16}{129}(\tilde{a}_0 - 2) + \tfrac{1}{129}(\tilde{a}_0 - 2) = 0,$$

whence

$$\tilde{a}_0 = -\tfrac{20}{33}.$$

Thus the approximation is

$$\begin{aligned} p_4(x) &= -\tfrac{10}{33} + \tfrac{32}{99} T_2(x) - \tfrac{2}{99} T_4(x) \\ &= -\tfrac{16}{99}(4 - 5x^2 + x^4). \end{aligned}$$

The solution of the problem is

$$y(x) = 1 - \mathrm{e}^{(1-x^2)/2}$$

and on $[-1, 1]$ the maximum error in using p_4 as an approximation to y occurs at $x = 0$, when

$$y(0) - p_4(0) \simeq 22 \times 10^{-4}.$$

One interesting feature of the method is that we can make a backward error analysis. The coefficients $\tilde{a}_r$ satisfy all the equations (14.50) with the exception of that with $r = 6$, which must be replaced by

$$\frac{1}{4.6}\,\tilde{a}_4 + \tilde{a}_6 - \frac{1}{4.6}\,\tilde{a}_8 = \frac{1}{4.6}\,\tilde{a}_4.$$

Now this equation would be obtained if a term $\tilde{a}_4 T_6/24$ were added to the right side of (14.47). We thus find that p_4 satisfies an equation of the form

$$p_4(x) + \int x p_4(x)\,\mathrm{d}x = \tfrac{1}{2}x^2 - \tfrac{1}{1188} T_6(x) + C.$$

We have perturbed (14.47) by an amount $T_6/1188$. □

The method is easily extended to second order equations of the form

$$q_1 y'' + q_2 y' + q_3 y = q_4, \tag{14.51}$$

where q_r, $r = 1, 2, 3, 4$, are polynomials. We integrate (14.51) twice to obtain

$$q_1 y + \int (q_2 - 2q_1')y\,\mathrm{d}x + \iint (q_3 - q_2' + q_1'')y\,\mathrm{d}x = \iint q_4\,\mathrm{d}x + C_1 x + C_2,$$

where C_1 and C_2 are arbitrary constants. We now proceed as before and obtain a system of linear equations in the Chebyshev coefficients. This time we omit the equations equating coefficients for both T_0 and T_1, as these involve the arbitrary constants. These equations are replaced by the two boundary conditions which are needed with the second order equation (14.51).

The method can also be used to deal with boundary conditions of a more general form than (14.42). For example, we could deal with the unlikely condition

$$\alpha y(-1) + \beta y(1) = s,$$

where $\alpha \neq 0$, $\beta \neq 0$ are constants. This will just give a linear equation in the a_r. For a second order problem there is no difficulty in dealing with conditions of the form

$$\alpha y(a) + \beta y'(a) = s.$$

If the interval on which we require a solution is not $[-1, 1]$ we must use a simple transformation of variable (see Problem 4.44). Polynomial terms in the differential equation will remain polynomials after such a transformation. Fox and Parker (1968) discuss the use of Chebyshev methods for differential equations and similar problems more fully. Much of the pioneering work on these methods was done by C. Lanzcos (1893–1974).

Problems

Section 14.1

14.1 Write a formal algorithm for obtaining the numerical solution of the boundary value problem (14.1) and (14.2) by a predictor–corrector method. Note that you should have three 'nested' iteration loops corresponding to:

(i) choice of starting conditions,
(ii) step-by-step advancement,
(iii) use of the corrector.

Section 14.2

14.2 Suppose that y satisfies

$$y'' = f(x, y)$$

and let

$$t(x_n) = \frac{1}{h^2}(y(x_{n+1}) - 2y(x_n) + y(x_{n-1}))$$
$$- \tfrac{1}{12}(f(x_{n+1}, y(x_{n+1})) + 10f(x_n, y(x_n)) + f(x_{n-1}, y(x_{n-1}))),$$

where $x_{n\pm1} = x_n \pm h$. If the eighth derivative of y exists and is continuous on $[x_{n-1}, x_{n+1}]$, show that

$$t(x_n) = -\tfrac{1}{240}h^4 y^{(6)}(x_n) + O(h^6).$$

(*Hint*: use the Taylor series for both y and y'' at $x = x_n$.)

14.3 Use the difference approximation (14.19) with $h = 0.2$ to solve

$$y'' = 1$$

with

$$y(0) = y(1) = 0.$$

Show that the calculated solution is identical to the exact solution

$$y(x) = (x^2 - x)/2.$$

14.4 Repeat Problem 14.3 with the boundary conditions replaced by

$$y'(0) = -\tfrac{1}{2}, \qquad y(1) = 0.$$

Use (14.9) to approximate to the first of these conditions.

14.5 If $v(x) \equiv 0$ in (14.4) and we scale the equation so that $u(x) \equiv 1$, the $N \times N$ coefficient matrix $\mathbf{A}$ in (14.8) is the tridiagonal matrix

$$\mathbf{A} = -\frac{1}{h^2}\begin{bmatrix} (2-h^2w_1) & -1 & & \mathbf{0} \\ -1 & (2-h^2w_2) & -1 & \\ \mathbf{0} \quad \ddots & \ddots & & \ddots \end{bmatrix} = -\frac{1}{h^2}\mathbf{B},$$

say. Show that if $\mathbf{z}$ is any N-dimensional (real) vector with components z_n then

$$\mathbf{z}^{\mathrm{T}}\mathbf{B}\mathbf{z} = z_1^2 + (z_1 - z_2)^2 + (z_2 - z_3)^2 + \cdots + (z_{N-1} - z_N)^2 + z_N^2 - h^2(w_1z_1^2 + w_2z_2^2 + \cdots + w_Nz_N^2).$$

Hence show that if $w(x) \leqslant 0$ on $[a, b]$, the matrix $\mathbf{B}$ is symmetric and positive definite (see § 9.8) and thus $\mathbf{A}$ is non-singular.

14.6 Given the non-linear boundary value problem

$$y'' = -2yy'$$

with

$$y(0) + y'(0) = 0,$$
$$y(1) = \tfrac{1}{2},$$

derive a system of non-linear equations for approximations y_r to $y(0.2r)$, $r = 0, 1, 2, 3, 4$, using a difference approximation of the form (14.16) with $h = 0.2$.

14.7 The equations (14.16) can be written in the form

$$F_n(y_{n-1}, y_n, y_{n+1}) = 0, \qquad n = 1, 2, \ldots, N,$$

with $y_0 = \alpha$, $y_{N+1} = \beta$. We can write these as

$$\mathbf{F}(\mathbf{y}) = \mathbf{0},$$

where $\mathbf{F}$ is an N-dimensional vector whose nth component is F_n. Show that the Jacobian matrix (see §12.2) of $\mathbf{F}$ is tridiagonal.

14.8 The boundary value problem on $[0, 1]$,

$$y'' = 4(y + x)$$

with

$$y(0) = 1, \qquad y(1) = e^2 - 1,$$

is to be solved using the difference approximation (14.19). Use Theorem 14.1 to determine what size of interval should be used if the global error is to be less than 10^{-2} at all grid points. (Note that the required solution of the differential equation is $y(x) = e^{2x} - x$.)

Section 14.3

14.9 A one-step method for first order initial value problems may be formed as follows. Suppose y_n is known.

(i) Advance one step using Euler's method (13.11) with step size h; let $y(x_{n+1}; h)$ be the resulting approximation to $y(x_{n+1})$.
(ii) Advance from y_n again, using two steps of Euler's method with step size $h/2$. Let $y(x_{n+1}; h/2)$ be the resulting approximation to $y(x_{n+1})$.
(iii) Extrapolate using

$$y_{n+1} = 2y(x_{n+1}; h/2) - y(x_{n+1}; h).$$

Show that the resulting method is precisely the modified Euler method (13.37).

14.10 Use Euler's method with extrapolation to the limit to obtain approximations to $y(0.8)$, where $y(x)$ is the solution of

$$y' = 1 - y$$

with

$$y(0) = 0.$$

Take $h = 0.4$, 0.2, 0.1 and 0.05. Compare your computed solutions with the exact solution $y(x) = 1 - e^{-x}$.

Section 14.4

14.11 Consider deferred correction for Euler's method. Show that, if

$$y' = f(x, y)$$

and $y^{(3)}(x)$ exists for $x_n < x < x_{n+1}$, then

$$\frac{y(x_{n+1}) - y(x_n)}{h} = f(x_n, y(x_n)) + \frac{h}{2}\, y''(x_n) + \frac{h^2}{6}\, y^{(3)}(\xi_n),$$

where $\xi_n \in (x_n, x_{n+1})$. Thus we can estimate the truncation error using (see §7.7)

$$\frac{h}{2}\, y''(x_n) \simeq \frac{h}{2}\, \frac{\Delta^2 y_{n-1}}{h^2} = \frac{1}{2h}\, \Delta^2 y_{n-1}.$$

Having computed approximations y_n, we seek new approximations z_n satisfying (cf. (14.35))

$$\frac{1}{h}\,(z_{n+1} - z_n) = f(x_n, z_n) + \frac{1}{2h}\, \Delta^2 y_{n-1}.$$

Apply this method to the initial value problem given in Problem 14.10, taking $h = 0.2$ and $0 \leqslant x \leqslant 0.8$. Note that the method requires y_{-1}, which must be calculated from

$$y_0 = y_{-1} + hf(x_{-1}, y_{-1}),$$

where $x_{-1} = x_0 - h = -h$.

14.12 Derive the formulas needed for applying the deferred correction process to the method (14.19) for boundary value problems of the form (14.17).

Section 14.5

14.13 Verify the formulas (14.37)–(14.40). (*Hints*: (14.37) may be proved by mathematical induction and the recurrence relation (4.55) written in the form

$$xT_r(x) = \tfrac{1}{2}(T_{r-1}(x) + T_{r+1}(x));$$

the other formulae may be proved using $T_r(x) = \cos r\theta$ where $\theta = \cos^{-1} x$. See also § 7.4.)

14.14 Use the method of § 14.5 to obtain an approximation $p_2 \in P_2$ to the solution of the initial value problem given in Example 14.5. Compare the accuracy of p_2 with that of the p_4 given in the example.

14.15 Use the method of § 14.5 to determine an approximation $p_3 \in P_3$ to the solution of the initial value problem

$$y' + y = x^2$$

with

$$y(0) = 1.$$

(See Example 14.3 for the exact solution.)

14.16 Show that we obtain the same approximation p_4 as in Example 14.5 if we replace the initial condition (14.46) by

$$y(-1) + y(1) = 0.$$

14.17 Use the method of §14.5 to determine an approximation $p_4 \in P_4$ to the solution of the boundary value problem on $[-1, 1]$

$$y'' = y$$

with $y(-1) = -1$, $y(1) = 1$. (The solution of the differential equation is $y(x) = \sinh x / \sinh 1$.)

Appendix

COMPUTER ARITHMETIC

A.1 Binary numbers

For reasons of economy in design, computers use *binary* representation of numbers instead of the *decimal* system adopted by humans. In the decimal system we use 10 as a *base* and there are 10 different digits, 0, 1, 2, 3, 4, 5, 6, 7, 8, 9. Numbers are represented as sequences of these digits. For example,

$$\begin{aligned} 739 &= 7 \times 10^2 + 3 \times 10^1 + 9 \times 10^0 \\ &= 7 \times 100 + 3 \times 10 + 9 \times 1 \end{aligned}$$

and

$$\begin{aligned} 910.68 &= 9 \times 10^2 + 1 \times 10^1 + 0 \times 10^0 + 6 \times 10^{-1} + 8 \times 10^{-2} \\ &= 9 \times 100 + 1 \times 10 + 0 \times 1 + 6 \times \tfrac{1}{10} + 8 \times \tfrac{1}{100}. \end{aligned}$$

In the binary system we use two as the base and there are two different digits, 0 and 1. For example,

$$11011 = 1 \times 2^4 + 1 \times 2^3 + 0 \times 2^2 + 1 \times 2^1 + 1 \times 2^0$$

and

$$11.101 = 1 \times 2^1 + 1 \times 2^0 + 1 \times 2^{-1} + 0 \times 2^{-1} + 1 \times 2^{-3}.$$

In binary, each digit is called a *bit*.

It is shown in number theory that every real number can be expressed as a finite or infinite sequence of digits, regardless of the base used in the representation (except for base 1). It is also shown that if the representation of a rational number is an infinite sequence of digits then these must include a pattern which repeats after a certain point. It is interesting to note that a number which can be represented by a finite sequence of digits in one representation may need an infinite sequence in another representation (see Problem A.1).

In this book we normally discuss decimal representations because of their familiarity.

A.2 Integers and fixed point fractions

All computers can deal with integers but, because a computer store must be finite, only a restricted range is available. Usually this is more than adequate and typically a machine will store integers of up to 32 bits of which one is kept for the sign so that the range is approximately $-2 \times 10^9 < n < 2 \times 10^9$. Arithmetic with such integers is exact provided the result is an integer in the permitted range. If the result is too large most computers can give some kind of 'overflow' indication.

Some types of processor deal only with *fixed point fractions* as an alternative to integers. These fractions are restricted to the range $-1 < x < 1$ and a finite number of digits, usually about 32 binary digits (approximately 9 decimal digits). If we think in terms of decimal numbers, a fixed point fraction consists of only a finite number t (say) of decimal digits after the decimal point. In general, we have to terminate the decimal representation of a number after t digits. To do this we *round off* the number by chopping the decimal representation after t digits and then adding 10^{-t} if the amount neglected exceeds $\frac{1}{2}10^{-t}$. (If the amount neglected is exactly $\frac{1}{2}10^{-t}$ we can avoid statistical bias by forcing the last digit in the rounded number to be even.) In binary fixed point representation with t digits, we replace 10^{-t} by 2^{-t}.

Addition and subtraction of two fixed point fractions does not introduce any error unless the result lies outside the permitted range. Multiplication of two t-digit numbers yields a $2t$-digit product which is said to be *double precision*, as it consists of twice as many digits as the operands, which are said to be *single precision*. We must usually round the product to single precision (or t digits) and thus, provided the result lies in the permitted range, an error of at most $\frac{1}{2}10^{-t}$ is introduced. Similarly, division gives an error of at most $\frac{1}{2}10^{-t}$. If $\text{fix}(a)$ represents the fixed point representation of a real number a, we have the following results.

For any a,

(i) $|\text{fix}(a) - a| \leq \frac{1}{2}10^{-t}$.

If a and b are any two fixed point fractions with t digits:

(ii) $\text{fix}(a \pm b) = a \pm b$,

(iii) $|\text{fix}(ab) - ab| \leq \frac{1}{2}10^{-t}$. (A.1)

(iv) $|\text{fix}(a/b) - a/b| \leq \frac{1}{2}10^{-t}$. (A.2)

You will notice that multiplication and division will in general introduce rounding errors. These results are only valid if all quantities lie in the range†

† Strictly, this range should be $-1 + \frac{1}{2}10^{-t} \leq x \leq 1 - \frac{1}{2}10^{-t}$.

$-1<x<1$. If working is to t *binary* digits, 10^{-t} must be replaced by 2^{-t}. If 'chopping' is used, without rounding up if the error exceeds $\frac{1}{2}10^{-t}$, the bounds must be made twice as large.

On some processors, integers are stored in exactly the same way as fixed point fractions. The main differences lie in the way that the binary digits are interpreted. For integers, the binary point is considered to be at the least significant end and for fractions it is at the most significant end. There is also a difference in the way that the double-length product is interpreted.

In dealing with decimal numbers (not necessarily fractions) in fixed point form, we sometimes refer to the number of *decimal places*, by which we mean the number of decimal digits appearing after the decimal point. Thus 3.460 is a number given to three decimal places. If this is an approximation to some value, the least significant zero indicates that it is correct to three decimal places whereas 3.46 would suggest it is correct to only two decimal places. In general, we say that two numbers agree to t decimal places if their absolute difference does not exceed $\frac{1}{2}10^{-t}$.

A.3 Floating point arithmetic

One difficulty with fixed point arithmetic is that we usually have to worry about scaling so that results of arithmetical operations lie in the permitted range. We can avoid nearly all of such problems with *floating point* representation of numbers. Each non-zero number is expressed in the form $a \times 10^p$ where a is a fraction in the range $1>|a|\geqslant 0.1$ and p is an integer called the *exponent*.† We now represent a number by a pair (a, p). For example,

$$\begin{aligned} 27.3 &= 0.273 \times 10^2 = (0.273,\ 2) \\ -832{,}000 &= -0.832 \times 10^6 = (-0.832,\ 6) \\ 0.000641 &= 0.641 \times 10^{-3} = (0.641,\ -3). \end{aligned}$$

We represent zero by the pair $(0, 0)$. In practice a, the main part of the number, must consist of a finite number of digits and we will assume that only t decimal digits are permitted. The exponent must also be restricted in range and we assume that $-N \leqslant p \leqslant N$ so that numbers expressed in floating point form must lie within $10^N > |x| \geqslant 10^{-N-1}$. We say that the floating point number consists of t digits, irrespective of the number of digits in the exponent. On a typical scientific calculator, $t \simeq 10$ and $N = 99$ so that a very wide range of numbers is available.

† There is no special name for a although some people misname this the 'mantissa'.

On a computer we work with powers of 2 and numbers are usually expressed in the form $a \times 2^p$ where $1 > |a| \geq \frac{1}{2}$. The IEEE† have set standards for binary floating point arithmatic. These include *single precision* when a consists of 24 bit (about 7 decimal places) and p has 8 bits (so that numbers as large as 10^{38} or as small as 10^{-38} may be represented). In *double pecision* a has 53 bits (about 15 decimal places) and p has 11 bits.

Suppose $x \neq 0$ is any real number which we wish to express in the nearest floating point form of t decimal digits. It is always possible to find p such that $x = b.10^p$ and $0.1 \leq |b| < 1$. We assume that p is in the permitted range $-N \leq p \leq N$. We now choose a to be a t-digit fraction obtained by rounding b, so that‡

$$|b - a| \leq \tfrac{1}{2} 10^{-t}$$

and

$$0.1 \leq |a| < 1.$$

The floating point number we choose in place of x is (a, p). Since $|b| \geq 0.1$,

$$|x| = |b| \times 10^p \geq 0.1 \times 10^p = 10^{p-1} \tag{A.3}$$

and we find that

$$\begin{aligned} |x - a \times 10^p| &= |b \times 10^p - a \times 10^p| \\ &= |b - a| \times 10^p \\ &\leq \tfrac{1}{2} 10^{-t} \times 10 |x|. \end{aligned}$$

Hence

$$\left| \frac{x - a \times 10^p}{x} \right| \leq \frac{1}{2} 10^{-t+1} \tag{A.4}$$

and, therefore, the relative error in replacing x by (a, p) is at most $\frac{1}{2} 10^{-t+1}$. We shall write $\mathrm{fl}(x)$ for the floating point approximation to x, so that $\mathrm{fl}(x) = (a, p)$.

All the basic arithmetic operations on floating point numbers may introduce a rounding error and, from (A.4), the following statements are valid for any non-zero real numbers x and y, provided all quantities lie in the permitted range.

$$\begin{aligned} &\text{(i) } \mathrm{fl}(x) = x(1 + \varepsilon), \\ &\text{(ii) } \mathrm{fl}(x \pm y) = (x \pm y)(1 + \varepsilon), \\ &\text{(iii) } \mathrm{fl}(xy) = xy(1 + \varepsilon) \\ &\text{(iv) } \mathrm{fl}(x/y) = (x/y)(1 + \varepsilon), \end{aligned} \tag{A.5}$$

---

† The Institute of Electrical and Electronic Engineers (USA).

‡ If $1 - \frac{1}{2} 10^{-t} \leq |b| < 1$ it is necessary to increase p by $+1$ and take $a = +0.1$ or -0.1.

where in each case $|\varepsilon| \leq \frac{1}{2} 10^{-t+1}$. The results (ii)–(iv) cannot be improved even if x and y are floating point numbers. If binary arithmetic is used, 10^{-t+1} must be replaced by 2^{-t+1}. The arithmetic in most computers is arranged so that rounding is optimal and thus the above conditions are satisfied. Sometimes 'chopping' without rounding is used, so that errors can be up to twice as large. Usually a computer will give an 'overflow' indication if the exponent of a result is too large, that is $p > N$. 'Underflow', when $p < -N$, usually means that the result is treated as zero. Note that, for $|x| \simeq |y|$, cancellation may give large relative errors when calculating $x \pm y$ using floating point representations (see Problem A.5).

We often refer to the first r *significant digits* of a number and by this we mean the first r digits in the main part of the floating point representation of the number. Thus the first three significant digits of 730,000 are 7, 3 and 0 and the first two significant digits of 0.000681 are 6 and 8. Two numbers are said to be the same (or agree) to r significant digits if, after rounding to r-digit floating point form, the numbers are identical. Thus 730,423 and 730,000 are the same to three significant decimal digits. We are concerned here with relative accuracy rather than absolute accuracy, which is usually expressed in terms of decimal places. For example, although 0.00126 and 0.0013 agree to four decimal places, they agree to only two significant digits.

Problems

Section A.1

A.1 In each of the following the number on the left is in decimal and that on the right is in binary. Verify that they are both the same number.

(i) $39.25 = 100\ 111.01$
(ii) $0.2 = 0.0011\ 0011\ 0011\ldots,$
(iii) $1 = 0.111\ 111\ 1\ldots$

Use (ii) and (iii) to verify that $0.2 \times 5 = 1$ by working in binary arithmetic.

Section A.2

A.2 Give examples of fixed point fractions a and b such that equality holds in (A.1) and (A.2).

A.3 Show by means of examples that, if x, y and z are any real numbers, then in general

$$\text{fix}(x+y) \neq \text{fix}(x) + \text{fix}(y)$$
$$\text{fix}(\text{fix}(x.y).z) \neq \text{fix}(x.\text{fix}(y.z)).$$

Show that the last relation can hold even if x, y and z are already fixed point fractions with t digits.

Section A.3

A.4 We do not obtain equality in (A.3) unless $|b| = 0.1$ in which case $\mathrm{fl}(x) = x$. Use this result to show that we must have strict inequality in (A.4).

A.5 As in Problem A.3, show that if x, y and z are t-digit floating point numbers, then in general

$$\mathrm{fl}(\mathrm{fl}(x+y)+z) \neq \mathrm{fl}(x+\mathrm{fl}(y+z))$$
$$\mathrm{fl}(\mathrm{fl}(xy).z) \neq \mathrm{fl}(x\,\mathrm{fl}(yz)).$$

A.6 Suppose $\mathrm{fl}(x) = x(1+\varepsilon_1)$ and $\mathrm{fl}(y) = y(1+\varepsilon_2)$, where x and y are real with $x > y > 0$. Show that if $\varepsilon_1 = -\varepsilon_2$ and $\mathrm{fl}(x) - \mathrm{fl}(y) = (x-y)(1+\delta)$ then

$$\delta = \left(\frac{x+y}{x-y}\right)\varepsilon_1.$$

Conclude that for some choices of x and y, the relative error δ in forming $x - y$, may be much larger than ε_1 or ε_2.

SOLUTIONS TO SELECTED PROBLEMS

Chapter 1

1.1 (ii) *Constructive proof*: $\cos(2\cos^{-1}x) = \frac{1}{2} \Rightarrow 2\cos^{-1}x = (2k \pm \frac{1}{3})\pi$, where k is an integer. Thus $x = \cos\frac{1}{6}\pi$, $\cos\frac{7}{6}\pi$ are two real roots.

Non-constructive proof: let $f(x) = 2\cos(2\cos^{-1}x) - 1$, when $f(0) = -3$ and $f(1) = +1$. Thus as the function is continuous (see § 2.2) and has opposite signs at the endpoints of $[0, 1]$, it has at least one zero in $[0, 1]$. Similarly, there is at least one zero in $[-1, 0]$.

1.4 Condition (i) on p. 6 is not satisfied. There is no solution if we change any one of the coefficients in the first two equations by an arbitrary small amount, keeping the others fixed.

Chapter 2

2.1 (i) $[\frac{3}{4}, \infty)$, (ii) $[0, \frac{2}{3}]$, (iii) $[1, \infty)$. Inverses exist for (ii) and (iii) only.

2.10 All exist; only (iii) and (iv) are attained.

2.13 Only (i) and (ii) are continuous. At $x = -1$, (iii) is not even defined; similarly for (iv) at $x = 0$.

2.15 If $|f(x) - f(x')| \leqslant L|x - x'|$ for all $x, x' \in [a, b]$ then Definition 2.10 holds if we choose $\delta = \varepsilon/L$.

2.17 (ii) We have

$$\frac{f(x+h) - f(x)}{h} = \frac{1}{h}\left[\frac{1}{x+h+1} - \frac{1}{x+1}\right] = \frac{-1}{(x+h+1)(x+1)}$$

and thus $f'(x) = -1/(x+1)^2$.

2.19 $|f(x)-f(x')|=||x|-|x'||\leqslant|x-x'|$.

2.21 Since $|u_n-\frac{1}{2}|=1/2n$, (u_n) converges to $\frac{1}{2}$.

Chapter 3

3.2 $\cos x$: $p_6(x)=1-\dfrac{1}{2!}x^2+\dfrac{1}{4!}x^4-\dfrac{1}{6!}x^6$.

$$(1+x)^{1/2}:\quad p_6(x)=1+\frac{1}{2}x-\frac{1}{2\cdot 4}x^2+\frac{1\cdot 3}{2\cdot 4\cdot 6}x^3-\frac{1\cdot 3\cdot 5}{2\cdot 4\cdot 6\cdot 8}x^4+\frac{1\cdot 3\cdot 5\cdot 7}{2\cdot 4\cdot 6\cdot 8\cdot 10}x^5-\frac{1\cdot 3\cdot 5\cdot 7\cdot 9}{2\cdot 4\cdot 6\cdot 8\cdot 10\cdot 12}x^6.$$

$$\log(1-x):\quad p_6(x)=-x-\tfrac{1}{2}x^2-\tfrac{1}{3}x^3-\tfrac{1}{4}x^4-\tfrac{1}{5}x^5-\tfrac{1}{6}x^6.$$

3.4 $n=9$.

3.6 $R_n(x)=(-1)^{n+1}(n+1)x\left(\dfrac{x-\xi}{1+\xi}\right)^n\dfrac{1}{(1+\xi)^2}$,

where $-1<a\leqslant x<\xi<0$. Thus

$$|R_n(x)|<\frac{(n+1)|a|^n}{(1+a)^2},\qquad \text{for } -1<a\leqslant x<0.$$

Since $(n+1)|a|^n\to 0$ as $n\to\infty$, we have uniform convergence.

3.9 The nth term in the series is $(-1)^{n+1}x^n/n$. For $|x|>1$, this term increases in magnitude as $n\to\infty$, showing that the series cannot be convergent. Combine this last result with that of Problem 3.8 to show that the radius of convergence is $+1$.

3.11 For $x>0$ use $|R_1(x)|=\frac{1}{8}x^2(1+\xi)^{-3/2}<\frac{1}{8}x^2=O(x^2)$, where $0<\xi<x$. For $x<0$ use $|R_1(x)|=|\frac{1}{4}x(x-\xi)(1+\xi)^{-3/2}|<\frac{1}{4}|x^2(1+x)^{-3/2}|=O(x^2)$, where $-1<x<\xi<0$.

3.13 By Taylor's theorem

$$\begin{aligned}
e^x&=1+x+\tfrac{1}{2}x^2e^{\xi_1}\\
e^{nx}&=1+nx+\tfrac{1}{2}n^2x^2e^{\xi_2}\\
(1+x)^n&=1+nx+\tfrac{1}{2}x^2n(n-1)(1+\xi_3)^{n-2}.
\end{aligned}$$

Thus $e^x\geqslant 1+x$, $e^x=1+x+O(x^2)$, $e^{nx}=(e^x)^n\geqslant(1+x)^n$ and $e^{nx}-(1+x)^n=O(x^2)$.

Chapter 4

4.1 $p_1(x) = 4x + 1$.

4.4 0.6381.

4.5 $p_1(x) = \frac{1}{3}(x+2)$, so $2^{1/2} \simeq 1\frac{1}{3}$, $3^{1/2} \simeq 1\frac{2}{3}$.

4.7 $p_2(x) = 2x^2 - 1$.

4.8 $p_3(x) = x^2 + 1$. (See Theorem 4.1.)

4.9 $p(x) = 4x^3 - 3x$.

4.10 Error lies between 0.003 and 0.006.

4.13 $h = 0.002$.

4.17 Estimate is $x = 1.11$.

4.19 $p_3(x) = -9 + (x+3)\cdot 7 + (x+3)(x+1)\cdot(-3) + (x+3)(x+1)x\cdot 1$
$= x^3 + x^2 - 2x + 3$.

4.24 Polynomial is $2x^2 + 3x + 1$.

4.25 $S(r) = 1 + (r-1)\cdot 4 + \frac{1}{2}(r-1)(r-2)\cdot 5 + \frac{1}{6}(r-1)(r-2)(r-3)\cdot 2$
$= \frac{1}{6}(2r^3 + 3r^2 + r)$.

4.27 Degree 3.

4.36 Write $y = T_n(x) = \cos(n \cos^{-1} x)$.
Then $T_m(T_n(x)) = T_m(y) = \cos(m \cos^{-1} y) = \cos(mn \cos^{-1} x) = T_{mn}(x)$.

4.40 $n = 7$.

4.42 $n = 3$, 5 and 7 respectively.

Chapter 5

5.7 Normal equations are $a_0 + \frac{1}{2}a_1 = \frac{2}{3}$, $\frac{1}{2}a_0 + \frac{1}{3}a_1 = \frac{2}{5}$, with solution $a_0 = \frac{4}{15}$, $a_1 = \frac{4}{5}$.

5.8 $2.97 - 2x$.

5.9 Show that if the ψ_r are linearly dependent, they do not form a Chebyshev set.

5.10 $4x - 5x^3 + x^5$ is zero for every $x \in \{-2, -1, 0, 1, 2\}$.

5.13 $1.06 + 1.05x + 0.95y$.

5.17 $|x| \sim \frac{1}{2}\pi - \dfrac{4}{\pi}\sum_{r=0}^{\infty} \dfrac{\cos(2r+1)x}{(2r+1)^2}.$

5.18 $x^2 \sim \dfrac{\pi^2}{3} + 4\sum_{r=1}^{\infty}(-1)^r \dfrac{\cos rx}{r^2}.$

5.29 Make the substitution $x = \cos\theta$.

5.33 $\cos^{-1} x \sim \frac{1}{2}\pi - \dfrac{4}{\pi}\sum_{r=0}^{\infty} \dfrac{T_{2r+1}(x)}{(2r+1)^2}.$

5.34 $(1-x^2)^{1/2} \sim -\dfrac{4}{\pi}\sum_{r=0}^{\infty}{}' \dfrac{T_{2r}(x)}{4r^2-1}.$

5.35 The orthogonal polynomials are $1, x, x^2 - \frac{5}{2}, x^3 - \frac{17}{5}x$.

5.38 $x + \frac{1}{8}$.

5.39 $\frac{1}{4} + \frac{1}{2}\sqrt{2} - \frac{1}{2}x$.

5.41 Error is not greater than $\frac{1}{48}\cdot 2^{-k}$. Approximations to $\sqrt{2}$, $\sqrt{3}$, $\sqrt{10}$ are $1\frac{3}{8}$, $1\frac{17}{24}$, $3\frac{1}{12}$ with errors less than $\frac{1}{24}$, $\frac{1}{24}$, $\frac{1}{12}$ respectively.

5.49 The last two terms can be 'economized'.

5.53 Follow Example 5.17. Take $X = \{-1, -0.5, 0.5, 1\}$ initially. One cycle of the algorithm, exchanging the two interior points, gives $X = \{-1, -0.44, 0.56, 1\}$ and $p(x) = 0.9890 + 1.1302x + 0.5540x^2$. Compare with the Chebyshev series for e^x, which begins $1.2661T_0(x) + 1.1303T_1(x) + 0.2715T_2(x)$.

Chapter 6

6.1 For $x \in [t_i, t_{i+1}]$, write $f(x) - S_n(x) = (x-t_i)(x-t_{i+1})f''(\xi_x)/2!$, where $\xi_x \in (t_i, t_{i+1})$, and consider the value at the midpoint $x = (t_i + t_{i+1})/2$.

6.6 On $[t_s, t_{i+1}]$ with $s \geqslant i$, the right side has only two non-zero terms, giving

$$\begin{aligned}(t_s - t_i)B^1_{s-1}(x) + (t_{s+1} - t_i)B^1_s(x) &= (t_s - t_i)\left(\frac{t_{s+1} - x}{t_{s+1} - t_s}\right) + (t_{s+1} - t_i)\left(\frac{x - t_s}{t_{s+1} - t_s}\right)\\ &= x - t_i.\end{aligned}$$

6.7 Rearrange the terms involving B_{i+1}^{k-1} as follows:

$$\left(\frac{k}{t_{i+k+1}-t_{i+1}}\right)\left(-\left(\frac{x-t_i}{t_{i+k+1}-t_i}\right)+\left(\frac{t_{i+k+2}-x}{t_{i+k+2}-t_{i+1}}\right)\right)B_{i+1}^{k-1}(x)$$
$$=\left(\frac{k}{t_{i+k+1}-t_{i+1}}\right)\left(-\left(\frac{x-t_{i+1}}{t_{i+k+2}-t_{i+1}}\right)+\left(\frac{t_{i+k+1}-x}{t_{i+k+1}-t_i}\right)\right)B_{i+1}^{k-1}(x).$$

6.9 Write

$$f_x[t_i,t_{i+1},t_{i+2}]=\frac{(x-t_i)_+}{(t_i-t_{i+1})(t_i-t_{i+2})}+\frac{(x-t_{i+1})_+}{(t_{i+1}-t_i)(t_{i+1}-t_{i+2})}$$
$$+\frac{(x-t_{i+2})_+}{(t_{i+2}-t_i)(t_{i+2}-t_{i+1})}.$$

This is zero for $x\leqslant t_i$. For $t_i<x\leqslant t_{i+1}$,

$$(t_{i+2}-t_i)f_x[t_i,t_{i+1},t_{i+2}]=\frac{x-t_i}{t_{i+1}-t_i}=B_i^1(x).$$

Finally, check the two intervals $t_{i+1}<x\leqslant t_{i+2}$ and $x>t_{i+2}$.

6.15 Write

$$S(x)=\sum_{r=-2}^{n-1}a_rB_r^2(x).$$

The given conditions imply that

$$(a_{i-2}+a_{i-1})/2=f(i),\qquad i=0,1,\ldots,n,$$

and $-a_{-2}+a_{-1}=f'(0)$. We thus obtain $a_{-2}=(-f'(0)+2f(0))/2$ and $a_{-1}=(f'(0)+2f(0))/2$. The remaining coefficients $a_0,a_1,\ldots,a_{n-1}$ are determined by a system of $n+1$ linear equations. The matrix is not strictly diagonally dominant but is non-singular.

6.18 For $n\geqslant 1$ write

$$T_{n+1}(x)=2^n(x-x_0)(x-x_1)\ldots(x-x_n)$$

and verify that

$$T'_{n+1}(x_i)=2^n\prod_{j\neq i}(x_i-x_j).$$

Next write

$$\frac{d}{dx}T_{n+1}(x) = \frac{d}{d\theta}\cos(n+1)\theta\,\frac{d\theta}{dx} = \frac{(n+1)\sin(n+1)\theta}{\sin\theta}.$$

Finally, observe that, for $x = x_i$ (zero of T_{n+1}), $\sin\theta = \sqrt{1-\cos^2\theta} = \sqrt{1-x_i^2}$ and $\sin(n+1)\theta = 1$.

6.21 Note that $H \in P_3$ and verify that $H(0) = f(0)$, $H(1) = f(1)$ and $H'(0) = f'(0)$, $H'(1) = f'(1)$.

6.25

$$R_{2,1}(x) = \frac{1+\frac{2}{3}x+\frac{1}{6}x^2}{1-\frac{1}{3}x}, \qquad R_{1,2}(x) = \frac{1+\frac{1}{3}x}{1-\frac{2}{3}x+\frac{1}{6}x^2},$$

$$R_{3,2}(x) = \frac{1+\frac{3}{5}x+\frac{3}{20}x^2+\frac{1}{60}x^3}{1-\frac{2}{5}x+\frac{1}{20}x^2}, \qquad R_{2,3}(x) = \frac{1+\frac{2}{5}x+\frac{1}{20}x^2}{1-\frac{3}{5}x+\frac{3}{20}x^2-\frac{1}{60}x^3}.$$

6.27 For $\tan^{-1}(x)$ we obtain

$$R_{3,2} = \frac{x+\frac{4}{15}x^3}{1+\frac{3}{5}x^2} \quad \text{and} \quad R_{5,4} = \frac{x+\frac{7}{9}x^3+\frac{64}{945}x^5}{1+\frac{10}{9}x^2+\frac{5}{21}x^4}.$$

With $x = 1/\sqrt{3}$ these give approximations to $\pi/6$. These give 3.14335 and 3.14160 respectively as approximations for π.

6.28

$$\frac{1+\frac{1}{2}x+\frac{1}{12}x^2}{1-\frac{1}{2}x+\frac{1}{12}x^2} = 1 + \frac{12}{x-6+}\,\frac{12}{x}.$$

6.30 Apply the Euclidean algorithm to the polynomial p and its derivative p', and note that a repeated zero of p is also a zero of p'. The final polynomial produced by the algorithm is the greatest common divisor of p and p' and its roots (if any) will be repeated zeros of p.

6.31 For $(1/x)\log(1+x)$ we obtain

$$R_{1,1}(x) = \frac{1+\frac{1}{6}x}{1+\frac{2}{3}x} = 1 - \frac{\frac{1}{2}x}{1+\frac{2}{3}x}$$

and

$$R_{2,2}(x) = \frac{1+\frac{7}{10}x+\frac{1}{30}x^2}{1+\frac{6}{5}x+\frac{3}{10}x^2} = 1 + \frac{-\frac{1}{2}x}{1+\frac{2}{3}x+}\,\frac{-\frac{1}{18}x^2}{1+\frac{8}{15}x}.$$

Chapter 7

7.2 We obtain $a_0 = a_3 = \frac{3}{8}$, $a_1 = a_2 = \frac{9}{8}$. (This is called the three-eighths rule.)

7.3 Follow the method of Problem 7.2 and obtain $b_1 = b_2 = \frac{3}{2}$.

7.4 It is sufficient to put $x_0 = 0$, $h = 1$ and show that the rule is exact for $f = 1$, x, x^2 and x^3.

7.8 Use the fact that $\sum w_i = b - a$.

7.19 The required interpolating polynomial for e^{-t^2} is $1 + 10t(e^{-0.01} - 1)$. Since

$$\left| \frac{d^2}{dt^2} e^{-t^2} \right| \leqslant 2$$

the error of the interpolating polynomial is less than 0.0025. Thus, on integrating, $F(x) \simeq x + 5x^2(e^{-0.01} - 1)$ with error less than 0.00017 on $[0, 0.1]$.

7.21 Choose $M = 4$. Then Simpson's rule on $[0, 4]$ with $h = 0.025$ will give adequate accuracy.

7.22 Error is bounded by $(b-a)^2 h^2 (M_x + M_y)/12$, where M_x is a bound for $|\partial^2 f/\partial x^2|$ and M_y is defined similarly.

7.23 For $h = 0.5$, 0.25 and 0.0125 we obtain 0.63765, 0.63951 and 0.63951 respectively.

7.25 Error is $\frac{1}{3} h^2 f^{(3)}(\xi_0)$, where $\xi_0 \in (x_0, x_2)$.

7.26 We obtain the differentiation rule of Problem 7.24.

7.31 $f'(x_0) \simeq (f(x_0 + h) - f(x_0 - h))/2h$.

7.32 We need to differentiate $E(h) = 2 \times 10^{-k}/h^2 + \frac{1}{12} h^2 M_4$.

7.33 For $h = 0.1$, 0.2, 0.3 we obtain −0.9600, −0.9675, −0.9656 respectively, with $h = 0.2$ yielding the best value. This agrees with the result of Problem 7.32, which predicts $h \simeq 0.22$.

Chapter 8

8.1 If the equation is $f(x) = 0$, $f(0) > 0$ and $f(1) < 0$ and since $f'(x) < 0$ on $[0, 1]$ there is exactly one root. Fourteen iterations are required if we use the midpoint of the final subinterval $[x_0, x_1]$ to estimate the root.

8.2 Root $\simeq 1.618$.

8.5 One iteration of the secant method gives $x_2 = 0.382095$. Then one iteration of Muller's method gives $x = 0.382686$, compared with the exact value $\cos(3\pi/8) \simeq 0.382683$.

8.6 The root is 0.091 to three decimal places; (i) requires ten iterations and (ii) only three iterations.

8.7 Root ≈ 0.732244. Note that $x_1 > x_0$ and, for $r \geqslant 1$, $x_{r+1} - x_r = \frac{2}{5}(2^{x_r} - 2^{x_{r-1}})$ and we see by induction that (x_r) is monotonic increasing. (Since, by induction, $x_r < 1$ for all r, the sequence (x_r) is also bounded above and so has a limit.)

8.9 We need to choose λ near to $\cos(0.5) \simeq 0.9$. (See (8.21).) With $\lambda = 0.9$ we obtain 0.510828, 0.510971, 0.510973 for x_1, x_2, x_3. The last iterate is correct to six decimal places.

8.12 We have $x_{10} = 0.64563$, $a_{11} = 0.09091$, $a_{12} = -0.08333$ so that $x^*_{12} = 0.69306$ which agrees with log 2 to four decimal places.

8.14 The equation is $ax^2 + bx + c = 0$. If $c = 0$ the roots are $x = 0$ and $-b/a$ and the iterative method finds the second root in one iteration if $x_0 \neq 0$.

8.15 With $x_0 = 0$, the next iterates are 0.696564, 0.732115 and 0.732244.

8.17 The pth root process is

$$x_{r+1} = \frac{1}{p}\left[(p-1)x_r + \frac{c}{x_r^{p-1}}\right].$$

8.18 We find (see Theorem 8.2):

(i) $f(a) < 0$ and $f(b) > \frac{1}{4}(a + c/a)^2 - c = \frac{1}{4}(a - c/a)^2 > 0$.
(ii) $f''(x) = 2$.
(iii) We need to show that $\frac{1}{2}(x + c/x) \in [a, b]$ for $x = a$ (which is easy) and $x = b$. For $x = b$, we find $\frac{1}{2}(b + c/b) > c^{1/2} > a$ and, since $b > \frac{1}{2}(a + c/a) > c^{1/2}$, we have $b^2 > c$ and thus $\frac{1}{2}(b + c/b) < b$.

8.19 Show that

$$x_{r+1} \pm c^{1/2} = \frac{1}{2x_r}(x_r \pm c^{1/2})^2.$$

8.20 We obtain a sequence which converges to $-c^{1/2}$.

8.22 If $g(x) = x - \lambda f(x)/f'(x)$, then

$$g'(x) = 1 - \lambda + \lambda \frac{f(x)f''(x)}{[f'(x)]^2}$$

and if $f(x) = (x - \alpha)^2 h(x)$, the numerator and denominator of the last term each has a factor $(x - \alpha)^2$. We find that

$$g'(\alpha) = 1 - \lambda + \tfrac{1}{2}\lambda = 0, \qquad \text{if } \lambda = 2.$$

8.23 Iterative process for finding $1/c$ is $x_{r+1} = 2x_r - cx_r^2$.

8.30 The closure condition of Theorem 8.3 is easily verified, using the given inequality. For the Lipschitz condition, if $g(x) = 2 - x^{-k}$, $g'(x) = kx^{-k-1}$ and on $[2 - 2^{1-k}, 2]$

$$x^k \geqslant (2 - 2^{1-k})^k = 2^k(1 - 2^{-k})^k \geqslant 2^{k-1} \geqslant k$$

for $k = 2, 3, \ldots$. Thus

$$|g'(x)| \leqslant \frac{1}{|x|} = 1/(2 - 2^{1-k}) < 1$$

and g satisfies a Lipschitz condition with $L < 1$. With $k = 6$ and $x_0 = 2$ we have $x_1 = 1.984$ which is correct to three decimal places.

8.31 By continuity of g' we can find $\delta > 0$ such that $|g'(x)| \leqslant L < 1$ for all $x \in [\alpha - \delta, \alpha + \delta] \subset I$. For closure,

$$|g(x) - \alpha| = |g(x) - g(\alpha)| = |x - \alpha| \cdot |g'(\xi)| < \delta.$$

8.32 We have

$$\psi'(y) = \frac{\mathrm{d}x}{\mathrm{d}y} = 1 \Big/ \frac{\mathrm{d}y}{\mathrm{d}x} = 1/g'(x).$$

Thus $|\psi'(y)| \geqslant 1/L > 1$.

8.33 If $g(x) = c/x^{p-1}$, $g'(x) = (1 - p)cx^{-p}$ and $|g'(c^{1/p})| = |1 - p| \geqslant 1$.

Chapter 9

9.3 $\mathbf{A} = [1 \;\; 0]$, $\quad \mathbf{B} = [0 \;\; 1]^{\mathrm{T}}$.

9.4 $\mathbf{A} = [1 \;\; 0]$, $\quad \mathbf{B} = [0 \;\; 1]$, $\quad \mathbf{C} = [1 \;\; 1]^{\mathrm{T}}$.

9.5 (i) Let $\mathbf{C} = \mathbf{AB}$, where $\mathbf{A}$ and $\mathbf{B}$ are lower triangular.

$$c_{ij} = \sum_{k=1}^{n} a_{ik} b_{kj} = \sum_{k=1}^{i} a_{ik} b_{kj}$$

as $a_{ik} = 0$ for $k > i$. Thus $c_{ij} = 0$ for $j > i$ as $b_{kj} = 0$ for $j > i \geqslant k$.

9.7 The ith column of $\mathbf{AD}$ is the ith column of $\mathbf{A}$ multiplied by d_i.

9.8 If $\delta = ad - bc \neq 0$, we obtain

$$e = d/\delta, \qquad f = -b/\delta, \qquad g = -c/\delta, \qquad h = a/\delta.$$

9.11 Need to interchange rows (or columns), e.g. rows 2 and 3.

$$\mathbf{x} = [-3\ \ 2\ \ -1\ \ 4]^{\mathrm{T}}.$$

9.16 $$\int \left[\sum_i c_i\psi_i\right]^2 \mathrm{d}x = \int \sum_j c_j\psi_j\left[\sum_i c_i\psi_i\right] \mathrm{d}x$$

$$= \sum_j c_j \int \psi_j \sum_i c_i\psi_i \, \mathrm{d}x = \sum_j c_j\left[\sum_i c_i \int \psi_i\psi_j \, \mathrm{d}x\right] = 0.$$

Finally use the argument at the end of § 9.3.

9.23 $$\mathbf{A} = \begin{bmatrix} 1 & 0 & 0 & 0 \\ 2 & 1 & 0 & 0 \\ -2 & 2 & 1 & 0 \\ 1 & 1 & -1 & 1 \end{bmatrix}\begin{bmatrix} 1 & 0 & 0 & 0 \\ 0 & 1 & 0 & 0 \\ 0 & 0 & -3 & 0 \\ 0 & 0 & 0 & 3 \end{bmatrix}\begin{bmatrix} 1 & 2 & -2 & 1 \\ 0 & 1 & 2 & 1 \\ 0 & 0 & 1 & -1 \\ 0 & 0 & 0 & 1 \end{bmatrix}.$$

9.28 The inverse matrix is

$$\mathbf{A}^{-1} = \begin{bmatrix} 95 & -28 & 18 \\ 10 & -3 & 2 \\ -8 & 2 & -1 \end{bmatrix}.$$

9.29 The (i, j)th element of $(\mathbf{AB})^{\mathrm{T}} = (j, i)$th element of $\mathbf{AB}$ = inner product of (jth row of $\mathbf{A}$) and (ith column of $\mathbf{B}$) = inner product of (ith row of $\mathbf{B}^{\mathrm{T}}$) and (jth column of $\mathbf{A}^{\mathrm{T}}$) = (i, j)th element of $\mathbf{B}^{\mathrm{T}}\mathbf{A}^{\mathrm{T}}$.

9.30 $(\mathbf{ABC})^{\mathrm{T}} = ([\mathbf{AB}]\mathbf{C})^{\mathrm{T}} = \mathbf{C}^{\mathrm{T}}[\mathbf{AB}]^{\mathrm{T}} = \mathbf{C}^{\mathrm{T}}\mathbf{B}^{\mathrm{T}}\mathbf{A}^{\mathrm{T}}$.

9.31 (ii) $$\begin{bmatrix} 1 & 1 \\ 1 & 1 \end{bmatrix}\begin{bmatrix} 1 & 0 \\ 0 & 0 \end{bmatrix} = \begin{bmatrix} 1 & 0 \\ 1 & 0 \end{bmatrix}.$$

9.33 $$\mathbf{M} = \begin{bmatrix} 2 & 0 & 0 \\ -1 & 4 & 0 \\ -2 & 2 & 1 \end{bmatrix}, \qquad \mathbf{x} = \begin{bmatrix} 2 \\ 1 \\ -1 \end{bmatrix}.$$

9.34 $$\mathbf{x}^{\mathrm{T}}\mathbf{A}^2\mathbf{x} = \mathbf{x}^{\mathrm{T}}\mathbf{A}^{\mathrm{T}}\mathbf{A}\mathbf{x} = (\mathbf{A}\mathbf{x})^{\mathrm{T}}\mathbf{A}\mathbf{x} = \mathbf{y}^{\mathrm{T}}\mathbf{y},$$

where $\mathbf{y} = \mathbf{Ax}$. As $\mathbf{A}$ is non-singular, $\mathbf{x} \neq \mathbf{0} \Rightarrow \mathbf{y} \neq \mathbf{0}$ and $\mathbf{y}^{\mathrm{T}}\mathbf{y} = y_1^2 + \cdots + y_n^2 > 0$ for $\mathbf{y} \neq \mathbf{0}$.

9.39 Suppose $\mathbf{b} \neq \mathbf{0}$ and $\boldsymbol{\Psi}^{\mathrm{T}}\mathbf{b} = \mathbf{0}$. The jth component of this last equation is

$$b_0\psi_0(x_{j-1}) + b_1\psi_1(x_{j-1}) + \cdots + b_n\psi_n(x_{j-1}) = 0$$

for $0 \leqslant j \leqslant N$. Thus $b_0\psi_0 + \cdots + b_n\psi_n$ has $N+1$ zeros, which is not possible for a Chebyshev set.

9.41 $\mathbf{x} = [2\ 3\ 4\ 5\ 6]^{\mathrm{T}}$.

Chapter 10

10.5 $\|\mathbf{x}\| = \|(\mathbf{x}-\mathbf{y}) + \mathbf{y}\| \leqslant \|\mathbf{x}-\mathbf{y}\| + \|\mathbf{y}\|$ by (10.9). Thus $\|\mathbf{x}\| - \|\mathbf{y}\| \leqslant \|\mathbf{x}-\mathbf{y}\|$. Similarly $\|\mathbf{y}\| - \|\mathbf{x}\| \leqslant \|\mathbf{x}-\mathbf{y}\|$.

10.8 $\mathbf{x} = \begin{bmatrix} 1 \\ 0 \end{bmatrix}, \quad \mathbf{y} = \begin{bmatrix} 0 \\ 1 \end{bmatrix}.$

10.10 $\mathbf{AA}^{-1} = \mathbf{I} \Rightarrow \|\mathbf{AA}^{-1}\| = \|\mathbf{I}\| = 1$.
By Lemma 10.2(v), $\|\mathbf{A}\| \cdot \|\mathbf{A}^{-1}\| \geqslant \|\mathbf{AA}^{-1}\| = 1$. For second part use $\mathbf{A}^{-1} - \mathbf{B}^{-1} = \mathbf{A}^{-1}(\mathbf{B}-\mathbf{A})\mathbf{B}^{-1}$ and Lemma 10.2(v).

10.11 $\mathbf{x}^{\mathrm{T}}\mathbf{x} = (\mathbf{Ty})^{\mathrm{T}}\mathbf{Ty} = \mathbf{y}^{\mathrm{T}}\mathbf{T}^{\mathrm{T}}\mathbf{Ty} = \mathbf{y}^{\mathrm{T}}\mathbf{y}$.

10.12 With $\mathbf{B} = \mathbf{A}^{\mathrm{T}}\mathbf{A}$, we have $\mathbf{x}^{\mathrm{T}}\mathbf{Bx} = \mathbf{x}^{\mathrm{T}}\mathbf{A}^{\mathrm{T}}\mathbf{Ax} = (\mathbf{Ax})^{\mathrm{T}}\mathbf{Ax} = \mathbf{y}^{\mathrm{T}}\mathbf{y} \geqslant 0$, where $\mathbf{y} = \mathbf{Ax}$.

10.13 Let $\mathbf{B} = \mathbf{I} + \mathbf{A}$, when $\|\mathbf{A}\|_1 = 0.9$. By Lemma 10.3, $\mathbf{B}$ is non-singular and $\|\mathbf{B}^{-1}\|_1 \leqslant 1/(1 - \|\mathbf{A}\|_1) = 10$.

10.15 $\|\mathbf{A}^r\| \leqslant \|\mathbf{A}\|^r$ so that by Lemma 10.2

$$\begin{aligned} \|p(\mathbf{A})\| &\leqslant a_0\|\mathbf{I}\| + a_1\|\mathbf{A}\| + \cdots + a_n\|\mathbf{A}^n\| \\ &\leqslant a_0 + a_1\|\mathbf{A}\| + \cdots + a_n\|\mathbf{A}\|^n = p(\|\mathbf{A}\|). \end{aligned}$$

10.17 $\left\| \dfrac{1}{n!}\mathbf{A}^n \right\| \leqslant \dfrac{1}{n!}\|\mathbf{A}\|^n \to 0 \quad \text{as} \quad n \to \infty,$

regardless of the value of $\|\mathbf{A}\|$. Thus

$$\lim_{n\to\infty} \left\| \frac{1}{n!}\mathbf{A}^n - \mathbf{0} \right\| = 0.$$

10.19 $k_1(\mathbf{A}) = k_\infty(\mathbf{A}) \simeq 12011$.

10.21 $\mathbf{A}^{-1} = \dfrac{1}{\lambda}\begin{bmatrix} 1 & -\lambda \\ -1 & 2\lambda \end{bmatrix}.$

For $|\lambda| \leqslant \frac{2}{3}$, $\|\mathbf{A}\|_\infty = 2$, $\|\mathbf{A}^{-1}\|_\infty = 2 + 1/|\lambda|$ and $k_\infty(\mathbf{A}) = 4 + 2/|\lambda|$. For $|\lambda| > \frac{2}{3}$, $\|\mathbf{A}\|_\infty = 3|\lambda|$, $\|\mathbf{A}^{-1}\|_\infty = 2 + 1/|\lambda|$ and $k_\infty(\mathbf{A}) = 3 + 6|\lambda|$.

10.24 Use (10.31) with $\boldsymbol{\delta}\mathbf{A} = \mathbf{0}$ and $\boldsymbol{\delta}\mathbf{b} = \mathbf{r}_0$.

10.25 0.498 × first equation is subtracted from the second equation.

$$\mathbf{r}_0 = -[0.00371 \;\; 0.001]^T, \qquad \boldsymbol{\delta}\mathbf{x}_0 = [0.140 \;\; -0.283]^T$$

so that $\mathbf{x}_1 = [0.971 \;\; -1.94]^T$. A further application of the process yields $\mathbf{x}_2 = [0.995 \;\; -1.99]^T$.

10.28 (i) $\mathbf{x}_3 = [-0.895 \;\; -3.865 \;\; -2.830]^T$
(ii) $\mathbf{x}_3 = [-0.978 \;\; -3.983 \;\; -2.990]^T$.

10.32 $\mathbf{M}_J = \begin{bmatrix} 0 & \frac{1}{2} \\ \frac{1}{2} & 0 \end{bmatrix}, \qquad \mathbf{M}_G = \begin{bmatrix} 0 & \frac{1}{2} \\ 0 & \frac{1}{4} \end{bmatrix}.$

Chapter 11

11.2 First matrix: there is one eigenvalue in each of the disks $|\lambda + 7| \leqslant 4$ and $|\lambda - 1| \leqslant 2$; the third is in the intersection of the disks $|\lambda - 6| \leqslant 2$ and $|\lambda - 1| \leqslant 6$. Second matrix: one eigenvalue is in the disk $|\lambda - 10| \leqslant 1$; two are in the disk $|\lambda - 4| \leqslant 2$.

11.3 The critical value is $\alpha = 1 - 1/\sqrt{2}$, and the set $|\lambda - 7| \leqslant 2\alpha$, with λ real, is contained in the interval $[6.4, 7.6]$.

11.5 We obtain $\mathbf{y}_{10} \simeq [0.445 \;\; 0.802 \;\; 1]^T$ and $\lambda_1 \simeq 8.05$.

11.7 In decreasing order of magnitude, the eigenvalues are $-(5+\sqrt{5})/2$, $-(3+\sqrt{5})/2$, $-(5-\sqrt{5})/2$ and $-(3-\sqrt{5})/2$. The initial vector happens to contain no contribution from the eigenvector corresponding to the dominant eigenvalue. Thus, without rounding error, the process converges to the eigenvalue of second largest modulus, $-(3+\sqrt{5})/2 \simeq -2.618034$.

11.9 After six iterations, (11.13) gives $\lambda_1 \simeq 8.0655$ and the Rayleigh quotient (11.14) gives $\lambda_1 \simeq 8.0489$, which is correct to four decimal places.

11.12 First factorize

$$\mathbf{A} - 8\mathbf{I} = \begin{bmatrix} 1 & 0 & 0 \\ -1/4 & 1 & 0 \\ -1/4 & -9/11 & 1 \end{bmatrix} \begin{bmatrix} -4 & 1 & 1 \\ 0 & -11/4 & 9/4 \\ 0 & 0 & 1/11 \end{bmatrix}.$$

(In practice, this is done 'in the computer'. The above factorization is given merely as a check.) With $\mathbf{y}_0 = [1 \;\; 1 \;\; 1]^T$ we obtain $\mathbf{y}_1 = [0.4400 \;\; 0.8000 \;\; 1]^T$, after normalization, and $\mathbf{y}_2 = [0.4451 \;\; 0.8020 \;\; 1]^T$. In both cases the Rayleigh quotient, using (11.21) and (11.22), gives $\lambda_1 \simeq 8.0489$.

11.16 We have $\lambda_1 \simeq 6.5$, $\lambda_2 \simeq 3.5$, $\lambda_3 \simeq 1.5$. Using Newton's method (11.28) we find that, more precisely, $\lambda_3 \simeq 1.8549$.

11.17 After four iterations, we obtain -3.618034.

11.18 Two iterations of Newton's method gives -7.029173.

11.20 Let **A** and **B** both be orthogonal. Then

$$(\mathbf{AB})^{\mathrm{T}}\,\mathbf{AB} = \mathbf{B}^{\mathrm{T}}\mathbf{A}^{\mathrm{T}}\mathbf{AB} = \mathbf{B}^{\mathrm{T}}\mathbf{B} = \mathbf{I}.$$

11.21 The required transformation **T** is

$$\mathbf{T} = \frac{1}{15}\begin{bmatrix} -3 & -6 & 6 & 12 \\ -6 & 13 & 2 & 4 \\ 6 & 2 & 13 & -4 \\ 12 & 4 & -4 & 7 \end{bmatrix}.$$

11.22 With

$$\mathbf{S} = \frac{1}{5}\begin{bmatrix} 5 & 0 & 0 \\ 0 & -3 & -4 \\ 0 & -4 & 3 \end{bmatrix}, \qquad \text{we have } \mathbf{S}^{\mathrm{T}}\mathbf{AS} = \frac{1}{25}\begin{bmatrix} 125 & -125 & 0 \\ -125 & 74 & 7 \\ 0 & 7 & 26 \end{bmatrix}.$$

Chapter 12

12.1 Closure condition is straightforward. Second, from (12.9),

$$|g_1(\mathbf{x}) - g_1(\mathbf{x}')| \leqslant \tfrac{1}{12}[|x_1 - x_1'| \cdot 2 + |x_3 - x_3'| \cdot 2] \leqslant \tfrac{1}{6}\|\mathbf{x} - \mathbf{x}'\|_1.$$

Similarly $|g_2(\mathbf{x}) - g_2(\mathbf{x}')| \leqslant \tfrac{1}{6}\|\mathbf{x} - \mathbf{x}'\|_1$ and $|g_3(\mathbf{x}) - g_3(\mathbf{x}')| \leqslant \tfrac{1}{3}\|\mathbf{x} - \mathbf{x}'\|_1$ and so $\|\mathbf{g}(\mathbf{x}) - \mathbf{g}(\mathbf{x}')\|_1 \leqslant \tfrac{2}{3}\|\mathbf{x} - \mathbf{x}'\|_1$, giving a contraction mapping.

12.2 If there exists a solution (x, y), then we see that $-\tfrac{1}{2} \leqslant x, y \leqslant \tfrac{1}{2}$. The closure and contraction mapping properties are easily verified, showing that the solution exists and is unique. With $x_0 = y_0 = 0$, we obtain $x = 0.49$, $y = 0.23$ in four iterations.

12.5

$$\mathbf{G} = \begin{bmatrix} \frac{1}{3} & \frac{1}{3} & \frac{1}{3} \\ \frac{1}{3} & \frac{1}{6} & \frac{1}{6} \\ \frac{1}{3} & \frac{1}{6} & \frac{1}{6} \end{bmatrix},$$

where **G** has elements g_{ij} given by (12.10). Thus from (12.12) $L = \sqrt{(2/3)}$, showing there is a contraction mapping. Note that in this case we cannot infer that a contraction mapping exists with either of the other two common norms.

12.7 Two iterations of Newton's method gives $x_1 = -0.82$, $x_2 = 1.91$.

12.10 The first two iterations give $x = \frac{1}{2}$, $y = \frac{1}{4}$ and $x = 0.4865$, $y = 0.2338$.

$$\mathbf{J}^{-1} = \frac{1}{1 + \frac{1}{4}\sin y \cos x}\begin{bmatrix} 1 & -\frac{1}{2}\sin y \\ \frac{1}{2}\cos x & 1 \end{bmatrix},$$

which exists for all x and y.

12.11 In this case $\mathbf{J} = \mathbf{A}$ and Newton's method is

$$\mathbf{x}_{r+1} = \mathbf{x}_r - \mathbf{A}^{-1}(\mathbf{A}\mathbf{x}_r - \mathbf{b})$$

which gives $\mathbf{x}_{r+1} = \mathbf{A}^{-1}\mathbf{b}$ and thus we do not obtain an iterative method.

12.13 With initial values $x = 2$, $y = -1$, Newton's method gives the solution $x = 1.709427$, $y = -1.441478$. The other solution is $x = -1.441478$, $y = 1.709427$.

Chapter 13

13.1 $L = 1$. With $y_0(x) = 1$,

$$y_n(x) = 1 + \frac{x^2}{2} + \frac{x^4}{2.4} + \cdots + \frac{x^{2n}}{2.4\ldots 2n}.$$

13.2 $L = 1$.

13.4 (i) $c_1 a^n$. (iv) $c_1 + c_2 2^n + c_3 n 2^n$.

13.5 (i) $c_1 2^n + c_2 n 2^n + 1$. (iii) $c_1 + c_2 n + \frac{1}{6} n^3$.

13.8 Difference equation: $y_{n+1} = y_n + anh^2 + bh$ with $y_0 = 0$. Solution (verify by induction): $y_n = \frac{1}{2} an^2 h^2 + bnh - \frac{1}{2} anh^2$.

13.12 Simple: $y_1 = 0.024$, $y_2 = 0.09488$, $y_3 = 0.2186$.
Classical: $y_1 = 0.02127$, $y_2 = 0.08969$, $y_3 = 0.2112$.
$y(0.6) = y(x_3) = 0.2112$.

13.13 $y_1 = 1.176819$, $y_2 = 1.367624$, $y_3 = 1.566211$, $y_4 = 1.764979$, $y_5 = 1.956291$.

13.16 $L_\phi = L = 1$. $|y''| = |2 - e^{-x}| \leqslant 2 - e^{-1} \simeq 1.6321$. Thus $|t(x; h)| = \frac{1}{2} h |y''(\xi)| \leqslant 0.8161 h$ and (13.52) becomes $|y(x_n) - y_n| \leqslant (e^{x_n} - 1)0.8161 h$.

13.17 $p = 2$, $L_0 = L = 3\sqrt{3}/8$, $L_1 = 2$, $L_\phi = 3\sqrt{3}/8 + h_0$, $N = \frac{1}{3}$.

$$|y(x_n) - y_n| \leqslant \frac{1}{3}\left(\frac{e^{L_\phi x_n} - 1}{L_\phi}\right)h^2.$$

13.20 $y_{n+1} = y_n + \frac{1}{12}h(23f_n - 16f_{n-1} + 5f_{n-2})$.

13.21 $y_5 = 0.6265$, $y(1.0) = 0.6321$.

13.22 We take starting values $y_0 = 1$ and y_1, y_2, y_3 as computed above in Problem 13.13 by the Runge–Kutta method. Then the Adams–Bashforth algorithm gives $y_4 = 1.764587$, $y_5 = 1.955417$.

13.23 $B_1 = 2$, $b_2 = \frac{5}{12}$, $L = 1$, $M_3 = 1$, $e_5 \leqslant 0.0569$.

13.25 $m = 0, 1$: $y_{n+1} = y_{n-1} + 2hf_n$.

$$m = 2: \quad y_{n+1} = y_{n-1} + \frac{h}{3}(7f_n - 2f_{n-1} + f_{n-2}).$$

13.26 See Problem 13.36(i).

13.28 $y_5 = 0.6334$.

13.29 We use $y_0 = 1$ and y_1, y_2 as computed in Problem 13.13 by the Runge–Kutta method. The Adams–Moulton method (13.84) with two corrections per step then gives $y_3 = 1.566238$, $y_4 = 1.765044$, $y_5 = 1.956389$.

13.30 $\delta = 0$, $|\gamma_{-1}| = \frac{1}{2}$, $|c_2| = \frac{1}{12}$, $M_3 = 1$, $C_0 = 1$, $L = 1$, $e_5 \leqslant 0.00679$.

13.36 (i) Stable for $h \leqslant 3/(2A)$. (ii) Unstable. (iii) One root of the characteristic equation is $z_2 = -1$ and the method is neither stable nor unstable. (It is said to be neutrally stable.).

13.40 (i) $y_{n+1} = (1 - Ah + \frac{1}{2}A^2h^2)y_n \Rightarrow y_n = (1 - Ah + \frac{1}{2}A^2h^2)^n y_0 \to 0$ if $h < 2/A$.

13.42
$$\begin{aligned} y_{1,n+1} &= y_{1,n} + h(x_n^2 y_{1,n} - y_{2,n}) \\ &\quad + \tfrac{1}{2}h^2[(1 + 2x_n + x_n^4)y_{1,n} - x_n(1 + x_n)y_{2,n}] \\ y_{2,n+1} &= y_{2,n} + h(-y_{1,n} + x_n y_{2,n}) \\ &\quad + \tfrac{1}{2}h^2[(-x_n(1 + x_n)y_{1,n} + (2 + x_n^2)y_{2,n}]. \end{aligned}$$

13.43 $y_5 = 0.8187$, $y_5' = -0.7301$. Exact solution $y(x) = (1 + x^2)^{-1}$.

13.46 Let $z = y'$, $w = z' = y''$. Corrector is

$$\begin{aligned} y_{n+1}^{(r+1)} &= y_n + \tfrac{1}{2}h(z_n + z_{n+1}^{(r)}) \\ z_{n+1}^{(r+1)} &= z_n + \tfrac{1}{2}h(w_n + w_{n+1}^{(r)}) \\ w_{n+1}^{(r+1)} &= w_n + \tfrac{1}{2}h(2w_n + 2w_{n+1}^{(r+1)} + x_n^2 y_n + x_{n+1}^2 y_{n+1}^{(r+1)} + 2 + x_n + x_{n+1}). \end{aligned}$$

Chapter 14

14.6

$$\begin{aligned} -1.6y_0 + 2y_1 - 0.08y_0^2 &= 0 \\ y_0 - 2y_1 + y_2 + 0.2y_1(y_2 - y_0) &= 0 \\ y_1 - 2y_2 + y_3 + 0.2y_2(y_3 - y_1) &= 0 \end{aligned}$$

$$y_2 - 2y_3 + y_4 + 0.2y_3(y_4 - y_2) = 0$$
$$y_3 - 1.9y_4 - 0.2y_3y_4 + 0.5 = 0.$$

14.8 $h < \sqrt{6}/10\mathrm{e} \simeq 0.0901$.

14.10

0.6400			
	0.5408		
0.5904		0.5512	
	0.5486		0.5509
0.5695		0.5509	
	0.5503		
0.5599			

$y(0.8) = 0.55067$.

14.11 $y_{-1} = -0.2500$, $y_0 = 0$, $y_1 = 0.2000$, $y_2 = 0.3600$, $y_3 = 0.4880$, $y_4 = 0.5904$; $z_1 = 0.1750$, $z_2 = 0.3200$, $z_3 = 0.4400$, $z_4 = 0.5392$.

14.12 $(z_{n+1} - 2z_n + z_{n-1})/h^2 = f(x_n, z_n) + \Delta^4 y_{n-2}/(12h^2)$.

14.15 $p_3(x) = (27T_0 - 19T_1 + 5T_2 + T_3)/22 = (2x^3 + 5x^2 - 11x + 11)/11$.

14.17 $p_4(x) = (51T_1 + 2T_3)/53 = (8x^3 + 45x)/53$.

REFERENCES AND FURTHER READING

Bartle, R. G. and Sherbert, D. R. (1992) *Introduction to Real Analysis* (second edition). Wiley, New York.

Cheney, E. W. (1966) *Introduction to Approximation Theory*. McGraw-Hill, New York.

Davis, P. J. (1976) *Interpolation and Approximation*, Dover, New York.

Davis, P. J. and Rabinowitz, P. (1984) *Methods of Numerical Integration* (second edition). Academic Press, New York.

Forsythe, G. E. and Moler, C. B. (1967) *Computer Solution of Linear Algebraic Systems*. Prentice-Hall, Englewood Cliffs, New Jersey.

Fox, L. (1957) *The Numerical Solution of Two-point Boundary Problems in Ordinary Differential Equations*. Oxford University Press, Oxford.

Fox, L. and Parker, I. B. (1968) *Chebyshev Polynomials in Numerical Analysis*. Oxford University Press, Oxford.

Haggerty, Rod (1993) *Fundamentals of Mathematical Analysis* (second edition). Addison-Wesley, Reading, Massachusetts.

Henrici, P. (1962) *Discrete Variable Methods in Ordinary Differential Equations*. Wiley, New York.

Hildebrand, F. B. (1974) *Introduction to Numerical Analysis* (second edition). McGraw-Hill, New York.

Isaacson, E. and Keller, H. B. (1966) *Analysis of Numerical Methods*. Wiley, New York.

Johnson, L. W., Riess, R. D. and Arnold, J. T. (1993) *Introduction to Linear Algebra* (third edition). Addison-Wesley, Reading, Massachusetts.

Lambert, J. D. (1991) *Numerical Methods for Ordinary Differential Equations*. Wiley, New York.

Powell, M. J. D. (1981) *Approximation Theory and Methods*. Cambridge University Press, Cambridge.

Ralston, A. and Rabinowitz, P. (1978) *A First Course in Numerical Analysis* (second edition). McGraw-Hill, New York.

Rivlin, T. J. (1981) *An Introduction to the Approximation of Functions*. Dover, New York.

Rivlin, T. J. (1990) *The Chebyshev Polynomials* (second edition). Wiley, New York.

Wilkinson, J. H. (1988) *The Algebraic Eigenvalue Problem* (new edition). Oxford University Press, Oxford.

INDEX